SENSORIMOTOR CONTROL
of Movement and Posture

ADVANCES IN EXPERIMENTAL MEDICINE AND BIOLOGY

Recent Volumes in this Series

Volume 499
FRONTIERS IN MODELING AND CONTROL OF BREATHING
Edited by Chi-Sang Poon and Homayoun Kazemi

Volume 500
BIOLOGICAL REACTIVE INTERMEDIATES VI: Chemical and Biological Mechanisms of Susceptibility to and Prevention of Environmental Diseases
Edited by Patrick M. Dansette, Robert Snyder, Marcel Delaforge, G. Gordon Gibson, Helmut Greim, David J. Jollow, Terrence J. Monks, and I. Glenn Sipes

Volume 501
BIOACTIVE COMPONENTS OF HUMAN MILK
Edited by David S. Newburg

Volume 502
HYPOXIA: From Genes to the Bedside
Edited by Robert C. Roach, Peter D. Wagner, and Peter H. Hackett

Volume 503
INTEGRATING POPULATION OUTCOMES, BIOLOGICAL MECHANISMS AND RESEARCH METHODS IN THE STUDY OF HUMAN MILK AND LACTATION
Edited by Margarett K. Davis, Charles E. Isaacs, Lars Å. Hanson, and Anne L. Wright

Volume 504
MYCOTOXINS AND FOOD SAFETY
Edited by Jonathan W. DeVries, Mary W. Trucksess, and Lauren S. Jackson

Volume 505
FLAVONOIDS IN CELL FUNCTION
Edited by Béla A. Buslig and John A. Manthey

Volume 506
LACRIMAL GLAND, TEAR FILM, AND DRY EYE SYNDROMES 3
Edited by Daniel A. Sullivan, Michael E. Stern, Kazuo Tsubota, Darlene A. Dartt, Rose M. Sullivan, and B. Britt Bromberg

Volume 507
EICOSANOIDS AND OTHER BIOACTIVE LIPIDS IN CANCER, INFLAMMATION, AND RADIATION INJURY 5
Edited by Kenneth V. Honn, Lawrence J. Marnett, Santosh Nigam, and Charles Serhan

Volume 508
SENSORIMOTOR CONTROL OF MOVEMENT AND POSTURE
Edited by Simon C. Gandevia, Uwe Proske, and Douglas G. Stuart

SENSORIMOTOR CONTROL of Movement and Posture

Edited by

Simon C. Gandevia
Prince of Wales Medical Research Institute
Sydney, New South Wales, Australia

Uwe Proske
Monash University
Melbourne, Victoria, Australia

and

Douglas G. Stuart
University of Arizona
Tempe, Arizona, USA

Kluwer Academic / Plenum Publishers
New York, Boston, Dordrecht, London, Moscow

Based on the Movement and Sensation International Symposium, held September 3–6, 2001, in Cairns, Australia

ISBN 0-306-47285-6

233 Spring Street, N.Y., NY 10013

http://www.wkap.nl/

10 9 8 7 6 5 4 3 2 1

A C.I.P. record for this book is available from the Library of Congress

Printed in the United States of America

AUTHORS

A

C. M. Adreani
University of California
Davis, CA 95616-8636 USA

J-M. Aimonetti
Centre National de la Recherche Scientifique (CNRS)
DPM-CNRS 31 chemin Joseph Aiguier
13402 Marseille cedex 20, France

A. Alaburda
MFI 12.5.9. The Panum Institute
University of Copenhagen
Blegdamsvej 3DK-2200N, Denmark

B. Alstermark
Department of Integrative Medical Biology
University of Umeå
S 901 87 Umeå, Sweden
Email: bror.alstermark@physiol.umu.se

J. Armand
Centre National de la Recherche Scientifique (CNRS)
Université de la Mediterranée
Faculté des Sciences du Sport
Marseille, France

B

R. W. Banks
University of Durham
Durham DH1 3LE
United Kingdom
Email: r.w.banks@durham.ac.uk

D. Bennett
University Centre for Neuroscience
University of Alberta
Edmonton ABTG6 2H9 Canada
Email: david.bennett@ualberta.ca

E. N. Benz
Rehabilitation Institute of Chicago & Department of Physical Medicine and Rehabilitation
Northwestern University Medical School
Chicago, IL 60611-3015 USA

G. S. Bewick
University of Aberdeen
Aberdeen AB25 2ZD UK
Email: g.s.bewick@abdn.ac.uk

M. D. Binder
Department of Physiology and Biophysics
University of Washington School of Medicine
Seattle, WA 98195 USA
Email: mdbinder@u.washington.edu

J. A. Brock
Prince of Wales Medical Research Institute
Randwick NSW 2031 Australia
Email: j.brock@unsw.edu.au

C. L. Brockett
Department of Physiology
Monash University VIC 3800 Australia
Email: camilla.brockett@med.monash.edu.au

I. E. Brown
A.E. Mann Institute for Biomedical Engineering
University of Southern California
Los Angeles, California 90089-1112 USA

J. Buchanan
Texas A&M University
College Station, Texas 77843 USA

D. Burke
Spinal Injuries Research Centre
Prince of Wales Medical Research Institute
Randwick NSW2031 Australia
Email: d.burke@unsw.edu.au

R. E. Burke
Laboratory of Neural Control
National Institute of Neurological Disorders and Stroke
National Institutes of Health
Bethesda, MD 20092-4455 USA
Email: reburke@helix.nih.gov

J. E. Butler
The Miami Project to Cure Paralysis
Department of Neurological Surgery
University of Miami School of Medicine
Miami, FL 33101 USA
Email: jbutler@miami.edu

C

M. B. Calford
School of Biomedical Sciences
University of Newcastle
Newcastle NSW 2308 Australia
Email: mike.calford@newcastle.edu.au

R. W. Carr
Prince of Wales Medical Research Institute
Randwick NSW2031 Australia
Email: rcarr@unsw.edu.au

O. Carter
Vision Touch and Hearing Research Centre
School of Biomedical Sciences
University of Queensland
St Lucia QLD 4072 Australia
Email: o.carter@vthrc.uq.edu.au

R. Chua
University of British Columbia
School of Human Kinetics
University of British Columbia
Vancouver BC V6T 1Z1 Canada
Email: rchua@interchange.ubc.ca

J. Cole
University of Southampton
Department of Clinical Neurosciences
University of Southampton and Poole Hospital
UK

J. G. Colebatch
Department of Neurology and Clinical School
Prince of Wales Hospital
Randwick NSW 2031 Australia
Email: j.colebatch@unsw.edu.au

D. F. Collins
Faculty of Physical Education and Recreation
University of Alberta
Edmonton, AB T6G 2H0 Canada
Email: dave.collins@ualberta.ca

R. Creath
Department of Kinesiology
University of Maryland
College Park, MD 20742 USA

D

B. L. Day
MRC Human Movement and Balance Unit
Institute of Neurology, Queen Square
London WC1N 3BG UK
Email: bday@ion.ucl.ac.uk

R. Davoodi
A.E. Mann Institute for Biomedical Engineering
University of Southern California
Los Angeles, California 90089-1112 USA

A. De Troyer
Brussels School of Medicine and Chest Service
Erasme University Hospital
Brussels 1070 Belgium
Email: a_detroyer@yahoo.fr

P. DiZio
Ashton Graybiel Spatial Orientation Laboratory & Center for Complex Systems
Brandeis University
Waltham, MA 02454 USA
Email: dizio@brandeis.edu

R. Durbaba
Division of Neuroscience
Imperial College School of Medicine
London W6 8RF UK
Email: r.durbaba@ic.ac.uk

M. Duxson
Department of Physiology
Otago School of Medical Sciences
University of Otago
PO Box 913
Dunedin, New Zealand
Email: marilyn.duxson@stonebow.otago.ac.nz

P. Dyhre-Poulsen
Institute of Medical Physiology and Institute of Anatomy
Panum Institute, University of Copenhagen
DK 2200 Copenhagen N Denmark
Email: dyhre@mfi.ku.dk

E

P. Ellaway
Division of Neuroscience
Imperial College School of Medicine
London W6 8RF UK
Email: p.ellaway@ic.ac.uk

F

J. B. Fallon
Department of Electrical Engineering and Computer Systems Engineering
Monash University
Clayton VIC 3800 Australia
Email: fallonj@mail.medoto.unimelb.edu.au

E. E. Fetz
Department of Physiology and Biophysics
University of Washington
Seattle, WA 98195-7290 USA
Email: fetz@u.washington.edu

R. C. Fitzpatrick
Prince of Wales Medical Research Institute
Randwick NSW 2031 Australia
Email: r.fitzpatrick@unsw.edu.au

H-J. Freund
Department of Neurology
Heinrich Heine University Duesseldorf
Duesseldorf 40225 Germany
Email: freund@rz.uni-duesseldorf.de

T. Fukunaga
Department of Life Sciences (Sports Sciences)
University of Tokyo
Komaba 3-8-9 Meguro Tokyo, Japan
Email: fukunaga@idaten.c.u-tokyo.ac.jp

G

S. C. Gandevia
Prince of Wales Medical Research Institute
Randwick NSW 2031 Australia
Email: s.gandevia@unsw.edu.au

M. Gorassini
University Centre for Neuroscience
University of Alberta
Edmonton AB TG6 2H9 Canada
Email: monica.gorassini@ualberta.ca

J. E. Gregory
Department of Physiology
Monash University VIC 3800 Australia
Email: ed.gregory@med.monash.edu.au

V. Gritsenko
Centre for Neuroscience
University of Alberta
Edmonton, Alberta T6G 2S2 Canada
Email: valeriya@ualberta.ca

M. Guerraz
Laboratoire "Sport, Performance
Santé" UFR STAPS, Montpellier, France

H

S. G. Hayes
University of California
Davis, CA 95616-8636 USA
Email: sghayes@ucdavis.edu

C. J. Heckman
Department of Physiology and Biophysics
University of Washington School of Medicine
Seattle, WA 98195 USA
Email: cjheckman@u.washington.edu

F. B. Horak
Neurological Sciences Institute
Oregon Health Sciences University
Beaverton, OR 97006 USA
Email: horakf@ohsu.edu

J. Hounsgaard
MFI 12.5.9. The Panum Institute
University of Copenhagen
Blegdamsvej 3DK-2200N Denmark
Email: j.hounsgaard@mfi.ku.dk

H. Hultborn
Department of Medical Physiology
Faculty of Health Sciences
University of Copenhagen
Blegdamsvej 3
Copenhagen DK-2200 Denmark
Email: h.hultborn@mfi.ku.dk

I

J. T. Inglis
School of Human Kinetics
University of British Columbia
Vancouver BC V6T 1Z1 Canada
Email: tinglis@interchange.ubc.ca

T. Isa
Department of Integrative Physiology
National Institute for Physiological Sciences
Myodaiji Okazaki 444-8585 Japan
Email: tisa@nips.ac.jp

J

J. Jeka
Department of Kinesiology
University of Maryland
College Park, MD 20742 USA
Email: jj96@umail.umd.edu

R. S. Johansson
Department of Integrative Medical Biology
Umeå University
SE 901 87 Umeå, Sweden
Email: roland.s.johansson@physiol.umu.se

K

H. Kanehisa
Department of Life Sciences (Sports Sciences)
University of Tokyo
Komaba 3-8-9 Meguro, Tokyo, Japan

R. Katz
Laboratoire de Neurophysiologie Clinique
Hôpital de la Salpêtrière
75651 Paris Cedex 13, France
Email: rose.katz@psl.ap-hop-paris.fr

M. P. Kaufman
Departments of Internal Medicine and Human Physiology
Division of Cardiovascular Medicine
University of California
Davis, CA 95616 8636 USA
Email: mpkaufman@ucdavis.edu

Y. Kawakami
Department of Life Sciences (Sports Sciences)
University of Tokyo
Komaba 3-8-9 Meguro, Tokyo, Japan
Email: kawakami@idaten.c.u-tokyo.ac.jp

P. M. Kennedy
School of Human Kinetics
University of British Columbia
Vancouver BC V6T 1Z1 Canada

D. Kernell
Department of Medical Physiology
University of Groningen
PO Box 196
9700 AD Groningen, The Netherlands
Email: dhkernell@hotmail.com

G. K. Kerr
School of Human Movement Studies
Queensland University of Technology
Brisbane QLD 4059 Australia
Email: g.kerr@qut.edu.au

P. A. Kirkwood
Sobell Department for Motor Neuroscience and Movement Disorders
Institute of Neurology
University College London
Queen Square, London WC1N 3BG UK
Email: pkirkwoo@ion.ucl.ac.uk

L

J. R. Lackner
Ashton Graybiel Spatial Orientation Laboratory & Center for Complex Systems
Brandeis University
Waltham, MA 02454 USA
Email: lackner@brandeis.edu

T. Lam
Centre for Neuroscience & Department of Physiology
University of Alberta
Edmonton AB T6G 2S2 Canada
Email: tlamb@ualberta.ca

N. Lan
A.E. Mann Institute for Biomedical Engineering
University of Southern California
Los Angeles, California 90089-1112 USA

R. N. Lemon
Sobell Department for Motor Neuroscience and Movement Disorders
Institute of Neurology
University College London
Queen Square, London WC1N 3BG UK
Email: rlemon@ion.ucl.ac.uk

G. E. Loeb
A.E. Mann Institute for Biomedical Engineering
University of Southern California
Los Angeles, California 90089-1112 USA
Email: gloeb@usc.edu

M

V. Macefield
Prince of Wales Medical Research Institute
Randwick NSW 2031 Australia
Email: vg.macefield@unsw.edu.au

M. A. Maier
INSERM U483
Université Pierre et Marie Curie
75005 Paris, France

V. Marchand-Pauvert
Service de Rééducation Neurophysiologie
Hôpital de la Salpêtrière
756551 Paris Cedex 13, France

Y. Matsuzaka
Department of Physiology
Tohoku University School of Medicine
Sendai 980-8575 Japan

P. B. C. Matthews
University Laboratory of Physiology
Parks Road, Oxford OX1 3PT UK
Email: peter.matthews@physiol.ox.ac.uk

C. Maurer
Neurological University Clinic
University of Freiburg
D 79106 Freiburg, Germany
Email: maurer@uni-freiburg.de

P. McNulty
Prince of Wales Medical Research Institute
Randwick NSW 2031 Australia
Email: p.mcnulty@unsw.edu.au

T. Mergner
Department of Neurology
University of Freiburg
D 79106 Freiburg, Germany
Email: mergner@uni-freiburg.de

S. Meunier
Laboratoire de Neurophysiologie Clinique
Hôpital de la Salpêtrière
75651 Paris Cedex 13, France
Email: meunier@chups.jussieu.fr

T. S. Miles
Department of Physiology
University of Adelaide
Adelaide SA 5005 Australia
Email: timothy.miles@adelaide.edu.au

D. L. Morgan
Department of Electrical and Computer Systems Engineering
Monash University VIC 3800 Australia
Email: david.morgan@eng.monash.edu.au

T. Muraoka
Department of Life Sciences (Sports Sciences)
University of Tokyo
Komaba 3-8-9 Meguro, Tokyo, Japan
Email: muraoka@idaten.c.u-tokyo.ac.jp

N

T. R. Nichols
Department of Physiology
Emory University
Atlanta, GA 30322 USA
Email: trn@physio.emory.edu

J. B. Nielsen
Department of Medical Physiology
University of Copenhagen
2200 Copenhagen N Denmark
Email: j.b.nielsen@mfi.ku.dk

M. A. Nordstrom
Department of Physiology
University of Adelaide
Adelaide SA 5005 Australia
Email: michael.nordstrom@adelaide.edu.au

P

A. Paul
Dept. of Anatomy and Structural Biology
Otago School of Medical Sciences
University of Otago
PO Box 913
Dunedin, New Zealand
Email: angelika.paul@stonebow.otago.ac.nz

K. G. Pearson
Centre for Neuroscience & Department of Physiology
University of Alberta
Edmonton, AB T6G 2S2 Canada
Email: kpearson@ualberta.ca

S. I. Perlmutter
Department of Physiology and Biophysics
University of Washington
Seattle, WA 98195 7290 USA

J-F. Perrier
MFI 12.5.9. The Panum Institute
University of Copenhagen
Blegdamsvej 3DK-2200N Denmark

R. J. Peterka
Neurological Sciences Institute
Oregon Health and Science University
Portland, Oregon USA

N. T. Petersen
Prince of Wales Medical Research Institute
Randwick NSW 2031 Australia
Email: nicolas@unsw.edu.au

J. D. Pettigrew
Vision Touch and Hearing Research Centre
School of Biomedical Sciences
University of Queensland
St Lucia QLD 4072 Australia
Email: j.pettigrew@vthrc.uq.edu.au

J. G. Pickar
University of California
Davis, CA 95616-8636 USA
Email: jgpickar@ucdavis.edu

E. Pierrot-Deseilligny
Service de Rééducation Neurophysiologie
Hôpital de la Salpêtrière
756551 Paris Cedex 13 France
Email: emmanuel.pierrot-deseilligny@chups.jussieu.fr

R. K. Powers
Department of Physiology and Biophysics
University of Washington School of Medicine
Seattle, WA 98195 USA
Email: rkpowers@u.washington.edu

A. Prochazka
Centre for Neuroscience
University of Alberta
Edmonton, Alberta T6G 2S2 Canada
Email: arthur.prochazka@ualberta.ca

U. Proske
Department of Physiology
Monash University VIC 3800 Australia
Email: uwe.proske@med.monash.edu.au

Y. Prut
Department of Physiology and Biophysics
University of Washington
Seattle, WA 98195 7290 USA

R

S. Rawlinson
Division of Neuroscience
Imperial College School of Medicine
London W6 8RF UK
Email: s.rawlinson@ic.ac.uk

K. Refshauge
School of Physiotherapy
Cumberland College of Health Sciences
University of Sydney
PO Box 170
Lidcombe NSW 2141 Australia
Email: k.refshauge@cchs.usyd.edu.au

B. Reid
University of Aberdeen
Aberdeen AB25 2ZD UK
Email: brian.reid@abdn.ac.uk

C. Richardson
University of Durham
Durham DH1 3LE UK
Email: a.c.richardson@durham.ac.uk

J. C. Rothwell
Sobell Department of Neurophysiology
Institute of Neurology
Queen Square, London WC1N 3BG UK
Email: j.rothwell@ion.ucl.ac.uk

M. J. Rowe
Department of Physiology and Pharmacology
University of New South Wales
Sydney NSW 2052 Australia
Email: m.rowe@unsw.edu.au

P. Rudomin
Department of Physiology, Biophysics and Neurosciences
Centro de Investigación y de Estudios Avanzados del Instituto Politécnico Nacional
México DF 07000 Mexico
Email: rudomin@fisio.cinvestav.mx

W. Z. Rymer
Rehabilitation Institute of Chicago & Department of Physical Medicine and Rehabilitation
Northwestern University Medical School
Chicago, IL 60611-3015 USA
Email: zevric@casbah.acns.nwu.edu

S

M. H. Schieber
Departments of Neurology & Neurobiology and Anatomy
University of Rochester School of Medicine and Dentistry
Rochester, New York 14642 USA
Email: mhs@cvs.rochester.edu

B. D. Schmit
Rehabilitation Institute of Chicago & Department of Physical Medicine and Rehabilitation
Northwestern University Medical School
Chicago, IL 60611-3015 USA

A. Schmied
Centre National de la Recherche Scientifique (CNRS)
DPM-CNRS 31 chemin Joseph Aiguier
13402 Marseille cedex 20
France
Email: schmied@dpm.cnrs-mrs.fr

K. Seki
Department of Physiology and Biophysics
University of Washington
Seattle, WA 98195-7290 USA

P. Sheard
Department of Physiology
Otago School of Medical Sciences
University of Otago, PO Box 913
Dunedin, New Zealand
Email: phil.sheard@stonebow.otago.ac.nz

K. Shima
Department of Physiology
Tohoku University School of Medicine
Sendai 980-8575 Japan

M. Simonetta-Moreau
Institut National de la Santé et de la Recherche Médicale (INSERM)
455CHUPurpan, Toulouse, France

E. B. Simonsen
Institute of Medical Physiology and Institute of Anatomy
Panum Institute
University of Copenhagen
DK 2200 Copenhagen N Denmark

T. Sinkjaer
Center for Sensory-Motor Interaction
University of Aalborg
9220 Aalborg, Denmark

D. G. Stuart
Department of Physiology
The University of Arizona College of Medicine
Tucson, AZ 85724-5051 USA
Email: dgstuart@u.arizona.edu

T

J. Tanji
Department of Physiology
Tohoku University School of Medicine
Sendai 980-8575 Japan
Email: tanjij@mail.cc.tohoku.ac.jp

A. Taylor
Division of Neuroscience
Imperial College School of Medicine
London W6 8RF UK
Email: ataylor@ic.ac.uk

J. L. Taylor
Prince of Wales Medical Research Institute
Randwick NSW 2031 Australia
Email: jl.taylor@unsw.edu.au

C. K. Thomas
The Miami Project to Cure Paralysis
Department of Neurological Surgery
University of Miami School of Medicine
Miami, FL 33101 USA
Email: cthomas@miami.edu

K. S. Türker
Department of Physiology
University of Adelaide
Adelaide SA 5005 Australia
Email: kemal.turker@adelaide.edu.au

V

J-P. Vedel
Centre National de la Recherche Scientifique (CNRS)
DPM-CNRS 31 chemin Joseph Aiguier
13402 Marseille cedex 20 France
Email: vedel@dpm.cnrs-mrs.fr

W

D. L. Wardman
Prince of Wales Medical Research Institute
Randwick NSW 2031 Australia
Email: d.wardman@unsw.edu.au

C. Wells
School of Human Kinetics
University of British Columbia
Vancouver BC V6T 1Z1 Canada

A. K. Wise
Department of Physiology
Monash University
Clayton VIC 3800 Australia
Email: wisea@mail.medoto.unimelb.edu.au

C. J. Worringham
School of Human Movement Studies
Queensland University of Technology
Brisbane QLD 4059 Australia

Y

S. Yakovenko
Centre for Neuroscience
University of Alberta
Edmonton, Alberta T6G 2S2 Canada
Email: sergiy@ualberta.ca

H-W. Yang
Department of Life Sciences
Chung Shan Medical and Dental College
Taichung 402 Taiwan

Z

I. Zijdewind
Department of Medical Physiology
University of Groningen
Groningen
The Netherlands
Email: i.zijdewind@med.rug.nl

FOREWORD

This collection of contributions on the subject of the neural mechanisms of sensorimotor control resulted from a conference held in Cairns, Australia, September 3-6, 2001. While the three of us were attending the International Union of Physiological Sciences (IUPS) Congress in St Petersburg, Russia, in 1997, we discussed the implications of the next Congress being awarded to New Zealand. We agreed to organise a satellite to this congress in an area of mutual interest - the neuroscience of movement and sensation.

Australia has a long-standing and enviable reputation in the field of neural mechanisms of sensorimotor control. Arguably this reached its peak with the award of a Nobel Prize to Sir John Eccles in 1963 for his work on synaptic transmission in the central nervous system. Since that time, the subject of neuroscience has progressed considerably. One advance is the exploitation of knowledge acquired from animal experiments to studies on conscious human subjects. In this development, Australians have achieved international prominence, particularly in the areas of kinaesthesia and movement control. This bias is evident in the choice of subject matter for the conference and, subsequently, this book. It was also decided to assign a whole section to muscle mechanics, a subject that is often left out altogether from conferences on motor control.

Cairns is a lovely city and September is a good time to visit it. Since we wanted to offer our international colleagues something more than just a high-standard conference, we picked Cairns as the venue. It meant that we would be close to one of the wonders of the world, the Great Barrier Reef. The organisers took the unprecedented decision to completely interrupt the conference for one day, while all conference delegates visited and enjoyed the marvels of the Great Barrier Reef. It generated an atmosphere of informality and shared adventure that helped to break down even the most intractable communication barriers between conference delegates. The free exchange of ideas led in several instances to the formulation of new consensus views and to agreement over the way forward in future experiments. The conference was brought to a dramatic end with a gala dinner at which delegates were introduced to aspects of Australian aboriginal culture.

In assembling the book we have chosen to slightly alter the order of presentations from that used at the conference. This was done in an attempt to bring related topics as close together as possible. Each section is preceded by a Preface. In the preface we make reference to particular chapters and use the names of the presenting authors from the conference rather than first-named authors of the chapters. Each chapter underwent a formal review process by experts in the field so, hopefully, the standard maintained during the conference is reflected in the level of scholarship achieved in this book.

The volume includes contributions by two authors who were unable to attend the Symposium: a chapter on presynaptic inhibition by Pablo Rudomin and one on motoneurones by Peter Matthews, who also wrote a Preface for the whole volume.

Organising a conference with 180 delegates, most of whom were coming from overseas, proved to be a mammoth job. It would not have been possible without considerable help from various quarters. We would like to thank in particular the various helpers in Sydney and Melbourne in Australia, and Tucson in the USA. The bulk of the organisational burden fell on the Sydney group, in particular, Jane Butler, Robert Gorman and Nicolas Petersen. Communication between the organisers and delegates, attendance at the reception desk and overseeing many crucial details at the conference were all competently and efficiently carried out by Mary Sweet and Emily Mifsud. We would also like to thank Emily for her huge input to the task of formatting chapters and assembling the book. We gratefully acknowledge financial support from the Prince of Wales Medical Research Institute, Sydney and Monash University, Melbourne. Additional financial support for American delegates was provided by the National Institutes of Health, Bethesda, MD, USA (NS 41876). Finally, we would like to acknowledge support from the IUPS and thank its representatives for assigning to the meeting the status of an official satellite symposium.

Simon C. Gandevia **Uwe Proske** **Douglas G. Stuart**

PREFACE

Those from the Antipodes have contributed nobly to the advance of neurophysiological understanding of movement and sensation, the basis of all human activity. In the first half of the last century their talent was largely exported, especially to Britain and the USA, and this has continued with several such emigrants returning home for the present meeting. Notable examples in other fields of science include Rutherford, a New Zealander working in Cambridge, who laid the groundwork for "splitting the atom" and Florey, an Australian working in Oxford, who showed that Fleming's curiosity of penicillin could be used to such wondrous effect in man. In the second half of the 20th century the balance shifted and most remained at home, developing many distinguished laboratories. This was facilitated by the dramatic growth in the ease and speed of travel and most recently by electronic communication; speaking personally, in 1965 I took over a month to reach Sydney from the UK by ship. Jack Eccles, whose contribution and influence was spelt out in a Poster, marked the turning point; educated in Melbourne, he travelled to Oxford in 1925 to work with Sherrington and established himself there apparently for ever. But he then returned to Australia in 1937, subsequently moved to New Zealand where he discovered the IPSP in 1951, and then came back to Australia to make Canberra a Mecca for a generation of neurophysiologists; finally, approaching 65, he postponed retirement by moving to the USA in 1966. The present symposium, held in Australia with a wide-ranging international attendance, helped celebrate this maturation of home-grown neuroscience; its organisers have a long record of distinguished contribution, 17 of the speakers were Australian as would have been equally appropriate if the meeting had been held somewhere else, and 33 of the 84 posters were Australian.

The format of the meeting was standard, with 10 two hour sessions ranging from sensory receptors, motoneurones, and interneurones to the motor cortex with the emphasis on the lower level mechanisms. Each session started with an overview by the chairman followed by a single major paper; four "discussants" then each gave shorter descriptions of their own work within the same area. The present volume gathers these papers together, with each section now briefly introduced by the organisers. Thus the volume ranges from the general to the specific. It thereby provides a welcome up-dating on a variety of topics. This is all most helpful; although the topics may sound familiar the focus of interest has shifted very considerably, with new findings leading to new thinking.

Certain particular advances may be high-lighted by comparing this volume with a comparable symposium on "Muscle afferents and the spinal control of movement" held 10 years earlier in Paris to which 19 of the present speakers also contributed. The muscle spindle remains of central interest, especially its role in walking. The underlying fusimotor drive for the cat gastrocnemius has now apparently been definitively fractionated, both temporally and into its functional components. The stretch reflex retains its importance in supporting extension during the stance phase, but tendon organ afferents are now thought to assist rather than antagonise the spindle afferents. Moreover, both types of afferent have been shown to take part in timing the switching from extension to flexion, and back again, in multi-sensory rule based operations instead of being simply responsible for "resistance reflexes". In addition, their signals are essential in enabling higher control centres to adaptively adjust their output to match the biomechanical properties of the limb, including the effect of changing gravitational fields and the effects of rotation (Coriolis forces). These Coriolis studies have also dealt a severe blow to servo type hypotheses of goal-directed movements in which the movement is specified by the balance of spindle or muscle activity required at the end point. Interest in proprioception continues unabated, but muscle receptors no longer reign in isolation; detailed information from cutaneous receptors has now been shown to contribute both to sensory awareness and to the up-dating of motor commands. Finally, the histology of the spindle continues to surprise with new evidence suggesting that the liberation of glutamate from synaptic type vesicles contributes to its firing.

The motoneurone has also moved on from ten years ago. Plateau potentials have come of age and are now seen as the outward sign of an extreme action of an omni-present mechanism for regulating MN responsiveness; there is intense investigation of the underlying metabotropic receptors which facilitate the voltage activated channels that generate the prolonged inward current. Human motor studies continue in health and disease, with modelling becoming inreasingly important for interpreting the findings. Synaptic noise has been recognised as crucial in triggering low-frequency firing, which occurs while the MN's mean membrane potential remains subthreshold, with firing continuing at the final equilibrium potential when the AHP is completed. The biomechanical properties of the motor unit remains of interest, with a new concentration on the effect of forcibly lengthening contracting muscle fibres (eccentric contraction). Interneurones have been successfully recorded from during voluntary hand movments in the monkey, but remain an enigma because of their great functional plasticity; moreover, violent debate continues on a largely anatomical matter, namely whether or not humans have a powerful C3\C4 propriospinal system mediating motor commands as in the cat. The "silent" vestibular system continues to be probed in man, and in particular how it interacts with other sensory inputs in controlling posture.

The classical study of the motor cortex continues with a mix of stimulation and recording, showing that simple ideas of a finely-grained topographical localisation of function become ever less tenable. Grossly separate motor areas are confirmed to differ in anatomical and functional organisation; but little emerges as to the precise parcellation of function between them, in the way things are known for many of the multiple cortical visual areas. The new light on the horizon is the recognition of a synchronisation of the firing of neurones in cortical motor areas that varies with the conditions; this occurs at 15-30 Hz and "paces" the motoneurones so that it can be detected in the EMG, giving a signal that is coherent with both electric and magnetic cortical recordings. The current challenge is to decide whether this of itself represents an important form of signal coding or is simply an epi-phenomenon arising from other mechanisms. This was approached from the standpoint of the human studies, without the related animal work being presented. It remains an important topic for the future, probably requiring the development of yet more detailed ways for the mathematical analysis of multiple recordings, accompanied by modelling.

In conclusion, contrasting this volume with its related predecessor of ten years ago shows a steady advance rather than dramatic breakthroughs. But the progress has been very real and cannot be ignored by those who teach the next generation and who should thus aim for a wider view than can be found in many textbooks. One forcible impression is that the current tools may be inadequate for understanding function much above the level of the motoneurone, and that particularly for human work even those in use require progressive refinement and testing, of which there are examples here. However, new tools have their dangers and their underlying basis must be subjected to continued critical thought and analysis. The current example is the widespread deployment of transcranial magnetic stimulation of the motor cortex, with the machinery so readily available and so easy to use. The oft-repeated statement that a simple entity entitled 'cortical excitability' is then being directly measured represents an almost wilful neglect of the underlying complexities; these are still being probed some 15 years after the inception of the technique. John Hunter's classical dictum "but why think. Why not try the experiment" can still be justified when approaching a bewilderingly complex situation; but once the initial results have been obtained they must be subjected to intense intellectual analysis, even more so than when an explicit hypothesis is being tested.

Peter B.C. Matthews

CONTENTS

SECTION VI: LOCOMOTION

SECTION VII: SUPRASPINAL CONTROL OF MOVEMENT

SECTION VIII: MECHANICS AND MOVEMENT

SECTION I

Impulse Initiation and Central Transmission from Muscle and Skin

This section considers properties of sensory receptors, aspects of their stimulus transduction, their response properties and functional roles. U. Proske (Chapter 1) discusses muscle receptors, in particular, the muscle spindle. He considers uncertainties involved in deciphering fusimotor activation patterns during combined γ_S and γ_D stimulation, as has been done recently for activity recorded during locomotion (see Chapter 38). The underlying receptor mechanisms remain under debate and there is uncertainty over response summation for a given set of conditions (Fallon et al., 2001). Proske also raises some points about signalling properties of tendon organs. It has become the popular view that any one tendon organ provides an inaccurate and incomplete signal of whole muscle tension and a functional role for tendon organs is derived from ensemble responses.

Repeated eccentric contractions of a muscle lead to damage in muscle fibres and the rise in tension effected by the injury contracture is signalled by tendon organs. When the whole muscle is subjected to eccentric contractions, with the exception of the 6-7 motor units with a specific excitatory action on one tendon organ, all of the muscle's tendon organs signal the presence of damage, that is, all but the one tendon organ whose motor units have not been stimulated. Thus specific information about local, regional tension changes can be provided by tendon organs. It remains uncertain whether the central nervous system makes use of this information.

On the subject of sensory transduction, R. Banks (Chapter 2) provided evidence of the presence of synaptic-like vesicles in the sensory terminals of rat muscle spindles. These vesicles recycle in an activity-dependent way, and release glutamate which has an autogenic excitatory effect on mechanosensory transduction. The findings recall the observation that axonal transport from the cell body to the distal terminals of the receptor is involved in the establishment and maintenance of typical response properties of muscle receptors (Proske and Luff, 1997). Conceivably, such transport could include synaptic-like vesicles.

Some exciting new observations are presented by R. Carr (Chapter 3) on aspects of sensory transduction in cold thermoreceptors in the cornea of the guinea pig. Extracellular recordings of nerve terminal impulses showed that tissue cooling reduced action potential amplitude, slowing of its time course, and an increase in frequency. Warming had the reverse effects. These changes may reflect the change in membrane potential. How cooling achieves its effects remains to be elucidated. One suggestion is that it acts on transmembrane sodium and potassium conductances that are temperature sensitive.

M. Kaufman (Chapter 4) summarises the properties of muscle receptors served by small-diameter afferents, conducting in the Gp III and Gp IV range. Both have aspects that would categorise them as polymodal receptors. Both respond to chemical and mechanical stimuli, although for mechanical stimuli this is normally in the noxious range for Gp IV. Responses, particularly for Gp IV, increase during ischemia. Infusing a stretch-activated channel blocker, gadolinium into the muscle reduced responses of Gp III afferents to contraction and stretch as well as suppressing the known reflex actions of these fibres, an increase in blood pressure, heart rate and respiration rate. Finally, it has recently been shown that both Gp III and Gp IV afferents can be activated during locomotion in the decerebrate cat evoked by midbrain stimulation. However the final word on the role of these receptors in regulation of skeletal muscle function has not been said.

D. Burke (Chapter 5) sounds a note of caution that many of us would do well to heed. He points out that there are important changes in axon excitability of human peripheral nerve following activity and that these changes are not the same for sensory and motor axons. So trying to control for the size of an afferent volley in the H-reflex by maintaining constancy of M-response amplitude turns out not to be reliable. Yet, at present, we have nothing better for this kind of experiment. It will be important, in the future, to introduce some simple practical guides for generating afferent and motor volleys of reliably reproducible size.

An important consideration is the reflex potency of particular groups of afferents. V. Macefield (Chapter 6) shows recordings of EMG in hand muscles triggered by activity in single, identified mechanoreceptors in the skin. The timing of activity peaks points to an oligosynaptic linkage. This linkage was absent for SAI afferents (Merkel receptors) and for muscle spindle endings. The latter result fits with previous results (Gandevia et al., 1986). While this observation illustrates the ability of cutaneous inputs to modify motor output, it raises the interesting question as to why the linkage is not demonstrable for the SAI afferents. Perhaps their role is more important at perceptual rather than reflex levels.

The issue of the central transmission of afferent information is raised by M. Rowe (Chapter 7). He documents the remarkable security of transmission across synapses in single dorsal column cells in the cat, security which exists whether the afferent innervates a joint, a muscle spindle, a Pacinian corpuscle or one of the slowly adapting cutaneous receptors. It is unclear whether properties of this linkage vary with the location of the peripheral receptors, although (in humans) the potency of segmental and supraspinal reflexes and the strength of percepts evoked by stimulation depends on the location of the input (e.g. finger tip versus forearm; see, Vallbo et al., 1984; Torebjork et al., 1987).

REFERENCES

Fallon, J. B., Carr, R. W., Gregory, J. E., and Proske, U., 2001, Summing responses of cat soleus muscle spindles to combined static and dynamic fusimotor stimulation, *Brain Research,* 888, 348-355.

Gandevia S.C., Burke D., and McKeon B., 1986, Coupling between human muscle spindle endings and motor units assessed using spike-triggered averaging, *Neuroscience Letters,* 71, 181-186.

Proske, U., and Luff, A. R., 1998, Mechanical sensitivity of muscle afferents in a nerve treated with colchicine. *Experimental Brain Research,* 119, 391-398.

Torebjork H.E., Vallbo A.B., and Ochoa J.L., 1987, Intraneural microstimulation in man. Its relation to specificity of tactile sensations, *Brain,* 110, 1509-1529.

Vallbo A.B., Olsson K.A., Westberg K.G., and Clark F.J., 1984, Microstimulation of single tactile afferents from the human hand. Sensory attributes related to unit type and properties of receptive fields, *Brain,* 107, 727-749.

1

SIGNALLING PROPERTIES OF MUSCLE SPINDLES AND TENDON ORGANS

Uwe Proske and John E. Gregory*

ABSTRACT

Some important issues for muscle receptors remain unresolved. For muscle spindles it is uncertain how responses to combined static and dynamic fusimotor stimulation may summate. Such summation may occur during certain phases of locomotion. Two mechanisms considered here include electrotonic spread of generator current between sources of impulse activity and mechanical interactions between contracting intrafusal fibres. For tendon organs, it remains uncertain what aspects of muscle tension they signal. Here they were tested for their ability to respond to rises in whole-muscle passive tension after eccentric contractions. It was found that only when motor units were contracted which had a direct action on a tendon organ did it signal the rise in tension. The finding raises questions about the role of tendon organs as monitors of muscle tension.

INTRODUCTION

In recent years studies of the motor system have paid less attention to muscle receptors than in the past. Perhaps that is because we believe that most of what we need to know about them has already been described. The aim of this account is to point out that there remain a number of areas of uncertainty in our understanding of the internal workings of muscle spindles and tendon organs and of the central processing of their afferent signals.

THE MUSCLE SPINDLE

Muscle spindles have been found in almost all skeletal muscles of mammals (Barker, 1974). Spindles lie alongside the ordinary muscle fibres and because of their location they are thought to signal muscle length. In the larger muscles they lie alongside only part

* Department of Physiology, Monash University, Melbourne, Victoria, 3800, Australia.
Email: uwe.proske@med.monash.edu.au

of the length of muscle fibres and here it is assumed that the length changes experienced by the spindle are representative of changes in the whole muscle (Proske et al., 2000).

Each spindle consists of a bundle of intrafusal fibres enclosed within a fluid-filled, connective tissue capsule. The middle of each intrafusal fibre is contacted by the spiral terminations of the Group Ia fibre, the primary endings. There are three kinds of intrafusal fibres, bag_1, bag_2 and chain. All receive terminations of the Ia fibre. The bag_2 and chain fibres may also receive secondary endings from a Group II fibre. A feature that distinguishes the spindle from other sensory receptors is that it has its own motor supply, the fusimotor innervation. Two kinds of fusimotor fibre have been identified, γ_S, static fusimotor neurons and γ_D, dynamic fusimotor neurons. The γ_D axons terminate on bag_1 fibres while γ_S axons terminate on bag_2 and chain fibres. Stimulation of γ_D axons produces a small increase in spindle firing at a set muscle length, but greatly increases the dynamic component of a response to a length change. γ_S stimulation produces a much larger increase in firing at a set length and leaves the dynamic response largely unchanged.

Studies of the responses of passive spindles, that is, in the absence of fusimotor activity, indicate that impulse activity in the primary ending, both background activity and activity in response to a stretch, arises largely in the bag_2 fibre (Proske et al., 1991; Gioux et al., 1991; Morgan et al., 1991). That is, the passive spindle, functionally speaking, over much of the muscle's working range behaves as though it is a single intrafusal fibre spindle. This is counter-intuitive in the sense that the intrafusal fibre responsible for enhancing the dynamic response during fusimotor activity, the bag_1 fibre, does not contribute to the dynamic response of the passive spindle.

Because the spindle is supplied by a large, branching sensory axon, it raises the question of how activity is generated in different branches. While branching of the afferent axon is quite common amongst mechanoreceptors, the issue is particularly important for the spindle where different terminals are under the influence of activity generated by different fusimotor fibres.

Observations of afferent responses to different patterns of fusimotor input have led to proposal of a model where activity generated in the bag_1 sensory terminals during dynamic fusimotor stimulation represents one source of afferent impulses that feeds in to the parent axon. The other source is the summed activity from bag_2 and chain fibres (Carr et al., 1996; Carr and Proske, 1996). The question raised by such an arrangement is how to account for the spindle's response to combined static and dynamic fusimotor inputs. There is evidence of overlapping static and dynamic fusimotor activity during certain phases of the locomotion cycle, so an answer to this question has some functional importance (Taylor et al., 2000; see also Taylor, Chapter 38).

The basic observation is that static and dynamic fusimotor responses measured at constant length show partial occlusion: that is, combined stimulation produces a response bigger than the larger individual response, but smaller than their sum. How can this be accounted for? Three possible mechanisms have been considered (Banks et al., 1997) (Fallon et al., 2001).

In the model, during combined fusimotor stimulation, impulse streams will arise from two sources. Normally the impulse generator with the higher mean rate will suppress the second generator, by re-setting (Horch et al., 1974). Therefore on combined stimulation of two, non-equal fusimotor inputs, the larger should entirely determine the discharge in the parent axon. That is not what is observed experimentally. One explanation for this result is probabilistic mixing (Eagles and Purple, 1974). In probabilistic mixing, if there is some irregularity in the two converging impulse streams, a longer interval in the dominant stream can be filled by impulses from the normally suppressed generator. Probabilistic mixing is a regularising process, leading to a small reduction in inter-impulse intervals and a reduction in variability. But we can dismiss probabilistic mixing as a significant mechanism for summation from combined fusimotor inputs because invariably the observed increase in combined rate is much larger than predicted by the statistical model (Carr and Proske, 1996).

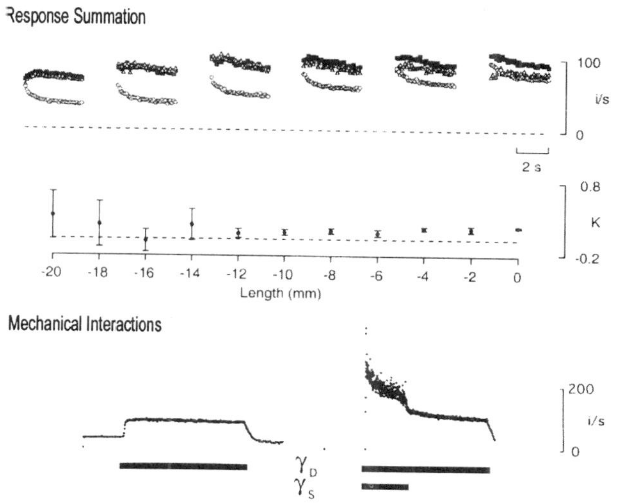

Figure 1. Response summation and mechanical interactions in muscle spindles. Upper panel, response summation for the primary ending of muscle spindle of the soleus muscle of the anaesthetised cat during stimulation (100 pps) of a static fusimotor fibre (open triangles), a dynamic fusimotor fibre (open circles) and both fibres together (filled squares). Action potentials shown as instantaneous frequency. Responses were recorded while the muscle was held at each of a number of lengths, given in mm below maximum body length, (L_m), which was assigned a value of zero. Summation of the two fusimotor inputs was expressed as K, the summation coefficient.

$$K = \frac{\text{Combined Response} - \text{Larger Individual Response}}{\text{Smaller Individual Response}}$$

K values (±S.E.M.) are given for the pooled data from 4 afferents. Dashed lines indicate zero. (Redrawn from Fallon et al., 2001.) Lower panel, mechanical Interactions. Instantaneous frequency plots of the responses of the primary ending of a cat soleus spindle to stimulation of a dynamic fusimotor fibre (γ_D) at 200 pps for 6 s (left-hand panel) and combined stimulation of both a static (γ_S) and a γ_D fibre (right-hand panel). The two γ fibres were stimulated together for 2 s, then γ_S stimulation was stopped while γ_D stimulation was continued for a further 4 s. Filled bars below the frequency records indicate periods of stimulation of fusimotor fibres. All measurements were made at L_m – 6 mm. (Redrawn from Carr et al., 1998.)

A second mechanism for summation of fusimotor responses is based on the idea that current from one impulse generating site can spread passively through the branching afferent tree to influence other generators (Banks et al., 1997). So summation of fusimotor responses would, in part, be the result of summation of receptor currents from different sources. It is envisaged that current summation can take place despite ongoing resetting. However, there is no direct evidence for such a mechanism. We have tried to obtain indirect support for the idea by studying fusimotor responses at different muscle lengths.

Our reasoning was that as a spindle is stretched to progressively longer lengths, fusimotor responses will grow according to the intrafusal length-tension relation. But since the branches of the afferent axon are unlikely to stretch very much during muscle lengthening, the electrotonic path length between impulse generators should remain about the same, so that summation should change in direct proportion to the increase in fusimotor responses. Yet at lengths beyond L_m −12, K, the summation coefficient, remained approximately constant (Fig. 1, upper panel). At shorter lengths, K tended to be larger and there was considerable variation in values between afferents. We conclude that spreading generator current, by itself, cannot adequately account for response summation.

That leaves the third possibility. Because the intrafusal fibres are bound together with connective tissue and elastic filaments (Cooper and Gladden, 1974), it provides the opportunity for mechanical interactions between contracting intrafusal fibres. The proposal here is that during combined fusimotor stimulation the impulses are all coming from one generator but the firing rate is higher than with either input on its own because one contracting intrafusal fibre is able to transmit some of its tension to its neighbours. We have been able to provide evidence for such a mechanism (Fig. 1, lower panel). When static and dynamic fusimotor fibres are stimulated together and then stimulation of the static axon is stopped, the persisting response is higher than for stimulation of the dynamic fusimotor fibre on its own.

We conclude that fusimotor response summation appears to involve two mechanisms, electrotonic spread of generator current and mechanical interactions between contracting intrafusal fibres. The third possibility of probabilistic mixing appears less important. Putting these influences on a quantitative basis will require a different kind of experiment to those described here. What does emerge, in view of the large variability in summation coefficients at different muscle lengths and between spindles, is that its precise value is unlikely to be important for central processing.

TENDON ORGANS

Tendon organs are distributed almost as widely in skeletal muscles as muscle spindles. There are occasional examples of muscles, like tenuissimus in the cat, which contain spindles but not tendon organs. A tendon organ consists of a connective tissue capsule which is penetrated by a large-diameter Ib axon. The axon branches and makes sprays of terminations on strands of tendon each of which is attached to a muscle fibre. Contraction of muscle fibres stretches the tendon strands and that, in turn, stretches the nerve endings (Fukami and Wilkinson, 1977). So tendon organs, like muscle spindles, are stretch receptors. However stretch threshold for tendon organs is much higher than for muscle spindles and tendon organs typically do not exhibit background activity. They are thought to be monitors of muscle tension. They respond preferentially to contraction of a small number of motor units that contribute muscle fibres to insert on sensory-innervated tendon strands within the receptor. The tendon organ appears to largely ignore tension changes in other parts of the muscle (Gregory et al., 1986).

The number of experimental manipulations that can be used to study tendon organs is more limited than for muscle spindles with their fusimotor supply. We have recently introduced a new experimental means of studying tendon organs based on their ability to signal tension changes in a muscle after exercise.

When a contracting muscle is stretched, an eccentric contraction, particularly if it is stretched to long muscle lengths, some parts of muscle fibres become damaged. During repeated eccentric contractions the damage extends to membranous elements of the excitation contraction-coupling system. That, in turn, leads to an uncontrolled release of Ca^{2+} into the sarcoplasm and development of a local injury contracture (Morgan and Allen, 1999). Such contractures are believed to be responsible for the observed rise in whole-muscle passive tension after exercise (Whitehead et al., 2001).

There are reports in the literature of disturbance of the sense of tension following eccentric exercise (Saxton et al., 1995; Brockett et al., 1997). This raised the question of whether changes in passive tension following eccentric exercise, or perhaps the exercise itself, led to a change in tendon organ responsiveness. We therefore embarked on a series of experiments in which we subjected the medial gastrocnemius muscle of the anaesthetised cat to eccentric contractions while measuring responses of tendon organs before and after the contractions.

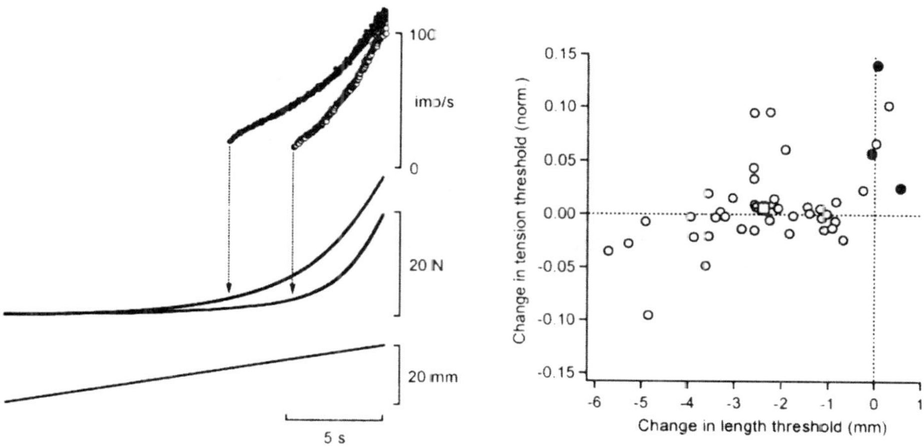

Figure 2. Effect of eccentric contractions on the response of tendon organs.
Left panel, response of a tendon organ of the medial gastrocnemius muscle of the anaesthetised cat to a slow, 20 mm stretch at 1 mm s^{-1}, across the full physiological range of the muscle, before (open symbols) and after (filled symbols) a series of 50 eccentric contractions of the muscle. Eccentric contractions were produced by repetitive stimulation of the muscle at 80 pps and simultaneous, 6 mm stretch at 50 mm s^{-1}. All measurements were made at the muscle's optimum length for active tension. Top trace instantaneous frequency display, middle trace tension, bottom trace length.
Right panel, Change in tension threshold for each of 40 tendon organ afferents (open circles), normalised to tension at maximum body length, plotted against the change in length threshold, after 50 eccentric contractions. Negative values indicate shorter length thresholds after the contractions. Open square, mean (± S.E.M.) value for the 40 afferents. Filled circles, 3 additional tendon organs whose motor units had not been subjected to eccentric contractions. (Redrawn from Gregory et al., 2001.)

In agreement with earlier reports (Whitehead et al., 2001), it was found that after 50 - 150 eccentric contractions, passive tension had increased over all of the muscle's working range. At the maximum physiological length of the muscle, L_{max}, passive tension had risen by an average of 33%. As a result, the muscle had to be stretched less to reach threshold for a particular tendon organ. Tension thresholds of tendon organs had changed very little. So, for example, after the contractions, the discharge of a tendon organ started at a length shorter by 3.4 mm (Fig. 2, left-hand panel). However the tension threshold of the tendon organ, 4.6 N, had not changed. Essentially the same pattern was seen with all 40 tendon organs (Fig. 2, right-hand panel). The mean change in length threshold (2.4 mm) was significant, while the mean tension threshold (4 N) had not changed significantly. These observations therefore confirmed that the small rise in passive tension produced by the eccentric contractions was able to be detected by tendon organs.

Length thresholds fell for nearly all of the tendon organs studied. This suggested that when a muscle was subjected to 50 or more eccentric contractions, the passive tension changes were wide-ranging, and not restricted to a particular part of the muscle. However, since tension was measured at the muscle tendon, it could be argued that passive tension had risen in only one part of the muscle and spread to tendon organs in the rest of the muscle. In additional experiments this possibility was eliminated.

On three separate occasions the usual complement of about 6 tendon organs was isolated in filaments of dorsal root, but for one of them the ventral root was subdivided, seeking individual motor units with strong excitatory actions on its response. These were most likely motor units which contributed muscle fibres to insert directly into the receptor. Typically, between 6 and 8 motor units could be isolated in this way. Then the remaining motor supply to the muscle was stimulated to produce eccentric contractions in the same way as had been done before. The isolated motor units were not stimulated. Subsequently, tendon organs showed the typical reduction in length threshold, indicative of a rise in whole-muscle passive tension. All, that is, except for the tendon organs whose complement of motor units had been separated out and not stimulated (Fig. 2, lower panel). The important conclusion from this observation is that after the eccentric contractions, not only were the passive tension changes in the muscle widespread, but tendon organs signalled the increase in tension only if it involved motor units with muscle fibres inserting directly into them. It indicated that the rise in passive tension had taken place in muscle fibres and not in extra-muscular connective tissue or tendon. Furthermore, it reinforced the view that a tendon organ signalled only local, regional tension changes and, under the conditions of our experiments, remained unresponsive to tension changes generated elsewhere in the muscle. The point is important because it has implications for ideas about the role of tendon organs as monitors of muscle tension. It has been suggested that because tendon organs respond both to forces acting in series and in parallel with the receptor, they are not suited to a role as absolute force sensors but are concerned primarily with signalling changes in force (Jami et al., 1992). However, it seems that when it comes to signalling small changes in passive tension in the muscle, each tendon organs responds only to the tension directly in series with it and remains unresponsive to other changes.

Following a series of eccentric contractions there is both a rise in passive tension in the muscle and a large fall in active tension. In our experiments tension fell by an average of 52%. As well as studying tendon organs during passive tension changes, their responses were measured during gradations of active tension, before and after the eccentric contractions. It was found that the discharge frequency : tension relationship had not changed after the contractions. In particular, the slope of the relation, an expression of tendon organ sensitivity to tension changes, had not changed. It suggested that the eccentric contractions had not altered the tendon organs' tension signalling capacity.

To conclude, this study reaffirms the view that tendon organs are regional tension sensors in that they signal the tension only in those muscle fibres inserting into them, whether it is active or passive. However, since these muscle fibres typically belong to different motor units, each with its fibres scattered through about one-third of the muscle (Burke and Tsairis, 1973), a tendon organ appears to sample tension generated across a wide area of muscle. While individual tendon organs may not accurately signal whole muscle tension, their ensemble discharge probably does (Prochazka and Gorassini, 1998). However, if the role of tendon organs is to provide the central nervous system only with information about whole muscle force, why are they so numerous, and why are they located where they are? Would not just a few receptors suffice, located out in the common tendon? That they are organised in the way they are suggests they have functions, yet to be understood, other than just signalling whole muscle force.

In view of our findings in this animal study, it seems unlikely that the disturbance in the ability of human subjects to accurately match levels of tension after eccentric exercise (Saxton et al., 1995; Brockett et al., 1997) results from changes in receptor properties. Rather, central mechanisms involving the sense of effort are more likely to be responsible.

ACKNOWLEDGEMENTS

This work was carried out with support from the Australian National Health and Medical Research Council.

REFERENCES

Banks, R.W., Hulliger, M., Scheepstra, K. A., and Otten, E., 1997, Pacemaker activity in a sensory ending with multiple encoding sites: the cat muscle spindle primary ending, *Journal of Physiology*, 498, 177-199.

Barker, D., 1974, *The morphology of muscle receptors*, Springer, Berlin.

Brockett, C., Warren, N., Gregory, J. E., Morgan, D. L., and Proske, U., 1997, A comparison of the effects of concentric versus eccentric exercise on force and position sense at the human elbow joint, *Brain Research*, 771, 251-258.

Burke, R. E., and Tsairis, P., 1973, Anatomy and innervation ratios in motor units of cat gastrocnemius, *Journal of Physiology*, 234, 749-765.

Carr, R. W., and Proske, U., 1996, Action of cholinesters on sensory nerve endings in skin and muscle, *Clinical Experimental Pharmacology and Physiology*, 23, 355-362.

Carr, R. W., Morgan, D. L., and Proske, U., 1996, Impulse initiation in the mammalian muscle spindle during combined fusimotor stimulation and succinyl choline infusion, *Journal of Neurophysiology*, 75, 1703-1713.

Carr, R. W., Gregory, J. E., and Proske, U., 1998, Summation of responses of cat muscle spindles to combined static and dynamic fusimotor stimulation, *Brain Research*, 800, 97-104.

Cooper, S., and Gladden, M. H., 1974, Elastic fibres and reticulin of mammalian muscle spindles and their functional significance, *Quarterly Journal of Experimental Physiology and Cognate Medical Sciences*, 59, 367-385.

Eagles, J. P., and Purple, R. L., 1974, Afferent fibers with multiple encoding sites, *Brain Research*, 77, 187-193.

Fallon, J. B., Carr, R. W., Gregory, J. E., and Proske, U., 2001, Summing responses of cat soleus muscle spindles to combined static and dynamic fusimotor stimulation (1), *Brain Research*, 888, 348-355.

Fukami, Y., and Wilkinson, R. S., 1977, Responses of isolated Golgi tendon organs of the cat, *Journal of Physiology*, 265, 673-689.

Gioux, M., Petit, J., and Proske, U., 1991, Responses of cat muscle spindles which lack a dynamic fusimotor supply, *Journal of Physiology*, 432, 557-571.

Gregory, J. E., Morgan, D. L., and Proske, U., 1986, The discharge of cat tendon organs during unloading contractions, *Experimental Brain Research,* 61, 222-226.
Gregory, J. E., Wood, S. A., and Proske, U., 2001, An investigation into mechanisms of reflex reinforcement by the jendrassik manoeuvre, *Experimental Brain Research,* 138, 366-374.
Horch, K. W., Whitehorn, D., and Burgess, P. R., 1974, Impulse generation in type I cutaneous mechanoreceptors, *Journal of Neurophysiology,* 37, 267-281.
Jami, L., Petit, J., Zytnicki, D., and Horcholle-Bossavit, G., 1992, Unloading: A reason why individual Golgi tendon organs cannot measure muscle force, in: *Muscle Afferents and Spinal Control of Movement*, L. Jami, E. Pierrot-Deseilligny and D. Zytnicki, eds., Pergamon Press, Oxford, pp. 71-76.
Morgan, D. L. and Allen, D. G., 1999, Early events in stretch-induced muscle damage, *Journal of Applied Physiology,* 87, 2007-2015.
Morgan, D. L., Proske, U., and Gregory, J. E., 1991, Responses of primary endings of cat muscle spindles to locally applied vibration, *Experimental Brain Research,* 87, 530-536.
Prochazka, A., and Gorassini, M., 1998, Ensemble firing of muscle afferents recorded during normal locomotion in cats, *Journal of Physiology,* 507, 293-304.
Proske, U., Gregory, J. E., and Morgan, D. L., 1991, Where in the muscle spindle is the resting discharge generated? *Experimental Physiology,* 76, 777-785.
Proske, U., Wise, A.K., and Gregory, J. E., 2000, The role of muscle receptors in the detection of movements, *Progress in Neurobiology,* 60, 85-96.
Saxton, J. M., Clarkson, P. M., James, R., Miles, M., Westerfer, M., Clark, S., and Donnelly, A. E., 1995, Neuromuscular dysfunction following eccentric exercise, *Medicine and Science in Sports and Exercise,* 27, 1185-1193.
Taylor, A., Durbaba, R., Ellaway, P. H., and Rawlinson, S., 2000, Patterns of fusimotor activity during locomotion in the decerebrate cat deduced from recordings from hindlimb muscle spindles, *Journal of Physiology,* 522, 515-532.
Whitehead, N., Weerakkody, N., Gregory, J., Morgan, D., and Proske, U., 2001, Changes in passive tension of muscle in humans and animals after eccentric exercise, *Journal of Physiology,* 533, 593-604.

2

EVIDENCE FOR ACTIVITY-DEPENDENT MODULATION OF SENSORY-TERMINAL EXCITABILITY IN SPINDLES BY GLUTAMATE RELEASE FROM SYNAPTIC-LIKE VESICLES

Robert W. Banks[1], Guy S. Bewick[2], Brian Reid[2], and Christine Richardson[1]

ABSTRACT

Sensory terminals of muscle spindles and similar mechanosensory neurons contain large numbers of 50 nm, "synaptic-like" vesicles (SLVs), about whose role very little is known. Using fluorescence microscopy, immunocytochemistry and electrophysiological recording, we present evidence that SLVs undergo a recycling process, and that they release glutamate that has an autogenic excitatory effect on mechanosensory transduction, probably involving a metabotropic receptor linked to phospholipase D. The rate of recycling of SLVs is activity dependent, at least in part, as shown by an increased rate of destaining of preparations labelled with FM1-43 during high-frequency, small-amplitude vibration. Immunogold labelling showed levels of glutamate-like reactivity in the sensory terminals at least as great as in probable Ia presynaptic terminals in the spinal cord. Exogenously applied glutamate has an excitatory effect on the spindle's response to stretch, which is blocked by 3,5-dihydroxyphenylglycine.

INTRODUCTION

It has been known for several decades that the sensory terminals of muscle spindles (and of similar primary afferent mechanosensory neurons) contain large numbers of small (50 nm), clear vesicles (Adal, 1969). Despite frequent acknowledgements of their obvious presence, the role of these "synaptic-like vesicles" (SLVs) has rarely been discussed and remains unknown. Whereas this indifference has been due, in part no doubt, to lack of evidence, it may also have been sustained by a preconceived notion that SLVs had no right to be at such an obviously non-synaptic site. However, it was already

[1] University of Durham, Durham, DH1 3LE, UK. Email: r.w.banks@durham.ac.uk
[2] University of Aberdeen, Aberdeen, AB25 2ZD, UK.

Sensorimotor Control of Movement and Posture
Edited by Gandevia *et al.*, Kluwer Academic/Plenum Publishers, 2002

known that the vesicles of both sensory and motor terminals are depleted by black widow spider venom (Queiroz and Duchen, 1982) and that spindle primary afferents are functionally blocked by tetanus toxin with a similar time course to its action on motor endings (Mizote and Takano, 1985), before the positive immunostaining reactions for synapsin I and synaptophysin confirmed the resemblance between the SLVs and the small vesicle population of chemical synapses (de Camilli et al., 1988).

Moreover, our own unpublished observations using the fluorescent styryl dye FM1-43 (Bewick and Betz) and electron microscopy (Banks and Stewart) provided evidence consistent with the hypothesis that the population of SLVs is constantly turned over, cyclicly fusing with and regenerating from the plasmalemma of the sensory terminals. If so, we reasoned that it might be possible to modulate the rate of turnover and that the SLVs might be releasing a neuroeffective substance (we deliberately avoid the use of the term "neurotransmitter" in this context). Knowing that the central Ia synapses are glutamatergic (Engberg et al., 1993; Walmsley and Bolton, 1994), we further reasoned by Dale's principle that the SLVs contain and release glutamate. Here we present preliminary evidence consistent with these hypotheses, based on fluorescence microscopy, immunocytochemistry and electrophysiology.

METHODS AND OBSERVATIONS

Fluorescence Microscopy

Hind foot lumbrical muscles were isolated from adult rats killed by cervical dislocation. Spindle afferent terminals were labelled by immersing the muscles in 5 μM FM1-43 (Molecular Probes) for 2 hrs in gassed (95% O_2-5% CO_2) physiological saline under moderate to maximal stretch. In contrast to the (motor) neuromuscular junctions, labelling proceeds without the need for stimuation of the nerve indicating a constitutive recycling of SLVs. Nevertheless, a similar intensity of labelling can be obtained in neuromuscular junctions after only 15 min stimulation of the nerve at 10 Hz (Betz et al., 1992). The muscles were then rinsed in 3 changes of dye-free saline and washed for a further 30-60 min in a bath of fresh saline before viewing or further processing. The spindles were observed using wide-field epifluorescence microscopy (MicroInstruments M2B microscope, fitted with a Zeiss 40x, 0.75 NA, water immersion objective). Images were captured with a MonoCoolview (PhotonicScience) digital camera and saved on a Macintosh 8500/150AV computer running Openlab software (Improvision). Individual spindles were mechanically stimulated by lightly placing at the spindle pole a blunt probe, vibrated at 200 Hz, ~50 μm amplitude by an N-802 piezo-electric drive (Vibrating Probe Company). In control experiments, the probe was placed in a spindle-free region of the muscle under otherwise identical conditions. Images were acquired between alternate 5 min periods of rest and stimulation. The mean fluorescence intensity of at least 5 labelled sites was determined for each image and corrected for variations in background labelling between preparations by subtracting the value for a nearby, non-labelled site.

Labelled terminals destained slowly (2.7% ± 0.9% S.E.M., n = 6, of initial intensity per 5 min period) at resting length (Fig.1, C-E), consistent with the SLVs labelled by endocytosis during the dye-incubation period now undergoing a subsequent round of exocytosis. Further, when spindles were stimulated by vibration at the pole, destaining increased markedly (27.6 ± 6.4%, $P < 0.01$, Student's t). The rate of dye loss returned to pre-stimulation values again (3.4 ± 2.7%; $P > 0.8$) on removal of mechanical stimulation, implying a corresponding return to basal levels of the rate of SLV exocytosis. Electroneurogram recording of afferent output in the muscle nerve during these experiments confirmed spindle excitation during destaining, whereas vibration applied in muscle regions devoid of spindles evoked neither increased spindle activity nor destaining.

Immunocytochemistry

Samples of extensor digitorum brevis muscles (EDB), cerebellum, and spinal cord were taken from adult ♀ albino rats following transcardial perfusion with 2% paraformaldehyde and 2.5% glutaraldehyde in 0.1M sodium cacodylate buffer pH 7.3. Blocks were post-fixed with cacodylate-buffered 1% OsO_4, dehydrated and embedded in Araldite CY212 (Agar Scientific). Ultrathin sections were collected on formvar-coated Ni grids, etched for 5 mins with saturated sodium meta-periodate, and incubated for 2 hrs at ambient temperature with a rabbit polyclonal antibody raised against glutamate conjugated with BSA (Storm-Mathisen et al., 1983), diluted 1:1000 in Tris/BSA pH 7.4. Secondary-antibody incubation (goat-anti-rabbit IgG labelled with 10 nm gold particles, diluted 1:20 in Tris/BSA pH 8.4, Sigma) was for 1 hr at ambient temperature. The cerebellum was used as a positive control; primary-antibody negative controls were also prepared. Image analysis was carried out using Scion Image (Scion Corporation). Sampled areas were delineated on digitised micrographs (taken at 10,000x) and the levels of immunoreactivity expressed as surface density (no. of particles μm^{-2}).

Two experiments have been carried our so far. In the first, glutamate-like immunoreactivity occurred at much greater levels in the spindle sensory terminals (7.65 ± 0.15 S.E.M. gold particles μm^{-2}, n = 22) than in the underlying intrafusal muscle fibres (2.08 ± 0.22, 22) or the more remote non-spindle (extrafusal) muscle fibres (1.21 ± 0.16, 9, $P < 0.001$ in each case, Student's t). Compared with other neuronal structures, these levels of terminal labelling were lower than, for instance, glutamatergic cerebellar structures (parallel fibres, 22.29 ± 1.58, 40; and mossy fibres, 23.91 ± 1.29, 15; also granule cell dendrites, 18.5 ± 1.17, 22). However, they were significantly higher than other neuronal (Purkinje dendritic spines; 5.2 ± 0.83, 40; $P < 0.05$) and non-neuronal, non-synaptic cerebellar structures (glial processes; 3.29 ± 0.3, 27; $P < 0.001$) and were very similar to those of the Golgi axon synaptic terminals (7.73 ± 0.7, 15) where glutamate is a precursor for GABA synthesis. In the second experiment (Fig. 1) very high labelling occurred in the terminals of a primary sensory ending (24.6 ± 1.7, 20), as compared to probable central terminals of Ia afferents on motoneuron dendrites (6.5 ± 0.5, 20; $P < 0.001$), or to the underlying dendrites themselves (3.0 ± 0.5, 19), whereas granule cell dendrites (18.4 ± 1.9, 20) and mossy fibres (27.9 ± 1.5, 20) of the cerebellum showed similar amounts of labelling as in the previous experiment.

Electrophysiology

Assuming that glutamate is present in SLVs and released from the sensory terminals by exocytosis, whose rate is modulated by activity, we next examined the effect of exogenously applied glutamate on spindle responses to maintained stretch to test the

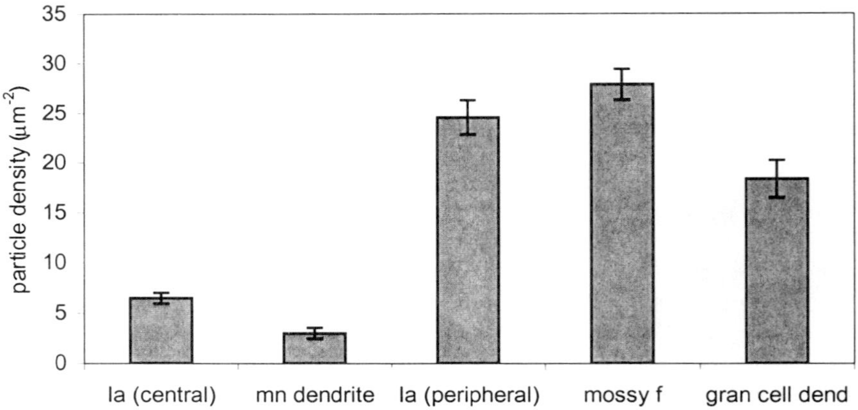

Figure 1. Glutamate-like immunoreactivity of various neuronal components from spinal cord, cerebellum, and extensor digitorum brevis of adult rat, estimated as particle density using immunogold labelling. Ia (central), probable Ia afferent presynaptic terminals on motoneuronal dendrites; mn dendrite, motoneuronal dendrite; Ia (peripheral), primary sensory ending terminals; mossy f, mossy fibre rosettes; gran cell dend, granule cell dendrites in synaptic glomerulus.

hypothesis that one of the functions of glutamate release is an autogenic support of sensory terminal excitability. Fourth (deep) lumbrical muscles were isolated from rat hind feet together with 1-1.5 cm of the nerve. The phalanx at the distal insertion was fixed to the bottom of a silicone rubber base in the organ bath, which was filled with physiological saline. The proximal, free tendon was attached to a 3-axis manual micromanipulator (Prior) and the nerve was placed in contact with a pair of silver-wire electrodes above the surface of the saline. Muscle length was adjusted to give minimal afferent discharge. Nerve impulses were amplified (CFP 8102 pre-amplifiers and Isleworth Electronics A103 amplifiers) and displayed on an oscilloscope (Gould 400). Signals were captured simultaneously on DAT tape (Bio-Logic 1204 DTR) and computer hard-drive (Strathclyde WCP software). Data were converted into Spike2 (CED) format for further processing. Responses to 3 consecutive 1 mm stretches applied manually were assessed for each condition (saline control, saline + 1 mM glutamate, etc.) by constructing spike-count histograms (200 ms bins). Four-second periods from the plateau output at the new resting length were used, so as to exclude the early, transitory peak during and shortly after the initial length increase. Differences between treatments were assessed by 2-way ANOVA; all differences were considered significant if $P < 0.05$.

Application of glutamate (0.1-1 mM) for 60-120 min (Fig. 2) resulted in the 4 s spike count increasing to an average of 75% above control values during the plateau of the stretch (3 trials each of 3 replicates, ANOVA; $P < 0.001$). It appeared that there was also an increased afferent discharge at resting length, but this was not quantified. Both of these effects were reversible on washing in glutamate-free saline, over a similar time course (60 -120 min). The long times required for these effects to be established or removed are likely to be due substantially, though perhaps not entirely, to the presence of the spindle capsule acting as a diffusion barrier.

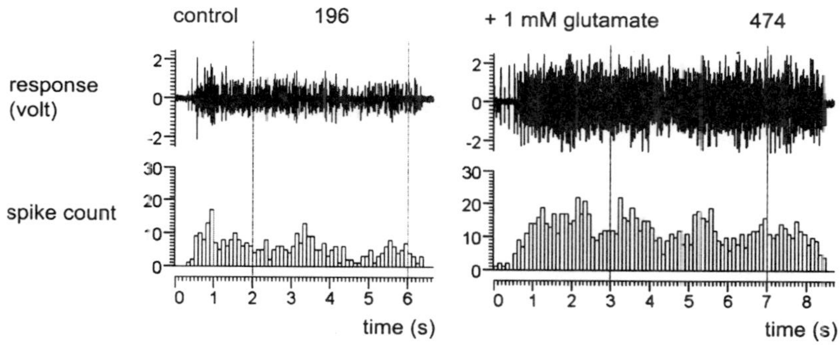

Figure 2. Electroneurograms (top) and derived spike count histograms (bottom) from 4th deep lumbrical isolated from adult rat hind foot. A 1 mm stretch was applied manually and the muscle then maintained at constant length between 1 and 6 s (left) or 2 and 8 s (right). The control response (left) shows much lower spindle firing rates than the same preparation about 2 hrs after exposure to 1 mM glutamate in the bathing solution. Numbers above each graph are the total spike counts for the 4 s periods between the vertical cursors.

The enhancement of spindle excitability by glutamate would be expected to be a specific, receptor-mediated effect. Ionotropic receptors (iGluRs) do not seem to be involved since 1 mM kynurenate, which blocks both AMPA and NMDA iGluRs, had little effect on the level of enhancement. Further addition of antagonists of Group I-III metabotropic receptors (mGluRs), including 0.5 mM MCPG, 50 nM CPPG, and 1 mM MAP4*, also had little effect. However, the inclusion in the cocktail of 3,5-dihydroxyphenylglycine (3,5-DHPG; 200 μM), a non-selective antagonist of phospholipase D activation by mGluR (Albani-Torregrossa et al., 1999), completely blocked any of the usual enhancement of activity induced by the addition of glutamate. Indeed, it actually reduced excitability to 43% *below* that in drug-free control preparations, even in the continued presence of 1 mM glutamate (2 trials, 3 replicates each, ANOVA; $P < 0.001$).

DISCUSSION

Our results, though preliminary, clearly demonstrate a major role for the SLVs in the control of excitability of spindle sensory endings. We have presented evidence that SLVs recycle to and from the plasmalemma, that the endings exhibit relatively high levels of glutamate-like immunoreactivity, and that exogenously applied glutamate has a reversible excitatory effect on spindle sensory endings, which can be blocked by 3,5-DHPG. We have further shown that the rate of SLV recycling is activity dependent by an, as yet, unknown mechanism. This implies a level of autogenic excitation, a proposition supported by the observation that 3,5-DHPG not only blocks the excitation by exogenous

* MCPG, (*S*)-α-methyl-4-carboxyphenyl-glycine; CPPG, (*RS*)-cyclopropyl-4-phosphonophenyl-glycine; MAP4, (*S*)-α-methyl-2-amino-4-phosphonobutyrate.

glutamate, but also inhibits spindle firing to rates well below drug-free control levels. We are not, of course, suggesting that such autogenic excitation is in any way the primary mechanism of sensory transduction, which presumably resides in stretch-sensitive ionotropic channels. Rather it appears that we are dealing with a modulatory effect of the kind now well known in many post-synaptic systems to be mediated by various second messengers and often activated by mGluRs (Hammond, 2001).

The precise nature and mechanism of that modulation has yet to be determined, and just how general it is among mechanosensory endings. The apparent involvement of phospholipase D indicates that phosphatidic acid and diacylglycerol may be acting as second messengers (Exton, 1997). Almost certainly Ca^{2+} will also prove to be involved: not only is there is a Ca^{2+} component to the receptor current (Hunt, Wilkinson, and Fukami, 1978), but several Ca^{2+}-binding proteins of the EF hand superfamily have also now been described as occurring in spindle sensory endings, including probable buffers (calbindin D-28k, calretinin) and switches (neurocalcin, frequenin) (for review, see Banks and Barker, in press). If the release mechanism of SLVs is similar to that of most presynaptic vesicles and is Ca^{2+}-dependent, this would readily account for the observed activity-dependency of SLV recycling.

REFERENCES

Adal, M. N., 1969, The fine structure of the sensory region of cat muscle spindles, *Journal of Ultrastructure Research,* 26, 332-354.

Albani-Torregrossa, S., Attucci, S., Marinozzi, M., Pellicciari, R., Moroni, F. & Pellegrini-Giampietro, D. E., 1999, Antagonistic pharmacology of metabotropic glutamate receptors coupled to phospholipase D activation in adult rat hippocampus: focus on (2R,1′S,2′R,3′S)-2-(2′-carboxy-3′-phenylcyclopropyl)glycine versus 3,5-dihydroxyphenylglycine, *Molecular Pharmacology,* 55, 699-707.

Banks, R. W., and Barker, D., 2002, The muscle spindle, in: *Myology,* 3rd edition, A. G. Engel and C. Franzini-Armstrong, eds., McGraw-Hill, New York, in press.

Betz, W. J., Mao, F., and Bewick, G. S., 1992, Activity-dependent fluorescent staining and destaining of living vertebrate motor nerve terminals, *Journal of Neuroscience,* 12, 363-375.

de Camilli, P., Vitadello, M., Canevini, M. P., Zanoni, R, Jahn R., and Gorio, A., 1988, The synaptic vesicle proteins synapsin I and synaptophysin (protein P38) are concentrated both in efferent and afferent nerve endings of the skeletal muscle, *Journal of Neuroscience,* 8, 1625-1631.

Engberg, I., Tarnawa, I., Durand, J., and Ouardouz, M., 1993, An analysis of synaptic transmission to motoneurons in the cat spinal cord using a new selective receptor blocker, *Acta Physiologica Scandinavica,* 148, 97-100.

Exton, J. H., 1997, Phospholipase D: enzymology, mechanisms of regulation, and function, *Physiological Reviews,* 77, 303-320.

Hammond, C., 2001, *Cellular and Molecular Neurobiology,* 2nd edition, Academic Press, London, pp. 314 326.

Hunt, C. C., Wilkinson, R. S., and Fukami, Y., 1978, Ionic basis of the receptor potential in primary endings of mammalian muscle spindles, *Journal of General Physiology,* 71, 683-698.

Mizote, M., and Takano, K., 1985, The response of cat muscle spindle primary endings to FM muscle vibration during fusimotor stimulation or following local injection of tetanus toxin, in: *The Muscle Spindle,* I. A. Boyd, and M. H. Gladden, eds. Macmillan, Basingstoke, pp. 365-369.

Queiroz, L. S., and Duchen, L. W., 1982, Effects of *Latrodectus* spider venoms on sensory and motor nerve terminals of muscle spindles, *Proceedings of the Royal Society of London. Series B: Biological Sciences,* 216, 103-110.

Storm-Mathisen, J., Leknes, A. K., Bore, A. T., Vaaland, J. L., Edminson, P., Haug, F.-M. S., and Ottersen, O. P., 1983, First visualization of glutamate and GABA in neurones by immunocytochemistry, *Nature,* 301, 517-520.

Walmsley, B., and Bolton, P. S., 1994, An *in vivo* pharmacological study of single group Ia fibre contacts with motoneurones in the cat spinal cord, *Journal of Physiology,* 481, 731-741.

3

ELECTROPHYSIOLOGY OF CORNEAL COLD RECEPTOR NERVE TERMINALS

Richard W. Carr and James A. Brock*

ABSTRACT

The mechanisms of sensory transduction in the fine nerve terminals of free nerve endings supplied by Aδ and C sensory axons are largely a matter of speculation. This is because the nerve terminals are small and inaccessible, particularly in intact tissues like skin. However, some of the difficulties associated with investigating the physiology of fine nerve terminals have recently been overcome using an *in vitro* preparation of the guinea-pig cornea that allows nerve terminal impulses (NTIs) to be recorded extracellularly from single polymodal and cold receptor nerve terminals. For cold receptors, the rate of spontaneously occurring NTIs is increased during cooling and decreased during heating. In addition, heating and cooling differentially modulate the shape of the recorded NTI. At the same temperature, NTIs are larger in amplitude and faster in time course during heating than those during cooling. The differential effect of heating and cooling on NTI shape is not considered to result simply from the temperature dependence of voltage-activated conductance kinetics or activity dependent changes in membrane excitability. Instead, changes in NTI shape may reflect changes in nerve terminal membrane potential that underlie the process of thermal transduction.

INTRODUCTION

Sensory receptors with a preferential sensitivity to cold or cooling stimuli are found in the epithelium of external tissues such as skin. Activation of cold receptors in the skin generates a sensation of cooling or 'freshness' and in the cornea these receptors may also contribute to painful sensations (Belmonte et al., 1997). Cold receptors typically have small diameter, thinly myelinated axons and, in most tissues, they have an ongoing impulse discharge that is modulated by temperature. Cooling increases their impulse activity and heating decreases it. However, the process responsible for the temperature sensitivity of cold receptors is largely unknown. This is primarily because their nerve

* Prince of Wales Medical Research Institute, Barker St., Randwick, Sydney, 2031, NSW, Australia.
Email: rcarr@unsw.edu.au

Sensorimotor Control of Movement and Posture
Edited by Gandevia *et al.*, Kluwer Academic/Plenum Publishers, 2002

terminals are small, less than 0.5 μm in diameter, and are difficult to access in tissues like skin. Recently, we have developed an extracellular recording technique that allows electrical activity to be recorded directly from identified sensory nerve terminals in guinea-pig cornea (Brock et al., 1998). Using this technique, it has been possible to investigate the mechanisms controlling the excitability of free sensory nerve endings, including those sensitive to cold.

The cornea is densely innervated by small-diameter, Aδ and C, sensory axons that form functional nerve terminals in the superficial layers of the epithelium (Belmonte et al., 1997). Three types of sensory receptor are found in the cornea, polymodal, mechano- and 'cold'-sensitive (Gallar et al., 1993), and all have unmyelinated axons in the epithelium (Whitear, 1960). By applying a small diameter suction electrode (~50 μm) to the surface of the cornea, nerve terminal impulses (NTIs) can be recorded extracellularly from each of these three functional classes (Brock et al., 1998). NTIs evoked by stimulating the ciliary nerves at the back of the eye, have conduction velocities in the range 0.3 to 2.7 m s^{-1}, at 31-32 °C, consistent with similar estimates made for corneal afferents using split fibre techniques (see Belmonte et al., 1997).

The observed pattern of corneal cold receptor NTI activity in the guinea-pig is similar to that previously reported for axonal recordings from cutaneous and corneal cold receptors in other species (cf. Darian-Smith et al., 1973; Belmonte et al., 1991). At ambient temperatures between 30 and 32 °C, guinea-pig corneal cold receptors have a regular ongoing NTI discharge comprising short high frequency bursts of impulses that occur at a mean rate between 5 and 20 impulses s^{-1}. This pattern of NTI activity is modulated by temperature (Fig. 1A). During heating from 31 to 37°C, the NTI rate is reduced and often silenced. In contrast, cooling from 37 °C increases NTI activity. Changes in temperature also produce changes in NTI shape. During heating, NTIs increase in amplitude and speed in time course, while during cooling, NTIs decrease in amplitude and slow in time course (Fig. 1B). Moreover, heating and cooling differentially affect NTI shape. When measured at the same absolute temperature, NTIs are larger in amplitude and faster in time course during heating compared with those during cooling (Fig. 1B).

The signal recorded extracellularly from a nerve terminal is proportional to net membrane current and is largely capacitive (Smith, 1988). Consequently, the recorded NTI is typically diphasic (positive-negative) with a more prominent positive component (see Fig. 1B and Brock et al., 2001). Signals of similar shape have been recorded extracellularly from the terminals of motor nerves where their configuration can be explained either by active impulse invasion (Katz and Miledi, 1965) or by passive invasion of the nerve terminal by an action potential that fails proximal to the nerve terminal (Dudel, 1963). The NTIs recorded from corneal cold receptors are thought to arise from passive invasion of the nerve terminal. This is because NTI time course is unaffected by locally applied lignocaine (10 mM), a compound that blocks regenerative Na^+ channels (Brock et al., 2001). However, cold receptor nerve terminals are likely to possess regenerative Na^+ channels since hyperpolarizing the nerve terminal membrane with extracellular current reveals a lignocaine-sensitive inward current. The simplest explanation for the lack of contribution of these Na^+ channels to the NTI, under normal conditions, is that the nerve terminal membrane is relatively depolarized. Consequently, most Na^+ channels in the nerve terminal region are inactivated and the nerve terminal acts as a passive electrical conductor.

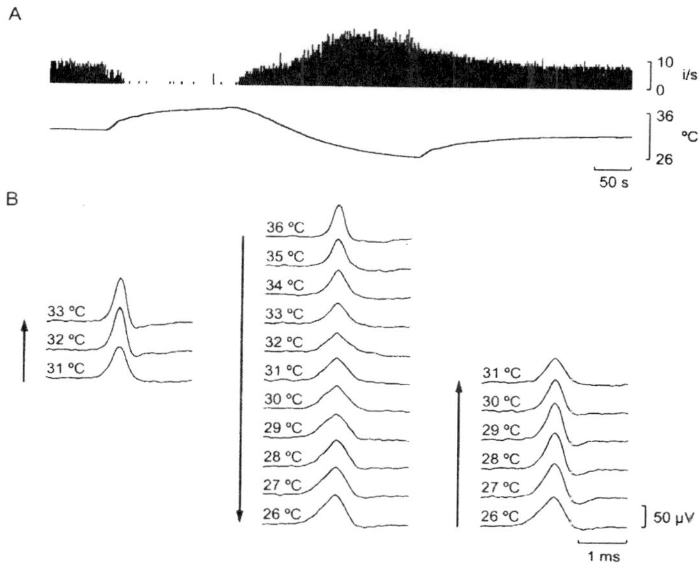

Figure 1. NTI activity recorded for a single corneal cold receptor during changes in temperature. Both the rate of NTI discharge (panel A, upper) and NTI shape (panel B) are dependent on temperature. In the upper portion of panel A, NTI rate is presented as impulse count per second. In panel B, NTIs are averages of 20 individual NTIs that occurred sequentially over a temperature range extending no more than ±0.1°C from the temperature specified. NTIs are shown during heating and cooling, indicated by the vertical arrows.

Assuming that the nerve terminal region of corneal cold receptors acts as passive electrical filter of changes in membrane potential produced by an action potential in the proximal axon, changes in NTI shape may then occur as a result of changes in the time course of the action potential or as a result of changes in the passive electrical properties of the nerve terminal itself. In addition, a change in the site of action potential failure in the proximal axon will modify the filtering of membrane potential changes by the nerve terminal as determined by the passive electrical properties of the nerve terminal membrane.

Temperature affects the kinetics of voltage-dependent conductances responsible for action potential conduction and therefore the rate of change of membrane potential. However, the temperature dependence of both regenerative Na^+ and rectifying K^+ conductances associated with action potential propagation is similar and both have a Q_{10} value of approximately 3 (Jack et al., 1975). Consequently, it is unlikely that temperature dependent changes in conductance kinetics can account for the differential effect of heating and cooling on NTI shape.

Changes in NTI shape produced by changing temperature are also associated with changes in NTI rate (Fig. 1). The differential effect of heating and cooling on NTI shape might therefore reflect activity-dependent changes in the excitability of the axon membrane. To investigate this possibility, the effect of stimulation frequency on the shape of NTIs, evoked by electrical stimulation of the ciliary nerves, was examined. At a constant temperature, trains of 50 stimuli at various rates between 10-60 Hz, produced a progressive increase in the amplitude and latency of the electrically evoked NTI. Both of these features are consistent with an activity-dependent hyperpolarization of the axonal membrane resulting from activation of the electrogenic Na^+–K^+-ATPase (Rang and Ritchie, 1968). However, when cold receptor activity increased during cooling, NTIs were reduced in amplitude and slowed in time course. Therefore, it is unlikely that the differential effects of heating and cooling on NTI shape result from an activity-dependent change in membrane excitability.

If it is assumed instead, that the nerve terminal membrane hyperpolarizes during heating and depolarises during cooling, the differential effects of heating and cooling on NTI shape may then reflect changes in the level of Na^+ channel inactivation in the nerve terminal. At a fixed ambient temperature, the nerve terminal is thought to be relatively depolarised (Brock et al., 2001), inactivating a proportion of Na^+ channels in the nerve terminal and also maintaining ongoing impulse activity. If the nerve terminal hyperpolarizes during heating, the rate of ongoing impulse activity will decrease and inactivation of a proportion of Na^+ channels will be relieved. As a result, the point at which the action potential fails will move closer to the nerve terminal causing the NTI to increase in amplitude and speed in time course. Conversely, if cooling depolarizes the nerve terminal, NTI activity would be expected to increase as would the degree of Na^+ channel inactivation proximal to the nerve terminal. Consequently, the site at which the action potential fails would move away from the nerve terminal, decreasing NTI amplitude and slowing NTI time course. Interestingly, two temperature regulated conductances have recently been described in a population of rat DRG neurons thought to be cold receptors: (1) a mixed-cationic inward current that is activated by cooling and sensitised by menthol (Reid and Flonta, 2001) and (2) a background K^+ conductance that increases with temperature and has a Q_{10} of approximately 7 (Reid and Flonta, 2001). Perhaps the combined action of these temperature sensitive conductances contributes to differential changes in membrane potential during heating and cooling.

REFERENCES

Belmonte, C. Gallar, J. Pozo, M. A., and Rebello, I., 1991, Excitation by irritant chemical substances of sensory afferent units in the cat's cornea, *Journal of Physiology,* 437, 709-725.

Belmonte, C. Garcia-Hirschfeld, J., and Gallar, J., 1997, Neurbiology of ocular pain, *Progress in Retinal Eye Research,* 16,117-156.

Brock, J. A. McLachlan, E., and Belmonte, C., 1998, Tetrodotoxin-resistant impulses in single nociceptor nerve terminals in guinea-pig cornea, *Journal of Physiology,* 489, 389-402.

Brock, J. A. Pianova, S., and Belmonte, C., 2001, Differences between nerve terminal impulses of polymodal nociceptors and cold sensory receptors of the guinea-pig cornea, *Journal of Physiology,* 533, 493-501.

Darian-Smith, I. Johnson, K. O., and Dykes, R., 1973, 'Cold' fiber population innervating palmar and digital skin of the monkey, *Journal of Neurophysiology,* 36, 325-346.

Dudel, J., 1963, Presynaptic inhibition of the excitatory nerve terminal in the neuromuscular junction of the crayfish, *Pflügers Archives,* 277, 537-557.

Gallar, J. Pozo, M. A. Tuckett, R. P., and Belmonte, C., 1993, Response of sensory units with unmyelinated fibres to mechanical, thermal and chemical stimulation of the cat's cornea, *Journal of Physiology,* 468, 609-622.

Katz, B., and Miledi, R., 1965, Propagation of electrical activity in motor nerve terminals, *Proceedings of the Royal Society B,* 161, 453-482.

Rang, H. P., and Ritchie, J. M., 1968, On the electrogenic sodium pump in mammalian non-myelinated nerve fibres and its activation by various external cations, *Journal of Physiology,* 196, 183-221.

Reid, G., and Flonta, M.-L., 2001, Cold current in thermoceptive neurons, *Nature,* 413:480.

Reid, G., and Flonta, M.-L., 2001, Cold transduction by inhibition of a background potassium conductance in rat primary sensory neurones, *Neuroscience Letters,* 297, 171-174.

Smith, D. O., 1988, Determinants of nerve terminal excitability, in: *Neurology and Neurobiology*, P. W. Lanfield and S. A. Deadwyler, ed., Alan Liss Inc., New York, pp. 411-438.

Whitear, M., 1960, An electron microscope study of the cornea in mice, with special reference to the innervation, *Journal of Anatomy,* 94, 387-409.

4

DISCHARGE PROPERTIES OF GROUP III AND IV MUSCLE AFFERENTS

Marc P. Kaufman, Shawn G. Hayes, Christine M. Adreani and Joel G. Pickar*

ABSTRACT

Stimulation of group III and IV muscle afferents has been shown to have important reflex effects on both the somatic and autonomic nervous systems. These include an inhibitory effect on alpha motoneurones, an excitatory effect on gamma motoneurones and an excitatory effect on the sympathetic nervous system. The purpose of this review is to describe the mechanical and metabolic stimuli that discharge group III and IV muscle afferents. Particular attention will be paid to the responses of these afferents to dynamic exercise induced by electrical stimulation of the mesencephalic locomotor region.

INTRODUCTION

Stimulation of group III and IV afferents with endings in the skeletal muscles of the limbs has been shown to have important reflex effects on both the somatic and autonomic nervous systems. For example, activation of group III and IV afferents has been shown to increase arterial blood pressure, heart rate and breathing (Kaufman and Forster, 1996). In addition, activation of group III and IV afferents in decerebrate cats has been shown to decrease the discharge of alpha motoneurones (Hayward et al., 1988) and increase the discharge of gamma motoneurones (Jovanovic et al., 1990). The aim of this account is to describe the metabolic and mechanical stimuli that discharge these thin-fiber muscle afferents.

* Division of Cardiovascular Medicine, Departments of Internal Medicine and Human Physiology, University of California, Davis, California, USA. Email: mpkaufman@ucdavis.edu

Sensorimotor Control of Movement and Posture
Edited by Gandevia *et al.*, Kluwer Academic/Plenum Publishers, 2002

DISCHARGE PROPERTIES OF GROUP III AFFERENTS

Group III afferents are thinly myelinated and conduct impulses between 2.5 and 30 meters per second. They are also referred to as Aδ fibres. One of the first to investigate the discharge properties of these afferents was Paintal, who found that group III endings located in the hindlimb muscles of dogs responded to punctate pressure (Paintal, 1960). Since then several laboratories have independently reached similar conclusions about the discharge properties of these thinly myelinated afferents. These are: (i) that about half respond to contraction, be it intermittent tetanic or maintained tetanic (static) contraction (Paintal, 1960; Ellaway et al., 1982; Kaufman et al., 1983; Mense and Stahnke 1983), (ii) that about half respond to intra-arterial injection of bradykinin, a potent algesic agent; (Kumazawa and Mizumura, 1977; Mense 1977; Kaufman et al., 1983), (iii) that many are discharged by non-noxious punctate pressure applied to their receptive fields (Paintal, 1960; Kumazawa and Mizumura, 1977; Kaufman et al., 1983; Mense and Stahnke, 1983; Kaufman et al., 1984b) and (iv) that their response to tendon stretch is variable - some investigators finding a frequent effect (Abrahams et al., 1984), others finding a moderate effect (Kaufman et al., 1983; Mense and Stahnke, 1983; Kaufman and Rybicki, 1987; Hayward et al., 1991) and still another finding an infrequent effect (Paintal, 1960).

Group III muscle afferents are believed to possess polymodal discharge properties because they respond to both chemical and mechanical stimuli (Kumazawa and Mizumura, 1977). Group III muscle afferents, for example, respond vigorously at the onset of tetanic contraction, with the first impulse often being discharged within 200 ms of the start of this manoeuvre (Kaufman et al., 1983). Moreover, group III afferents increase their responses to tetanic contractions as the peak tension developed by the working muscle increases (Kaufman et al., 1983; Mense and Stahnke, 1983; Hayward et al., 1991). In addition, group III afferents usually decrease their discharge rate during a static contraction as the tension developed by the working muscle decreases (Kaufman et al., 1983; Hayward et al., 1991). Finally, group III afferents are capable of synchronizing their discharge with a constantly oscillating stimulus, be it either 5 Hz twitch contraction (Kaufman et al., 1984b) or intermittent tetanic contraction (Mense and Stahnke, 1983).

The mechanical sensitivity of group III afferents might be their most relevant discharge property when assessing their contribution to the reflex autonomic responses to exercise. For example, the pressor, cardioaccelerator and ventilatory responses to tetanic contraction have been shown to be reduced by about half when stretch activated channels in group III afferents were blocked by gadolinium infusion into the arterial supply of the triceps surae muscles (Hayes and Kaufman, 2001; Fig. 1). In addition, gadolinium infusion prevented most of the responses of group III afferents to tetanic contraction and to tendon stretch, the latter being a pure mechanical stimulus (Hayes and Kaufman, 2001).

Despite their obvious mechanical sensitivity, group III afferents might also respond to the metabolic products of muscular contraction. For example, they are stimulated by bradykinin (Kumazawa and Mizumura, 1977; Mense, 1977; Kaufman et al., 1983), potassium (Hník et al., 1969; Kumazawa and Mizumura, 1977; Mense, 1977; Rybicki et al., 1985; Thimm and Baum, 1987), arachidonic acid (Rotto and Kaufman, 1988), a thromboxane A_2 mimetic, (Kenagy et al., 1997) and lactic acid (Thimm and Baum 1987; Rotto and Kaufman, 1988; Sinoway et al., 1993). Of these metabolites, only potassium (Rybicki et al., 1985) and lactic acid (Thimm and Baum, 1987; Sinoway et al., 1993) have been shown, as of yet, to stimulate group III afferents in concentrations that may approximate those occurring in a contracting muscle. Sensitivity to metabolic products of contraction might explain why some group III afferents display a secondary response to static contraction at a time when the working muscle is fatiguing (Kaufman et al., 1983). In addition, histamine and serotonin stimulate group III afferents, but the required doses were quite high and the percentage of afferents stimulated was small (Mense, 1977). Also, it is unclear whether interstitial concentrations of histamine and serotonin increase above resting levels when skeletal muscle is contracted.

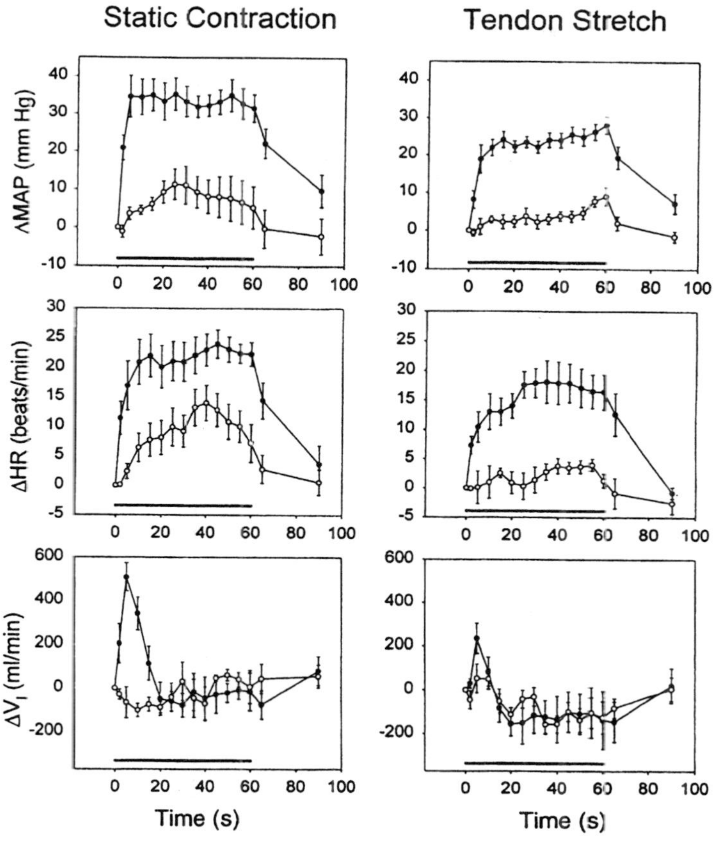

Figure 1. Time course of pressor, cardioaccelerator and ventilatory responses to static contraction (left) and tendon stretch (right) before (•) and 60 min after (○) gadolinium injection into the femoral artery (n = 7). Circles represent means, and brackets represent ±SE. Circles at *time 0* represent baseline (i.e., before the manoeuvre). First circle after *time 0* represents the mean 2 s after the manoeuvre started, and every circle afterwards represents the mean at 5-s intervals starting from *time 0*. Time period when the muscles were contracted or stretched is represented by the filled horizontal bar. During contraction and stretch, all means for ΔMAP and ΔHR before gadolinium injection were significantly greater ($P < 0.05$) than their corresponding means 60 min after gadolinium injection. Also, during contraction, means for the change in minute volume of inspiratory ventilation (ΔV_I) at 2, 5, 10, and 15 s before gadolinium were significantly greater ($P < 0.05$) than corresponding means 60 min after gadolinium injection. During stretch, however, only the mean ΔV_I value at 5 s before gadolinium was significantly greater ($P < 0.05$) than its corresponding mean 60 min later.

DISCHARGE PROPERTIES OF GROUP IV AFFERENTS

Group IV afferents are unmyelinated and conduct impulses at less than 2.5 meters per second. They are also referred to as C fibres. In many respects, group IV muscle afferents possess different discharge properties than do group III muscle afferents. Although both respond to tetanic contraction, group IV afferents often do not discharge vigorously at the onset of contraction as do group III afferents. Instead, group IV afferents usually display obvious responses to contraction with latencies of 5-30 seconds. (Kaufman et al., 1983; Mense and Stahnke, 1983). Nevertheless, group IV afferents sometimes generate one to two impulses at the onset of contraction, a response suggesting that they might possess a minimal mechanical sensitivity.

About half of the group IV afferents respond more to a tetanic contraction while the muscles are ischaemic than to a static contraction of equal magnitude while the muscles are freely perfused (Bessou and LaPorte 1958; Iggo, 1961; Mense and Stahnke, 1983; Kaufman et al., 1984a). By contrast, only about 10-12% of the group III afferents responded more to tetanic contraction while the circulation to the muscle was occluded than to a contraction while the muscle was freely perfused (Mense and Stahnke, 1983; Kaufman et al., 1984a). Therefore, group IV afferents appear to be primarily responsible for the ischaemia-induced amplification of the reflex pressor response to tetanic (i.e., static) contraction which is seen in anesthetized preparations. In addition, group IV afferents are much less sensitive than are group III afferents to either probing their receptive fields or to tendon stretch. Most often, group IV afferents require noxious pinching of their receptive fields to discharge them, and frequently do not respond to even noxious tendon stretch (Kniffki et al., 1978; Mense and Stahnke, 1983; Kaufman et al., 1984a; Kaufman and Rybicki, 1987).

Group IV afferents are stimulated by the same substances that have been shown to stimulate group III afferents. Hence, intra-arterial injection of bradykinin (Mense and Schmidt, 1974; Mense 1977; Kaufman et al., 1983) potassium (Hník et al., 1969; Mense 1977; Kaufman and Rybicki, 1987), lactic acid (Graham et al., 1986; Thimm and Baum, 1987; Rotto and Kaufman, 1988) and arachidonic acid (Rotto and Kaufman, 1988) have been shown to stimulate at least half of the group IV afferents tested. In addition, prostaglandin E_2 and a thromboxane A_2 mimetic also stimulate about half the group IV afferents (Mense, 1981; Kenagy et al., 1997). The responses of group III afferents to PGE_2 is not known as of yet. Histamine and serotonin stimulated only a few group IV afferents and the threshold doses appeared to be high (Mense, 1977; Kniffki et al., 1978).

The responses of group III and IV afferents to tetanic contraction can be affected markedly by the nature of their chemical environment. For example, intraarterial injection of bradykinin increased the responses of both group III and IV afferents to intermittent tetanic contraction of the triceps surae muscles (Mense and Meyer, 1988). Likewise, intra-arterial injections of arachidonic acid increased the responses of group III but not those of group IV afferents to maintained tetanic contraction (Rotto et al., 1990a,b). Decreasing the concentrations of various metabolites in the muscle also have an effect on the afferents' responses to contraction. Hence, indomethacin and aspirin, compounds which decrease the muscle's ability to synthesize prostaglandins and thromboxanes, decreased the responses of group III and IV afferents to maintained tetanic contraction (Rotto et al., 1990a,b). Likewise, sodium dichloroacetate, a compound which decreased the concentration of lactic acid in the muscle, decreased the responses of group III afferents to maintained tetanic contraction (Sinoway et al., 1993). Specifically, dichloroacetate decreased the sensitivity of the group III afferents to the first ten sec of the contraction period. The effect of dichloroacetate on the responses of group IV afferents to contraction appears to be minimal.

RESPONSES OF GROUP III AND IV AFFERENTS TO LOCOMOTION

Tetanic muscular contractions, whether maintained (Kaufman et al., 1983; Kaufman et al., 1984a) or intermittent (Mense and Stahnke, 1983) have provided important information about the discharge properties of group III and IV muscle afferents. For example, group III afferents are more sensitive to mechanical stimuli than group IV afferents. In addition, the responses of group IV afferents to contraction are amplified by ischaemia, whereas the responses of group III afferents have not. This finding has led to the suggestion that group IV afferents are more sensitive to ischaemic metabolites than are group III afferents. To obtain this information, investigators have contracted hindlimb muscles by electrical stimulation of motor axons in either the ventral roots or peripheral nerves. However, this method of stimulation recruits alpha motoneurones (with respect to their conduction velocities) from "fast to slow", whereas exercise recruits motoneurones from "slow to fast" (e.g. Henneman et al., 1965). Moreover, electrical stimulation activates motoneurones synchronously (i.e., with a constant interspike interval) whereas exercise activates motoneurones asynchronously (i.e., irregularly).

Clearly electrical stimulation of alpha motoneurones is not a satisfactory method of determining if group III and IV muscle afferents are stimulated by a true form of exercise. A superior method of simulating exercise is the locomotion evoked in decerebrate unanesthetized cats by activation of either the hypothalamic or mesencephalic locomotor regions. Locomotion evoked by stimulation of brainstem regions in decerebrated unanesthetized cats is quite similar to locomotion in intact, conscious cats. Rhythmic alternating limb movements are elicited by stimulation of the mesencephalic (MLR) or hypothalamic locomotor regions and a variety of gaits may be evoked by changing the intensity of the electrical pulses activating these midbrain or diencephalic areas (Shik and Orlovsky ,1966; Wetzel and Stuart, 1976;). A most important feature of this preparation is that activation of either the mesencephalic or hypothalamic locomotor region recruits alpha motoneurones from slow to fast (Zajac and Faden, 1985; Hoffer et al., 1987a,b). This order of recruitment appears to be identical to the order of recruitment of alpha motoneurones in intact conscious cats performing treadmill exercise (Hoffer et al., 1981; Hoffer et al., 1987a, b).

We recorded the impulse activity of group III and IV muscle afferents while we stimulated the MLR in decerebrate unanesthetized cats. Specifically, we found that 19 of 25 group III afferents with endings in the triceps surae muscles responded to the dynamic exercise induced by MLR stimulation. Many of these group III afferents discharged in synchrony with the rhythmic contractions of the triceps surae muscles (Pickar et al., 1994; Adreani et al., 1997). Occlusion of the arterial supply of the triceps surae muscles increased the responses to dynamic exercise of 11 of the 25 group III afferents (Adreani and Kaufman, 1998). The occlusion-induced increase in responsiveness to exercise might be due to a sensitization of group III afferents by prostaglandins (Rotto et al., 1990b) and bradykinin (Mense and Meyer ,1988).

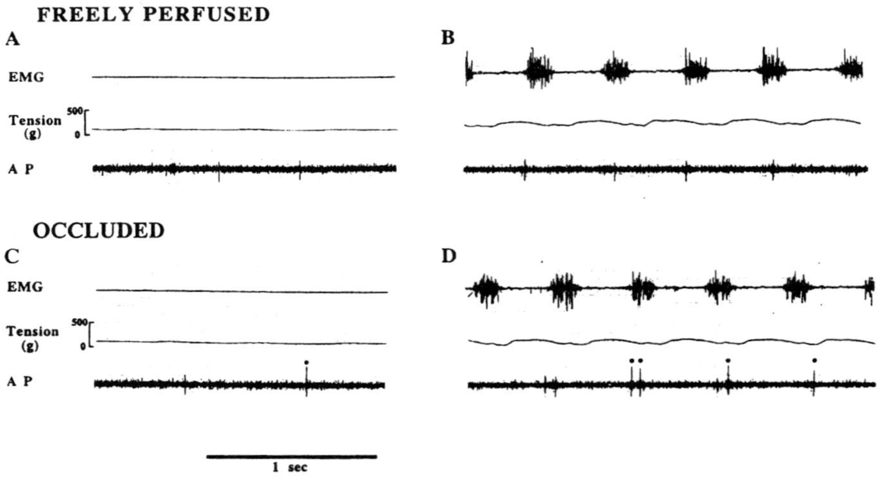

Figure 2. Original trace of 2 group IV afferents (unmarked AP have a CV = 2.4 m/s; AP with (•) have a CV = 1.7 m/s) with femoral artery freely perfused (A and B) and with femoral artery occluded (C and D). Afferent with unmarked AP had similar discharge rates during rest and exercise whether femoral artery was occluded or was freely perfused. Afferent with AP marked by (•) did not discharge when femoral artery was freely perfused but did discharge during rest and exercise when femoral artery was occluded.

We also found that 16 of 19 group IV afferents with endings in the triceps surae muscles responded to dynamic exercise induced by MLR stimulation (Adreani et al., 1997; Adreani and Kaufman, 1998). Nine of the 19 responded more to exercise when the triceps surae muscles were "ischaemic" (i.e., femoral artery occluded) than when the muscles were freely perfused (Fig. 2). Finally, the "level" of dynamic exercise that stimulated these thin fiber muscle afferents was modest. Specifically, the oxygen consumption of the triceps surae muscles increased by only 2.5 times (Adreani et al., 1997), whereas this muscle group can increase its oxygen consumption by 10 times (Bockman, 1983).

In summary, group III and IV muscle afferents appear to be activated by low to moderate levels of exercise. This finding is surprising and contradictory to previous speculation that group III and IV muscle afferents evoke reflex effects only at severe or painful levels of exercise. The precise role played by these thin fiber afferents in regulating skeletal and autonomic function during exercise is unclear and remains to be determined.

ACKNOWLEDGEMENTS

This work was supported by NIH grant HL 30710. We thank Ms. Angela DiStefano for typing this manuscript.

REFERENCES

Abrahams, V. C., Lynn, B., and Richmond, F. J. R., 1984, Organization and sensory properties of small myelinated fibres in the dorsal cervical rami of the cat, *Journal of Physiology,* 347, 177-187.

Adreani, C. M., Hill, J. M., and Kaufman, M. P., 1997, Responses of group III and IV muscle afferents to dynamic exercise, *Journal of Applied Physiology,* 86, 1811-1817.

Adreani, C. M., and Kaufman, M. P., 1998, Effect of arterial occlusion on responses of group III and IV afferents to dynamic exercise, *Journal of Applied Physiology,* 84(6), 1827-1833.

Bessou, P., and LaPorte, Y., 1958, Activation des fibres afférentes amyélinisées de petit calibre, d'origine musculaire (fibre du groupe III), *Comptes Rendus des Seances de la Societe de Biologie et de Ses Filiales,* 152, 1587-1590.

Bockman, E. L., 1983, Blood flow and oxygen consumption in active soleus and gracilis muscles in cats, *American Journal of Physiology,* 244, H546-H551.

Ellaway, P. H., Murphy, P. R., and Tripathi, A., 1982, Closely coupled excitation of gamma-motoneurons by group III muscle afferents with low mechanical threshold in a cat, *Journal of Physiology,* 331, 481-498.

Graham, R., Jammes, Y., Delpierre, S., Grimaud, C., and Roussos, Ch., 1986, The effects of ischemia, lactic acid and hypertonic sodium chloride on phrenic afferent discharge during spontaneous diaphragmatic contraction, *Neuroscience Letters,* 67, 257-262.

Hayes, S., and Kaufman, M. P., 2001, Gadolinium attenuates exercise pressor reflex in cats, *American Journal of Physiology,* 280, H2153-H2161.

Hayward, L., Breitbach, D., and Rymer, W. Z., 1988, Increased inhibitory effects on close synergists during muscle fatigue in the decerebrate cat, *Brain Research,* 440, 199-203.

Hayward, L., Wesselmann, U., and Rymer, W. Z., 1991, Effects of muscle fatigue on mechanically sensitive afferents of slow conduction velocity in the cat triceps surae, *Journal of Neurophysiology,* 65, 360-370.

Henneman, E., Somjen, G., and Carpenter, D. O., 1965, Functional significance of cell size in spinal motoneurons, *Journal of Neurophysiology,* 28, 560-580.

Hník, P., Hudlická, O., Kucera, J., and Payne, R., 1969, Activation of muscle afferents by nonproprioceptive stimuli, *American Journal of Physiology,* 217, 1451-1458.

Hoffer, J. A., Loeb, G. E., Marks, W. B., O'Donovan, M. J., Pratt, C. A., and Sugano, N., 1987a, Cat hind limb motoneurons during locomotion. I. Destination, axonal conduction velocity, and recruitment threshold, *Journal of Neurophysiology,* 57, 510-529.

Hoffer, J. A., O'Donovan, M. J., Pratt, C. A., and Loeb, G. E., 1981, Discharge patterns of hindlimb motoneurons during normal cat locomotion, *Science,* 213, 466-468.

Hoffer, J. A., Sugano, N., Loeb, G. E., Marks, W. B., O'Donovan, M. J., and Pratt, C. A., 1987b, Cat hind limb motoneurons during locomotion. II. Normal activity patterns, *Journal of Neurophysiology,* 57, 530-553.

Iggo, A., 1961, Non-myelinated afferent fibers from mammalian skeletal muscle, *Journal of Physiology,* 155, 52-53.

Jovanovic, K., Anastasijevic, R., and Vuco, J., 1990, Reflex effects on gamma fusimotor neurones of chemically induced discharges in small diameter muscle afferents in decerebrate cats, *Brain Research,* 521, 89-94.

Kaufman, M. P., and Forster, H. V., 1996, Reflexes controlling circulatory, ventilatory and airway responses to exercise. In 'Handbook of Physiology, Section 12: Exercise: Regulation and Integration of Multiple Systems. II. Control of Respiratory and Cardiovascular Systems', L. B. Rowell and J. T. Shepherd, eds., Oxford University Press, New York, pp. 381-447.

Kaufman, M. P., Longhurst, J. C., Rybicki, K. J., Wallach, J. H., and Mitchell, J. H., 1983, Effects of static muscular contraction on impulse activity of groups III and IV afferents in cats, *Journal of Applied Physiology,* 55, 105-112.

Kaufman, M. P., and Rybicki, K. J., 1987, Discharge properties of group III and IV muscle afferents: their responses to mechanical and metabolic stimuli, *Circulation Research,* 61, 160-165.

Kaufman, M. P., Rybicki, K. J., Waldrop, T. G., and Ordway, G. A., 1984a, Effect of ischemia on responses of group III and IV afferents to contraction, *Journal of Applied Physiology,* 57, 644-650.

Kaufman, M. P., Rybicki, K. J., Waldrop, T. G., Ordway, G. A., and Mitchell, J. H., 1984b, Effects of static and rhythmic twitch contractions on the discharge of group III and IV muscle afferents, *Cardiovascular Research,* 18, 663-668.

Kenagy, J., VanCleave, J., Pazdernik, L., and Orr, J. A., 1997, Stimulation of group III and Iv afferent nerves from the hindlimb by thromboxane A_2, *Brain Research,* 744, 175-178.

Kniffki, K.-D., Mense, S., and Schmidt, R. F., 1978, Responses of group IV afferent units from skeletal muscle to stretch, contraction and chemical stimuli, *Experimental Brain Research,* 31, 511-522.

Kumazawa, T. N., and Mizumura, K., 1977, Thin-fibre receptors responding to mechanical, chemical and thermal stimulation in the skeletal muscle of the dog, *Journal of Physiology*, 273, 179-194.
Mense, S., 1977, Nervous outflow from skeletal muscle following chemical noxious stimulation, *Journal of Physiology*, 267, 75-88.
Mense, S., 1981, Sensitization of group IV muscle receptors to bradykinin by 5-hydroxytryptamine and prostaglandin E-2, *Brain Research*, 225, 95-105.
Mense, S., and Meyer, H., 1988, Bradykinin-induced modulation of the response behavour of different types of feline group III and IV muscle receptors, *Journal of Physiology*, 398, 49-63.
Mense, S. and Schmidt, R. F., 1974, Activation of group IV afferent units from muscle by algesic agents, *Brain Research*, 72, 305-310.
Mense, S., and Stahnke, M., 1983, Responses in muscle afferent fibers of slow conduction velocity to contractions and ischemia in the cat, *Journal of Physiology*, 342, 383-397.
Paintal, A. S., 1960, Functional analysis of group III afferent fibres of mammalian muscles, *Journal of Physiology*, 152, 250-270.
Pickar, J. G., Hill, J. M., and Kaufman, M. P., 1994, Dynamic exercise stimulates group III muscle afferents, *Journal of Neurophysiology*, 71, 753-760.
Rotto, D. M., Hill, J. M., Schultz, H. D., and Kaufman, M. P., 1990a, Cyclooxygenase blockade attenuates the responses of group IV muscle afferents to static contraction, *American Journal of Physiology*, 259, H745-H750.
Rotto, D. M., and Kaufman, M. P., 1988, Effects of metabolic products of muscular contraction on the discharge of group III and IV afferents, *Journal of Applied Physiology*, 64, 2306-2313.
Rotto, D. M., Schultz, H. D., Longhurst, J. C., and Kaufman, M. P., 1990b, Sensitization of group III muscle afferents to static contraction by products of arachidonic acid metabolism, *Journal of Applied Physiology*, 68, 861-867.
Rybicki, K. J., Waldrop, T. G., and Kaufman, M. P., 1985, Increasing gracilis interstitial potassium concentrations stimulates group III and IV afferents, *Journal of Applied Physiology*, 58, 936-941.
Shik, M. L., and Orlovsky, G. N., 1966, Locomotion of the diencephalic cat elicited by stimulation of the pyramids, *Biofizika*, 13, 143-152.
Sinoway, L. I., Hill, J. M., Pickar, J. G., and Kaufman, M. P., 1993, Effects of contraction and lactic acid on the discharge of group III muscle afferents in cats, *Journal of Neurophysiology*, 69, 1053-1059.
Thimm, F., and Baum, K., 1987, Response of chemosensitive nerve fibers of group III and IV to metabolic changes in rat muscles, *Pflügers Archives*, 410, 143-152.
Wetzel, M. C., and Stuart, D. G., 1976, Ensemble characteristics of cat locomotion and its neural control, *Progress in Neurobiology*, 7, 1-98.
Zajac, F. E., and Faden, J. S., 1985, Relationship among recruitment order, axonal conduction velocity and muscle-unit properties of type-identified motor units in cat plantans muscle, *Journal of Neurophysiology*, 53, 1303-1322.

5

EFFECTS OF ACTIVITY ON AXONAL EXCITABILITY: IMPLICATIONS FOR MOTOR CONTROL STUDIES

David Burke*

ABSTRACT

In many motor control studies, such as those involving H reflexes in human subjects, stimuli are delivered to nerves in which axons are or have recently been active. It is assumed that a constant stimulus intensity ensures that the same population of axons has been activated, particularly when there is little change in the M wave. However, the neurophysiological literature is replete with evidence that activity depresses axonal excitability. Indeed, some motor control studies have exploited this phenomenon to stimulate specific afferent populations selectively. This chapter will review studies performed in human subjects demonstrating that a voluntary contraction can have profound effects on the excitability of the active axons and on their ability to transmit impulses. These data need to be considered when interpreting human reflex studies.

AXONS HYPERPOLARIZE WHEN THEY ARE ACTIVE

When axons are stimulated repetitively, they developed a positive afterpotential, the size of which depends on the impulse load, i.e., stimulus rate and the duration of the stimulus train (Gasser, 1935; Bergmans, 1970). The positive afterpotential reflects axonal hyperpolarization that is largely driven by the electrogenic Na^+/K^+ pump. The pump is stimulated because activity results in the accumulation of Na^+ ions intracellularly and K^+ ions extracellularly, but it exchanges 3 Na^+ ions for 2 K^+ ions, thus producing membrane hyperpolarization.

There is evidence for greater expression of the hyperpolarization-activated cation conductance (I_H) on human sensory axons than human motor axons (Bostock et al., 1994). This 'inwardly rectifying' conductance produces accommodation to hyper-

* Spinal Injuries Research Centre, Prince of Wales Medical Research Institute, University of New South Wales, and Department of Neurology, Prince of Wales Hospital, Sydney, Australia.
Email: d.burke@unsw.edu.au

polarizing shifts in membrane potential, and its function is presumably to help maintain membrane potential during periods of intense impulse activity. It is likely that, for equivalent impulse loads, human motor axons hyperpolarize more than human sensory axons (Vagg et al., 1998), presumably because of the difference in expression of I_H.

ACTIVITY-DEPENDENT HYPERPOLARIZATION HAS BEEN EXPLOITED IN MOTOR CONTROL EXPERIMENTS

Prolonged vibration can be used to produce axonal hyperpolarization confined to the active group Ia afferents (Coppin et al., 1970), a technique that then allows selective stimulation of group Ib axons that normally have thresholds close to those of Ia afferents. The axonal hyperpolarization does not greatly alter the spindle response to physiological stimuli (such as quick stretch and vibration), and group Ia volleys can still be produced by mechanical stimuli. These techniques have been exploited by Jankowska and

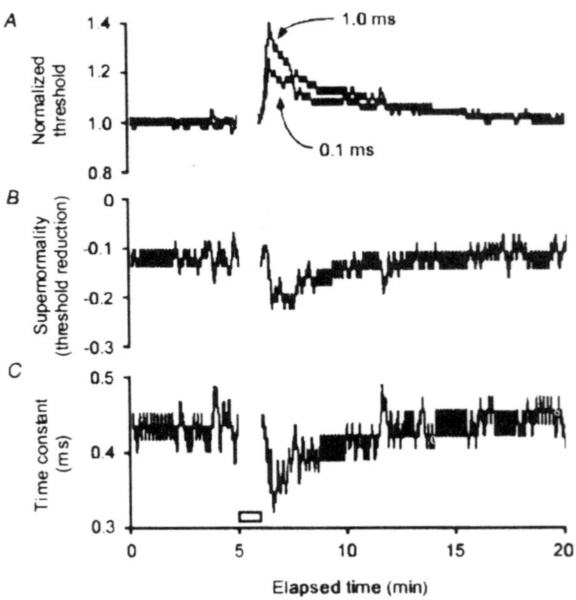

Figure 1. Changes in excitability in a single subject when the median nerve was stimulated at the wrist. CMAP of 70% max was tracked over the thenar muscles, using increments and decrements in stimulus intensity of 2%. At 5 min, a 60 s voluntary contraction was performed by the subject, as indicated by the open rectangle. *A*, threshold changes measured using test stimuli of 1.0 and 0.1 ms duration, normalized to 1.0. *B*, threshold changes during the supernormal period measured using a 0.1 ms test stimulus, delivered 7 ms after a supramaximal conditioning stimulus. Supernormality is expressed as the decrease in current below the control threshold necessary to produce the target CMAP. *C*, strength-duration time constant, calculated from the threshold changes in *A*. The discrete steps in each of the traces result from the tracking steps, and the graded post-contraction change is the result of tracking with small steps (2%). The changes in excitability are maxmal immediately after the end of the contraction. (From Vagg et al., 1998, with permission.)

colleagues (Fetz et al., 1979) to demonstrate convergence of group Ia and group Ib afferents on common interneurons responsible for non-reciprocal group I inhibition (formerly called 'Ib' inhibitory interneurons).

NATURAL ACTIVITY SIGNIFICANTLY INCREASES THE THRESHOLD OF THE HUMAN AXONS

If normal subjects perform a maximal voluntary contraction of the thenar muscles for 1 min, the threshold for the active motor axons increases by ~40%, and this decrease in excitability takes 15-20 min to return to normal (Figs. 1 and 2). The increase in threshold is accompanied by an increase in latency, a decrease in strength-duration time constant and an increase in supernormality (Fig. 1; data for a single subject). In myelinated axons, strength-duration time constant is a nodal property, while changes in supernormality reflect changes in the resistance of the internodal membrane. The activity-dependent changes in these indices indicate that the increase in threshold was due to axonal hyperpolarization.

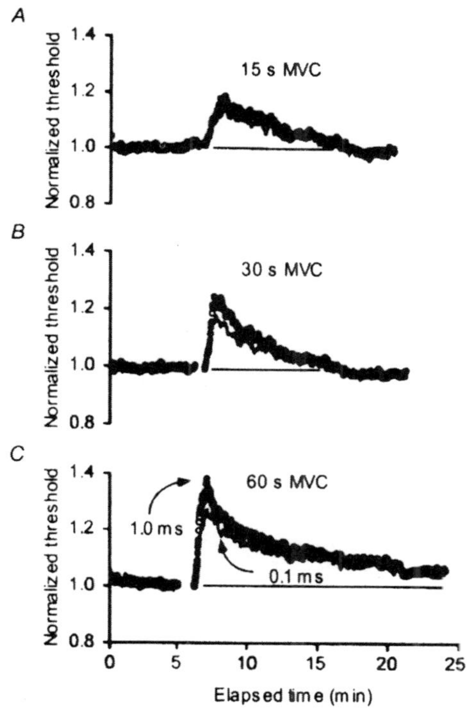

Figure 2. Dependence of the decrease in excitability on the duration of contraction. Threshold changes, measured using 1.0 (open symbols) and 0.1 ms (filled symbols) test stimuli, continuous line, respectively for contractions lasting 15, 30 and 60s. Each trace represents mean data for six subjects. (From Vagg et al., 1998, with permission.)

Contractions of only 15-s duration can increase threshold by 15-20% with return to control levels after 5-10 min (Fig. 2; mean data for 6 subjects). These are profound changes in excitability: a 40% increase in threshold implies that the stimulus intensity would need to be increased by 40% to activate the same population of axons, and inevitably axons that had not been active in the contraction might then also be activated. (Indeed, this was the rationale for the technique developed by Coppin et al., 1970.)

In these studies the response to a constant stimulus intensity decreased by >50% and recovered with the same time course as threshold. These findings indicate that, if a manoeuvre or conditioning stimulus produced a sustained discharge (or a sustained increase in discharge) of axons, the same stimulus intensity would not excite the same population of axons before, during and after the maneuver and, importantly, the effects would be long-lasting.

In addition, the hyperpolarization causes conduction slowing, by some tens of microseconds per node. Considering the number of nodes of Ranvier in long human nerves this would amount to >1.0 ms over for a limb segment of 0.5-1.0 m. This degree of slowing could be misinterpreted as indicating that a reflex change involved an interneuronal delay.

HOW RELIABLE IS THE M WAVE AS A CONTROL THAT THE STIMULUS RECRUITED THE SAME NUMBER OF AFFERENT AXONS?

Unfortunately, this control, though better than none, is flawed because the excitability properties of sensory and motor axons differ. There are differences between sensory and motor axons not only in I_H, as discussed above, but also their strength-duration properties (Mogyoros et al., 1996). Indeed, in comparison with □ motor axons, it is the lower rheobase of group Ia axons rather than any small difference in axonal diameter that confers a lower threshold when test stimuli of 1-ms duration are used. Motor and sensory axons would not undergo the same changes in excitability (and thereby threshold) even if they conducted identical impulse trains (Vagg et al., 1998). In addition, axonal hyperpolarization shortens strength-duration time constant (as in Fig. 1) and increases rheobase, changing the difference between the properties of sensory and motor axons. These hyperpolarization-induced changes in strength-duration properties can be produced by natural activity (Vagg et al., 1998; Kuwabara et al., 2001).

ACTIVITY-DEPENDENT HYPERPOLARIZATION REDUCES THE SAFETY MARGIN FOR IMPULSE CONDUCTION

The hyperpolarization produced by activity is insufficient to jeopardize impulse conduction because the normal safety margin for action potential generation is ~5:1 at nodes of Ranvier in the mid-axon. An increase in threshold of 40% is relatively small compared to this. However, it could be important when the safety margin is lowered by pathology or is normally low – i.e., at axonal branch points and nerve terminals.

When there is axonal pathology, the hyperpolarization that accompanies a maximal voluntary contraction can precipitate conduction block at sites of impaired safety margin for impulse conduction. This has been demonstrated recently in patients with a chronic demyelinating peripheral nerve disease (Cappelen-Smith et al., 2000). These findings are probably applicable to patients with similar pathology involving axons within the central nervous system, particularly multiple sclerosis and spinal cord injury, and studies of strength and fatigue in such patients may need to be interpreted judiciously.

Recent studies have demonstrated impairment of the refractory period of transmission between axon and muscle following maximal voluntary contraction for 1 min (Kuwabara et al., 2001). There was transmission failure between mid-axon and muscle of impulses separated by 2-3 ms, an effect that lasted a few minutes. Whether this occurred at branch points (as postulated) or more distally could not be established. Such short interspike intervals are not normally encountered with human motor units, and this finding is unlikely to reflect a limitation on motor behavior. However, in pathology, when other factors intervene, this might not be so. It is possible that a similar limitation on transmission occurs at central synapses.

CONCLUSIONS

Whenever axons are active they undergo changes in excitability and regain their 'resting' level slowly. This chapter has focused on the hyperpolarization that follows the conduction of long impulse trains. However, there are significant fluctuations in excitability following a single action potential and following brief impulse trains, though these subside within 100-200 ms. The message is that activity affects the stimulus required to activate an axon again. This may be important in motor control studies (i) following voluntary activity, (ii) when conditioning and test stimuli are delivered to the same nerve, or (iii) when transcranial stimuli are delivered at the same scalp site.

REFERENCES

Bergmans, J., 1970, *The Physiology of Single Human Nerve Fibres*, Vander, Louvain.

Bostock, H., Burke, D., and Hales, J. P., 1994, Differences in behaviour of sensory and motor axons following release of ischaemia, *Brain*, 117, 225-234.

Cappelen-Smith. C., Kuwabara, S., Lin, C. S.-Y., Mogyoros, I., and Burke, D., 2000, Activity-dependent hyperpolarization and conduction block in chronic inflammatory demyelinating polyneuropathy, *Annals of Neurology*, 48, 826-832.

Coppin, C. M., Jack, J. J., and MacLennan, C. R., 1970, A method for the selective electrical activation of tendon organ afferent fibres from the cat soleus muscle, *Journal of Physiology*, 210, 18P-20P.

Fetz, E. E., Jankowska, E., Johannisson, T., and Lipski, J., 1979, Autogenetic inhibition of motoneurones by impulses in group Ia muscle spindle afferents, *Journal of Physiology*, 293, 173-195.

Gasser, H.S., 1935, Changes in nerve-potentials produced by rapidly repeated stimuli and their relation to the responsiveness of nerve to stimulation, *American Journal of Physiology*, 111, 35-50.

Kuwabara, S., Lin, C. S.-Y., Mogyoros, I., Cappelen-Smith, C., and Burke, D., 2001, Voluntary contraction impairs the refractory period of transmission in healthy human axons, *Journal of Physiology*, 531, 265 275.

Mogyoros, I., Kiernan, M. C., and Burke, D., 1996, Strength-duration properties of human peripheral nerve, *Brain*, 119, 439-447.

Vagg, R., Mogyoros, I., Kiernan, M. C., and Burke, D., 1998, Activity-dependent hyperpolarization of human motor axons produced by natural activity, *Journal of Physiology*, 507, 919-925.

REFLEXES IN THE HAND: STRONG SYNAPTIC COUPLING BETWEEN SINGLE TACTILE AFFERENTS AND SPINAL MOTONEURONES

Penelope McNulty and Vaughan Macefield*

ABSTRACT

We investigated whether the discharge of single afferents could modulate ongoing EMG in muscles acting on the hand. Recordings were made from 129 single low threshold mechanoreceptors via tungsten microelectrodes inserted into the median or ulnar nerves at the wrist in awake human subjects. Reflex modulation of ongoing EMG was observed for 9 of 26 FA I, 1 of 5 FA II, 18 of 39 SA II and 2 ectopically active afferents via spinally mediated, presumably oligosynaptic, pathways. Two reflex responses were found: a single excitatory response which arose from irregularly firing afferents (FA I, FA II), and a cyclic EMG modulation associated with regularly discharging afferents (SA II) and one FA I afferent firing in regular bursts. No strong synaptic coupling could be detected from the discharge of any of the 21 SA I, 2 joint or 34 muscle spindle afferents. These results highlight the important role of certain types of single cutaneous mechanoreceptors in sensorimotor control of the hand.

INTRODUCTION

Peripheral feedback is an important element in fine motor control of the hand. The glabrous skin of the human hand, which is the primary source for tactile information, contains four low-threshold mechanoreceptors: rapidly adapting type I and II (FA I and FA II, respectively) and slowly adapting type I and II afferents (SA I and SA II) (Knibestöl and Vallbo, 1970). Muscle spindles give rise to two afferent endings, primary (or group 1a) and secondary (group II) (Boyd, 1962). Afferents can be type-identified based on their discharge patterns, receptive field properties and response and adaptation to external mechanical stimulation (Knibestöl, 1973; Johnson et al., 2000). Activation of populations of either cutaneous or muscle spindle afferents results in a short latency

* Prince of Wales Medical Research Institute & Univ. of New South Wales, Sydney, NSW 2031, Australia. Email: vg.macefield@unsw.edu.au

facilitation of the motoneurone pool of muscles acting on the hand where there is ongoing EMG (Caccia et al., 1973; Buller et al., 1980; Datta and Stephens, 1981; Darton et al., 1985; Evans et al., 1989; Deuschl et al., 1995; Macefield et al., 1996). However, it was not known whether the activity of single afferents was able to produce a similar modulation of associated ongoing EMG. In this chapter we have combined the results of two series of experiments (McNulty et al., 1999; McNulty and Macefield, 2001).

TACTILE AFFERENTS

Ninety-one single low threshold tactile afferents were recorded via insulated microelectrodes inserted percutaneously into either the median or ulnar nerves at the wrist in awake human subjects. A further two joint afferents and two ectopically active (presumed cutaneous) afferents were recorded. EMG was recorded from muscles acting on the hand, namely thenar muscles, first dorsal interosseous (FDI), adductor digiti minimi (ADM), flexor pollicis longus (FPL) and flexor digitorum superficialis (FDS). The receptive field of each afferent was activated by a rigid probe while subjects performed a weak voluntary contraction (<20% MVC). Unitary afferent recordings were confirmed by spike superimposition. Autocorrelograms of the neural signal were constructed in 5 ms bins to illustrate the temporal profile of the afferent discharge pattern. Data were analysed by spike-triggered averaging from the neural spike to reveal any events in the associated EMG that were time-locked to the afferent discharge.

High frequency bursts of irregular discharges were recorded from 26 FA I afferents. This characteristic bursting pattern reflects the movement of the stimulating probe across the receptive field of the receptor end organ and can be seen in the raw data and afferent autocorrelogram (Fig. 1). The EMG associated with the activity of nine FA I afferents revealed a single excitatory peak time-locked to the receptor discharge. The EMG response, measured peak to peak, represents a modulation of between 4-23% of the total averaged myoelectrical activity. A typical example of an FA I afferent mediated facilitation of the motoneurone pool is illustrated in Figure 1*C*.

A highly variable discharge rate typifies the activity of FA II afferents, five of which were recorded. Only one of these units was associated with a short latency facilitation of the ongoing EMG with a 50 ms latency to the peak of the response. The mean firing rate of this unit was 125.7 Hz with a coefficient of variation of 0.63. While the excitatory reflex response for this unit was not large, comprising only 1.3% of the total EMG activity, it was present when averages were constructed from different portions of the raw data. Conversely, none of the 21 SA I afferents recorded showed any evidence of modulation of ongoing EMG. The discharge variability of these units was less than that of the FA II afferent demonstrating a reflex response, with coefficients of variation ranging between 0.34-0.62. Thus, the absence of a reflex response associated with the discharge of single SA I tactile afferents was not due to their irregular discharge pattern.

SA II afferents typically fire with a regular discharge in response to sustained indentation, as seen in Figure 2. The averaged EMG revealed a series of peaks the periodicity of which, measured between subsequent positive peaks, matched the mean interspike interval of the afferent discharge. Thus for every neural spike there was a corresponding facilitation of the motoneurone pool. When all 15 SA II afferents associated with a cyclic modulation of the ongoing EMG were analysed in this way, a high correlation was found between the afferent discharge rate and the modulation of the averaged EMG (r=0.97).

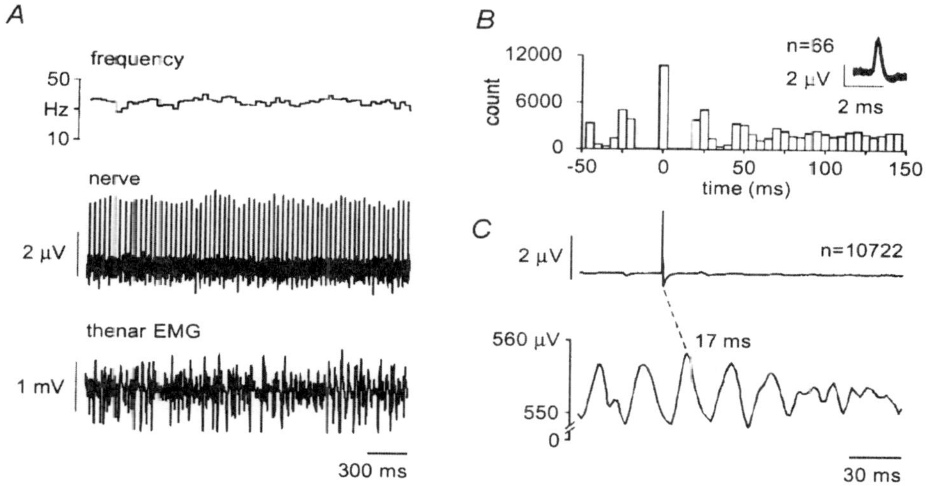

Figure 1. Short-latency excitation in an FA I afferent, with receptive field on middle phalanx of digit II. *A*, Raw data with instantaneous frequency in top trace, neurogram in second trace and FDS EMG below; *B*, autocorrelogram of the neural spikes illustrating the temporal profile of the afferent discharge; inset: superimposed spikes; *C*, averaged data from 9718 sweeps, with averaged afferent spike in top trace, and averaged EMG below with a single excitatory short latency reflex response, dashed line highlights the peak response latency (36 ms). (Redrawn from McNulty et al., 1999.)

A further two short-latency reflex responses were observed in the ongoing EMG associated with the ectopic discharge of two presumably cutaneous afferents (see Macefield, 1998). Immediately preceding the location of these spontaneously active discharges, the tip of the electrode was located in the nerve fascicle supplying the skin of the lateral hemidigit of the fourth finger. Although we suspect that these were cutaneous afferents, we cannot exclude the possibility that these discharges arose from joint afferents. However, two joint afferents were recorded and analysed, neither of which revealed any evidence of a strong synaptic coupling to the spinal motoneurones. A larger sample is required before further conclusions can be drawn.

The range of response latencies in the averaged EMG was 17 to 98 ms. While this range seems broad it reflects the range of conduction velocities for cutaneous afferents reported in the literature (Knibestöl, 1973; Johansson and Vallbo, 1983; Mackel, 1988; Macefield et al., 1989; Kakuda, 1992) and suggests a spinally mediated, oligo-synaptic pathway. Previous reports of cutaneomuscular hand reflexes have been elicited by activation of populations of afferents (see Introduction) and in some cases by electrical stimulation of whole digital nerves (Caccia et al., 1973; Datta and Stephens, 1981; Evans et al., 1989). The latencies of the evoked reflex responses in such cases represent a synchronised volley via either all, or a large proportion, of the available afferents and hence provide little variation between studies or methods of activation.

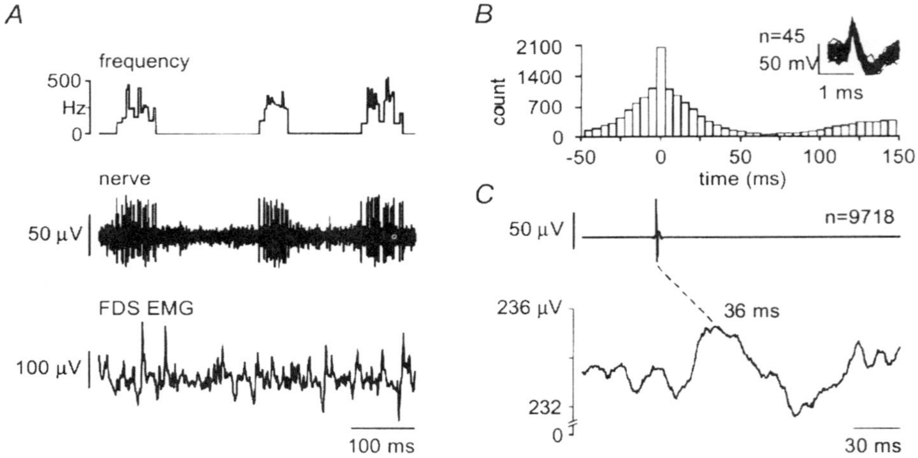

Figure 2. Cyclic response in an SA II afferent with a regular discharge. *A*, Raw data, receptive field in the nail bed of digit I: top trace instantaneous frequency of the afferent discharge, the neurogram in the second trace and thenar EMG below. *B*, Autocorrelogram of the afferent discharge; inset – superimposed afferent spikes from Panel A. *C*, Averaged data from 10722 sweeps with afferent in top trace, and averaged EMG showing cyclic fluctuations below. The dotted line highlights the short-latency reflex with the first peak at 17 ms and the second at 44 ms. (Reprinted from McNulty et al., 2001.)

Two kinds of reflex responses were noted: a single excitatory peak (Fig. 1) and a cyclic modulation of the ongoing EMG (Fig. 2). The single parameter determining the nature of the reflex response is the pattern of the afferent discharge. Regularly firing afferents produce a cyclic modulation and those with an irregular firing rate are associated with a single excitatory peak. However, an afferent discharging irregularly, but in regularly occurring bursts, can also produce a cycle modulation in the averaged EMG (Fig. 3). Fifteen of the eighteen SA II afferents associated with a reflex response evoked a cyclic modulation of the ongoing EMG. The remaining three afferents were associated with a single excitatory peak. Two of these afferents discharged with a 'sliding regularity' so that any 1 s epoch appeared regular, but a 300 s epoch demonstrated a monotonic change in firing rate reflecting an altered level of stimulus indentation. The mechanical threshold of the third afferent was low enough to encode the smallest changes in stimulus intensity as a change in firing rate. Thus all three did not discharge with a regular output, typical of SA II afferents.

The absence of a response in the SA I receptors cannot be attributed to their irregular firing rate as the coefficient of variation for all 21 of these units was less than that of the FA II afferent that was associated with a single short latency reflex modulation. Whether these afferents require spatial and/or temporal summation to modulate spinal motoneurones (as do the muscle spindle endings, see below), or synapse with more inhibitory interneurones is unclear. While it is known that populations of primary and secondary muscle spindle endings can have a facilitatory effect on the motoneurone pool, there is no evidence to support SA I afferents exerting a similar effect. It may be that their primary role is to transmit tactile information to the cortex as rapidly as possible while other cutaneous mechanoreceptors, which also rapidly relay similar tactile signals centrally but with a poorer resolution, interact with the motor system via a stronger net synaptic input to spinal motoneurones.

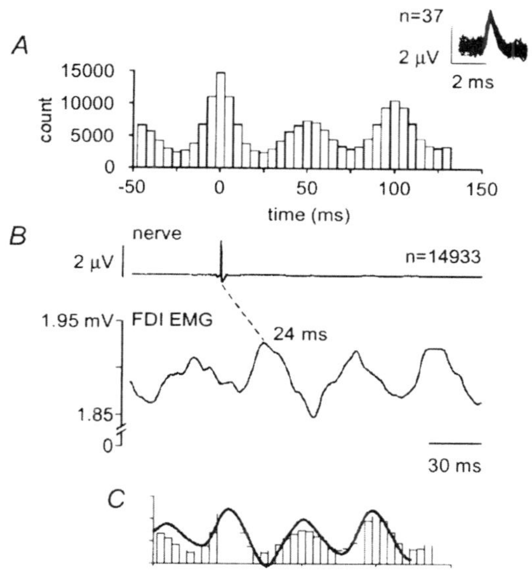

Figure 3. Cyclic response in an irregularly firing afferent discharging in regular bursts. Averaged data for an FA I afferent located in the middle of the thenar eminence. *A* Autocorrelogram of the afferent discharge with inset of superimposed spikes. *B* Averaged neural spike from 14933 sweeps above and averaged EMG from first dorsal interosseous below. *C* Superimposition of the averaged EMG trace (more highly filtered) on the autocorrelogram with a -25 ms shift (reflecting the phase lag of the reflex latency) to emphasise the correlation between afferent bursting pattern and averaged EMG. (Reprinted from McNulty et al., 2001.)

MUSCLE SPINDLES

Thirty four muscle spindle afferents were recorded. The search for these afferents was not undertaken during voluntary contraction so that the potential sampling of Golgi tendon organs was avoided. Experiments were conducted as for the tactile mechanoreceptors, with EMG being recorded from FDI, thenar muscles and adductor digiti minimi. Afferents were either activated by passive muscle stretch or their firing rate was increased in this manner for those that were fusimotor driven. Three spindles endings were unloaded by voluntary contractions and so were not considered further. Of the remaining 31 afferents, responses typical of both primary (n=12) and secondary (n=19) spindle endings were encountered in both homonymous and heteronymous muscles. No EMG modulation was evident following activation of any muscle spindle ending, whether primary or secondary. Averages were computed from large numbers of spike triggers so that the absence of a response was not due to insufficient averaging triggers.

Like SA I afferents, muscle spindles endings may require temporal and/or spatial summation, as postulated by Gandevia and colleagues (1986), despite the demonstration of monosynaptic contacts between Ia afferents and most of the α-motoneurones of the homonymous motoneurone pool (Mendell and Henneman, 1968). When measured in the same body segment, the conduction velocity of human muscle and cutaneous afferents is similar (Macefield et al., 1989), while spindle afferents in the cat (Hunt and Kuffler, 1951; Brown and Iggo, 1967) and monkey (Lindblom, 1965; Cheney and Preston, 1976) conduct approximately twice as fast as their respective low-threshold tactile afferents. Thus, it is possible that the role of cutaneous mechanoreceptors in humans has become more significant than that of muscle spindles.

CONCLUSION

We have demonstrated reflex modulation of the motoneurone pool of muscles acting on the hand by the input of single FA I, FA II and SA II cutaneous mechanoreceptors. The latencies involved suggest an oligosynaptic, spinally mediated pathway. There was no evidence of strong synaptic coupling between single SA I cutaneous, joint or muscle spindle afferents and spinal motoneurones. These results highlight the importance of single tactile afferents in fine motor control of the human hand and suggest that tactile afferents can assist in regulating the amount of EMG necessary to successfully manipulate held objects, possibly to prevent slipping and dropping of objects, or the use of excessive force.

ACKNOWLEDGEMENTS

This work was carried out with support from the National Health and Medical Research Council (of Australia).

REFERENCES

Boyd, I. A., 1962, The structure and innervation of the nuclear bag muscle fibres system and the nuclear chain muscle fibre system in mammalian muscle systems, *Philosophical Transactions of the Royal Society of London. Series B. Biological Sciences,* 245, 81-136.

Brown, A. G., and Iggo, A., 1967, A quantitative study of cutaneous receptors and afferent fibres in the cat and rabbit, *Journal of Physiology,* 193, 707-733.

Buller, N. P., Garnett, R., and Stephens, J. A., 1980, The reflex responses of single motor units in human hand muscles following muscle afferent stimulation, *Journal of Physiology,* 303, 337-349.

Caccia, M. R., McComas, A. J., Upton, A. R., and Blogg, T., 1973, Cutaneous reflexes in small muscles of the hand, *Journal of Neurology Neurosurgery and Psychiatry,* 36, 960-977.

Cheney, P. D., and Preston, J. B., 1976, Classification and response characteristics of muscle spindle afferents in the primate, *Journal of Neurophysiology,* 39, 1-8.

Darton, K., Lippold, O. C., Shahani, M., and Shahani, U., 1985, Long-latency spinal reflexes in humans, *Journal of Neurophysiology,* 53, 1604-1618.

Datta, A. K., and Stephens, J. A., 1981, The effects of digital nerve stimulation on the firing of motor units in human first dorsal interosseous muscle, *Journal of Physiology,* 318, 501-510.

Deuschl, G., Feifel, E., Guschlbauer, B., and Lücking, C. H., 1995, Hand muscle reflexes following air puff stimulation, *Experimental Brain Research,* 105, 138-146.
Evans, A. L., Harrison, L. M., and Stephens, J. A., 1989, Task-dependent changes in cutaneous reflexes recorded from various muscles controlling finger movement in man, *Journal of Physiology,* 418, 1-12.
Gandevia, S. C., Burke, D., and McKeon, B., 1986, Coupling between human muscle spindle endings and motor units assessed using spike-triggered averaging, *Neuroscience Letters,* 71, 181-186.
Hunt, C. C., and Kuffler, S. W., 1951, Stretch receptor discharges during muscle contraction, *Journal of Physiology,* 113, 298-315.
Johansson, R. S., and Vallbo, Å. B., 1983, Tactile sensory coding in the glabrous skin of the human hand, *Trends in Neurosciences,* 6, 27-32.
Johnson, K. O., Yoshioka, T., and Vega-Bermudez, F., 2000, Tactile functions of mechanoreceptive afferents innervating the hand, *Journal of Clinical Neurophysiology,* 17, 539-558.
Kakuda, N., 1992, Conduction velocity of low-threshold mechanoreceptive afferent fibers in the glabrous and hairy skin of human hands measured with microneurography and spike-triggered averaging, *Neuroscience Research,* 15, 179-188.
Knibestöl, M., 1973, Stimulus-response functions of rapidly adapting mechanoreceptors in human glabrous skin area, *Journal of Physiology,* 232, 427-452.
Knibestöl, M., and Vallbo, Å. B., 1970, Single unit analysis of mechanoreceptor activity from the human glabrous skin, *Acta Physiologica Scandinavica,* 80, 178-195.
Lindblom, U., 1965, Properties of touch receptors in distal glabrous skin of the monkey, *Journal of Neurophysiology,* 28, 966-985.
Macefield, G., Gandevia, S. C., and Burke, D., 1989, Conduction velocities of muscle and cutaneous afferents in the upper and lower limbs of human subjects, *Brain,* 112, 1519-1532.
Macefield, V. G., 1998, Spontaneous and evoked ectopic discharges recorded from single human axons, *Muscle Nerve,* 21, 451-468.
Macefield, V. G., Rothwell, J. C., and Day, B. L., 1996, The contribution of transcortical pathways to long-latency stretch and tactile reflexes in human hand muscles, *Experimental Brain Research,* 108, 147-154.
Mackel, R., 1988, Conduction of neural impulses in human mechanoreceptive cutaneous afferents, *Journal of Physiology,* 401, 597-615.
McNulty, P. A, and Macefield, V. G., 2001, Modulation of ongoing EMG by different classes of low-threshold mechanoreceptors in the human hand, *Journal of Physiology,* 537, 1021-1032.
McNulty, P. A., Türker, K. S., and Macefield, V. G., 1999, Evidence for strong synaptic coupling between single tactile afferents and motoneurones supplying the human hand, *Journal of Physiology,* 518, 883-893.
Mendell, L. M., and Henneman, E., 1968, Terminals of single Ia fibers: distribution within a pool of 300 homonymous motor neurons, *Science,* 160, 96-98.

7

THE SYNAPTIC LINKAGE FOR TACTILE AND KINAESTHETIC INPUTS TO THE DORSAL COLUMN NUCLEI

Mark J. Rowe*

ABSTRACT

Our sensory abilities in touch and kinaesthesia depend upon approximately eight major classes of receptors and sensory nerve fibres. When *individual* fibres of these different kinaesthetic and tactile fibre classes are selectively activated in conscious human subjects by means of the intraneural microstimulation procedure there are quite marked differences observed among the classes in their capacity to generate perceptual responses. These differences may be attributable to differential transmission characteristics for different fibre classes at synaptic junctions within the sensory pathways. To test this hypothesis we have employed a paired, simultaneous recording paradigm in which we have examined the efficacy of transmission across the dorsal column nuclei (DCN) in a one-to-one synaptic linkage between single, identified tactile or kinaesthetic afferent fibres and their central DCN target neurons. These studies demonstrate a remarkable security of transmission for all fibre classes examined. For all classes, the minimum sensory input, a single impulse in one sensory fibre, can generate spike output from DCN target neurons. The results demonstrate that the differential capacities of various tactile and kinaesthetic fibre classes to generate perceptual responses when activated singly, do not appear to be explicable in terms of systematic differences in DCN transmission characteristics.

INTRODUCTION

Receptors and Sensory Nerve Classes responsible for Tactile and Kinaesthetic Sensation

Tactile sensation in the human fingertips and palms depends upon three or four major classes of tactile receptors and sensory nerve fibres. These include the Pacinian

* Department of Physiology and Pharmacology, University of New South Wales, Sydney N.S.W. 2052, Australia. Email m.rowe@unsw.edu.au

corpuscles and their *PC* sensory nerve fibres; the Meissner corpuscles and their associated *RA (rapidly adapting)* class of afferent fibres; the Merkel receptors and their associated *SAI (slowly adapting type I)* nerve fibres, and the Ruffini endings that are associated with *SAII* afferent nerve fibres. In the glabrous skin of the limb extremities the latter class seems to be peculiar to human beings, as only the first three classes appear to be present in the macaque and marmoset monkeys (Talbot et al., 1968; Coleman et al., 2001) and in the cat (Jänig et al., 1968; Iggo and Ogawa, 1977; Ferrington and Rowe, 1980).

In the hairy regions of skin, the principal tactile afferent fibres include the *SAI* class together with an *SAII* class, associated with Ruffini receptors, and a major group whose endings are associated with hair follicles, the *HFA* afferent fibres.

While there is general agreement over the presence of these different tactile classes in glabrous and hairy skin, there is some variation in the nomenclature. In particular, Johansson and Vallbo, who have been responsible for the most detailed observations on human tactile afferent classes use an *FAI* and *FAII (fast adapting types I* and *II)* designation for the *RA* and *PC* fibre classes (Johansson and Vallbo, 1979; Vallbo and Johansson, 1984).

For the kinaesthetic sense the principal contributions arise from muscle spindle receptors and from receptors in the joint capsules, although cutaneous mechanoreceptors, in the vicinity of joints, principally associated with the *SAII* fibres, are sensitive to skin stretch when flexion-extension movements occur at the joint, and also contribute to our sense of limb position (Gandevia and McCloskey, 1976; McCloskey, 1978; Edin and Abbs, 1991; see Gandevia et al. in this volume).

Selective Activation of Single Tactile or Kinaesthetic Afferent Fibres in Conscious Human Subjects

The advent of microneurography in the late 1960s enabled researchers to record from single afferent or efferent nerve fibres in a conscious human subject attending to a particular sensory or motor task (Vallbo and Hagbath, 1968). An important extension of this technique entailed the conversion of the microneurography electrode from the recording mode to the stimulation mode, enabling the researcher to use weak electrical stimuli to *selectively* activate an identified tactile or kinaesthetic afferent fibre and determine whether there were perceptual consequences for the subject (Ochoa and Torebjörk, 1983; Vallbo et al., 1984). The quite striking finding in these studies has been that marked differences are observed from one fibre class to another in their capacities to generate a perceptual response when individual fibres of those classes are selectively activated.

Intraneural Microstimulation of Single PC, RA, SAI and SAII Afferent Fibres

When trains of weak electrical stimuli (20-100 Hz) are used with the intraneural microstimulation procedure to activate single *PC* or *RA* fibres, the attending human subject almost always experiences a sense of flutter or vibration depending upon the stimulation frequency in the train (Ochoa and Torebjörk, 1983; Vallbo et al., 1984). In contrast, when single *SAI* or *SAII* fibres are activated with the equivalent stimulus trains, there are very different perceptual consequences. First, *SAII* fibre activation rarely elicits any perceptual response. Second, although a majority of *SAI* fibres (~75%, Ochoa and

Torebjörk, 1983) generate a percept on being activated singly, the experience is one of steady pressure or deformation in the skin region approximating the *SAI* fibre's receptive field, in contrast to the vibrotactile percept generated by the equivalent train of stimuli delivered to single *RA* or *PC* fibres (Ochoa and Torebjörk, 1983; Vallbo et al., 1984; Macefield et al., 1990).

Intraneural Microstimulation of Single Kinaesthetic Afferent Fibres

Differences have also emerged between the two principal kinaesthetic fibre classes in their capacities for generating perceptual responses when activated singly. Almost all joint afferent fibres arising from the interphalangeal and metacarpophalangeal joints gave rise to a deep sensation associated with their associated joint capsule, or a sense of movement in the joint, whereas the activation of single muscle spindle afferents from various forearm muscles was almost always without effect (Macefield et al., 1990). A similar failure to generate a percept was observed for single spindle afferents associated with upper arm muscles (Ni et al., 1998).

Hypothesis for the Differential Capacities of Individual Fibres to Generate Percepts

As individual afferent fibres of the different classes respond in an apparently identical manner to trains of the intraneural microstimuli it appears that the differential capacities of different fibre classes to generate a percept when activated singly must lie within the central nervous system (Rowe, 2001). One of the hypotheses that we have examined is that there may be systematic differences from one fibre class to another in their transmission characteristics across synaptic junctions within the somatosensory pathways in the cat. The studies were predicated on the assumption that any systematic differences among the different somatosensory afferent fibre classes in their transmission characteristics would apply across mammalian species or, at least, would apply for both the cat and primate species, including human beings. Some support for this similarity across species comes at the peripheral level of the tactile and kinaesthetic systems where it appears that the same basic classes of fibres are present in the cat and in primates, and where *a particular class*, e.g. the *PC* class, shares an identity of functional characteristics across these species.

EXPERIMENTAL METHODS

Paired Recording Analysis of Transmission between Single Sensory Fibres and their Central Target Neurones

The transmission characteristics for single identified tactile or kinaesthetic afferent fibres at the synaptic relay in the dorsal column nuclei (*DCN*) were analyzed by paired simultaneous recording from an *individual* afferent fibre in an *intact* peripheral nerve

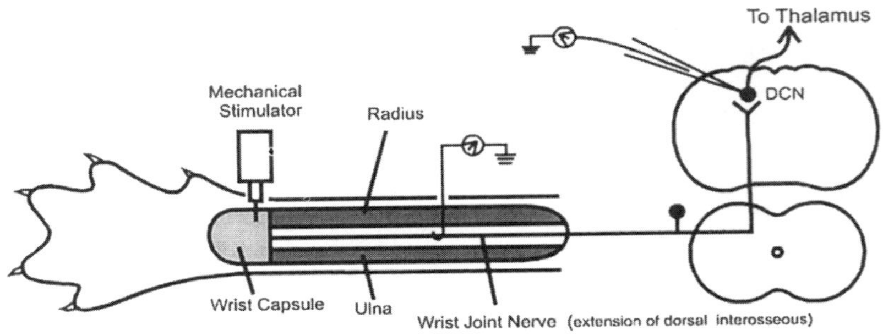

Figure 1. Schematic representation of the preparation used for paired, simultaneous recording from single afferent fibres in the wrist joint nerve and their target neurones in the cuneate division of the dorsal column nuclei (DCN).

fascicle, and from the central target neurone of that afferent fibre. With this approach we were able to examine the efficacy of transmission in a one-to-one linkage between an identified primary afferent fibre and one of its central target neurones of the *DCN*. The experimental arrangement is shown in Figure1 for the analysis of transmission characteristics for single joint afferent fibres whose activity could be monitored from the intact wrist joint nerve. The nerve, which is a distal extension of the dorsal interosseous nerve of the forearm, was freed by removal of forearm extensor muscles (Coleman et al., 1998) over a distance up to ~4 cm and covered by liquid paraffin in a pool created by the forearm skin flaps. It was then possible, with a silver or platinum hook electrode under the nerve, to monitor the activity of any active Group II fibre while the nerve remained in continuity with the central nervous system. A second electrode, a microelectrode, was used to record simultaneously from neurons within the cuneate division of *DCN* that were the central targets of the monitored wrist joint afferent fibre. We have employed similar paired recording preparations to examine the transmission characteristics for other identified classes of tactile or kinaesthetic afferent fibres (Ferrington et al., 1987a,b; Rowe, 1990, Vickery et al., 1994; Gynther et al., 1995; Rowe, 2001; Zachariah et al., 2001).

RESULTS

Transmission Characteristics for Single PC Afferent Fibres

The first class of afferent fibre for which we successfully applied the paired recording paradigm in the anaesthetized cat was the *PC* fibre class, using the hindlimb interosseous nerve (*IN*), which supplies an array of PC receptors on the hindlimb interosseous membrane (Ferrington et al., 1987a,b). These receptors show the same mechano-sensitivity, including vibrational sensitivity, as cutaneous as Pacinian corpuscles. Individual *PC* fibres were activated by means of vibration applied with a 200 μm diameter probe to individual corpuscles associated with the interosseous membrane. Thirty five PC fibre-gracile neurone pairs were isolated in which the activity

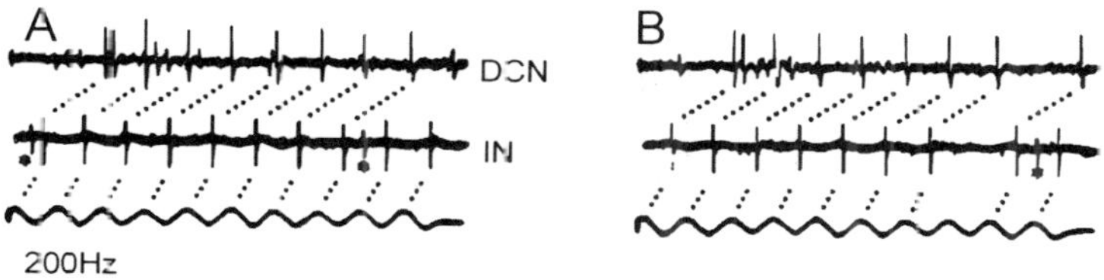

Figure 2. Single sensory nerve impulses in a PC fibre of the interosseous nerve (IN) evoke spike output with high security from a gracile neurone of dorsal column nuclei (DCN; uppermost traces). The lowest trace shows the stimulus train of ten cycles of 200 Hz vibration at an amplitude of <2 μm. (From Ferrington et al., 1987a.)

in a single fibre could generate spike output from the central neurone (Fig. 2). Furthermore, for a great many of these pairs, the potency of the one-to-one synaptic linkage was quite remarkable, as shown in the paired recording traces of Figure 2A-D in which a single *PC* fibre was activated in the IN by a ten-cycle train of vibration at 200 Hz (<2μm). In each set, A-D, a pair of spikes was generated in the gracile neurone by the first *PC* fibre spike, demonstrating amplification in the linkage, and a sustained gracile response obtained to almost every occurrence of subsequent *PC* spikes. Evidence in support of the monitored *PC* fibre accounting entirely for the observed gracile neurone responses was given in the original papers (Ferrington et al., 1987a), and included the fact that the gracile spike response was dependent upon the prior occurrence of the *observed PC* fibre spike and followed it at a fixed latency. When the *PC* fibre failed to respond on some cycles of the vibration, for example on the fourth cycle in *B*, the eight in *C*, and the ninth in *D*, there was a correlated absence of response in the gracile neurone.

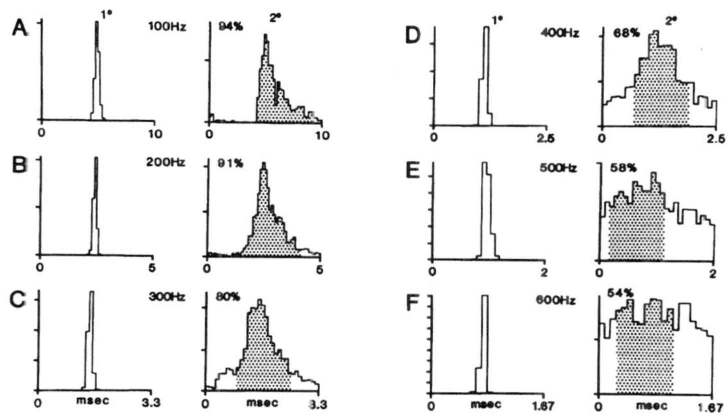

Figure 3. Cycle histogram pairs showing the distribution of impulse activity within the vibration cycle period for the responses of a PC fibre-gracile neurone pair (1° and 2°) at six frequencies of vibration (from 100 to 600 Hz). (From Ferrington et al., 1987a.)

Examination of the paired responses in traces such as Figure 2 shows that the *PC* fibre responds to the vibratory stimulus with a precise, temporally patterned sequence of spikes that reflects accurately the periodicity of the vibratory stimulus. However, it is almost equally striking that the gracile neurone displays tightly phaselocked, patterned activity that is also able to reflect the periodicity of the vibration, as the interspike intervals approximate the 5 ms cycle period of the 200 Hz vibratory stimulus.

As the neural code for the frequency parameter of vibrotactile stimuli appears to depend upon an impulse pattern code of this type, it appears from response traces such as those of Figure 2 that, in the process of synaptic transmission within the dorsal column nuclei, the central neurone retains reliable information about the frequency parameter of the stimulus even in response to a selective input from a *single PC* sensory fibre. These synaptic transmission characteristics at this central synapse are consistent with the ability of human subjects to recognize the vibrotactile character of the input when single *PC* fibres are driven at frequencies of 20-100 Hz with the intraneural microstimulation procedure (Ochoa and Torebjörk, 1983; Vallbo et al., 1984). In fact, when quantitative analysis is conducted on the impulse patterning in gracile responses to vibratory inputs carried over single *PC* fibres we find that tightly phaselocked patterns can be retained over a vibration frequency bandwidth extending up to ~400 Hz (Ferrington et al., 1987a). This is demonstrated in Figure 3 in the paired cycle histograms constructed from responses of a representative *PC* fibre-gracile neurone pair to six different vibration frequencies in the range 100-600 Hz. The cycle histograms, which have an analysis time on the abscissa corresponding to the cycle period of the vibration (10,5,3.3 ms etc) display the probability of impulse occurrence throughout the period of the vibration cycle. Those on the left hand side show that the impulse occurrences in the *PC* fibre are tightly confined within little more than ~10% of the vibration cycle, even at the highest frequencies of 400, 500 and 600 Hz, reflecting the metronome-like precision in the impulse patterning of the *PC* fibres in response to vibration. Although there is a greater dispersion of impulse occurrences in the gracile responses in the right hand histograms, these central neurone responses remain well phaselocked even at vibration frequencies up to 400 Hz. The percentage values on each of these histograms provide a quantitative measure of phaselocking and confirm that transmission across this synapse for *PC* fibre inputs preserves temporal information that could permit the vibratory character of the input to be signalled for frequencies beyond 400 Hz (Ferrington et al., 1987a).

Analysis of Transmission Characteristics across the DCN Synapse for Inputs from Single SAI and SAII Afferents

Single *SAI* or *SAII* tactile fibres supplying the forearm hairy skin could be identified, selectively activated, and their activity monitored in fine fascicles of the lateral branch of the superficial radial nerve (see Vickery et al., 1994; Gynther et al., 1995; Rowe, 2001) and their central actions on target neurones of the cuneate nucleus examined a the paired recording arrangement. Individual *SAI* and *SAII* fibres were also found to have extremely secure synaptic linkages with their cuneate target neurones when activated by steady skin displacement, with evidence for amplification in the central response (Vickery et al., 1994; Gynther et al., 1995; Rowe, 2001). This behaviour was, of course, consistent with the capacity of SAI fibres for generating a perceptual response, that of steady pressure, following microstimulation in conscious human subjects. However, as *single SAII* fibres can also display a similar high security of transmission in their synaptic linkage to

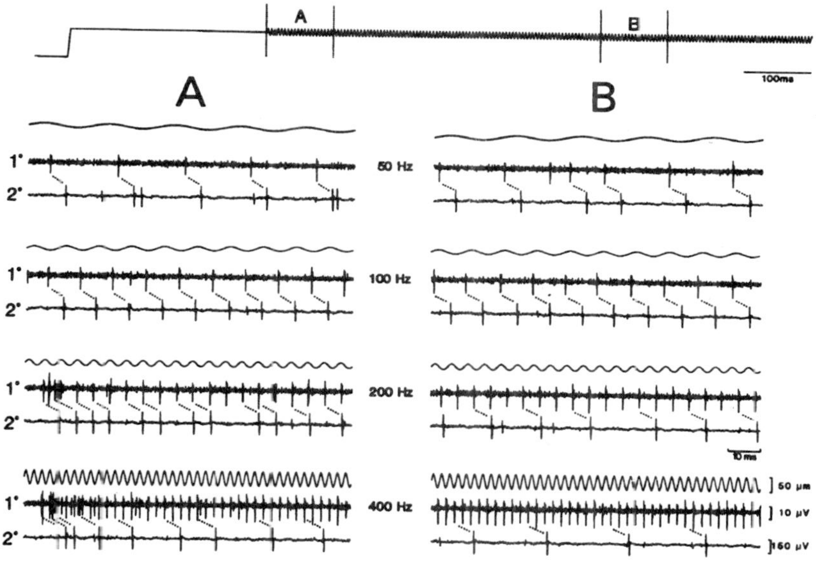

Figure 4. Impulse patterning in the cuneate response to a 1:1 response level in its SAI input is shown by the impulse records for the vibration-induced responses of the SAI (1°)-cuneate (2°) pair. Dashed lines link the cuneate impulses to the SAI impulses presumed to have caused them. The top trace in each record is a replica of the vibration stimulus waveform (250 μm probe, pre-indentation 200 μm). (From Vickery et al., 1994.)

cuneate neurones it appears that the failure of single SAII fibres to elicit perceptual responses in the human microneurography experiments cannot be explained in terms of a less secure synaptic linkage, at least at the level of the dorsal column nuclei (Gynther et al., 1995; Rowe, 2001).

As single SAI fibres, activated by trains of electrical pulses (20-100 Hz) in the human microneurography experiments, generate only a sense of static pressure, we employed vibrotactile stimuli over a range of frequencies, to examine the capacity of the *SAI*-cuneate neurone linkage to retain the temporally-precise, phaselocked pattern of spike activity in the transmission process to the output neurone. The unequivocal result of this analysis is reflected in the data of Figure 4, where the 100 ms long segments of paired *SAI*-cuneate traces show that the cuneate neurone can respond (at least at 50 and 100 Hz) with a faithful and precise impulse pattern reflecting that in the incoming *SAI* spikes, and can therefore reflect the vibratory stimulus periodicity. Furthermore, when the response patterning is quantified by means of the cycle histogram analyses described in Figure3, the phaselocking in the *SAI*-induced cuneate responses was found to be no poorer than that in DCN neurons driven by single *PC* fibres in response to vibration (Vickery et al., 1994; Rowe, 2001). We may therefore conclude that the failure of *SAI* fibres to generate a vibrotactile percept in the microneurography experiments cannot be explained by poor transmission security and degradation of the temporal pattern of incoming activity at this first central relay nucleus. It was also quite striking that, in response to vibrotactile stimuli, the *SAII*-cuneate synaptic linkage could also retain a precision of temporal patterning in the output spikes that was indistinguishable from that observed for the *DCN* linkages formed by single *PC* or *SAI* fibres (Gynther et al., 1995; Rowe, 2001).

Transmission Security Across the DCN Synaptic Linkage for Single Kinaesthetic Afferent Fibres

The extension of these studies to the two principal classes of kinaesthetic afferent fibres has been based upon the paired recording arrangement illustrated in Figure1 for the study of transmission characteristics for single joint afferent fibres arising from the wrist joint (Coleman et al., 1998) and, in the case of muscle spindle afferents, has been based upon the use of a very fine nerve in the cat forearm that supplies the indicis proprius muscle (Mackie and Rowe, 1997). In each case, selective activation of single kinaesthetic afferent fibres has been achieved most effectively by focal stimulation of either the joint capsule, or the muscle over the presumed site of individual spindles. Preliminary evidence for secure transmission between single afferents of both these kinaesthetic classes and their target cuneate neurons has been reported (Rowe, 2001).

SUMMARY AND CONCLUSIONS

The different capacity of the various tactile and kinaesthetic fibre classes to generate a perceptual response when activated singly in human microneurography experiments does not appear to be explicable in terms of differential transmission characteristics across the first synaptic relay site in the principal tactile and kinaesthetic sensory system that traverses the dorsal column-lemniscal pathway. In particular, the failure of individual SAII fibres and muscle spindle afferent fibres to generate a perceptual response upon activation with microneurography, does not appear to be attributable to a breakdown in synaptic transmission security at the level of the dorsal column nuclei. It may be presumed that for sensations to be generated by inputs arising in either of these fibre classes requires the concurrent activation of at least a small population of each afferent class before a percept will ensue. Whether this is related to a need for adequate spatial summation to take place at thalamic or cortical levels before transmission of the SAII or spindle inputs is effective as these levels, remains uncertain. An alternative hypothesis is that individual spindle afferents and SAII afferents (and, in a proportion of cases, at least, some SAI fibres which fail to generate a perceptual response) do not diverge in their ascending projection paths to the same extent that single PC, RA, joint afferent and the majority of SAI fibres do, and that for the 'critical mass' of cortical tissue to be activated by these inputs for the perceptual threshold to be crossed, may require peripheral activation of an 'ensemble' of each of these classes.

ACKNOWLEDGEMENTS

Work form the author's laboratory has been supported by grants from the National Health and Medical Research Council of Australia and the Australian Research Council.

REFERENCES

Coleman, G. T., Zhang, H. Q., Mackie, P. D., and Rowe, M. J., 1998, An intact peripheral nerve preparation for examining the central actions of single kinaesthetic afferent fibres arising in the wrist joint, *Primary Sensory Neuron,* 3, 61-70.

Coleman, G. T., Bahramali, H., Zhang, H. Q. and Rowe, M. J., 2001, Characterization of tactile afferent fibers in the hand of the marmoset monkey, *Journal of Neurophysiology,* 85, 1793-1804.

Edin, B. B., and Abbs, J. H., 1991, Finger movement responses of cutaneous mechanoreceptors in the dorsal skin of the human hand, *Journal of Neurophysiology,* 65, 657-670.

Ferrington, D. G. and Rowe, Mark J., 1980, Functional capacities of tactile afferent fibres in neonatal kittens, *Journal of Physiology,* 307, 335-353.

Ferrington, D. G., Rowe, M. J., and Tarvin, R. P., 1987a, Actions of single sensory fibres on cat dorsal column nuclei neurones: vibratory signalling in a one-to-one linkage, *Journal of Physiology,* 386, 293-309.

Ferrington, D. G., Rowe, M. J., and Tarvin, R. P., 1987b, Integrative processing of vibratory information in cat dorsal column nuclei neurones driven by identified sensory fibres, *Journal of Physiology,* 386, 311-331.

Gandevia, S. C., and McCloskey, D. I., 1976, Joint sense, muscle sense, and their combination as position sense, measured at the distal interphalangeal joint of the middle finger, *Journal of Physiology,* 260, 387-407.

Gynther, B. D., Vickery, R. M., and Rowe, M. J., 1995, Transmission characteristics for the 1:1 linkage between slowly adapting type II fibers and their cuneate target neurons in cat, *Experimental Brain Research,* 105, 67-75.

Iggo, A., and Ogawa, H., 1977, Correlative physiological and morphological studies of rapidly adapting mechanoreceptors in cat's glabrous skin, *Journal of Physiology,* 266, 275-296.

Jänig, W., Schmidt, R. F., and Zimmermann, M., 1968, Single unit responses and the total afferent outflow from the cat's foot pad upon mechanical stimulation, *Experimental Brain Research,* 6, 100-115.

Johansson, R. S., and Vallbo, Å. B., 1979, Detection of tactile stimuli, Thresholds of afferent units related to psychophysical thresholds in the human hand, *Journal of Physiology,* 297, 405-422.

Macefield, G., Gandevia, S. C., and Burke, D., 1990, Perceptual responses to microstimulation of single afferents innervating joints, muscles and skin of the human hand, *Journal of Physiology,* 429, 113-129.

Mackie, P. D., and Rowe, M.J., 1997, An intact peripheral nerve preparation for monitoring inputs from single muscle afferent fibres, *Experimental Brain Research,* 113, 186-188.

McCloskey, D. I., 1978, Kinesthetic sensibility, *Physiological Reviews,* 58, 763-820.

Ni, S., Wilson, L.R., Burke, D., and Gandevia S., 1998, Microneurographic stimulation of muscle spindle afferents in the radial nerve fails to elicit proprioceptive sensation, *Proceedings of the Australian Neuroscience Society,* 9, 183.

Ochoa, J., and Torebjork, E., 1983, Sensations evoked by intraneural microstimulation of single mechanoreceptor units innervating the human hand, *Journal of Physiology,* 342, 633-654.

Rowe, M. J., 1990, Impulse patterning in central neurons for vibrotactile coding, in: *Information Processing in Mammalian Auditory and Tactile Systems,* Rowe, M. J., and Aitkin, L.M., ed., Wiley-Liss, New York, pp. 111-125.

Rowe, M.J., 2001, Security of central tranmission for individual tactile and kinaesthetic sensory nerve fibres, in: *Somatosensory Processing: from Single Neuron to Brain Imaging,* Rowe, M.J., and Iwamura, Y., ed., Harwood Academic, Sydney, pp. 77-100.

Talbot, W. H., Darian-Smith, I., Kornhuber, H. H., and Mountcastle, V. B., 1968, The sense of flutter-vibration: comparison of the human capacity with response patterns of mechanoreceptive afferents from the monkey hand, *Journal of Neurophysiology,* 31, 301-334.

Vallbo, Å. B., and Hagbarth, K. E., 1968, Activity from skin mechanoreceptors recorded percutaneously in awake human subjects, *Experimental Neurology,* 21, 270-289.

Vallbo, Å. B., and Johansson, R. S., 1984, Properties of cutaneous mechanoreceptors in the human hand related to touch sensation, *Human Neurobiology,* 3, 3-14.

Vallbo, Å. B., Olsson, K. Å., Westberg, K. G., and Clark, F. J., 1984, Microstimulation of single tactile afferents from the human hand: Sensory attributes related to unit type and properties of receptive fields, *Brain,* 107, 727-749.

Vickery, R. M., Gynther, B. D., and Rowe, M. J., 1994, Synaptic transmission between single slowly adapting type I fibres and their cuneate target neurones in cat, *Journal of Physiology,* 474, 379-392.

Zachariah, M. K., Coleman, G. T., Mahns, D. A., Zhang, H. Q., and Rowe, M. J., 2001, Transmission security for single, hair follicle-associated tactile afferent fibres and their target cuneate neurones in the cat, *Journal of Neurophysiology,* 86, 900-911.

SECTION II

Proprioception in Limb Movements

In this Section, S. Gandevia (Chapter 8) reviews a number of aspects of proprioception. The first and very important point was that in designing our experiments we, legitimately, try to simplify the situation to obtain simple, unambiguous answers to our questions. In proprioception it has meant focussing attention on one or other of the several sensory inputs coming from skin, muscle or joint which are thought to contribute to kinaesthesia. We then attempt to extend the findings to more "natural" situations by invoking terms such as "parallel processing" and "redundancy". Here one is reminded of active touch, where the sensory experience from the inputs in combination is quite distinct from that felt from the movement alone or the touch alone. So we should be alert to this in proprioception and consider outcomes beyond a simple facilitatory action.

A second important point brought up, relates to possible roles of proprioceptors in kinaesthetic sensations other than joint position and movement and including aspects of body image. An example is proprioceptive acuity at the elbow joint which is greater with the forearms and hands in their normal working position, compared with the arms raised, or to the side (Gooey et al., 2000). Within a broader context, Gandevia describes changes in perceived size of body parts following their anaesthesia. Sometimes other body parts become involved. This relates to the common experience of distortion of the lips and face after dental anaesthesia. The change in perceived size of a body part, whose afferent input has not been interrupted but whose cortical field is adjacent to that of the anaesthetised part is not so surprising. Since we know that interruption of sensory input from one area leads to "invasion" of that area by cortical representations from other areas, the distortion is likely to be an expression of that expansion (see Calford, Chapter 51). More mysterious is expansion of the perceived size of the body part anaesthetised, in this case the thumb. One way to explain it would be to postulate that, based on learning, it is the boundaries between cortical areas, representing body parts, that determine the perceived size of the part. Removal of the boundary by removal of afferent input is interpreted as an undefined expansion of the boundary. Perhaps the new size of the thumb relates to the cortical area its afferent representation would occupy if it had access without the competitive influences from other areas.

J. Lackner (Chapter 9) describes experiments where subjects sat in a darkened room, carrying out pointing movements, while the room rotated. The Coriolis forces generated by the rotation led subjects to deviate from the target position in their initial attempts. However, they rapidly corrected. When motion of the room was stopped subjects made errors in a direction opposite to that seen previously.

It was concluded that during reaching movements in the absence of vision, a motor program was selected, based on previous experience. During repeated trials the program was iteratively updated, most probably by muscle afferent signals. Final positional information was provided by tactile cues from the pointing finger. Such a model undermines current ideas based on the equilibrium point hypothesis, where length-tension properties of the muscles are specified at the outset and updating of information is not necessary (Bizzi and Abend, 1983).

Labyrinth defective subjects continued to make end-point errors without signs of adaptation because in making their movements they approached the target more vertically than control subjects and so did not receive proper positional feedback of final target position. Cerebellar patients showed no adaptations of movements or end points, nor did they show any after-effects. Significantly, both normal and labyrinth-defectives could 'feel' the presence of the Coriolis force during initial trials while cerebellar patients felt nothing. It suggested that the sensory input during a movement which is used to update it reaches consciousness and involves the cerebellum.

Experiments reported by G. Kerr (Chapter 10) are directly relevant to the propositions put by Lackner. Subjects' passive arms were extended at different speeds and for different distances. Velocity perception was more accurate when distance and timing cues were also available. The data suggested that information about velocity alone could, in the absence of other cues, result in large performance errors in any predictive model of movement control. The conclusion highlighted the need for the additional cues.

A. Wise (Chapter 11) makes the important point that whenever a position or movement matching task is carried out on a passive limb, it is important to define the condition of the test muscle, whether its spindles are in a sensitive or insensitive state. This is because spindles, that is, intrafusal fibres, exhibit thixotropy, a dependence of passive properties on the previous history of contraction and length changes. A second point was that both movement and positional errors increased when the reference arm was undergoing a co-contraction. This was attributed to inaccuracies in the postulated central subtraction process required to separate stretch-evoked from fusimotor-evoked spindle activity. Recordings from animals also showed that sensitivity of the passive spindle to small movements was greater than when it was under static or dynamic fusimotor control.

A discussion of the effects of joint pathology on proprioception is initiated by K. Refshauge (Chapter 12). Important considerations are the type of injury, whether it has only affected the articular surfaces and whether joint mechanics had been altered. In the latter circumstances, proprioception is more likely to be modified. Another consideration is effect, direct or reflex, of any inflammation. Activity in γ motoneurones to hip flexor spindles goes up during an inflammation (He et al., 1988). This might be expected to reduce proprioceptive acuity (see Chapter 11). It was concluded that detailed, quantitative studies are needed to more precisely define changes in proprioception about a diseased joint.

REFERENCES

Bizzi, E., and Abend, W., 1983, Posture control and trajectory formation in single- and multi-joint arm movements, in: *Motor Control Mechanisms in Health and Disease - Advances in Neurology,* Volume 39, J.E. Desmedt, ed., Raven Press, New York, pp. 31-45.

Gooey, K., Bradfield, O., Talbot, J., Morgan, D. L., and Proske, U., 2000, Effects of body orientation, load and vibration on sensing position and movement at the human elbow joint, *Experimental Brain Research*, 133, 340-348.

He, X., Proske, U., Schaible, H. G., and Schmidt, R. F., 1988, Acute inflammation of the knee joint in the cat alters responses of flexor motoneurons to leg movements, *Journal of Neurophysiology*, 59, 326-340.

8

PROPRIOCEPTION: PERIPHERAL INPUTS AND PERCEPTUAL INTERACTIONS

Simon C. Gandevia[1], Kathyrn M. Refshauge [2] and David F. Collins[1]

ABSTRACT

Much emphasis has been placed on the specific role of specific inputs from muscle, joint and cutaneous afferents in the detection of movement. However, particularly for the hand, multiple inputs from the moving part are likely to be important. This chapter reviews some recent studies which examine the co-operative interaction between the various proprioceptive channels. Proprioceptive control of movement must also take account of the length of the various limb segments, a variable which is independent of muscle lengths and joint angles. Evidence is presented that body image can be affected by the tonic discharge of non-muscle receptors.

INTRODUCTION

Proprioception represents a group of sensations that encompasses first, detection of the position and movement of the joints, second, the sensations of force and heaviness accompanying contractions, third, sensations related to the conscious distribution and timing of motor commands, and fourth, sensation of the body's orientation and that of its segments (for reviews which trace the development of these ideas (see McCloskey, 1978; Matthews, 1988; Gandevia, 1996). Much of research in this field has centred on the mechanisms subserving the sensations of joint position and movement. Because of this emphasis, and because clinical testing has relied so heavily on simply waggling a joint, the majority of studies have examined the detection of movements applied to the passive (although not necessarily relaxed) subject.

Historically, the interest in the neural mechanisms underlying movement sensation has focused on attempts to establish one class of receptor as the main or even exclusive one for a particular proprioceptive function. Hence, prior to the 1970s physiologists strongly favoured the view that muscles were insentient and that therefore joint receptors sent the main signals for the detection of joint movement. Subsequently, studies by

[1] Prince of Wales Medical Research Institute, Randwick, NSW, 2031, Sydney, Australia.
Email: s.gandevia@unsw.edu.au
[2] University of New South Wales, Sydney, NSW, 2031, Australia.

Matthews and colleagues (Goodwin et al., 1972) and later many others (e.g. Roll and Vedel, 1982; Lackner, 1988; Cordo, 1990) showed that activation of muscle spindle receptors by vibration caused proprioceptive illusions consistent with perceived muscle lengthening. The pendulum swung heavily in favour of muscle spindles as occupying "pride of place" for movement detection (Matthews, 1988; McCloskey, 1999). In this debate the cutaneous receptors were given little attention. Eventually application of the technique of microneurography revealed that for the ventral surface of the hand most cutaneous afferent classes were activated by active and passive movement of the digits (Hulliger et al., 1979; Burke et al., 1988). However, there had already been a result showing that some cutaneous receptors in the hand signaled movement across the angular range (Knibestol, 1975). Subsequently the proprioceptive capacity of cutaneous afferents innervating the skin over the dorsum of the hand (Edin and Abbs, 1991; Edin, 1992; Grill and Hallett, 1995), and over the thigh (Edin, 2001) was demonstrated. Indeed, the sensitivity (measured in terms of joint motion appeared similar for muscle spindle and slowly adapting cutaneous afferents (Grill and Hallett, 1995), a finding which is unlikely to apply only to the metacarpophalangeal joint. Of course, the discharge of an afferent to a particular stimulus does not establish that the central nervous system uses that signal to provide a conscious appreciation of it. The afferent recordings reaffirmed, if such was needed, that multiple inputs responded to joint movement and hence redundancy in the central processing systems was likely. Furthermore, none of the proprioceptive candidates supplied an uncomplicated 'potentiometer-like' signal of joint motion. Thus joint receptors can be affected by contraction of muscles pulling on joint capsules, slowly adapting cutaneous receptors may respond to motion at more than one joint (Edin and Abbs, 1991) or in more than one direction (Hulliger et al., 1979; Burke et al., 1988), and muscle spindle afferents (apart from being prone to potentially large history-dependent changes, and falling slack at some lengths) respond, depending on their anatomy, when all joints they cross are moved in isolation. The one exception may be the large subset of joint receptors which respond as detectors of the limits of natural and unnatural movements (Burke et al., 1988).

This chapter considers initially some recent studies that have focused on the interaction between the different inputs for proprioception. We accept that various skin, joint and muscle receptors will fire in response to movement and consider how this may operate when all 'channels' are operative. Any interactions observed under these circumstances are likely to be important simply because under normal conditions all proprioceptive channels are available.

ILLUSORY MOVEMENTS AND ARTIFICIAL STIMULATION OF RECEPTORS

Since the pioneering studies of Goodwin and colleagues (Goodwin et al., 1972) one of the tools of trade of the psychophysicist interested in human proprioception has been to activate a group of receptors or afferents with relatively little effect on the other proprioceptive inputs and to quantify the evoked sensory consequences (e.g. illusory movement or changes in position). Thus vibration has been used to stimulate muscle spindle receptors, local skin movement to active cutaneous inputs, and various forms of surface and intraneural stimulation to activate large-diameter afferents. It is unlikely that the stimuli applied are absolutely selective for a particular proprioceptive channel but nonetheless the results have been sufficiently clear that this has not yet been a critical issue.

Recent studies have considered the capacity of cutaneous afferents in the hand to provide proprioceptive information. Edin and Johansson found that movements of the skin applied directly by the experimenter to parts of an anaesthetized finger were transmitted proximally to sentient regions and subjects reported illusory movement at the metacarpophalangeal joint (Edin and Johansson, 1995). At the same time, Collins and Prochazka applied stretch through tapes attached to skin over the dorsum of the hand and found that more than half the subjects reported illusory finger movements (Collins and Prochazka, 1996). However their overall conclusion was that movement illusions were evoked more reliably by muscle afferent input than by cutaneous inputs. Of course, a drawback to all studies which 'hijack' one afferent channel and examine the sensory consequences is that the other channels continuously signal the absence of any change. Thus the presence of an evoked illusory movement is evidence for the channel's direct contribution to proprioception but the absence of an illusion or an inconsistent one is not evidence against a contribution - the artificial input may not have been sufficiently 'natural' for correct central interpretation.

To examine the interaction between cutaneous and muscle afferents for the detection of movement, we used tendon vibration to provide a signal that the fingers were moving and then added various stimuli to the skin on of the hand (Collins et al., 2000). Vibration was applied (100 Hz) over the tendons of the finger extensors at wrist level and either electrical stimulation (200 Hz at twice sensory threshold via small surface electrodes) or skin stretch (via tapes attached to the skin) was applied. Stimuli were delivered repeatedly in phasic bursts of about 4-s duration and subjects tracked any perceived movements of the stationary experimental hand by movement of the contralateral one. Several findings emerged. First, not surprisingly, bursts of vibration alone produced illusory flexion and extension at the metacarpophalangeal joint, usually of all four fingers. The vibration was adjusted to evoke illusions of about 10 degrees in amplitude. Second, stretch or electrical stimulation of the skin on the dorsum of the hand commonly evoked illusions. However, illusory motion occurred in all subjects with skin movement applied over the dorsal and ventral surface of the skin around a particular metacarpophalangeal joint. These illusory movements were about 5 degrees. Third, the combination of skin stretch associated with one finger and vibration over finger extensors produced a significant 'focusing 'of the illusory movements to the joint under the stimulated finger. The illusory movements were larger at this finger joint, close to the simple summation of the effects of cutaneous and muscle stimulation alone. Sometimes the effects were not simply additive because combined stimulation evoked a large illusory movement at the metacarpophalangeal joint of the finger receiving the cutaneous input while the illusory motion was smaller at adjacent the metacarpophalangeal of adjacent fingers (Fig. 1). Subjects were sometimes unaware that the localized cutaneous stimuli had produced any changes in the vibration-induced illusions of movement. This focusing effect did not occur with stimulation at a more remote site (the finger tip) and was more prominent with skin stretch than electrical stimulation, presumably because the latter input less resembled that activated in natural movement.

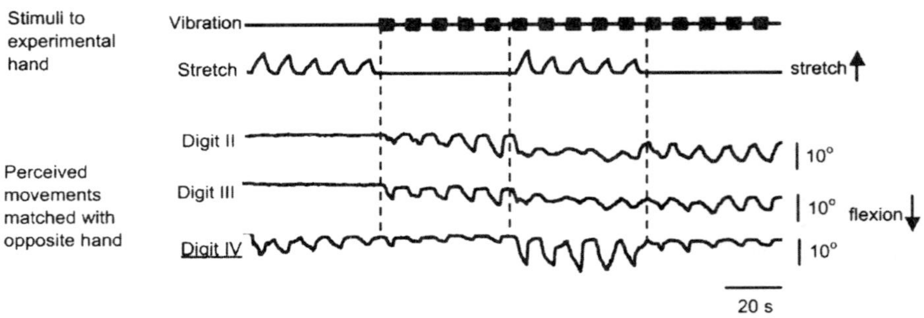

Figure 1. Illusory movements evoked when skin stretch is applied over the dorsal aspect of the hand and the dorsal and ventral sides of the metacarpophalangeal joint (MCP) of the ring finger (IV) in one subject. The top two traces show the time course of the stimuli: skin stretch and vibration (over the extensor tendons of the fingers at wrist level). The bottom three traces show the resulting illusory movements of the MCP joints as indicated by the voluntary movements of the matching hand. Flexion at the MCP joints is shown as a downward deflection. (From Collins et al., 2000.)

POSSIBLE 'FACILITATION' OF MUSCLE AFFERENT INPUTS

Possible interactions between the various proprioceptive channels have been raised to explain the often-reported deterioration in detection of movements at a particular finger when its digital nerves are blocked (e.g. Brown et al., 1954; Merton, 1964; Gandevia and McCloskey, 1976; Refshauge et al., 1998). Indeed the impairment can be so profound that it was possible to propose that *all* movement detection was lost, hence supporting the physiological view at the time that only joint receptors were responsible for detection (Merton, 1964). The simple position adopted by Goodwin and colleagues (Goodwin et al., 1972) was that the residual sensation indicated the role of intramuscular receptors which were proximal and unaffected by the nerve block. However it seemed hard to reconcile this finding with the potent illusory movements evoked by muscle receptor activation (e.g. Goodwin et al., 1972; Roll and Vedel, 1982; Gandevia, 1985; Cordo, 1987) and the equally marked deterioration in movement detection when muscle receptors were disengaged by changing the position of the fingers adjacent to the one at which movements were applied (Gandevia and McCloskey, 1976).

This impasse was seemingly resolved when it was proposed that there might be facilitation at the level of detection between the proprioceptive inputs from muscle receptors in the forearm and the local inputs from the finger being moved. This view was summarized by McCloskey as follows: "Possibly, any cutaneous contribution is one of facilitation of the central action of intramuscular and joint receptors, rather than a specific individual role" (McCloskey, 1999). Hence removal of muscle or digital nerve inputs (considered to be from joint receptors at the time) would reduce detection. Evidence for such facilitation was sought by elimination of the digital nerve inputs from the fingers next to the ones at which proprioception was being tested: if detection of movements deteriorated then it was assumed that such an effect would also operate for the digital inputs from within the test finger. Although initial tests of this proved successful (Gandevia and McCloskey, 1976), the numbers of subjects were very small (usually 3-5 per study) and although some subsequent studies appeared to corroborate the results (Clark et al., 1985) others did not (Ferrell and Smith, 1988). The outcome was unimpressive. Consequently, we have recently revisited this issue and tested the detection of movements at the proximal joint of the index finger in large number of subjects when the inputs from one or two of the adjacent digits were removed by anaesthesia. There was no significant deterioration in this proprioceptive task (Gandevia et al., 2001). This result makes it hard to sustain the view that the deterioration in movement detection at joints in the hand during removal of the joint and local cutaneous input reflects loss of some general facilitation. A more parsimonious interpretation is that the anaesthesia simply removes some inputs that are normally contributing to detection. This view gives both muscle and non-muscle proprioceptive inputs an important role in movement detection. Even within the distal joints of the hand, the relative contribution of the various muscle and non-muscle inputs may vary: at the thumb the input from the long flexor of the thumb seems very important (Refshauge et al., 1998).

Another interaction involving muscle afferent (and other proprioceptive) signals is their central interpretation during a muscle contraction. Several factors must come in to play. During a voluntary contraction tendon slackness is removed, intrafusal contraction secondary to fusimotor drive will alter spindle afferent discharge (to a degree depending on alpha-gamma co activation and the speed and strength of contraction), and corollary discharges will be available centrally to cancel or modify the interpretation of spindle signals. Cancellation of the reafferent spindle signal by a motor command corollary has been postulated to explain the reduction in vibration-induced illusions of movement as contraction intensity increases or during fatigue (McCloskey, 1973). Other explanations such as a reduced afferent input evoked by the vibration during contraction need more consideration. There is currently an unexpected controversy about whether detection of movements is impaired when added during a voluntary contraction. Studies in which a digit was anaesthetized (removing joint and cutaneous inputs) have supported the view that a contraction can restore some of the lost ability to detect applied movements (e.g. (Gandevia et al., 1983). However, this view has been challenged with the finding that movement detection thresholds at the elbow increase during co-contraction (a manoeuvre designed to remove slack from muscle spindles; Wise et al., 1998). For further discussion see Chapter 12 (see also Proske et al., 1999). In addition, the detection of twitch contractions superimposed during voluntary movements is impaired compared to detection during non-movement controls (Collins et al., 1998). It is unclear whether this relates to the unphysiological task given to the subject, but still the result may bring the literature on muscle afferents into line with that for cutaneous afferents. It has long been known that voluntary contraction reduces the detectability and perceived intensity of cutaneous stimuli.

ASPECTS OF THE BODY IMAGE

While much attention has focused on the mechanisms subserving movement detection under passive and to a lesser extent active conditions, accurate movements require more than just a knowledge of joint angles, muscle lengths and stretch on the skin. It is necessary to know the length of the various segments involved in the movement. This involves use of the various input signals and knowledge of goal-directed movements to synthesize a 'map' of the skeleton. Recent studies have emphasized the lability of various cortical and subcortical maps of the body surface. For example, following dorsal rhizotomy in monkeys affecting the input from the whole arm, the cortical areas previously devoted to the arm eventually become 'invaded' by inputs related to the face (Pons et al., 1991), for review see (Kaas, 1991). Even in acute studies of deafferentation there are marked changes in the properties of cortical cells (e.g. Kelahan and Doetsch, 1984; Calford and Tweedale, 1988; Calford and Tweedale, 1991). Furthermore, studies in humans have shown that the properties of the motor cortex change when tested with transcranial magnetic stimulation (e.g. Brasil-Neto et al., 1993; Kew et al., 1994; McNulty et al., 2002), and that somatosensory maps alter (Yang et al., 1994; Kew et al., 1997).

We recently used a combination of psychophysical techniques to demonstrate that during large changes in the peripheral input from the digits (including the thumb) there were clear changes in the perceived size of the body part (Gandevia and Phegan, 1999). For example, complete anaesthesia of the thumb with a digital nerve block increased its size by about 70%. This effect did not involve adjacent fingers or digits on the contralateral side. To establish that these illusions were sufficient to alter movement control subjects traced the shapes of the affected parts – these drawings revealed the same perceptual changes. The responses to thumb anaesthesia were accompanied by significant increases in perceived size of the (unanaesthetized) lips.

This illusory enlargement of an anaesthetized part is paradoxical as it accompanied complete *removal* of the normal afferent input from the part. Subjects often volunteered that the thumb felt larger during anaesthesia with no recognition of the incompatibility of their report with the input from the part. Another common response was that perceived enlargements of parts of the face also occurred with dental anaesthesia.

These perceptual distortions presumably result from convergence and divergence in somatosensory pathways with thumb and lip areas being adjacent within both primary and secondary somatosensory cortex (e.g. Robinson and Burton, 1980; Cusick et al., 1989). This convergence may be present not only at the level of inputs to the neurons which are 'subthreshold' (e.g. Zarzecki et al., 1993) because some cells in the primate somatosensory cortex discharge to both facial and thumb inputs (e.g. Robinson and Burton, 1980). Finally this aspect of proprioception is interesting because it is presumably susceptible to all the factors controlling 'excitability' along somatosensory pathways. Furthermore, this information about segment length must be built into the functional 'maps' or 'frames of reference' that are used to guide movement (e.g. Soechting and Flanders, 1989).

CONCLUSION

This chapter has focused on some integrative aspects of proprioception with emphasis on the inputs from skin, joint and muscle afferents involved in detection and monitoring of movement and on the operation of central maps which contribute to labile perceptions of the body image. As indicated at the outset, much argument has previously surrounded the struggle to establish one proprioceptive channel as *the* prime one for particular proprioceptive functions (e.g. movement detection). The time has come to explore integrative aspects of how the channels interact and how their operation differs at various joints, or more usefully, parts, of the body. However it will not be easy to adapt all the techniques used to explore these issues for the hand to more proximal joints.

ACKNOWLEDGEMENTS

This work was funded by the National Health and Medical Research Council of Australia (#3206).

REFERENCES

Brasil-Neto, J. P., Valls-Sole J., Pascual-Leone, A., Cammarota, A., Amassian, V.E., Cracco, R., Maccabee, P., Cracco, J., Hallett, M., and Cohen, L. G., 1993, Rapid modulation of human cortical motor outputs following ischaemic nerve block, *Brain,* 116, 511-525.

Brown, K., Lee J., and Ring, P. A., 1954, The sensation of passive movement at the metatarso-phalangeal joint of the great toe in man, *Journal of Physiology,* 12, 38-58.

Burke, D., Gandevia, S. C., and Macefield, G., 1988, Responses to passive movement of receptors in joint, skin and muscle of the human hand, *Journal of Physiology,* 402, 347-361.

Calford, M. B., and Tweedale, R., 1988, Immediate and chronic changes in responses of somatosensory cortex in adult flying-fox after digit amputation, *Nature,* 332, 446-448.

Calford, M. B., and Tweedale, R., 1991, Acute changes in cutaneous receptive fields in primary somatosensory cortex after digit denervation in adult flying fox, *Journal of Neurophysiology,* 65, 178-187.

Clark, F. J., Burgess, R.C., Chapin, J. W., and Lipscomb, W. T., 1985, Role of intramuscular receptors in the awareness of limb position, *Journal of Neurophysiology,* 54, 1529-1540.

Collins, D. F., Cameron, T., Gillard, D. M., and Prochazka, A., 1998, Muscular sense is attenuated when humans move, *Journal of Physiology,* 508, 635-643.

Collins, D. F., and Prochazka, A., 1996, Movement illusions evoked by ensemble cutaneous input from the dorsum of the human hand, *Journal of Physiology,* 496, 857-871.

Collins, D. F., Refshauge, K. M., and Gandevia, S. C., 2000, Sensory integration in the perception of movements at the human metacarpophalangeal joint, *Journal of Physiology,* 529, 505-515.

Cordo, P. J., 1987, Mechanisms controlling accurate changes in elbow torque in humans, *Journal of Neuroscience,* 7, 432-442.

Cordo, P. J., 1990, Kinesthetic control of a multijoint movement sequence, *Journal of Neurophysiology,* 63, 161-172.

Cusick, C. G., Wall, J. T., Felleman, D. J., and Kaas, J. H., 1989, Somatotopic organization of the lateral sulcus of owl monkeys: area 3b, S-II, and a ventral somatosensory area, *Journal of Comparative Neurology,* 282, 169-190.

Edin, B., 2001, Cutaneous afferents provide information about knee joint movements in humans, *Journal of Physiology,* 531, 289-297.

Edin, B. B., 1992, Quantitative analysis of static strain sensitivity in human mechanoreceptors from hairy skin, *Journal of Neurophysiology,* 67, 1105-1113.

Edin, B. B., and Abbs, J. H., 1991, Finger movement responses of cutaneous mechanoreceptors in the dorsal skin of the human hand, *Journal of Neurophysiology,* 65, 657-670.

Edin, B. B., and Johansson, N., 1995, Skin strain patterns provide kinaesthetic information to the human central nervous system, *Journal of Physiology,* 487, 243-251.

Ferrell, W. R., and Smith, A., 1988, Position sense at the proximal interphalangeal joint of the human index finger, *Journal of Physiology,* 399, 49-61.

Gandevia, S. C., 1985, Illusory movements produced by electrical stimulation of low-threshold muscle afferents from the hand, *Brain,* 108, 965-981.

Gandevia, S. C., 1996, Kinesthesia: roles for afferent signals and motor commands, in: *Handbook of Physiology. Section 12. Exercise: Regulation and Integration of Multiple Systems,* L. B. Rowell, and J. T. Shepherd, eds., Oxford University Press, Oxford, pp. 128-172.

Gandevia, S. C., Collins, D. F., and Refshauge, K. M. Detection of finger movements is not facilitated by cutaneous fedback from adjacent digits. IUPS Satellite Symposium abstracts, 2001, P26.

Gandevia, S. C., Hall, L. A., McCloskey, D. I., and Potter E. K., 1983, Proprioceptive sensation at the terminal joint of the middle finger, *Journal of Physiology,* 335, 507-517.

Gandevia, S. C., and McCloskey, D. I., 1976, Joint sense, muscle sense, and their combination as position sense, measured at the distal interphalangeal joint of the middle finger, *Journal of Physiology,* 260, 387-407.

Gandevia, S. C., and Phegan, C. M., 1999, Perceptual distortions of the human body image produced by local anaesthesia, pain and cutaneous stimulation, *Journal of Physiology,* 514, 609-616.

Goodwin, G. M., McCloskey, D. I., and Matthews, P. B. C., 1972, The contribution of muscle afferents to kinaesthesia shown by vibration induced illusions of movement and by the effects of paralysing joint afferents, *Brain,* 95, 705-748.

Grill, S. E., and Hallett, M., 1995, Velocity sensitivity of human muscle spindle afferents and slowly adapting type II cutaneous mechanoreceptors, *Journal of Physiology,* 489, 593-602.

Hulliger, M., Nordh, E., Thelin, A. E., and Vallbo, Å.B., 1979, The responses of afferent fibres from the glabrous skin of the hand during voluntary finger movements in man, *Journal of Physiology,* 291, 233-249.

Kaas, J. H., 1991, Plasticity of sensory and motor maps in adult mammals, *Annual Review of Neuroscience,* 14, 137-167.

Kelahan, A. M., and Doetsch, G. S., 1984, Time-dependent changes in the functional organization of somatosensory cerebral cortex following digit amputation in adult raccoons, *Somatosensensory Research,* 2, 49-81.

Kew, J. J., Halligan, P. W., Marshall, J. C., Passingham, R. E., Rothwell, J. C., Ridding, M. C., Marsden, C. D., and Brooks, D.J., 1997, Abnormal access of axial vibrotactile input to deafferented somatosensory cortex in human upper limb amputees, *Journal of Neurophysiology,* 77, 2753-2764.

Kew, J. J., Ridding, M. C., Rothwell, J. C., Passingham, R. E., Leigh P. N., Sooriakumaran, S., Frackowiak, R. S., and Brooks, D. J., 1994, Reorganization of cortical blood flow and transcranial magnetic stimulation maps in human subjects after upper limb amputation, *Journal of Neurophysiology,* 72, 2517-2524.

Knibestöl, M., 1975, Stimulus-response functions of slowly adapting mechanoreceptors in the human glabrous skin area, *Journal of Physiology,* 245, 63-80.

Lackner, J. R., 1988, Some proprioceptive influences on the perceptual representation of body shape and orientation, *Brain* 111, 281-297.

Matthews, P. B. C., 1988, Proprioceptors and their contribution to somatosensory mapping: complex messages require complex processing, *Canadian Journal of Physiology and Pharmacology,* 66, 430-438.

McCloskey, D. I., 1973, Differences between the senses of movement and position shown by the effects of loading and vibration of muscles in man, *Brain Research,* 61, 119-131.

McCloskey, D. I., 1978, Kinesthetic sensibility, *Physiological Reviews,* 58, 763-820.

McCloskey, D. I., 1999, Kinesthesia, kinesthetic perception, in: *Encyclopedia of Neuroscience,* G. Adelman, and B. H. Smith, eds., Elsevier, Amsterdam, pp. 995-998.

McNulty, P., Macefield, V. G., Taylor, J. L., and Hallett, M., 2002, Cortically evoked neural volleys to the human hand are increased during ischaemic block of the forearm, *Journal of Physiology*, 538, 279-288.

Merton, P. A., 1964, Human position sense and sense of effort, *Symposia of the Society for Experimental Biology*, 18, 387-400.

Pons, T. P., Garraghty, P. E., Ommaya, A. K., Kaas, J. H., Taub, E., and Mishkin, M., 1991, Massive cortical reorganization after sensory deafferentation in adult macaques, *Science* 252, 1857-1860.

Proske, U., Wise, A. K., and Gregory, J. E., 1999, Movement detection thresholds at the human elbow joint, *Progress in Brain Research,* 123, 143-147.

Refshauge, K. M., Kilbreath, S. L., and Gandevia, S. C., 1998, Movement detection at the distal joint of the human thumb and fingers, *Experimental Brain Research,* 122, 85-92.

Robinson, C. J., and Burton, H., 1980, Somatotopographic organization in the second somatosensory area of M. fascicularis, *Journal of Comparative Neurology,* 192, 43-67.

Roll, J. P., and Vedel, J. P., 1982, Kinaesthetic role of muscle afferents in man, studied by tendon vibration and microneurography, *Experimental Brain Research,* 47, 177-190.

Soechting, J. F., and Flanders, M., 1989, Sensorimotor representations for pointing to targets in three-dimensional space, *Journal of Neurophysiology,* 62, 582-594.

Wise, A. K., Gregory, J. E., and Proske, U., 1998, Detection of movements of the human forearm during and after co-contractions of muscles acting at the elbow joint, *Journal of Physiology,* 508, 325-330.

Yang, T. T., Gallen, C. C., Ramachandran, V. S., Cobb, S., Schwartz, B. J., and Bloom, F. E., 1994, Noninvasive detection of cerebral plasticity in adult human somatosensory cortex, *Neuroreport,* 5, 701-704.

Zarzecki, P., Witte, S., Smits, E., Gordon, D. C., Kirchberger, P., and Rasmusson, D. D., 1993, Synaptic mechanisms of cortical representational plasticity: somatosensory and corticocortical EPSPs in reorganized raccoon SI cortex, *Journal of Neurophysiology,* 69, 1422-1432.

ADAPTATION TO CORIOLIS FORCE PERTURBATION OF MOVEMENT TRAJECTORY

Role of proprioceptive and cutaneous somatosensory feedback

James R. Lackner and Paul DiZio*

ABSTRACT

Subjects exposed to constant velocity rotation in a large fully-enclosed room that rotates initially make large reaching errors in pointing to targets. The paths and endpoints of their reaches are deviated in the direction of the transient lateral Coriolis forces generated by the forward velocity of their reaches. With additional reaches, subjects soon reach in straighter paths and become more accurate at landing on target even in the absence of visual feedback about their movements. Two factors contribute to this adaptation: first, muscle spindle and golgi tendon organ feedback interpreted in relation to efferent commands provide information about movement trajectory, and second, somatosensory stimulation of the fingertip at the completion of a reach provides information about the location of the fingertip relative to the torso.

INTRODUCTION

The contribution of muscle spindle afferents to the conscious appreciation of limb position was first shown by Goodwin, McCloskey, and Matthews in 1972. They found that resisting the displacement of a limb moving under the action of a tonic vibration reflex led to illusory motion and displacement of the limb in the direction that would be associated with lengthening of the vibrated muscle. Their study along with subsequent observations was also key in revealing that position sense is computed by interpreting spindle output in relation to motor commands (cf. Matthews, 1988; Gandevia, 1996). In the present paper, we provide evidence that muscle spindle signals also figure prominently in the adaptive control mechanisms underlying limb trajectory formation and

* Ashton Graybiel Spatial Orientation Laboratory & Center for Complex Systems Brandeis University Waltham MA 02454, USA. Email: lackner@brandeis.edu

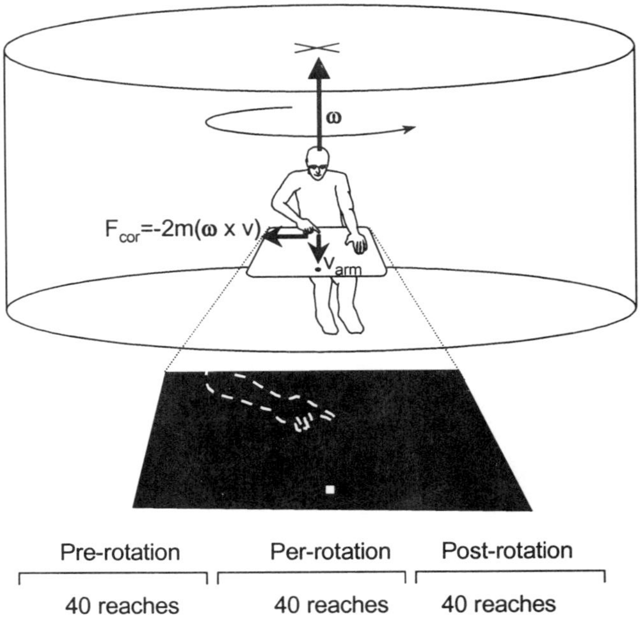

Figure 1. Rotation conditions. The subject in darkness reaches to point at a target that is extinguished as his hand lifts to point. Forward velocity of the arm generates a rightward Coriolis force on the arm.

regulation. We also show that the role of spindle signals in trajectory control and adaptation is complemented by tactile somatosensory afferents that specify limb position in relation to the torso when the hand makes contact at the end of a reaching movement.

CORIOLIS FORCE INFLUENCES ON REACHING MOVEMENTS

The experimental paradigm used in our studies is depicted in Figure 1 (Lackner and DiZio, 1994). A subject is illustrated seated at the center of a large, fully-enclosed, circular room that can be rotated. When the room has been turning at constant velocity for a minute the subject will feel completely stationary because the angular acceleration sensitive semicircular canals of his inner ear will have returned to resting discharge levels and he has no visual or somatosensory cues to indicate that he is rotating. If the subject makes a forward reaching movement to point to a target appearing on the surface in front of him, an inertial Coriolis force will be generated on his arm. This Coriolis force (F_{cor}) is proportional to the angular velocity of room rotation, ω, the forward velocity of the arm, *v*, in the plane of rotation, and the mass of the arm, *m*, with $F_{cor} = -2m(\omega \times v)$. For example, if the test room is rotating counterclockwise (CCW), the Coriolis force

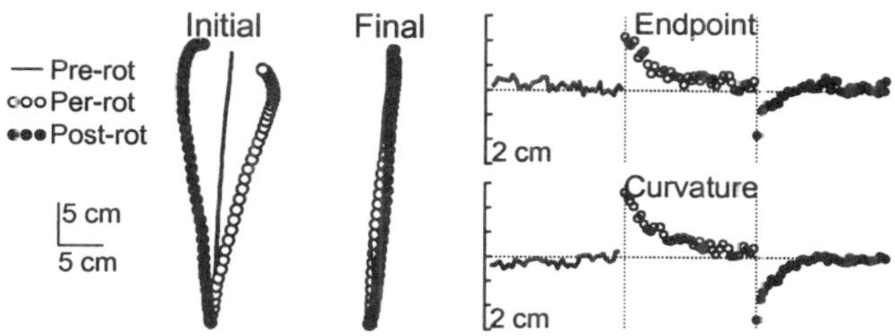

Figure 2. Left panel shows overhead view of average initial and final pre-, per-, and post-rotation movements. Right panel shows average movement endpoints and curvatures for all movements.

generated will act to displace the subject's arm rightward. Unlike mechanical forces that are often used to perturb reaching movements, e.g. manipulanda controlled by torque motors, Coriolis forces act without localized contact on the arm. As a consequence, no tactile cues are imposed on the arm that can provide information about the magnitude, direction, and duration of the perturbation.

Because Coriolis forces on the arm are dependent on the arm's linear velocity relative to the test chamber, they will mirror the forward velocity profile of the arm and be absent at the beginning and end of a reach. This also makes Coriolis forces an ideal tool for evaluating equilibrium point theories of movement control, which posit that movement trajectory is generated by an evolving series of commands specifying the length-tension properties of the muscles and that detailed monitoring of ongoing limb position is not necessary (Bizzi et al., 1992; Feldman, 1966, 1986). Such theories hold that transient perturbations of movement trajectory will not affect the pre-programmed endpoint of the movement. The endpoint is thought to be determined by the final length-tension properties of the agonist and antagonist muscles specified and these are thought not to be altered by a temporary perturbation. Experiments using mechanical perturbations delivered by manipulanda to disturb reaching movements have provided support for these theories (Bizzi et al., 1992; Feldman, 1986, Feldman et al., 1998).

Figure 2 shows average data of eleven subjects who made reaching movements to a light emitting diode target embedded in the underside of the smooth Plexiglas work surface (cf. Figure 1). When a subject lifted his finger to point, the target was extinguished and the reach was completed in total darkness. Thus, the subjects never received visual feedback about their movement onset, trajectory, or terminus. As can be seen, pre-rotation the subjects were very accurate. The endpoints of the first per-rotation reaches (made during 10 rpm, constant velocity, CCW rotation) were displaced, however, in the direction of the lateral Coriolis force generated by the forward velocity of the arm. With repeated per-rotation reaches, subjects became progressively more accurate and were back to baseline accuracy within 15 reaches despite the absence of visual feedback.

The trajectory curvature of the reaches (defined as the maximum horizontal distance between a straight line joining the start and endpoint of the movement and the actual movement path) showed the same pattern. The pre-rotation movements were essentially straight; the initial per-rotation movements were deviated in the direction of the Coriolis force but the subsequent per-rotation movements became progressively straighter. The initial post-rotation movements had a trajectory curvature mirror image to the initial per-rotation movements. An "overhead view" of the first pre-, per-, and post-rotation movements (averaged across subjects) and the final pre-, per-, and post-rotation movements are also presented in Figure 2. The existence of aftereffects in the post-rotation period indicates that the central nervous system has constructed a "model" of the Coriolis forces associated with the per-rotation movements and is still compensating for them although it is no longer appropriate. These anticipatory compensations are why movement trajectories become straight again during the per-rotation period. Such findings are totally contradictory to equilibrium point theories of movement control because these theories would not predict movement errors, adaptive changes, or aftereffects in our experimental situation.

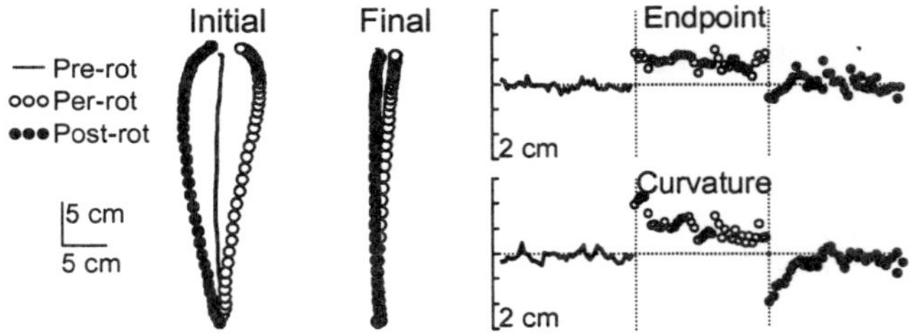

Figure 3. Movement characteristics when subjects end their movements in the air.

DISSOCIATION OF TRAJECTORY AND ENDPOINT ADAPTATION

If we expose subjects to the same experimental paradigm but have them point just above the location of the target so that their reaches end in the air, they exhibit the same pattern of per-rotation trajectory curvature and adaptation and post-rotation aftereffects. However, they do not show the same pattern of per-rotation endpoint adaptation as subjects who touch down at the end of their movements. Instead, they exhibit only partial endpoint adaptation, indicating that terminal hand contact is important for full adaptation to Coriolis forces (see Fig. 3).

To explore the factors that contribute to endpoint and trajectory adaptation and to determine why they are dissociable, we tested labyrinthine defective and cerebellar subjects using the same pre-, per-, and post-rotation paradigm. The subjects in these experiments always made hand contact with the target board surface at the end of their movements. The rotation rate was always 10 rpm, CCW. We tested the labyrinthine-defective subjects in part because equilibrium point theorists had suggested that subjects in our test situation might sense body rotation during the per-rotation period and "correct" for it, thus changing their reaching pattern. This argument does not cohere conceptually for at least two reasons: (i) it should not be necessary by these theories to "correct" for self-rotation, and (ii) if corrections were made, then our subjects should have been reaching more accurately than they were. Nevertheless, it was of interest to test vestibular loss patients because they have an impaired sense of orientation to the vertical and possibly of adaptive capacity that could affect their performance. All of them had normal tonic vibration reflexes. We tested the cerebellar patients because of possible degradations in adaptive control of movement owing to alterations in muscle spindle gain and feedback. Each patient group was tested along with age-matched, normal control subjects.

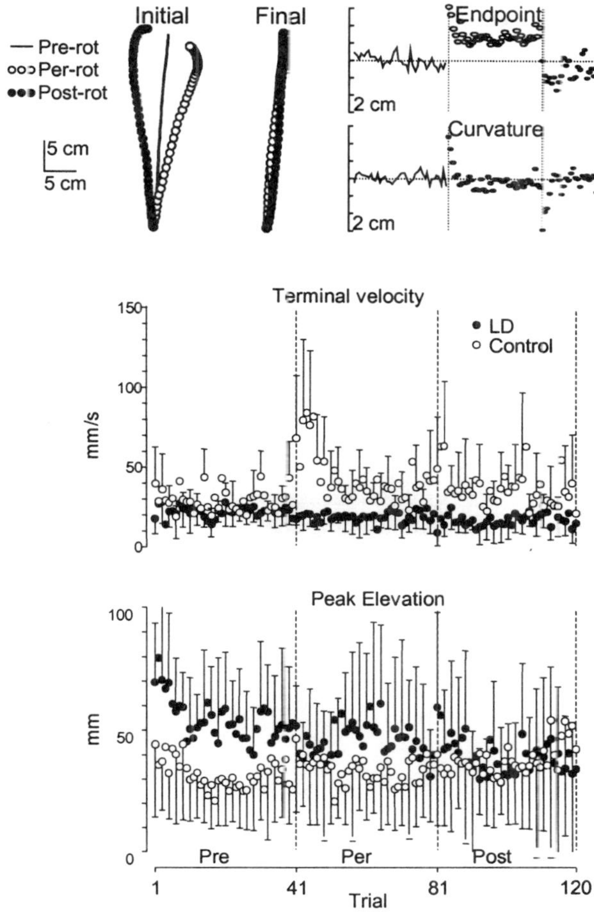

Figure 4. Upper panel shows movement paths, endpoints, and curvature for labyrinthine-defective (LD) subjects. Bottom panel shows movement terminal velocities and peak elevations for LD subjects and controls.

Figure 4 (top panel) presents the experimental findings from the labyrinthine defective subjects. These subjects all showed normal trajectory adaptation during the per-rotation period. Their per-rotation reaches were initially curvilinear but subsequent reaches became progressively straighter just like those of normal subjects. They also showed trajectory curvature aftereffects. However, unlike the normal subjects, they failed to show movement endpoint adaptation. During the per-rotation period, their reaches remained deviated in the direction of the transient Coriolis forces acting during the reaches. Thus, although their reaches straightened out, they reached straight to the wrong place.

As a consequence of this pattern, we wondered if there was something different about the way labyrinthine-defective subjects ended their reaching movements in darkness compared with normal subjects. We therefore examined the position and velocity profiles of the reaching movements of the two subject groups (Figure 4, lower two panels). Several features stand out from this analysis. Compared with the control subjects, the labyrinthine-defective subjects reached higher in the air, moved slower overall, and approached the target surface both more slowly and more vertically. This reaching strategy minimized shear and normal forces on landing thus avoiding the possibility of jamming the fingertip into the surface. We discuss elsewhere the possible reasons for this strategy which partly relate to the labyrinthine-defective subjects having a less precise knowledge of their torso position relative to the target surface (DiZio and Lackner, 2001).

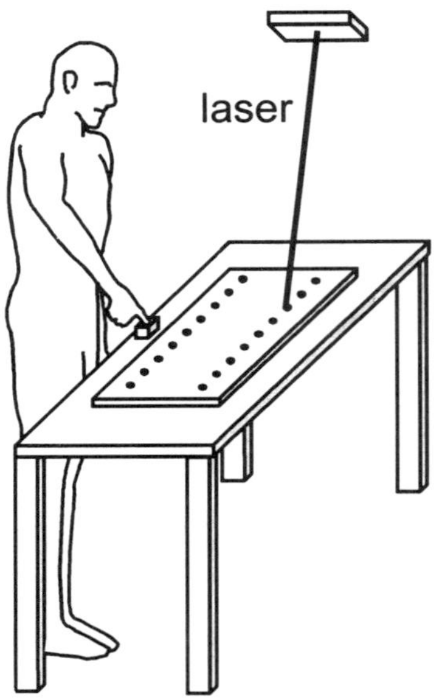

Figure 5. Experimental set up for studying impact forces on fingertip during reaching movements.

FINGERTIP CONTACT FORCES CODE HAND POSITION

The differences between labyrinthine-defective and intact control subjects in movement endpoint adaptation and in finger touchdown suggested impact patterns as being key to whether adaptation of movement terminus would occur. The landing pattern in normal subjects was clearly providing some form of positional information. To identify the nature of this information, we developed the simple experimental set-up shown in Figure 5. The experiment is done in a normal stationary environment. A subject stands in front of a force plate at waist level and attempts to reach and touch targets that are laser projected upon it. When the subject's finger lifts off a central start button near the body midline, the target is extinguished and the movement is completed in total

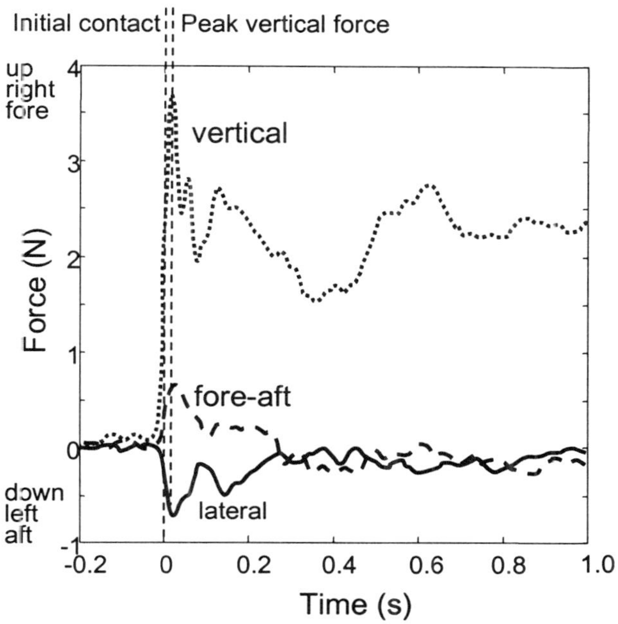

Figure 6. Typical landing forces on the fingertip at the end of a reach.

darkness. (We have evaluated conditions with visual feedback and have used fast as well as natural speed reaching movements but the basic pattern of results described below remains the same across all conditions.) Subjects were instructed to reach in a manner that was natural and comfortable for them.

Figure 6 shows a typical trace for the finger landing on a target location positioned to the right of the subject's midline. Several features stand out. On impact, lateral, fore-aft, and normal forces are present and all rapidly rise to a maximum. Within 50 msec, the lateral and fore-aft shear forces rapidly diminish and after ≈ 100 msec only a normal force remains. This is true regardless of target position. When a three dimensional impact force vector is plotted for each target position, it is oriented toward a point near the midline of the shoulder of the pointing arm. This pattern means that the impact forces on the finger at touch down provide a mapping of finger position relative to the torso. This positional information is absent in the labyrinthine-defective subjects because they end their reaches with the finger landing virtually normal to the surface.

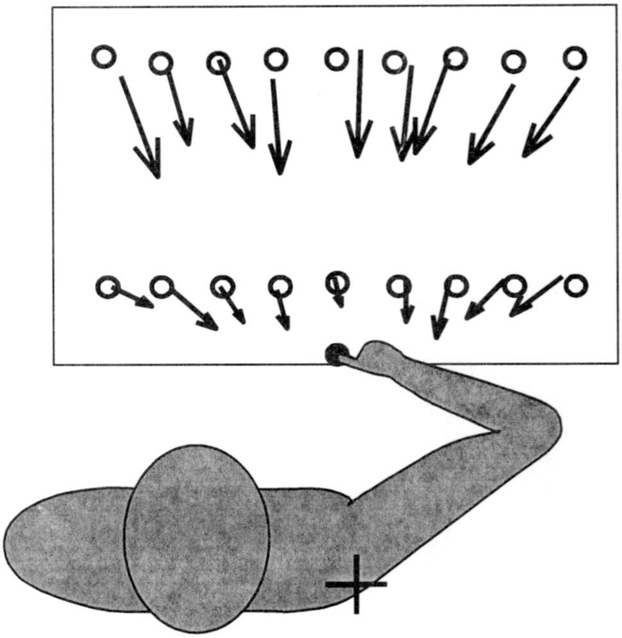

Figure 7. Fingertip reaction force vectors for landing on different targets are oriented to a location on the shoulder of the pointing arm.

Figure 7 shows the impact vector map for reaches to the various target locations. In addition to the impact vectors specifying direction, the precise region of the fingertip stimulated also codes finger position relative to the torso. When the subject reaches to more distal compared with more proximal targets the area of the fingertip in contact with the surface moves more proximal. When the subject reaches to the right of the body midline the area of the index finger stimulated is more medial toward the middle finger because the finger is more supinated, and for reaches to the left more lateral. Thus, the region of the finger contacted also provides a body relative specification of finger position. Recent physiological evidence indicates that different classes of tactile afferents may be biased to code different directions of fingertip stimulation, e.g. SAI for distal tangential force components, SAII for proximal components (Birznieks et al., 2001). Thus, both the fingertip region and the receptor population activated are providing spatially relevant information about hand position.

MUSCLE SPINDLES AND TRAJECTORY ADAPTATION

Both labyrinthine-defective and normal subjects sensed the Coriolis forces acting on their arm during their initial per-rotation reaches. They felt their arm was being deviated from its intended path. However, as their movement trajectories became progressively straighter over the course of subsequent reaches, the Coriolis forces although undiminished in strength, felt progressively smaller until the subjects no longer perceived them at all. During their initial post-rotation reaches, the subjects felt their arm being deviated from its intended trajectory. Here the self-generated compensation for an expected Coriolis force which was no longer appropriate was sensed as an external force, even though no external force was present.

Four of the five cerebellar patients we tested have damage implicating muscle spindle projection regions of the cerebellum. Although all five could reach without difficulty, none of the five showed normal adaptation of movement trajectory or endpoint during the per-rotation period. Their per-rotation reaches were deviated in the direction of the rightward acting Coriolis forces, had rightward curvature, and ended to the right of the target. This pattern persisted virtually without alteration for the entire per-rotation series of 40 reaches. The fifth subject showed partial trajectory adaptation in that his reaches became somewhat less curved but the endpoint errors remained. None of the cerebellar subjects exhibited post-rotation aftereffects. Notably, none of these subjects sensed the presence of Coriolis forces during their per-rotation reaches and they were totally unaware that their reaches were being deviated in path and endpoint from their intended trajectory and terminus.

The responses of our normal, labyrinthine-defective, and cerebellar subjects to Coriolis force perturbations of their reaching movements reveals the importance both of muscle spindle signals (and potentially Golgi tendon organs as well) in the monitoring and adaptive control of movement trajectory and of terminal somatosensory feedback from the fingertip in the adaptive control of movement endpoint. Normal and labyrinthine-defective subjects all adapted movement curvature in response to Coriolis forces. The normal subjects failed to show endpoint adaptation when they reached above the target location and the labyrinthine-defective subjects all of whom had abnormal patterns of movement termination also failed to adapt movement endpoint. By contrast, cerebellar subjects showed degraded ability to adapt either trajectory or endpoint and did not sense that their per-rotation reaches were deviated by the Coriolis forces generated by the forward velocity of their reaches.

CONCLUSION

The critical role of spindle activity in relation to efferent commands for the computation of position sense has been recognized since the classical observations of Goodwin, McCloskey and Matthews (1972). The present findings indicate that similar computations figure in the adaptive control of limb trajectory and that adaptive accommodations are automatically introduced when the path of the limb deviates from the goal intended. This means that, contrary to equilibrium point theories of movement control, the central nervous system must be continuously monitoring and controlling the path of the limb in relation to the intended goal. In addition, we have found that trajectory and endpoint control are dissociable with endpoint being coded at the end of a reach by the impact forces on the fingertip and the region of the fingertip stimulated. These impact vectors subside to a small vertical force within 100 msec, thus positional information is only transiently present and must be used immediately to update motor control, held in memory, or lost.

REFERENCES

Birznieks, I., Jenmalm, P., Goodwin, A. W., and Johansson, R. S., 2001, Encoding of direction of fingertip forces by human tactile afferents, *Journal of Neuroscience,* 21, 8222-8237.

Bizzi, E., Hogan, N., Mussa-Ivaldi, F. A., and Giszter, S., 1992, Does the nervous system use equilibrium-point control to guide single and multiple joint movements? *Behavioural Brain Science,* 15, 603-613.

DiZio, P., and Lackner, J. R., 2001, Coriolis force induced trajectory and endpoint errors in the reaching movements of labyrinthine defective subjects, *Journal of Neurophysiology,* 85, 784-789.

Feldman, A. G., 1966, Functional tuning of the nervous system during control of movement or maintenance of a steady posture--III. Mechanographic analysis of the execution by man of the simplest motor tasks, *Biofizika,* 11, 667-675.

Feldman, A. G., 1986, Once more on the equilibrium-point hypothesis (*lambda* model) for motor control, *Journal of Motor Behavior,* 18, 17-54.

Feldman, A. G., Ostry, D. J., Levin, M. F., Gribble, P. L., and Mitnitski, A. B., 1998, Recent tests of equilibrium point hypothesis (*lambda* model), *Motor Control,* 2, 26-42.

Gandevia, S. C., 1996, Kinesthesia: roles for afferent signals and motor commands, in: *Handbook on Integration of Motor, Circulatory, Respiratory and Metabolic Control during Exercise.* L.B. Rowell and J.T. Shephard, eds., American Physiological Society, pp. 128-172.

Goodwin, G. M., McCloskey, D. I., and Matthews, P. B. C., 1972, The contribution of muscle afferents to kinaesthesia shown by vibration-induced illusions of movement and by the effects of paralysing joint afferents, *Brain,* 95, 705-748.

Lackner, J. R., and DiZio, P., 1994, Rapid adaptation to Coriolis force perturbations of arm trajectory, *Journal of Neurophysiology,* 72, 299-313.

Matthews, P. B. C., 1988, Proprioceptors and their contribution to somatosensory mapping: complex messages require complex processing, *Canadian Journal of Physiology and Pharmacology,* 66, 430-438.

10

VELOCITY PERCEPTION AND PROPRIOCEPTION

Graham K. Kerr and Charles J. Worringham*

ABSTRACT

Proprioceptive information related to position and velocity reaches conscious perception and appears to be processed simultaneously via separate perceptual channels. Determining the fidelity of velocity perception is difficult, however, because of the interaction of distance and timing information in its derivation. There is also a complex relationship between receptor discharge and kinematic variables and thresholds for detection of proprioceptive information are increased during movement. Experiments determined the discrimination thresholds for different movement velocities when i) movement distance remained constant and duration varied; ii) movement duration remained constant and distance varied; iii) movement distance and duration were randomly varied. Discrimination thresholds increased as movement velocity increased. Velocity perception was more accurate when distance and timing cues were available, particularly at higher velocities. These results indicate that all available cues are used to make judgements of movement velocity.

INTRODUCTION

Proprioceptive information provides the central nervous system (CNS) with knowledge of the shape, size and mass of body segments and allows the CNS to determine the orientation, position and velocity of our body and limbs during movement (Gandevia, 1995; Kerr and Marshall, 1995). For the purposes of this chapter a broad definition of proprioception has been adopted which includes information derived from sensory receptors in muscles, tendons, joints and skin (Cordo et al., 1994; Gandevia, 1995).

Proprioception is crucial for coordinated movement. When there is a deficit in proprioception, controlled movements are impossible without continuous visual guidance (Ghez et al., 1995), maintenance of force or position is severely impaired, and a tremor develops (Marsden et al., 1984). In extreme cases, functional movement is restored only

* School of Human Movement Studies, Queensland University of Technology, Brisbane, Queensland, Australia. Email: g.kerr@qut.edu.au

after many years of continued practice (Cole, 1991). Despite the importance of being able to restore functional movement in people who have proprioceptive deficits, however, our understanding of how proprioception is used to control movement is far from complete. There is a lack of knowledge in understanding of *when* and *how* specific sources of proprioceptive information contribute to the preparation and performance of movement. This Chapter outlines how proprioceptive information related to movement velocity is perceived and how this perception is dependent upon associated movement cues.

Kinematic Variables and Proprioceptive Information

Sensory information arising from receptors in the muscles, tendons, joints and skin plays a role in controlling limb movements (Matthews, 1972; McCloskey, 1978; Gandevia and Burke, 1992). Position and velocity information may be derived from cutaneous mechanoreceptors and muscle spindles. Both signal information related to kinematics, in particular position and velocity information, although cutaneous afferents have less directional sensitivity than muscle spindles (Grill and Hallett, 1995). Muscle spindles discharge in relation to movement onset, velocity transitions, muscle length changes, and movement cessation during passive movements (Grill and Hallett, 1995). During active contractions and voluntary movement the relationship between kinematic variables and muscle spindle discharge is more complex (Burke et al., 1978; Rymer and D'Almeida, 1980), and depends on the history of muscle activation (Edin and Vallbo, 1988; Gregory et al., 1990). The non-linear output of muscle spindles (Hasan and Houk, 1975) further complicates the information conveyed to the CNS.

If proprioceptive information is used for controlling arm movements then the signals conveyed by proprioceptors must vary throughout the movement. During discrete movements the discharge from muscles spindles and cutaneous receptors is dependent upon the phase and velocity of movement (Grill and Hallett, 1995). However, in any movement the muscles acting about a joint will be subject to different length changes and velocities of stretch. Thus the information from individual muscles, and even different spindles within each muscle, will vary considerably throughout the course of a movement (Burgess et al., 1982). Furthermore, muscles that span more than one joint are subject to length changes from quite different actions, thus creating a potential ambiguity in the proprioceptive information. This is further complicated if the afferent connections between different muscle groups are considered. For example, afferents from wrist muscles project onto motoneurons of muscles acting at the elbow (Cavallari and Katz, 1989). Movement involving muscles at one joint can therefore influence the pattern of activation of muscles acting about other joints and, consequently, the afferent information arising from them (Kasai et al., 1992). Thus further processing by the CNS is required to extract meaningful information related to joint position and joint velocity from the proprioceptors .

Perceiving Proprioceptive Information

How proprioceptive information is used during movement is uncertain because thresholds for detection of cutaneous stimuli are increased (Williams et al., 1998) and sensation from muscle afferents is reduced (Collins et al., 1998). It is apparent, however, that people are able to interpret proprioceptive information related to position and velocity (Cordo et al., 1994; Kerr et al., 1994). Both reach conscious perception and appear to be processed simultaneously via separate perceptual channels (Sittig et al., 1985; 1987). Position and velocity may be used by the CNS to predict the future position of the arm so that a coordinated action (opening of the hand) can be initiated (Cordo et al., 1994). Arm position is also perceived in advance of its actual position (Kerr et al., 1994; Cordo et al., 1995b). This perceptual bias is accentuated by muscle contractions that assist movement, and attenuated by muscle contractions that resist movement (Kerr et al., 1999). Position and velocity information derived from proprioceptors may therefore be incorporated within internal predictive models and used to compensate for time delays inherent in the nervous system. Thus, although there is no clear answer as to how both joint position and velocity are represented in proprioceptor discharge, it is evident that information related to these kinematic variables is available to and used by the nervous system to control movement.

Velocity Perception

Reliable knowledge of limb velocity is required for the accurate performance of certain types of movement. However, there have been few reports about the resolution of velocity perception related to proprioception, although an increase in joint velocity improves the perception of movement onset (Taylor and McCloskey, 1992). When subjects attempt to match the speed of their passively moved elbow with their contralateral elbow they are accurate to within ±5°/s for velocities ranging from 15°/s to 78°/s (Kerr et al., 1994). However, thresholds for detection of velocity changes during passive ramp movements (10°/s to 35°/s) of the index finger are 10°/s (Grill et al., 1994). Discrimination thresholds between pairs of 30°/s and 50°/s passive shoulder extension movements have been reported as 2°/s and 10°/s respectively (Lonn et al., 2001). Although these studies have used different methods to determine the accuracy with which velocity is perceived they indicate that this perception is velocity dependent and that there may be quite large inaccuracies in perception, especially at higher velocities.

When determining velocity perception one important consideration is that perception of any aspect of movement is based on a mixture of different movement cues embedded within proprioceptive information. These include timing, location, distance and velocity information. Thus determining the ability of the nervous system to consciously discriminate between different velocities is difficult because of the interaction of distance and timing information in the derivation of velocity. For example, an increase in velocity can occur by decreasing movement duration, increasing distance or both.

The following experiments outline how the perceptual resolution of velocity was determined during passive rotations of the elbow when discriminations were required to be made between pairs of different constant velocities. A series of experiments determined the discrimination thresholds for different velocities when i) movement distance remained constant and duration varied; ii) movement duration remained constant and distance varied; iii) movement distance and duration were randomly varied. These experiments controlled for influences of distance and time and therefore determined whether the perception of joint velocity was independent of movement duration and distance.

METHODS

Subjects' (36) right elbows were passively extended at a constant velocity (15, 30, 45, 60, 75°/s) over a fixed distance (30°; reference movement). The arm was returned to a start position and moved at a different constant velocity (comparison movement) over the same *distance* (Expt. 1: velocity changed by alteration in duration) or the same *duration* (Expt. 2: velocity changed by alteration in distance). For Experiment three the arm was returned to a randomly varied start position and moved at a different constant velocity over a randomly varied distance (22° -26°). An adaptive forced-choice staircase procedure was used in which the initial difference between the two velocities (±170%) was altered (±80%) on subsequent trials until there was convergence on a velocity level (Falmagne, 1985; Brenner and van der Berg, 1994). Subjects were asked to attend to velocity irrespective of time and distance and to state whether the comparison movement was *slower* or *faster* than the reference movement.

Presentation of the different velocities and the different instruction sets was counterbalanced across subjects. Three repetitions were obtained for each velocity and instruction set for each subject. Thresholds were calculated averaging the velocity difference between the reference and comparison movement for the last three trials that represented the convergence level. Statistical analysis of the data was by repeated measures analysis of variance and Tukey's post-hoc tests.

RESULTS

The thresholds for detecting velocity differences were significantly greater for the Variable Time and Distance condition (Expt. 3) compared to the Constant Distance (Expt. 1) and Constant Duration (Expt. 2) experiments ($F\ (2,33) = 7.005$, $p < 0.05$). These thresholds were velocity dependent and increased linearly as the reference velocity increased ($F\ (4,132) = 78.4462$, $p < 0.05$). Post-hoc tests revealed that the threshold at 15°/s was significantly lower than the thresholds at all other velocities and that the thresholds at 30 °/s and 45°/s were significantly lower than thresholds at 60 °/s and 75°/s.

Figure 1 illustrates the increase in thresholds from 4.5°/s (Expts. 1, 2 and 3) at the 15°/s reference velocity to 10°/s (Expts. 1 and 2) and 16°/s (Expt. 3) at the 75°/s reference velocity. The differences in threshold between Experiment 3 and Experiments 1 and 2 resulted in a significant interaction ($F\ (8,132) = 3.063$, $p < 0.05$). Post-hoc tests showed that the differences between Experiment 3 and Experiments 1 and 2 occurred at 45 °/s, 60 °/s and 75°/s. At reference velocities of 45°/s and greater, thresholds were therefore higher when there was greater uncertainty about movement duration and distance.

These results show that velocity thresholds increased as a function of movement velocity. Moreover velocity perception was more accurate when distance and timing cues were also available, particularly at higher velocities (>45°/s).

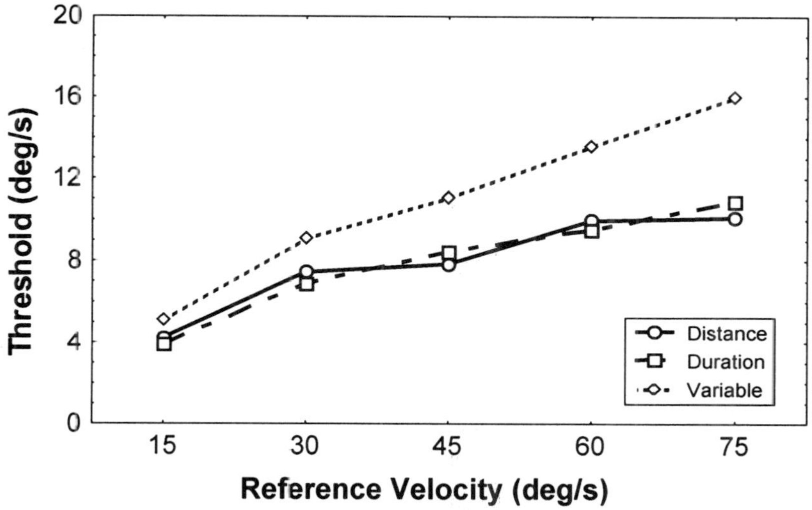

Figure 1. Velocity thresholds. Velocity thresholds for each of the movement velocities for Experiment 1 (Constant Distance), Experiment 2 (Constant Duration) and Experiment 3 (Variable Distance and Duration).

DISCUSSION

The present experiments determined the perceptual resolution of velocity during passive elbow rotations of the elbow at different velocities. They also determined whether the perception of joint velocity was independent of movement duration and distance by controlling for these influences.

Velocity thresholds were lower for the experiments in which subjects were able to use cues of distance (Expt. 1) and duration (Expt. 2) to discriminate between two different movement velocities. This difference was manifest at velocities of 45°/s and greater. These results indicate that perception of velocity is not independent of distance and duration when these cues are available. All available cues are used to make judgements of movement velocity, even when instructions are made to ignore them. These results are counter to those of Lonn and colleagues (2001) who reported that subjects did not use movement distance or time as primary discrimination cues.

Two previous experiments have determined thresholds for detecting velocity differences during passive finger movements (Grill et al., 1994) and between successive passive shoulder extension movements (Lonn et al., 2001). Both of these papers have reported similar threshold differences to those obtained in the present experiment for the 45°/s reference movement (11°/s threshold). However, the 8°/s threshold obtained at 30°/s in the present experiment is considerably higher than the 2°/s threshold reported by Lonn et al. (2001, their Fig. 6) for this velocity. At least some of these differences may be attributable to the different joints examined and, in the case of Grill et al. (1994), the different experimental protocol. However, the similar thresholds for the 45°/s reference movement indicates that there may be some common limits in the fidelity with which movement velocity is consciously perceived, irrespective of the joint used.

The CNS could sense velocity directly or could differentiate the afferent information that signals joint position. However, even if all or some of the position and velocity signals were used, it would still be necessary for the CNS to somehow extract this information from the proprioceptors distributed within the muscles, tendons, joints and skin. Even if muscle spindles alone were used to derive a measure of joint position and velocity all of the elbow flexor muscles would be stretched by different amounts during elbow extension and contribute to ambiguity in these measures. The discrimination thresholds obtained in the present experiment therefore indicate the fidelity of these measures because they define the bandwidth for consciously perceiving differences in movement velocity.

The discrimination thresholds indicate the level to which velocity has to change for a person to consciously perceive a difference. If these thresholds are indicative of the accuracy of velocity information conveyed by proprioceptors to the CNS then the use of velocity in forward models and predictive control could result in large performance errors if proprioceptive information alone is used. This is despite the fact that the CNS has access to different sources of afferent information to derive measures of position and velocity. Correct detection of joint angle and velocity is necessary for coordinating action, as has been demonstrated for a hand-opening (frisbee) task during passive elbow movements (Cordo et al., 1994). Therefore, if a person whose arm is moving at 75°/s does not detect a velocity difference of 10°/s, they will incur an error of at least ±10 degrees at the end of a one-second movement. However, given that people are very accurate on the frisbee task (±1°) it is likely that the proprioceptive information in these tasks is processed subconsciously and therefore not subject to the limitations implied by the present discrimination thresholds.

Finally, it is worth considering that for proprioceptive information to be useful for controlling arm movements, information about the position and orientation of the hand in space and the velocity at which it is moving must be provided to the CNS. Ultimately, this proprioceptive information must be integrated within a central representation of the body if the end effector is going to arrive at the correct position at the correct velocity and at the correct time. This is a difficult task because no single proprioceptor directly senses hand position or velocity (see Gandevia et al., Chapter 8). Rather these percepts must be constructed from the many and varied proprioceptors distributed throughout the arm. Modelling approaches directed at answering this question have indicated that ensemble discharge patterns of muscle spindles from many muscles are highly correlated with the tangential velocity of the hand (Wallace and Kerr, 1996a,b). These results indicate that the neural organization underlying the processing of peripheral afferent signals may contribute to the direct coding of the position and velocity of the hand in space. However, determining this remains a challenge for future microneurographic experiments.

ACKNOWLEDGEMENTS

This work was carried out with support from the Australian Research Council.

REFERENCES

Brenner, E., and Van den Berg, A.V., 1994, Judging object velocity during smooth pursuit eye movements, *Experimental Brain Research*, 99, 316-324.

Burgess, P. R., Wei, J. Y., Clark, F. J., and Simon, J., 1982, Signalling of kinesthetic information by peripheral sensory receptors, *Annual Review of Neuroscience*, 5, 171-187.

Burke, D., Hagbarth, K. E., and Lofstedt, L., 1978, Muscle spindle responses in man to changes in load during accurate position maintenance, *Journal of Physiology*, 276, 159-164.

Cavallari, P., and Katz, R., 1989, Pattern of projections of group I afferents from forearm muscles to motoneurones supplying biceps and triceps muscles in man, *Experimental Brain Research*, 78, 465-478.

Cole, J., 1991, Pride and a daily marathon, Duckworth, London.

Collins, D., Cameron, T., Gillard, T.M., and Prochazka, A., 1998, Muscular sense is attenuated when humans move, *Journal of Physiology*, 508, 635-643.

Cordo, P, Carlton, L, Bevan, L, Carlton, M, and Kerr, G.K., 1994, Proprioceptive triggering in movement sequences: The role of velocity input, *Journal of Neurophysiology*, 71, 1848-1861.

Cordo, P., Gurfinkel, V., Bevan, L., and Kerr, G.K., 1995, Proprioceptive consequences of tendon vibration during movement, *Journal of Neurophysiology*, 74, 1675-1688.

Edin, B.B., and Vallbo, A.B., 1988, Stretch sensitization of human muscle spindles, *Journal of Physiology*, 400, 101-111.

Falmagne, J., 1985, *Elements of Psychophysical Theory*, Clarendon Press, Oxford.

Gandevia, S., 1995, Kinesthesia: roles for aferent signals and motor commands, In: *Handbook of physiology, Section 12, Exercise: regulation and integration of multiple systems, Part 1, Neural control of movement*, J. Smith, ed., American Physiological Society, Oxford University Press, New York, pp. 128-172.

Gandevia, S.C., and Burke, D., 1992, Does the nervous system depend on kinesthetic input to control natural limb movements? *Behavioral and Brain Science*, 15, 615-633.

Ghez, C., Gordon, J., and Ghilardi, M., 1995, Impairments of reaching movements in patients without proprioception. II. Effects of visual information on accuracy, *Journal of Neurophysiology*, 73, 361-372.

Goodwin, G., Mccloskey, D.I., and Matthews, P., 1972, The contribution of muscle afferents to kinaesthesia shown by vibration induced illusions of movement and by the effects of paralyzing joint afferents, *Brain*, 95, 705-748.

Gregory, J.E., Mark, R.F., Morgan, D.L., Patak, A., Polus, B., and Proske, U., 1990, Effects of muscle history on the stretch reflex in cat and man, *Journal of Physiology*, 424, 93-107.

Grill, S.E., and Hallett, M., 1995, Velocity sensitivity of human muscle spindle afferents and slowly adapting type II cutaneous mechanoreceptors, *Journal of Physiology*, 489, 593-602.

Grill, S.E., Hallett, M., Marcus, C., and McShane, L., 1994, Disturbances in kinaesthesia in patients with cerebellar disorders, *Brain*, 117, 1433-1447.

Hasan, Z., and Houk, J.C., 1975, Transition in sensitivity of spindle receptors that occurs when muscle is stretched more than a fraction of a millimeter, *Journal of Neurophysiology*, 38, 673-689.

Kasai, T., Kawanishi, M., and Yahagi, S., 1992, The effects of wrist-muscle vibration on human voluntary elbow flexion-extension movements, *Experimental Brain Research*, 90, 217-220.

Kerr, G.K., Bevan, L., Gurfinkel, V., and Cordo, P., 1994, Effects of vibration on the perception of dynamic position and velocity of the passively moving elbow, *Proceedings of the Australian Neuroscience Society, 14th Annual Meeting*, 5, 195.

Kerr, G.K., and Marshall, R.N., 1995, Adaptation of arm movements to altered loads: Implications for sensorimotor transformations, In: *Motor control and sensory-motor integration: Issues and directions*, D.J. Glencross and J.P. Piek, eds., North Holland, pp. 321-339.

Kerr, G.K., Leonard, C., and Worringham, C., 1999, Effects of muscle contraction on the perception of dynamic position during passive movement of the elbow, *Proceedings of the Australian Neuroscience Society, Vol 10, 19th Annual Meeting.*

Lonn, J., Djupsjobacka, M., and Johansson, H., 2001, Replication and discrimination of limb movement velocity, *Somatosensory and Motor Research*, 18, 76-82.

Marsden, C., Rothwell, J., and Day, B., 1984, The use of peripheral feedback in the control of movement, *Trends In Neuroscience*, 7, 253-257.

Matthews, P.B.C., 1972, Mammalian Muscle Receptors and their Central Actions, Arnold, London.

Mccloskey, D.I., 1978, Kinesthetic sensibility, *Physiological Reviews*, 58, 763-820.

Rymer, W.Z., and d'Almeida, A., 1980, Joint position sense effects of muscle contraction, *Brain*, 103, 1-22.

Sittig, A., Denier van der Gon, J., and Gielen, C., 1985, Separate control of arm position and velocity demonstrated by vibration of muscle tendon in man, *Experimental Brain Research*, 60, 445-453.

Sittig, A., Denier van der Gon, J., and Gielen, C., 1987, The contribution of afferent information on position and velocity to the control of slow and fast human forearm movements, *Experimental Brain Res*earch, 67, 33-40.

Taylor, J.L., and McCloskey, D.I., 1992, Detection of slow movements imposed at the elbow during active flexion in man, *Journal of Physiology*, 457, 503-513.

Wallace, K.R., and Kerr, G.K., 1996a, A numerical simulation of muscle spindle ensemble encoding during planar movement of the human arm, *Biological Cybernetics*, 75, 339-350.

Wallace, K.R., and Kerr, G.K., 1996b, A numerical simulation of muscle spindle ensemble encoding during planar movement of the human arm: correlation sensitivity and parameter dependence, *Biological Cybernetics,* 75, 351-359

Williams, S., Shenasa, J., and Chapman, E., 1998, Time course of movement-related gating of tactile detection in humans. I. Importance of stimulus location, *Journal of Neurophysiol*ogy, 79, 947-963.

11

THE EFFECT OF MUSCLE CONTRACTION ON KINAESTHESIA

Andrew K. Wise[1] and James B. Fallon[2]

ABSTRACT

Kinaesthesia is our conscious awareness of body position and movement. Experiments are described examining kinaesthetic acuity in human subjects. The results showed a reduced ability to detect limb movement and match limb position during co-contraction of elbow extensors and flexors compared to when these muscles were relaxed. We also report results from animal experiments showing a reduction in muscle spindle stretch sensitivity during fusimotor and skeletomotor activation, a factor that might contribute to the decreased kinaesthetic acuity observed during muscle contraction.

INTRODUCTION

It is believed that feedback derived from a number of receptors located in muscle, joints and skin is essential for normal kinaesthesia. Signals from these receptors provide us with a dynamic update of limb movement and position and contribute to an internal schema, a conscious representation of our body necessary for normal motor functioning. The work of Goodwin et al. (1972) showed that signals arising from muscle spindles provided kinaesthetic information. They were able to elicit illusions of limb movement and altered position by vibrating muscle, a stimulus known to excite muscle spindles. The illusions reported by the subjects were in a direction of limb movement that would stretch the vibrated muscle, an observation consistent with the idea that muscle lengthening can be signalled by an increase in firing rate from muscle spindles. If agonist/antagonist muscles were vibrated concurrently and at the same frequency, no illusion was evoked (Gilhodes et al., 1986). It was only when the vibration frequency of one muscle was different to the other that movement illusions were reported suggesting that inputs from both muscles were involved in the illusion.

[1] Department of Physiology, Monash University, Clayton, VIC, 3800, Australia.
Email: wisea@mail.medoto.unimelb.edu.au
[2] Department of Electrical Engineering, Monash University, Clayton, VIC, 3800, Australia.

Muscle spindles have their own motor supply, the fusimotor system. When we voluntarily contract our muscles we not only activate our skeletal muscle but we also co-activate the fusimotor system (Vallbo, 1970) which can alter the firing behaviour and stretch sensitivity of spindle afferents (Matthews, 1962). This poses a potential problem when deriving kinaesthetic information from muscle spindles. If signals arising from spindles are influenced by changes in muscle length and also by fusimotor activity then how does the central nervous system interpret information arising from these receptors? One theory suggests that the fusimotor-evoked spindle activity is subtracted from the total signal so that only muscle-length-related activity reaches consciousness (see McCloskey, 1981). However, to complicate matters further, fusimotor activation also acts to alter muscle spindle stretch sensitivity, a factor that needs to be taken into account when determining muscle length information from spindles under fusimotor control.

MOVEMENT DETECTION

A common method used to measure kinaesthetic acuity is to impose limb movements and ask the subject to report the direction of movement. Most studies using this method have examined kinaesthesia when the muscles acting on the target limb were relaxed. However, we normally function by contracting our muscles to move limbs and maintain posture, and examination of kinaesthesia under these conditions has more functional relevance. Studies that have measured movement detection at the elbow joint have generally found improved acuity during isometric contraction of biceps muscles (Colebatch et al., 1987; Taylor et al., 1992) when compared to thresholds for the relaxed limb (Hall et al., 1983). However, kinaesthetic acuity during co-contraction of agonist and antagonist muscles has received less attention.

In experiments by Wise et al. (1998) movement detection thresholds were measured at the elbow with muscles relaxed and also during co-contraction of flexors and extensors. For the relaxed condition movement detection thresholds were measured after a brief co-contraction of the elbow extensors and flexors which was used to control muscle history effects (Wise et al., 1996). For thresholds measured during co-contraction, subject's were required to maintain a co-contraction of elbow extensors and flexors at 15% of their maximum. Biceps and triceps EMG was measured to quantify the level of contraction and the experimental set-up is shown in the left panel of Figure 1. Movement detection thresholds were measured for flexion and extension.

The average data from seven subjects are shown in the right panel of Figure 1. There was considerable variation in thresholds between subjects and therefore the data has been displayed with values normalised to the average of thresholds measured for the relaxed condition. There was no difference between extension and flexion thresholds so they were combined for each group. The results indicated that during co-contraction of biceps and triceps muscles kinaesthetic acuity decreased significantly (ANOVA, $p<0.01$), with thresholds approximately 2.4 times larger than those measured for the relaxed limb. In other words, during muscle co-contraction larger limb movements were required to detect the direction of the movement.

It is proposed that the reduced kinaesthetic acuity observed during co-contraction is a result of significant fusimotor drive to biceps and triceps muscles. During co-contraction the spindle discharge will be dominated by fusimotor activity, particularly for the slow movements used in these experiments. An additional consequence of fusimotor activity is that muscle spindle stretch sensitivity is likely to be altered, a point which will be discussed later.

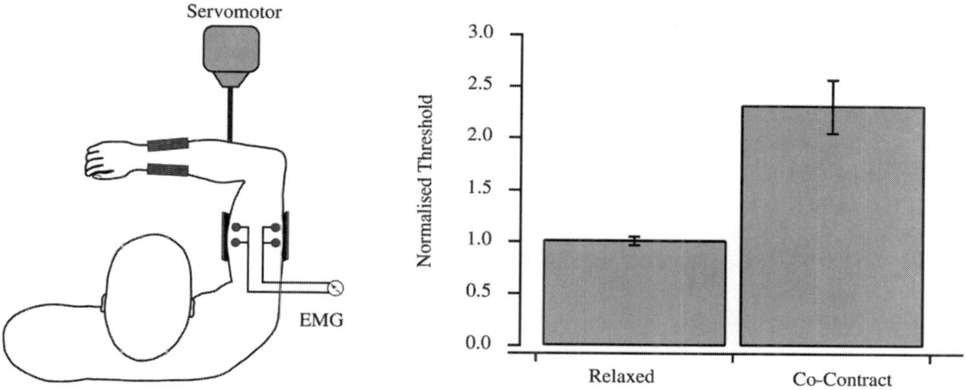

Figure 1.
Left panel, blindfolded subjects were required to indicate the direction of small rotations applied to the right forearm by a servomotor. Movement detection thresholds were measured for elbow extension and flexion under relaxed and co-contraction conditions (15% MVC co-contraction of biceps and triceps muscles).
Right panel, average thresholds measured for seven subjects. There was no difference between extension and flexion thresholds and values have been normalised to the average thresholds measured for the relaxed condition. Thresholds measured during co-contraction of elbow muscles were significantly greater than those measured for the relaxed arm. (Data from Wise et al., 1998.)

Movement detection thresholds measured during isometric contraction of biceps (Taylor et al., 1992) were lower than those measured during co-contraction. This finding could relate to differences in methods, for instance, the method of threshold determination or the acceleration of the limb for each test movement. Alternatively, these results may reflect a difference in the two motor tasks, that is, contraction of one muscle group is quite different to co-contraction of agonist and antagonist muscles. To resolve this issue movement detection thresholds need to be measured in the same subject, under the same conditions, using both forms of muscle contraction.

POSITION MATCHING

Kinaesthesia encompasses the sense of limb position. There is evidence that the sense of movement and position is mediated by separate lines of information (Horch et al., 1975). Alternatively, limb position could be determined by integration of movement signals. Is position sense, like movement sense, affected by muscle co-contraction? Previous studies have found that during isometric contraction of one muscle group matching errors increased proportionately with contraction force (Rymer et al., 1980). Another study reported that the ability to match elbow angle was reduced by changing the load on the arm, whereby an increase in the motor-drive, both skeletal and fusimotor, was required to move the matching arm into position (Gooey et al., 2000). We recently measured the ability of human subjects to match elbow angles to determine how muscle co-contraction would affect limb position sense.

Using the experimental set-up illustrated in the upper panel of Figure 2, each trial began with the subject's left arm at either an extended or flexed position and their right arm at the start position. The subject's right arm was then positioned at one of three randomly selected reference positions (A, B or C). Three methods of positioning the subject's right arm were used: voluntary positioning, relaxed positioning, and co-contraction positioning. The voluntary positioning (Vol) required the subject to move their arm to one of the three reference positions. The relaxed positioning (Rel) required the subject to remain relaxed while the experimenter positioned their arm and then perform a brief co-contraction of their right biceps and triceps muscles. The co-contraction condition (Cont) required the subject to maintain a 15% MVC co-contraction of biceps and triceps of their right arm while it was positioned, and then continue the co-contraction throughout the duration of the trial. Once the right arm was positioned the left arm was then rotated at 5°/s while the subject either kept their left arm relaxed (Rel) or maintained a 15% MVC co-contraction of left elbow muscles (Cont). A total of six untrained subjects were used. Each subject was blindfolded and instructed to indicate when the angle of their left arm matched the angle of the right arm (i.e. the elbows were bent to the same angle). The difference between the matched angle (left arm) and the reference angle (right arm) was recorded and values normalised to the average mismatch angle for the (Rel-Rel) condition. The resulting mismatch values were then analysed using a multi-factorial ANOVA and the pooled results for all subjects are shown in the lower panel of Figure 2.

The type of condition used for the right arm (Vol, Rel or Cont) had a significant effect on the subject's ability to match. Muscle co-contraction produced significant under estimation of elbow angle compared to either the relaxed or voluntary condition. Under estimation of the reference elbow angle indicated that subjects thought their right arm was more flexed than it was. The co-contraction did not lead to an increase in the variability of the match, as measured by the SEM of the match value, but simply an apparent shift into flexion. The trend of co-contraction resulting in the arm position being perceived as further flexed than its actual position was also evident during the left arm co-contraction condition, although in this case the result was not significant. During co-contraction of the left arm only, subjects over estimated the reference elbow position as they perceived the left arm to be more flexed than it actually was.

We propose that the errors in limb position sense produced during muscle co-contraction were a result of fusimotor-evoked spindle activity. There was a systematic shift in position sense and matching errors were biased in the direction of flexion of the limb during co-contraction. Elbow flexion would be signalled by a reduction of the muscle spindle activity in biceps and / or an increase in spindle activity in triceps. A possible explanation for the observed shift in matching with co-contraction might be that triceps was activated more vigorously than biceps, resulting in a larger increase in fusimotor-evoked activity in this muscle. However, analysis of the EMG showed that the activation level of biceps was significantly greater than triceps, which would result in a perceived extension of the elbow joint if this theory were correct. Therefore, the illusion of extra flexion is not simply explained by the different contraction levels but is likely to involve some errors in the proposed subtraction of the fusimotor-evoked spindle activity to recreate the muscle length signal.

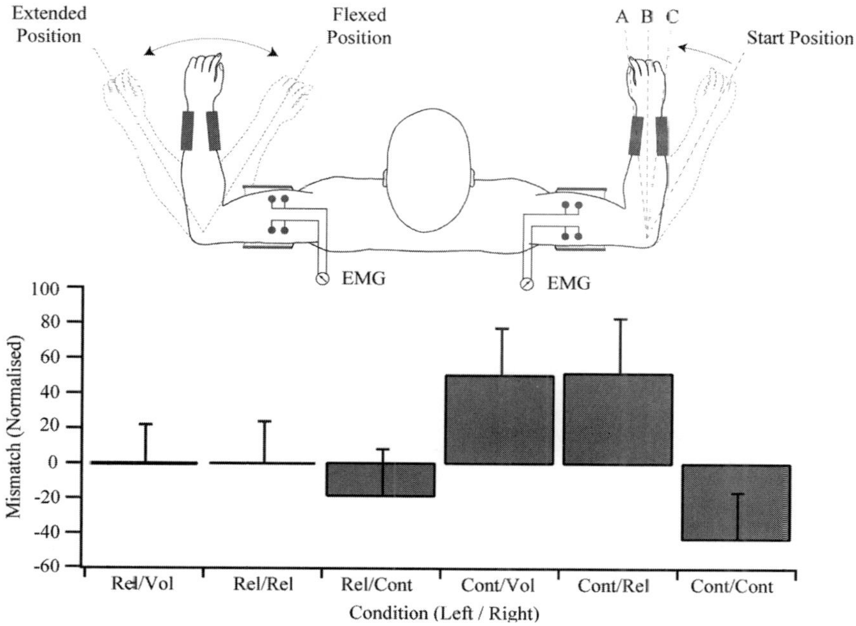

Figure 2.
Upper Panel, the right arm of a blindfolded subject was placed at one of three different positions (A, B or C) from an initial starting position using three different conditions, Voluntary (Vol), Relaxed (Rel) or Co-contraction (Cont). Once the right arm was positioned, the left arm was rotated from either a flexed or extended position at a constant velocity of 5°/s. Two different conditions were used for the left arm, either Relaxed (Rel) or Co-contraction (Cont). The subject was required to indicate when the position of their left arm matched their right arm. The mismatch angle was measured and normalised to the mismatch of the Pas/Pas condition.
Lower panel, average normalised results from six subjects. There was a systematic increase in matching errors during co-contraction of biceps and triceps muscles. Errors were biased in the direction of flexion for the right arm co-contraction condition (left arm matched more flexed than right arm). Matching errors during co-contraction of the left arm only were in the opposite direction (left arm matched more extended) implying that subjects perceived their left arm to be in a position more flexed than it actually was. Interestingly, this effect was not evident when the co-contraction condition was used for both arms.

MUSCLE SPINDLE RESPONSES DURING CONTRACTION

One possible explanation for the larger movement detection thresholds and position matching errors seen during muscle contraction might relate to changes in the response properties of muscle spindles during fusimotor activation. Co-activation of the fusimotor system during contraction may have led to a decrease in the ability of muscle spindles to signal muscle length changes. To examine this possibility the stretch sensitivity of muscle spindle primary endings in the soleus muscle of the anaesthetised cat was measured in response to small, slow movements (Wise et al., 1999). Spindle stretch responses were measured for the 'Passive' spindle (without fusimotor stimulation) and for the 'Active' spindle during combinations of fusimotor and skeletomotor stimulation.

The left panel of Figure 3 shows an example of the responses from a muscle spindle primary ending. Spindle activity, represented as instantaneous rate, was measured in the presence or absence of muscle stretch for the Passive condition and the Active condition. The Active condition involved stimulation (70pps) of single identified fusimotor axons, static, dynamic or both, and for four spindles, responses were measured during combined fusimotor and skeletomotor stimulation.

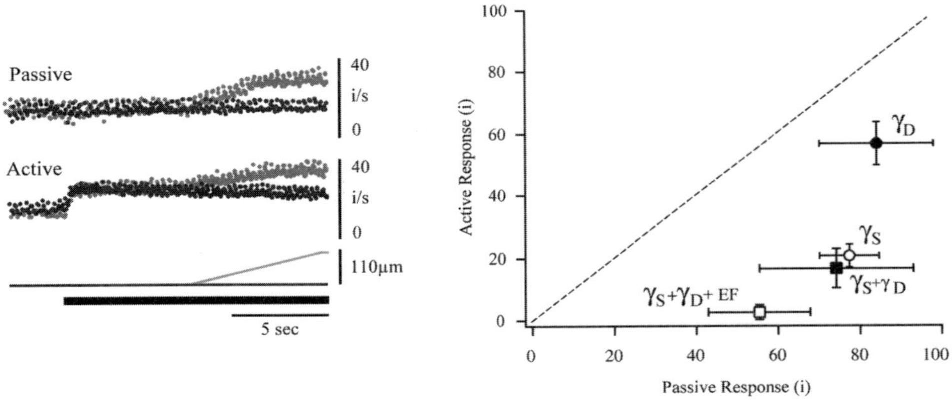

Figure 3.
Left panel, example of responses from a muscle spindle primary ending for the Passive and Active conditions. Active condition: measurements were made during stimulation of single fusimotor axons, indicated by the bar below the muscle length trace. In this example the Active response was measured during stimulation of a single dynamic fusimotor axon and produced a smaller stretch evoked response than the Passive condition. Stretch evoked responses were calculated by subtracting the response measured in the absence of stretch (black markers) from the response measured during muscle stretch (grey markers).
Right panel, average data from 16 units. Active responses were measured during stimulation of dynamic (γ_D), static (γ_S), combined ($\gamma_S+\gamma_D$) and with the addition of extrafusal tension ($\gamma_S+\gamma_D$+EF). All combinations of stimulation tended to reduce the stretch sensitivity of the spindle compared to its passive values. (Data from Wise et al., 1999.)

The right panel of Figure 3 shows the average Active stretch responses plotted against their Passive values. Stimulation of single static fusimotor axons produced a significant decrease in spindle stretch sensitivity ($p<0.001$). Stimulation of dynamic fusimotor axons, which is known to increase the spindle sensitivity for stretches of relatively large amplitude, decreased the stretch response for the small movements used here. However, this effect failed to reach significance. Combinations of static and dynamic fusimotor stimulation, in the presence or absence of skeletomotor activity, also reduced spindle stretch sensitivity. These results indicate that for small movements, of the kind used in the human movement detection experiments, all forms of fusimotor and/or skeletomotor stimulation tended to reduce spindle stretch responses. It means that during muscle contraction larger muscle stretches are required to elicit increases in spindle firing rate similar to that observed during stretch of a relaxed muscle. The implications of this finding to the human data is that if a simple subtraction process is used to determine muscle length information during fusimotor activity the resultant length signal would be smaller than the stretch response of a relaxed muscle.

CONCLUSION

The conclusion from these studies is that we are less capable of deriving muscle length information from sensory receptors during muscle co-contraction. We propose that a factor contributing to this finding is a reduction in muscle spindle stretch sensitivity during muscle contraction. However, a number of additional factors may be involved. One possibility is the loss of muscle length information in the process used to account for fusimotor-induce spindle activity. In addition there might be a general attenuation, or gating, of sensory inflow during muscle contraction. The reduction in kinaesthetic acuity observed during co-contraction might also reflect this particular motor task. Movement of a relaxed limb should increase the discharge from spindles in the lengthening muscle and decrease the discharge from spindles in the shortening muscle. There is a clear asymmetry when comparing signals from these muscles. However, during muscle co-contraction, with significant fusimotor activity to both muscles, spindle discharge will be dominated by the fusimotor drive. Slow limb movement will not produce the same decrease in spindle firing rate seen with a passive spindle and a lengthening response may be reduced by the fusimotor activity. Therefore, if comparison of the sensory inflow from agonist and antagonist muscles is used to compute the direction of limb movement then these differences would be less pronounced during muscle co-contraction.

ACKNOWLEDGMENTS

This work was carried out with support from the Australian NH&MRC and we would like to thank Uwe Proske and Ed Gregory for their contributions.

REFERENCES

Colebatch, J. G., and McCloskey, D. I., 1987, Maintenance of constant arm position or force: reflex and volitional components in man, *Journal of Physiology*, 386, 247-261.

Gilhodes, J. C., Roll, J. P., and Tardy-Gervet, M. F., 1986, Perceptual and motor effects of agonist-antagonist muscle vibration in man, *Experimental Brain Research*, 61, 395-402.

Goodwin, G. M., McCloskey, D. I., and Matthews, P. B., 1972, Proprioceptive illusions induced by muscle vibration: contribution by muscle spindles to perception? *Science*, 175, 1382-1884.

Gooey, K., Bradfield, O., Talbot, J., Morgan, D. L., and Proske, U., 2000, Effects of body orientation, load and vibration on sensing position and movement at the human elbow joint, *Experimental Brain Research*, 133, 340-348.

Hall, L. A., and McCloskey, D. I., 1983, Detections of movements imposed on finger, elbow and shoulder joints, *Journal of Physiology*, 335, 519-533.

Horch, K. W., Clark, F. J., and Burgess, P. R., 1975, Awareness of knee joint angle under static conditions, *Journal of Neurophysiology*, 38, 1436-1447.

Matthews, P. B. C., 1962, The differentiation of two types of fusimotor fibre by their effect on the dynamic response of muscle spindle primary endings, *Quarterly Journal of Experimental Physiology*, 47, 324-333.

McCloskey, D. I., 1981, Corollary discharges: motor commands and perception, in: *Handbook of Sensory Physiology: The Nervous System: Motor Control*. J. M. Brookhart, ed., pp. 1415.

Rymer, W. Z., and D'Almeida, A., 1980, Joint position sense: the effects of muscle contraction, *Brain*, 103, 1-22.

Taylor, J. L., and McCloskey, D. I., 1992, Detection of slow movements imposed at the elbow during active flexion in man, *Journal of Physiology,* 457, 503-513.
Vallbo, Å. B., 1970, Discharge patterns in human muscle spindle afferents during isometric voluntary contractions, *Acta Physiologica Scandinavica,* 80, 552-556.
Wise, A. K., Gregory, J. E., and Proske, U., 1996, The effects of muscle conditioning on movement detection thresholds at the human forearm, *Brain Research,* 735, 125-130.
Wise, A. K., Gregory, J. E., and Proske, U., 1998, Detection of movements of the human forearm during and after co-contractions of muscles acting at the elbow joint, *Journal of Physiology,* 508, 325-330.
Wise, A. K., Gregory, J. E., and Proske, U., 1999, The responses of muscle spindles to small, slow movements in passive muscle and during fusimotor activity, *Brain Research,* 821, 87-94.

12

PROPRIOCEPTION AND JOINT PATHOLOGY

Kathryn M. Refshauge*

ABSTRACT

Many problems consequent to joint pathology are attributed to impaired proprioception. However, proprioception is not consistently impaired by pathology, even pathology that severely affects the articular surface, and intriguingly, there may be a deficit in some aspects of proprioception but not others. Furthermore, deficits in proprioception appear to affect specific planes of movement in joints with more than one degree of freedom. For such knowledge to be useful for rehabilitation we need further insight into issues including: the magnitude of proprioceptive impairment that is clinically relevant, normal variability in sensibility for each proprioceptive task, and the relationship between proprioception and function.

INTRODUCTION

Proprioception encompasses a diverse group of sensations, including sense of movement, joint position sense, sensations related to muscle force and to muscle contractions, and sensations of body posture and size (Gandevia, 1996). A wide variety of tests has therefore been used to examine some of these sensations after injury. However, it is unclear whether there is any relationship between these tests and therefore whether one test is adequate to make judgements about general proprioceptive status after injury. A lack of relationship could potentially account for the conflicting results found for the effect of injury on proprioception.

Contributing Afferent Classes

Most work on patient populations has been directed at the sensations of movement and position. These sensations have been studied in few musculoskeletal injuries, however, a theoretical framework can be developed from which to predict the effect on proprioception of those pathologies not yet studied. It is currently understood that the three classes of afferent fibre (cutaneous, muscle and tendon, and joint capsule and

* School of Physiotherapy, The University of Sydney, Sydney, NSW 2141 Australia and Prince of Wales Medical Research Institute, Sydney, NSW 2031, Australia. Email: k.refshauge@fhs.usyd.edu.au

Sensorimotor Control of Movement and Posture
Edited by Gandevia *et al.*, Kluwer Academic/Plenum Publishers, 2002

ligament afferent fibres) all contribute to proprioceptive sensibility, but their relative importance is unclear. Muscle afferents may provide the most useful proprioceptive information at the large joints of the body. Joint receptors play a less important proprioceptive role in these large joints, instead signalling extremes of range to prevent joint damage, increase in pressure from intra-articular swelling, and perhaps nociception (e.g. Proske et al., 1988). The role of the cutaneous receptors is unclear, but is unlikely to be affected by most musculoskeletal pathology. It could be argued, therefore, that if joint receptors play a minor role in proprioception, injury to the joint structures would not cause a functionally relevant loss of proprioceptive acuity. However, the contribution of each afferent class to proprioceptive sensibility may not be the same at all joints, and because in many human joints the muscles insert into the joint capsule, "normal" proprioception may require "normal" input from all afferent classes.

Normal Acuity

The effect of injury on proprioception should be interpreted in light of normative data and an understanding of the factors that normally affect proprioceptive acuity. Goldscheider (1889) was the first to study thresholds for perception of movement, and measured acuity at most human joints using crude test methods. Since then, various workers have studied movement perception at individual joints (Fig. 1). Proprioceptive acuity improves with increasing velocity, decreases with increasing age and, in general, very small movements can be detected at all joints, except at the finger and the toe (Fig. 1A, B). It is not yet known what magnitude of decrement in acuity is of clinical relevance, particularly because normative data are somewhat variable.

INJURY

Acute Pathology

The effect of injury on proprioception probably depends on a number of factors including the anatomical structures affected and the phase of the injury. Most studies have focussed on proprioceptive impairment in the chronic phase of the pathology rather than the acute phase. However, the little work available on acute injury suggests that proprioception could be enhanced in an acutely inflamed, painful and swollen joint. Experimentally induced swelling in the dog and cat, even with small amounts of fluid, caused increased discharge frequency in joint receptors and recruitment of afferents that were previously silent within the medial articular nerve (Ferrell et al., 1986). Also, experimental distension of healthy human joints improved some aspects of proprioception but not others (Ferrell, et al., 1986). The effect of pain on proprioception has not been studied, although the increased discharge in the pain state may enhance proprioception sensibility. These findings suggest that at least some aspects of proprioception might be enhanced in acutely swollen joints.

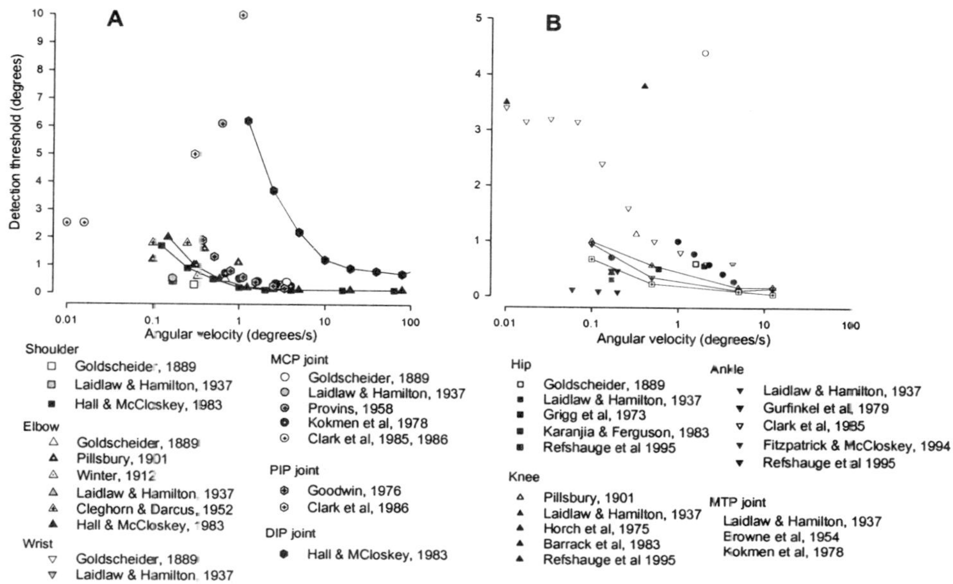

Figure 1. Summary of reported detection thresholds for sense of movement. Data are shown here from all comparable work. The left panel (A) shows data from joints in the upper limb, and the right panel (B) presents data from joints in the lower limb. (From Refshauge et al., 1995.)

Chronic Pathology

Osteoarthritis

Osteoarthrits (OA) impairs joint position sense at the knee (e.g. Barrett, 1991), despite the lack of correlation between proprioceptive acuity and severity of disease (Koralewicz and Engh, 2000). Joint position sense is significantly worse in OA knees than in either the contralateral unaffected knee or healthy controls (Koralewicz and Engh, 2000; Sharma, 1999). Interestingly, an isolated finding suggests that proprioceptive acuity in both knees in subjects with unilateral OA was worse than in knees of healthy controls (Koralewicz and Engh, 2000).

Joint Replacements

Total replacement of the joint for OA has generally, but not consistently, been shown to improve proprioception at both the knee and the hip. Different types of knee replacement improved joint position sense although it is unclear whether acuity was restored to normal healthy levels (Warren et al., 1993, cf. Barrett et al., 1991). However, such an improvement has not always been found after total knee arthroplasty, some authors finding no improvement (Barrack et al., 1983) and others finding that, in fact, proprioceptive acuity in replaced knees was worse than in unreplaced knees (Fuchs et al.,

1999). Even in those subjects in whom proprioceptive acuity deteriorated after surgery, replaced knees had an excellent clinical outcome (Pap et al., 2000) and retained normal performance in balance tasks (McChesney and Woollacott, 2000), suggesting that a minor deficit in proprioception is not necessarily related to loss of function. After total joint replacement at the hip, joint position sense and sense of movement returned to the level of age-matched controls in the abduction-adduction plane (Stender and Drowatzky, 1994) and to the level of the contralateral unaffected limb in the flexion-extension plane (Ishii et al., 1999).

The mechanism by which either the deficit or any subsequent restoration of normal acuity occurred is unclear. It is possible that joint receptors provide redundant input at the large hip and knee joints, because their removal in total joint arthroplasty did not consistently impair acuity. Alternatively, the joint receptors may have provided important abnormal input, and surgical removal of this abnormal input enhanced coherence of the residual signals. It is also possible, however, that either joint instability or the deformity from the pathology prior to surgery led to abnormal muscle input, and the surgery restored normal muscle alignment, resulting in normalised proprioceptive signals.

Instability

The effect of instability on proprioception has been studied in different joints for a variety of pathologies, including anterior shoulder instability, anterior cruciate ligament (ACL) insufficiency at the knee, and recurrent ankle inversion sprains.

Shoulder Instability. A deficit has consistently been found in both the senses of movement and position sense in unstable shoulders compared with the healthy contralateral shoulder and with controls (e.g. Lephart et al., 1994), although Jerosch and Thorwesten (1998) found joint position sense to be affected in some positions only. After surgical repair, acuity returned to the level of the healthy contralateral shoulder (Warner et al., 1996). Although many studies found the improvement in the early post-operative period, Machner et al (1998) found that recovery was gradual, and normal acuity was not achieved until 18 months after repair. Other proprioceptive tasks, such as an active pointing and matching task (Forwell and Carnahan, 1996.), and reflex contraction latency were not affected (Wallace et al., 1997).

Anterior Cruciate Ligament Injury (ACL). Proprioception after ACL injury has been extensively investigated. Various proprioceptive tasks, including sense of movement and joint position sense have been measured in ACL deficient knees and after surgical reconstruction.

Sense of movement was impaired in ACL deficient knees compared with both the contralateral uninjured knee and a healthy control group (e.g. Roberts et al., 1999), and the deficit was significantly worse in the mid-range (Borsa et al., 1997). Again, an isolated finding suggests that in subjects with unilateral ACL insufficiency there is a deficit in the healthy contralateral knee compared with age-matched controls (Roberts et al., 1999). These proprioceptive deficits were not restored by rehabilitation (Fremery et al., 2000), although Friden et al. (1997) found that the deficit resolved in the chronic phase. The effect of surgical reconstruction is unclear, there being evidence that reconstruction restored sense of movement to normal levels (Risberg et al., 1999), improved acuity but not to normal levels (Roberts et al., 2000), or improved acuity only at end of range but not in the mid range (Fremery et al., 2000), although the reverse was found by Lephart et al. (1992).

Joint position sense was found to be unaffected compared with healthy controls by most authors (e.g. Roberts et al., 1999 cf. Fischer-Rasmussen, 2000). Reconstruction conferred no proprioceptive benefit (Co et al., 1993), probably because there was no deficit to address. However, joint position sense when tested with a visual analogue scale (VAS) is less clear: acuity was significantly worse in ACL deficient knees than healthy control knees by some (Barrett, 1991), but not others (Roberts et al., 1999).

When both sense of movement and joint position sense were measured in the same subjects, there was no relationship between the sensations. Sense of movement was impaired in ACL deficient knees compared with both healthy controls (Friden et al., 1997) and the contralateral unaffected knee (Corrigan et al., 1992), while joint position sense was unaffected even in the presence of instability (Roberts et al., 1999).

Thus, in the ACL deficient knee there appears to be a deficit in sense of movement but not joint position sense. These findings support McCloskey's hypothesis (1973) that signals are processed differently for position sense and sense of movement. The deficit in sense of movement may arise from the loss of input from the ACL, which is thought to have a strong proprioceptive role via the fusimotor system (Johansson et al., 1991), or from abnormal input from muscles (e.g. hamstrings) that have adaptively lengthened (Beard et al., 1993). The effect of reconstructive surgery on proprioception is generally unclear. Any improvement noticed may be due to the increased stability of the joint (Iwasa et al., 2000), although proprioceptive acuity correlated poorly with mechanical stability (Barrett, 1991). Alternatively, improved joint kinematics may normalise input from the muscles, or the repair or replacement of the ACL may normalise input from joint receptors. Of particular interest is that proprioception correlated highly with patient satisfaction concerning surgical outcome (Barrett, 1991) despite the lack of clarity about the impact of proprioception on function (Barrett, 1991 cf. Beard, 1993). Finally, Corrigan and colleagues (1992) found that proprioceptive acuity correlated well with hamstrings:quadriceps strength ratio, leading the authors to suggest that either quadriceps atrophy or hamstrings development may be an adaptive mechanism to maximise proprioceptive signalling.

Ankle Inversion Sprain. Sense of movement and joint position sense have been examined in both the inversion-eversion and plantarflexion-dorsiflexion planes in subjects with recurrent ankle sprain. Sense of movement was shown to be impaired in the inversion-eversion plane after recurrent sprain compared with healthy controls by both studies to investigate this issue (Lentell, et al., 1995; Refshauge, et al., 2000b) although there was no correlation with mechanical stability (Konradsen, et al., 1998). However, in the plantarflexion-dorsiflexion plane, sense of movement was impaired by some authors (Forkin et al., 1996) but not others (Refshauge et al., 2000a).

Findings for joint position sense in the inversion-eversion plane are unclear: a significant impairment was found by some (e.g. Konradsen and Magnusson, 2000) although not by others (Boyle and Negus, 1998). Joint position sense in the plantarflexion-dorsiflexion plane has not been investigated at the ankle after sprain.

These findings suggest that proprioception is usually impaired in the inversion-eversion plane, but that the deficit is not generalised for either plane of movement or proprioceptive task, but rather is specific to the anatomical structures involved in the sprain. These structures include the lateral ligament, the peroneal retinaculum and perhaps the peroneal muscles. These structures resist inversion and not plantarflexion forces. Damage to these structures may therefore cause impairment only in proprioceptive tasks specific to the inversion-eversion plane of movement.

CONCLUSIONS

The clinical literature is generally inconclusive, nevertheless conclusions can be tentatively proposed by considering all findings together. In general, proprioception is worse in joints with degenerative disease or joints that are unstable, and is commonly improved, sometimes to normal, after surgical replacement or repair. Interestingly, different aspects of proprioception appear not to be related, but rather specific impairments may exist according to the structures affected by the pathology. Therefore, identification of a deficit is likely to depend on the proprioceptive variable measured.

It is interesting that, even when a proprioceptive deficit exists for a patient group, up to half the subjects may not have a deficit (e.g. Lentell et al., 1995). That is, pathology of a joint structure such as the ACL or the lateral ligament complex of the ankle, does not necessarily incur a proprioceptive deficit unless other conditions that are as yet unknown, exist. It may be that damage to a ligament does not impair proprioception if the joint remains stable with normal kinematics. However, given the lack of correlation between mechanical stability and proprioception, at least at the knee and ankle, it may be *functional* instability (giving way of the joint, and the perception of instability) that is the major concomitant factor. If this were the case, we could predict for injuries not yet studied that if the injury involves only an articular structure, but does not disturb the mechanics of the joint, then proprioception is unlikely to be affected. Finally, given the lack of demonstrated relationship between proprioceptive sensations, performance on one proprioceptive test cannot be generalised to tests of other proprioceptive modalities.

REFERENCES

Barrack, R. L., Skinner, H. B., Cook, S. D., and Haddad, R. J., 1983, Effect of articular disease and total knee arthroplasty on knee joint-position sense, *Journal of Neurophysiology*, 50, 684-687.

Barrett, D. S., 1991, Proprioception and function after anterior cruciate reconstruction, *Journal of Bone and Joint Surgery,* 73B, 833-837.

Beard, D. J., Kyberd, P. J., Fergusson, C. M., and Dodd, C. A. F., 1993, Proprioception after rupture of the anterior cruciate ligament, *Journal of Bone and Joint Surgery,* 75B, 311-315.

Borsa, P. A., Lephart, S. M., Irrgang, J. J., Safran, M. R., and Fu, F. H., 1997, The effects of joint position and direction of joint motion on proprioceptive sensibility in anterior cruciate ligament-deficient athletes, *American Journal of Sports Medicine,* 25, 336-340.

Boyle, J., and Negus, V., 1998, Joint position sense in the recurrently sprained ankle, *Australian Journal of Physiotherapy,* 44, 159-163.

Co, F. H., Skinner, H. B., and Cannon, W. D., 1993, Effect of reconstruction of the anterior cruciate ligament on proprioception of the knee and the heel strike transient, *Journal of Orthopaedic Research,* 11, 696-704.

Corrigan, J. P., Cashman, W. F., and Brady, M. P., 1992, Proprioception in the cruciate deficient knee, *Journal of Bone and Joint Surgery,* 74B, 247-250.

Ferrell, W. R., Nade, S., and Newbold, P. J., 1986, The interrelation of neural discharge, intra-articular pressure, and joint angle in the knee of the dog, *Journal of Physiology,* 373, 353-365.

Forkin, D., Koczur, C., Battle, R., and Newton, R. A., 1996, Evaluation of kinesthetic deficits indicative of balance control in gymnasts with unilateral chronic ankle sprains, *Journal of Orthopaedic and Sports Physical Therapy,* 23, 245-250.

Forwell, L. A., and Carnahan, H., 1996, Proprioception during manual aiming in individuals with shoulder instability and controls, *Journal of Orthopaedic and Sports Physical Therapy*, 23, 111-119.

Fremery, R., Lobenhoffer, P., Zeichen, J., Skutek, M., Bosch, U., and Tscherne, H., 2000, Proprioception after rehabilitation and reconstruction in knees with deficiency of the anterior cruciate ligament: a prospective, longitudinal study, *Journal of Bone and Joint Surgery*, 82, 801-806.

Friden, T., Roberts, D., Zatterstrom, R., Lindstrand, A., and Moritz, U., 1997, Proprioception after an acute knee ligament injury: a longitudinal study on 16 consecutive patients, *Journal of Orthopaedic Research*, 15, 637-644.

Fuchs, S., Thorwesten, L., and Niewerth, S., 1999, Proprioceptive function in knees with and without total knee arthroplasty, *American Journal of Physical Medicine and Rehabilitation*, 78, 39-45.

Ishii, Y., Tojo, T., Terajima, D., Terashima, S., and Bechtold, J. E., 1999, Intracapsular components do not change hip proprioception, *Journal of Bone and Joint Surgery*, 75B, 546-550.

Iwasa, J., Ochi, M., Adachi, N., Tobita, M., Katsube, K., and Uchio, Y., 2000, Proprioceptive improvement in knees with anterior cruciate ligament reconstruction, *Clinical Orthopaedics and Related Research*, 381, 168-176.

Johansson, H., Sjolander, P., and Sokja, P., 1991, A sensory role for cruciate ligaments, *Clinical Orthopaedics*, 268, 161-175.

Konradsen, L., and Magnusson, P., 2000, Increased inversion angle replication error in functional ankle instability, *Knee Surgery, Sports Traumatology, Arthroscopy*, 8, 246-251.

Konradsen, L., Olesen, S., and Hansen, H. M., 1998, Ankle sensorimotor control and eversion strength after acute ankle inversion injuries, *American Journal of Sports Medicine, 26*, 72-77.

Koralewicz, L. M., and Engh, G. A., 2000, Comparison of proprioception in arthritic and age-matched normal knees, *Journal of Bone and Joint Surgery*, 82A, 1582-1588.

Lentell, G., Baas, B., Lopez, D., McGuire, L., Sarrels, M., and Snyder, P., 1995, The contributions of proprioceptive deficits, muscle function, and anatomic laxity to functional instability of the ankle, *Journal of Orthopaedic and Sports Physical Therapy*, 21, 206-215.

Lephart, S. M., Warner, J. P., Borsa, P. A., and Fu, F. H., 1994, Proprioception of the shoulder in normal, unstable and postsurgical individuals, *Journal of Shoulder and Elbow Surgery*, 3, 371-380.

McChesney, J. W., and Woollacott, M. H., 2000, The effect of age-related declines in proprioception and total knee replacement on postural control, *Journals of Gerontology, Series A, Biological Sciences and Medical Sciences*, 55, M658-666.

Pap, G., Meyer, M., Weiler, H. T., Machner, A., and Awiszus, F., 2000, Proprioception after total knee arthroplasty: a comparison with clinical outcome, *Acta Orthopaedica Scandinavica*, 71, 153-159.

Proske, U., Schaible, H.-G., and Schmidt, R. F., 1988, Joint receptors and kinaesthesia, *Experimental Brain Research*, 72, 219-224.

Refshauge, K. M., Kilbreath, S. L., and Raymond, J., 2000, The effect of recurrent ankle sprain and taping on proprioception at the ankle, *Medicine and Science in Sports and Exercise*, 32, 10-15.

Refshauge, K. M., Kilbreath, S. L., Raymond, J., Heijnen, I., and Pengel, L., 2000, *Inversion-eversion proprioception in recurrent ankle sprains - the effect of taping*, Paper presented at the 6th International Physiotherapy Congress.

Risberg, M. A., Beynnon, B. D., Peura, G. D., and Uh, B. S., 1999, Proprioception after anterior cruciate ligament reconstruction with and without bracing, *Knee Surgery, Sports Traumatology, Arthroscopy*, 7, 303-309.

Roberts, D., Friden, T., Stomberg, A., Lindstrand, A., and Moritz, U., 2000, Bilateral proprioceptive defects in patients with a unilateral anterior cruciate ligament reconstruction: a comparison between patients and healthy individuals, *Journal of Orthopaedic Research*, 18, 565-571.

Roberts, D., Friden, T., Zatterstrom, R., Lindstrand, A., and Moritz, U., 1999, Proprioception in people with anterior cruciate ligament-deficient knees: comparison of symptomatic and asymptomatic patients, *Journal of Orthopaedic and Sports Physical Therapy*, 29, 587-594.

Sharma, L., 1999, Proprioceptive impairment in knee osteoarthritis, *Rheumatic Diseases Clinics of North America*, 25, 299-314.

Stender, B. L., and Drowatzky, J. N., 1994, Joint position sense in subjects with total hip replacements: The possible role of muscle afferents, *Clinical Kinesiology*, 48, 10-14.

Wallace, D. A., Beard, D. J., Gill, R. H., Eng, B., and Carr, A. J., 1997, Reflex muscle contraction in anterior shoulder instability, *Journal of Shoulder and Elbow Surgery*, 6, 150-155.

Warner, J. J. P., Lephart, S. M., and Fu, F. H., 1996, Role of proprioception in pathoetiology of shoulder instability, *Clinical Orthopaedics and Related Research*, 330, 35-39.

SECTION III

Afferent Contributions to Balance and Posture

To introduce the subject J. G. Colebatch (Chapter 13) provides an interesting, historically-oriented account of developments in our understanding of labyrinth function. He points out that much of the detailed knowledge of this system and its neural substrate derived from animal studies. However, problems in maintenance of posture in a four-legged animal are likely to differ from those in an upright human. Indeed, it has been suggested that human postural control is learned, not reflexive in origin. Furthermore, in humans vestibulo-ocular reflexes are stronger than vestibulo-collic reflexes. A better understanding of labyrinthine function in humans has been hampered by the lack of availability, until recently, of simple, non-invasive tests. While it is relatively straightforward to test semi-circular canal function using caloric stimulation and sudden head movements in the appropriate plane, testing otolith function has proved more difficult. Loud sounds have recently been proposed as a means of stimulating the saccule. This kind of approach is important because of the clinical implications and because it promises to improve our understanding of human labyrinth function.

T. Inglis (Chapter 14) discusses one particular sensory channel in the control of balance, somatosensory information arising from the sole of the foot. He points out that cooling of the sole increases body sway and increases the response to galvanic vestibular stimulation (GVS). Measurement of vibration thresholds indicate a range of values depending on stimulus frequency and the age of the subjects, with values lowest in the ball and arch of the foot. Microneurographic recordings indicate the presence of the expected four classes of cutaneous mechanoreceptor afferents. A question posed by this work was, what proportion of the afferents was truly cutaneous and how many were ligamentous (plantar fasciae) or even muscular in origin? Particularly the higher vibration frequencies could recruit afferents of non-cutaneous origin. Do these, too, play a part in the control of balance? It seems likely.

The question of what GVS actually does is addressed by R. Fitzpatrick (Chapter 15). Based on the structural arrangement of the otoliths and semicircular canals, it was argued that the effect of GVS on the utricle is likely to be greater than on the saccule. This is because of the asymmetric striola in the utricle, leading to perception of a laterally-directed acceleration. For the semi-circular canals, the net effect of GVS is to indicate head roll, as signalled by both vertical canals. The largest vestibulo-ocular response to canal stimulation is a torsional response with head-roll. Postural adjustment consists, predominantly, of body sway towards the side of the head with the anode. The direction of head movement is independent of head-on-trunk position. These predictions are well borne out by available observations.

B. L. Day (Chapter 16) proposes that for normal operation of the system all three major sensory input channels, vestibular, visual and somatosensory, have to be functional. Postural responses can be triggered by disturbances of each channel in isolation. The question is posed, how do these interact? Studies of a subject lacking somatosensory functions showed that with their eyes shut, galvanic vestibular stimulation (GVS) generated head and body sway responses, ten times larger than in normal subjects. Even with the eyes open, responses were larger. It points to the importance of the somatosensory system in body correctional responses to vestibular stimulation. In normal subjects, degrading the quality of the visual information produced graded responses to GVS. It is argued that when one sensory channel is disturbed, there is a gain change in the vestibular-postural reflex as well as a delayed, feedback adjustment of the response. The gain change, which manifests itself earlier, is the result of a sensitivity-tuning of the system based on the amount and quality of information available from other channels.

Further experimental observations on subjects with vestibular loss are provided by F. Horak (Chapter 17). Such subjects showed larger head accelerations and displacements during walking. The vestibular system was not used for postural responses to a sudden slip or trip since these subjects showed normal responses. But they did show disruption of upper body control when trying to balance over a narrow support base. They lacked an adequate 'hip-strategy'. During forward-backward or rotational movements at increasing sinusoidal frequencies, the vestibular-loss subjects had difficulty in uncoupling head and trunk movement from movement of the hips, but they showed reasonable control of their centre of mass. When subjects closed their eyes, or they were unable to steady themselves with finger pressure on a support, their performance deteriorated, particularly at higher frequencies. It remains uncertain why some subjects are able to compensate better than others.

A global model, incorporating the vestibular system as part of a postural control strategy is put forward by T. Mergner (Chapter 18). He introduces the concept of sensory fusion. The information provided by two different sensory systems, each signalling its own distinct response to environmental stimuli, is combined to provide a third sensation, unique in its own right. Simple postural adjustments can be achieved by vestibulo-spinal and cervico-spinal reflexes. However such a system does not cope well with external disturbing forces. Additional sensory cues are provided by somatosensory inputs signalling centre of pressure on the feet and proprioceptive inputs providing joint angle information. Internal representations of external stimuli are used as set points for local proprioceptive loops, as a means of achieving a particular posture. The internal representation can be applied to different joints depending on the momentary demands of body geometry. Some support for this kind of model is provided by observations on normal subjects and on subjects with vestibular loss.

13

CONSEQUENCES AND ASSESSMENT OF HUMAN VESTIBULAR FAILURE

Implications for Postural Control

James G. Colebatch*

ABSTRACT

Labyrinthine afferents respond to both angular velocity (semicircular canals) and linear acceleration (otoliths), including gravity. Given their response to gravity, the otoliths are likely to have an important role in the postural functions of the vestibular apparatus. Unilateral vestibular ablation has dramatic effects on posture in many animals, but less so in primates. Nevertheless, bilateral vestibular lesions lead to disabling symptoms in man related to disturbed ocular and postural control and impaired perception of slopes and accelerations. While seimicircular canal function can be assessed through its effects on vestibular ocular reflexes, assessment of otolith function in man has traditionally been much more difficult. Recent definition of a short latency vestibulocollic reflex, activated by sound and appearing to arise from the saccule, shows promise as a new method of non-invasive assessment of otolith function.

INTRODUCTION

Much of the basic definition of vestibular function occurred in the late 19th century. Marked postural alterations as well as effects on the eyes were recognised as characteristic consequences of labyrinthine lesions (Camis, 1930). Unilateral lesions led to rotation of the trunk and head to the side of the lesion. In the case of the rabbit, Magnus (1926) stated that neck rotation following a unilateral labyrinthine lesion persisted for as long as the animal lived. The effects of unilateral and bilateral labyrinthectomy were found to be less marked in primates than in other animals (Dow, 1938) and unilateral lesions in man had few overt postural effects (Cairns and Brain,

* Department of Neurology and Clinical School, Prince of Wales Hospital, High Street, Randwick NSW 2031, Sydney, Australia. Email: j.colebatch@unsw.edu.au

1933). Sherrington regarded the role of the labyrinth as akin to that of muscle receptors in maintaining posture. Whereas muscle receptors maintained the posture of a limb, labyrinthine receptors acted to orient the animal with respect to gravity. He was also aware of the "eyeball reflex" which maintained the direction of gaze despite movement of the head (Sherrington, 1947). The postural role of the labyrinth was evident in its role in righting reflexes. A striking example is the ability of a (blind-folded) cat to right itself mid air, a well-known natural phenomenon (Fig. 1). The pattern of neural projections from the vestibular nuclei are consistent with a role in oculomotor and postural control with pathways ascending to the oculomotor nuclei, projections to the cerebellum and descending pathways to the spinal cord (Brodal, 1981). The principle descending projections travel in the medial (MVST) and lateral vestibulospinal tracts (LVST). The LVST travels ipsilaterally for the length of the cord and originates from the lateral vestibular nucleus. The MVST descends to the cervical and thoracic levels and originates from the lateral, medial and descending nuclei (Wilson and Peterson, 1978). Intracellular recordings have confirmed that, following stimulation of the vestibular nuclei, the largest EPSPs and IPSPs are evoked in cervical motoneurones (Wilson and Yoshida, 1969). Selective stimulation of individual semicircular canal afferents produces characteristic neck movements which approximate the plane of the stimulated canal, similar to the eye movements evoked by the same stimuli (Suzuki and Cohen, 1964)

Figure 1. Cat's ability to right themselves is regarded as an immutable part of life as we know it. (Reproduced with permission from Cathy Wilcox, SMH/The Age.)

THE VESTIBULAR APPARATUS AND POSTURE

It has long been recognised that the vestibular apparatus consists of two main classes of receptors: the three semicircular canals (lateral, superior and posterior) and the two otolith organs (utricle and saccule). The natural stimuli which excite these two classes of receptors are very different. The semicircular canals are excited by angular velocity or acceleration, whereas the otoliths respond to linear acceleration, including gravity. With considerable prescience, Breuer in 1875 wrote that "labyrinthine reflexes can be divided into two classes: those that are reactions to movements (functions of the canals) and those which are a reaction to position (function of the otoliths) (quoted in Camis, 1930). Thus vestibular influences may act to stabilise a given position in response to imposed disturbances (phasic reflexes) or may act to define the desired position (set point) with respect to gravity. The receptors capable of mediating responses specifically dependent upon gravity are the otoliths, which are characterised by their responses to tilt (Fernandez et al., 1972). The initial reflex response to falling in both man and cats appears to be mediated by the otoliths (Melvill et al., 1971; Watt, 1976). Selective lesions of the individual otolith organs are technically difficult (Uchino et al., 1997) and this has prevented corresponding behavioural studies in animals. Magnus (1926) deduced that the saccules were responsible for achieving a symmetrical position of the head in the vertical plane (roll plane) and the utricle for establishing head position in pitch. The short latency responses to selective stimulation suggest the opposite: saccular stimulation has effects consistent with a role in maintaining the head in pitch and utricular with a role in responses to head tilt (Uchino et al., 1997). Wilson and Peterson (1978) illustrate the function of vestibulospinal and vestibulocollic reflexes in quadrupeds with the example of an animal suddenly pitched forwards. Under these circumstances semicircular canal receptors are activated by the angular acceleration of the head and otolith receptors are activated by linear acceleration of the head. Phasic vestibulospinal reflexes cause contraction of the dorsal neck muscles, extension of the forelimbs and flexion of the hindlimbs, thereby acting to maintain the position of the animal's head (and postural stability). Similar responses can be seen in humans on all fours (Roberts, 1995).

In bipedal posture, as typical a characteristic of humans as is fractionated movements of the fingers, the role of the forelimbs is likely to be very different from quadripedal stance. Responses in four-legged animals may therefore be poor models for postural control in upright man. Roberts (1995) considers human posture primarily a learned "skill" rather than a series of reflexes per se. He gives an example of forced hopping in which a hop occurs with a lateral tug but not if the surface tilts to the same degree. Such responses seem unlikely to be explicable in terms of classical reflex mechanisms and are probably better viewed as "learned" or "automatic" responses. While Roberts emphasises the importance of subjects' recognition of the "gestalt" for postural responses, his observations may be analogous to the motor strategies and rules proposed for gait (Prochazka, 1996; Prochazka and Yakovenko, Chapter 41). Vestibular reflexes do show considerable alterations in response to changes in postural task (Welgampola and Colebatch, 2001) but short-latency vestibular influences may be of less importance in postural control than the information the discharges provide about the subject's position and motion. For control of certain aspects of head posture, even though vestibulocollic reflexes are weak, the performance of subjects with intact vestibular apparatuses is superior to those with severe vestibulopathy (Guitton et al., 1986).

SYMPTOMS OF VESTIBULAR FAILURE IN HUMANS

Early observations of the effects of vestibular failure in man were made in deaf mutes. James (1882) observed that many of his deaf-mutes reported no vertigo during or after rotation, became disoriented under water and did not report any sensation of sea-sickness in response to rough weather (Hess, 1996). Barany found no clinical difference between deaf-mutes with preserved vestibular function and those without it, leading him to conclude initially that bilateral absence of vestibular function in humans was of no consequence (Hess, 1996). In the 1930s and '40s the effects of vestibular failure in humans came to be recognised. A particularly clear report was published by JC, a doctor treated with streptomicin for presumed tuberculous arthritis. JC's (1952) report included postural disturbances, particularly when vision was excluded, ocular disturbances ("oscillopsia") and perceptual changes, the latter including difficulty in detecting movement and sloping ground. The difficulty in detecting slopes may be related to the lack of response shown by patients with vestibular failure to tilts of their support surface (Martin, 1967). JC noted that "if I turned my head from side to side while looking forward I had the sensation that the room turned around me, rather than I was turning around in the room" and "I learned that I must stand still in order to read the lettering of a sign". Gait was an effort and more dependent upon visual input – "it was imperative to restrain the impulse to look up when an airplane passed overhead until I could brace myself against a railing or a tree". A particular difficulty was trying to walk in the dark: "I still find myself almost helpless in the dark". Despite this he was subsequently able to live a near-normal life, including playing tennis socially. He did not again suffer from seasickness and was unaware of acceleration in a lift (Hess, 1996). These very precise observations give us important insight into the role of the vestibular apparatus in man. They indicate that complex motor performance is possible in the absence of vestibular function. Vestibular input appears to be important under specific circumstances; for posture, for example, when vision is not available.

ASSESSMENT OF VESTIBULAR FUNCTION IN MAN

Assessment of vestibular function in man has traditionally relied mainly upon measures of canal function acting upon the eyes. Vestibulocollic reflexes in intact humans are weaker than in the cat and much weaker than vestibulo-ocular reflexes (Guitton et al., 1986; Gresty, 1987). Both caloric stimulation and rotatory tests measure the vestibulo-ocular reflex evoked from the lateral canal (Fetter, 2000). Conventional clinical examination of patients with bilateral vestibular failure may reveal little beyond a minor degree of gait ataxia, but abnormalities of both ocular and postural control can be shown with specific testing. A sudden imposed transient head rotation activating the lateral canal is an important bedside and laboratory test of vestibular function (Halmagyi and Curthoys, 1988). Transient head rotations applied in the planes of the other canals can be used to assess their function (Cremer et al., 1998). Camis (1930) stated that ocular counter-rolling was the only means then available of measuring the effect of saccular and utricular stimulation. Measurement of otolith function has remained problematical: many initial abnormalities may attenuate with increasing time following a lesion and are insensitive to bilateral abnormalities (Halmagyi et al., 1990). The subjective visual vertical is a method of assessing unilateral utricular impairment (Halmagyi and Curthoys, 1999). Recently Colebatch et al. (1994) reinvestigated the effects of loud click stimulation on vestibular receptors. They were able to show a short latency projection to the sternocleidomastoid (SCM) muscle in the neck that was vestibular-dependent and which they proposed might be the result of saccular stimulation. Single unit studies of vestibular nerve afferents support a role for the saccule in the response to loud sounds (McCue and Guinan, 1995; Murofushi et al., 1995). The reflex appears to be mediated by a brief period of inhibition (Colebatch and Rothwell, 1993). Very similar effects can be shown using intracellular recording and selective saccular stimulation in the cat (Kushiro et al., 1999), suggesting that this pathway has been preserved despite the different role of the SCM with respect to gravity in the two species. Additional methods of evoking vestibular-dependent reflexes in this same pathway have since been published (Halmagyi et al., 1995; Watson and Colebatch, 1998). The functional role of these short latency pathways may not be major (Wilson and Schor, 1999) but it is now possible to assess both canal and otolith function in man non-invasively. Saccular function, so assessed, is often preserved despite impairment of lateral canal function (Halmagyi and Colebatch, 1995) and vestibular failure should not be assumed unless both types of test show absent responses. Better delineation of vestibular lesions should allow more accurate definition of vestibular function in humans. In particular it may be possible to deduce the physiological roles of the semicircular canals and otoliths from the deficits that are associated with their selective impairment.

REFERENCES

Brodal A., 1981, *Neurological Anatomy*, 3rd ed., Oxford University Press, New York, pp. 470-488.

Cairns, H., and Brain, W. R., 1933, Aural vertigo: treatment by division of the eighth nerve. *Lancet* i: 946-952.

Camis, M., 1930, *The Physiology of the Vestibular Apparatus*, Translated by Creed, R.S., Oxford University Press, Oxford, pp. 47-68.

Colebatch, J. G., Halmagyi, G. M., and Skuse, N. F., 1994, Myogenic potentials generated by a click-evoked vestibulocollic reflex, Journal of Neurology, Neurosurgery and Psychiatry, 57, 190-197.

Colebatch, J. G. and Rothwell, J. C., 1993, Vestibular-evoked EMG responses in human neck muscles, *Journal of Physiology*, 473, 18P.

Cremer, P. D., Halmagyi, G. M., Aw, S. T., Curthoys, I. S., McGarvie, L. A., Todd, M. J., Black, R. A., and Hannigan, I. P., 1998, Semicircular canal plane head impulses detect absent function of individual semicircular canals, *Brain*, 121, 699-716.

Dow, R. S., 1938, The effects of unilateral and bilateral labyrinthectomy in monkey, baboon and chimpanzee, *American Journal of Physiology*, 121, 392-399.

Fernandez C., Goldberg J. M., and Abend, W. K., 1972, Responses to static tilts of peripheral neurons innervating otolith organs of the squirrel monkey, *Journal of Neurophysiology*, 35, 978-997.

Fetter, M., 2000, Assessing vestibular function: which tests, when? *Experimental Brain Research*, 247, 335-342.

Gresty, M., 1987, Stability of the head: studies in normal subjects and in patients with labyrinthine disease, head tremor, and dystonia, *Movement Disorders*, 2, 165-185.

Guitton, D, Kearney, R. E., Werely, N., and Peterson, B. W., 1986, Visual, vestibular and voluntary contributions to human head stabilisation, *Experimental Brain Research*, 64, 59-69.

Halmagyi, G. M., and Colebatch, J. G., 1995, Vestibular evoked myogenic potentials in the sternomastoid muscle are not of lateral canal origin, *Acta Oto-Laryngologica, Supplement*, 520, 1-3.

Halmagyi, G. M., and Curthoys, I. S., 1988, A clinical sign of canal paresis, *Archives of Neurology*, 45, 737-739.

Halmagyi, G. M., and Curthoys, I. S., 1999, Clinical testing of otolith function, *Annals of the New York Academy of Sciences*, 871, 195-204.

Halmagyi, G. M , Curthoys, I. S., and Dai, M. J., 1990, Diagnosis of unilateral otolith hypofunction, *Neurologic Clinics*, 8, 313-329.

Halmagyi G. M., Yavor R. A., and Colebatch J. G., 1995, Tapping the head activates the vestibular system: a new use for the clinical tendon hammer, *Neurology*, 45, 1927-1929.

Hess, K., 1996, Vestibulotoxic drugs and other causes of acquired bilateral peripheral vestibulopathy, in: *Disorders of the Vestibular System*, R.W. Baloh, G.M. Halmagyi, eds., Oxford University Press, New York, pp. 360-373.

J.C., 1952, Living without a balancing mechanism, *New England Journal of Medicine*, 246, 458-460.

Kushiro, K., Zakir, M., Ogawa, Y., Sato, H., and Uchino, Y., 1999, Saccular and utricular inputs to sternocleidomasotid motoneurons of decerebrate cats, *Experimental Brain Research,* 126, 410-416.
Magnus, R., 1926, Some results of studies on the physiology of posture, *Lancet* ii: 531-536, 585-588.
Martin, J.P., 1967, *The Basal Ganglia and Posture*, Pitman, London, pp. 36-51.
McCue, M. P., and Guinan, J. J. Jnr, 1995, Spontaneous activity and frequency selectivity of acoustically responsive vestibular afferents in the cat, *Journal of Neurophysiology,* 74, 1563-1572.
Melvill Jones G., and Watt, D. G. D., 1971, Muscular control of landing from unexpected falls in man, *Journal of Physiology,* 219, 729-737.
Murofushi, T., Curthoys, I. S., Topple, A. N., Colebatch, J. G., and Halmagyi, G. M., 1995, Responses of guinea pig primary vestibular neurons to clicks, *Experimental Brain Research,* 103, 174-178.
Prochazka, A., Proprioceptive feedback and movement regulation, in: *Exercise: Regulation and Integration of Multiple Systems, Handbook of Physiology, Section 12,* L.B. Rowell, J.T. Shepherd, eds, Oxford University Press, New York, pp. 89-217.
Roberts, T. D. M., 1995, *Understanding Balance*, Chapman & Hall, London, pp. 94-174.
Sherrington, C. S., 1947, *The Integrative Action of the Nervous System*, 2nd ed., Yale University Press, New Haven, pp. 335-344.
Suzuki, J. I., and Cohen, B., 1964, Head, eye, body and limb movements from semicircular canal nerves, *Experimental Brain Research,* 10, 393-405.
Uchino, Y., Sato, H., Sasaki, M., Imagawa, M., Ikegami, H., Isu, N., and Graf, W., 1997, Sacculocollic reflex arcs in cats, *Journal of Neurophysiology,* 77, 3003-3012.
Watson, S. R. D., and Colebatch, J. G., 1998, Vestibulocollic reflexes evoked by galvanic stimulation in man, *Journal of Physiology,* 513, 587-597.
Watt, D. G. D., 1976, Responses of cats to sudden falls: an otolith-originating reflex assisting landing, *Journal of Neurophysiology,* 39, 257-265.
Welgampola, M. S., and Colebatch, J. G., 2001, Vestibulospinal reflexes: quantitative effects of sensory feedback and postural task, *Experimental Brain Research,* 139, 345-353.
Wilson, V. J., and Peterson, B. W., 1978, Peripheral and central substrates of vestibulospinal reflexes, *Physiological Reviews,* 58, 80-105.
Wilson, V.J., and Schor, R.H., 1999, The neural substrate of the vestibulocollic reflex: what needs to be learned, *Experimental Brain Research,* 129, 483-493.
Wilson, V.J., and Yoshida, M., 1969, Comparison of effects of stimulation of Deiter's nucleus and medial longitudinal fasciculus on neck, neck, forelimb, and hindlimb motoneurons, *Journal of Neurophysiology,* 32, 743-758.

THE ROLE OF CUTANEOUS RECEPTORS IN THE FOOT

J. Timothy Inglis, Paul M. Kennedy, Cari Wells, and Romeo Chua*

ABSTRACT

Cutaneous receptors in the foot sole appear to contribute to the control of human stance and locomotion. Two approaches were undertaken to establish the characteristics of the receptors in the sole. Psychophysical vibrotactile thresholds (range 25 - 400 Hz) were determined across the unloaded sole in young and elderly subjects. Thresholds were lower in the ball and arch of the sole, than in the heel and toe regions. Elderly subjects demonstrated significantly elevated thresholds for high-frequency vibration. Secondly, microneurographic recordings were made from skin afferents of the unloaded sole in young subjects. Results indicated that while similar types of cutaneous receptors exist in the sole of the foot and hand, there appear to be differences in receptor density and distribution. Our results demonstrate that cutaneous afferent inputs from the foot sole provide useful information for the control of posture and locomotion.

INTRODUCTION

The task of maintaining an upright standing posture in the human involves a complex sensorimotor control system, with somatosensory, vestibular, and visual sensory information all contributing to the control of stance and locomotion (Inglis et al., 1994). While there is little doubt that it is the successful integration of all these inputs that leads to optimal balance control in standing, it has been suggested that somatosensory information from the lower limb appears to play a rather dominant role (Inglis et al., 1994; Fitzpatrick et al., 1994). The exact source of this essential somatosensory information remains to be determined, but recent evidence is accumulating that the glabrous cutaneous receptors from the foot sole may contribute significantly.

A number of different lines of research support this notion. Cooling of the sole of the foot, thereby reducing the input of the cutaneous information from the sole, is associated with an increase in postural sway during quiet stance (Orma, 1957; Asai et al., 1992).

* School of Human Kinetics, The University of British Columbia, Vancouver, British Columbia, Canada.
Email: tinglis@interchange.ubc.ca

Postural sway induced following galvanic vestibular stimulation (Magnusson et al., 1990b) is likewise increased following foot sole cooling. Finally, compensatory stepping reactions following a sudden postural perturbation (i.e. movement of the supporting surface) are also dramatically affected by reduced plantar support information (Perry et al., 2000).

While reduction of cutaneous foot sole information is associated with postural instability, low amplitude vibratory activation (0.2-0.5 mm, 20-80 Hz) of foot sole cutaneous afferents leads to directionally specific postural sway (Kavounoudias et al. 1998, 1999). For example, bilateral vibration of the metatarsal footpads caused sway in a backward direction. Interestingly, the velocity of the induced postural sway depended on the frequency of the vibratory stimulation, with larger velocities of sway following higher frequencies of skin vibration. Recent work by Mergner and colleges has also found that mechanical stimulation of the plantar skin, in the range of natural postural sway (i.e. 0.1-0.4 Hz), evokes sway that is highly correlated with the cutaneous stimuli (Maurer et al., 2001).

While the above studies illustrate the importance of cutaneous information in the control of standing balance, our knowledge of the nature and type of information that is available from the foot sole is largely based on indirect evidence. It seems logical that cutaneous afferents from the foot sole could code for changes in foot pressure by acting as a pressure "map". This map could monitor, much like a sensory array, the pressure changes encountered with movements of the center of pressure across the foot sole during either stance or locomotion. A few studies that have investigated vibrotactile psychophysical thresholds for the plantar skin (Kekoni et al., 1989; Nurse and Nigg, 1999) have found evidence of regional differences across the foot sole, hinting that such a map may exist. However, to assess further the role of cutaneous mechanoreceptors in bipedal stance it seems essential to understand the distribution and behaviour of the sensory receptors of the plantar skin in humans. Microneurographic recordings of cutaneous afferents from the peripheral nerves have provided direct analysis of the functional properties of skin receptors in response to various stimuli (Johansson and Vallbo, 1983; Edin, 2001). However, there are only a limited number of microneurography based studies related to the lower limb, and these have examined the hairy skin of the calf and the glabrous skin isolated to the lateral border of the foot (Vallbo & Hagbarth, 1968; Ribot et al., 1989; Edin, 2001; Trulsson, 2001). Consequently, there is no information about the characteristics of the mechanoreceptors specific to the total foot sole, their distributions, regional variations or densities, or their response characteristics.

To understand the potential contribution of plantar cutaneous afferents, and to see if this region could potentially code for information useful for postural control, two directions of research were undertaken. In the first experiment, psychophysical vibration thresholds from 55 sites across the foot sole were determined using a range of different vibrational frequencies to evaluate the nature of the regional variations in sensitivity and to establish if evidence exists for a foot sole sensor map. As a second aspect of this experiment, both young and elderly subjects were investigated, since vibrotactile thresholds are known to become elevated with aging, and this may be related to some of the postural disturbances and tendency toward increased falling in the older population. In a second experiment, microneurographic recordings were made from isolated cutaneous afferents that originated in the glabrous skin of the foot sole, in an attempt to clarify what types of receptors are found in the foot sole, their distributions, the level of their physiological activation thresholds, and the nature of their receptive fields.

METHOD

A total of 12 subjects volunteered for the first experiment (6 young; mean age = 26 years; 6 older; mean age = 89 years). For the second experiment, 31 microneurographic recording sessions were performed (mean age 30 years). For all experiments subjects were prone lying with their right foot supported with the ankle in neutral and the foot sole surface unloaded. The clinical ethics board at the University of British Columbia approved the following experimental procedures.

For experiment 1, the vibrotactile stimulus (1.5 mm diameter cylindrical probe attached to a DC motor) was applied to the sole at 55 locations equally distributed about the foot sole (see Fig. 1). The method of limits was used to determine the psychophysical thresholds at each of 4 frequencies (25, 50, 250, and 400 Hz) and at each location. The thresholds were determined using the up and down staircase method, with the last 8 of 10 up and down pairs determining each threshold value. Stimuli were presented for 1s bursts of signal, followed by a 1 s silent period, with the amplitude of the sine wave increasing in 0.025μm steps. Amplitude was increased until the participant closed a hand switch indicating that they had felt the stimulus. To start the down portion of the staircase, the amplitude of the first stimulus burst was set to 1.5 the amplitude of the threshold amplitude from the previous up portion of the staircase. The 55 sites were tested in serial order, however order was counterbalanced between subjects to eliminate serial effects. All 4 frequencies were tested at one site before moving to a new site. Trials were blocked by frequency, however the order of presentation of the 4 blocks was randomized.

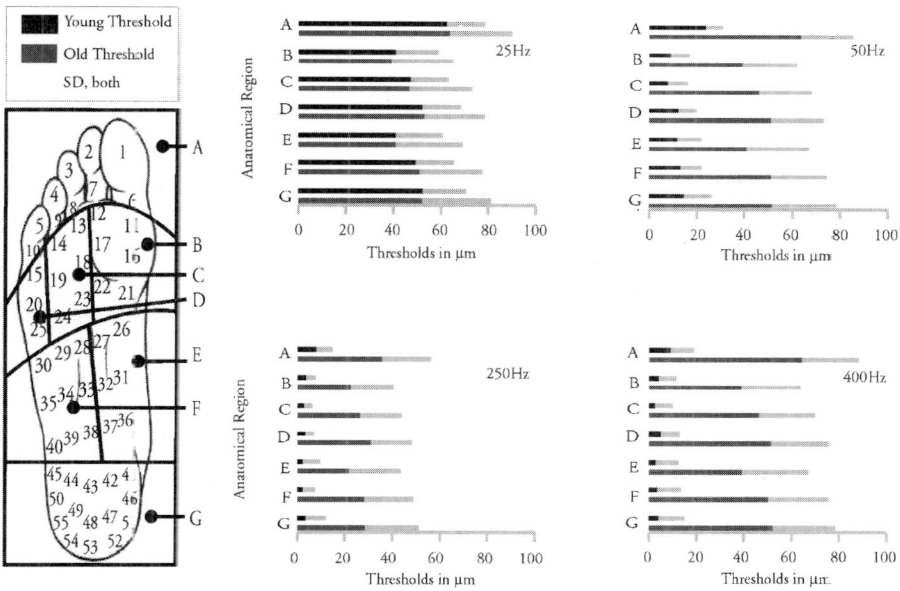

Figure 1. *Left*, an outline of the foot-sole showing the 55 points where vibration thresholds were tested, and the 7 anatomical regions into which the points were segregated. *Right*, plots showing the relationship between anatomical region (x-axis) and vibrotactile threshold (y-axis) in young and old subjects at 4 different frequencies. Thresholds were measured in μm.

For experiment 2, a sterile insulated tungsten microelectrode (0.2 mm diameter, 65-mm length, impedance 50-360 [mean 150] kΩ at 1 kHz in situ) was inserted manually into the tibial nerve at the level of the popliteal fossa. A subcutaneous electrode of similar make, placed approximately 10 mm proximally, served as a reference. By manually manipulating the recording electrode, single unit activity from low-threshold cutaneous afferents originating in the glabrous skin of the foot sole were isolated. The neurogram was amplified 10-25K and band-passed filtered between 0.3 and 10 KHz, and then analogue-to-digitally converted at a sample rate of 25-50 KHz (1401-Micro interface). Single-unit morphology was determined on-line using an oscilloscope, and verified off-line using Spike2 software (Inglis et al., 1996). Physiological threshold and receptive field sizes were determined and afferents were classified using Semmes-Weinstein monofilaments (Stoelting Co, USA) in the range of 0.5 to 5000 mN of force. A monofilament of 4-5 times threshold force was used to outline the receptive fields. Afferents were classified using the same method as previously documented (Johansson and Vallbo, 1983) into either slowly or rapidly adapting receptors and then subclassified into type I and type II afferents as follows: Slowly adapting type I (SAI); Slowly adapting type II (SAII); Fast adapting type I (FAI) Fast adapting type II (FAII)

RESULTS

To facilitate analysis of the many different regions, and so comparisons could be made with earlier reports (Kekoni et al., 1989) the 55 locations were initially grouped into 7 anatomical areas: toes, medial ball, mid-ball, lateral ball, medial arch, lateral arch and heel, and overall means and standard deviations for each sub-region were calculated for comparison. Also, grand mean thresholds (all 55 regions) were calculated separately for each frequency and age group. Figure 1 shows that the ordinal pattern of threshold level amongst the regions is age and frequency invariant. The highest thresholds were in the toes (A), followed by the heel (G) and lateral ball (D); lateral arch (F); mid-ball (C); medial arch (E); and medial ball (B). This pattern appeared to remain throughout each of the four frequencies. However, the overall pattern did seem to be different depending on age and frequency. For grand overall means, at 25 Hz, thresholds for young and old subjects were not statistically different from each other (y=40.4 μm, O=49.6 μm). At 50 Hz, thresholds decreased significantly for the younger group (13.1μm), but not for the older group (49.2 μm). At 250 Hz, both age groups exhibited their lowest thresholds, however younger thresholds (3.6 μm) were significantly lower than older thresholds (27.8 μm). At 400Hz, the thresholds for the younger subjects stayed low (4.4 μm), while the elderly thresholds increased (49.1 μm) to a similar level to what they were at 250 Hz. Regional differences demonstrated that for both young and old, the ball and the arch of the foot had lower thresholds than the heel and toes. These regional differences suggested that perhaps there is potentially a variation in innervation profiles for the different regions of the foot, suggestive of the existence of a sensory map, with a change in these profiles with aging.

In experiment 2, 106 foot sole cutaneous afferents were classified according to their type, physiological threshold, and receptive field. Based on previously established criteria (Johansson and Vallbo, 1983), there were 15 SAIs, 16 SAIIs, 60 FAIs, and 15 FAIIs. The receptors had increased activation thresholds, larger receptive

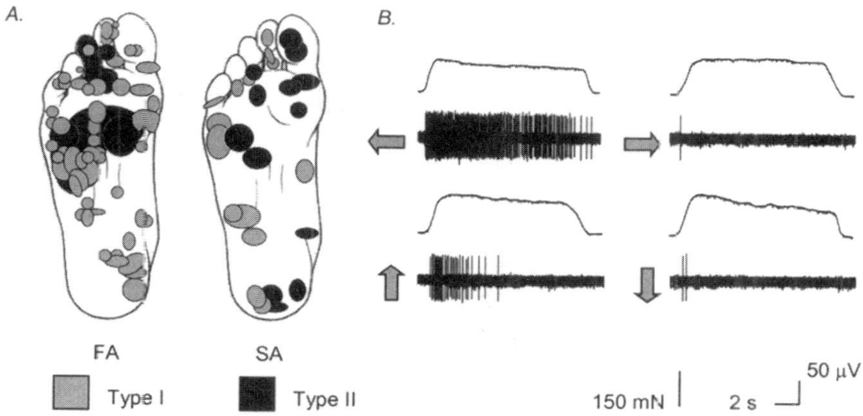

Figure 2. *A*, the positions and sizes of the receptive fields for each receptor type on the foot sole are shown above. *B*, Directional sensitivity of a SAII receptor was measured in response to repeatable skin stretch. The direction of skin stretch on the foot sole is represented by the arrow. The arrow pointing to the left indicates that the direction of stretch was toward the lateral border of the foot sole. The level of stretch applied to the skin and the accompanying discharge activity are illustrated.

field profiles, and clearly a different distribution pattern than that observed in the human hand (Johansson and Vallbo, 1983). Instead of the concentration of type I receptors in the distal phalangeal regions, the slowly adapting receptors appeared to be located more along the borders of the foot, with the fast adapting receptors having a more random distribution throughout the sole. Of note, none of the slowly adapting receptors demonstrated a background discharge. Interestingly, most of the SA II receptors did demonstrated a dramatic directional skin stretch sensitivity, as illustrated in Figure 2B by a SAII's response to 4 different directions of skin stretch.

DISCUSSION AND CONCLUDING REMARKS

Results from the psychophysical experiments show clearly that there is a regional variation in vibrotactile threshold across the human foot sole, and that these regional differences do not seem to be frequency dependent. Surprisingly, for the sole, the toe region demonstrated the highest regional thresholds, in contrast to the hand where the fingers have been shown to have the lowest regional threshold (Johansson and Vallbo, 1983). This regional variation was the same in the sole of the older subjects, but the actual thresholds for high frequency stimulation were dramatically elevated as compared to younger subjects. It seems possible that the regional variations could be due to different receptor densities, and/or different receptor physiological thresholds, within the regions of the sole. It is also possible that with normal aging there is a loss of the rapidly adapting receptors, or perhaps a change in their coding behaviour, that is responsible for the decreased ability to sense higher frequency vibrations.

The microneurographic experiments demonstrated that these regional variations are more than likely not dependent on receptor density differences across regions. There was no evidence of a regional variation in receptor density or of a decreased concentration of type I receptors in the toes, suggesting that the psychophysical variations were not due to a sensor map that is dependent on receptor concentrations. However, since both the physiological thresholds and the receptive field sizes were much larger in the sole than has been described for the hand, it is possible that the regional variations are due to the response characteristics of the individual receptors rather than their distributions. Variations between the younger and older subjects likewise could be due to this, or to variations in the skin stiffness or its mechanical behaviour changing with aging.

The results of both experiments demonstrate that the foot sole receptors can provide rich tactile information, and that this information is probably regionally specific. Issues for future study will need to include microneurographic recordings of afferents in young and elderly subjects, under different conditions of foot sole loading.

ACKNOWLEDGEMENTS

The Natural Science and Engineering Research Council of Canada supported this work.

REFERENCES

Asai, H., Fujiwara, K., Toyama, H., Yamashina, T., Tachino, K., and Nara, I., 1992, The influence of foot soles cooling on standing postural control analyzed by tracking the center of foot pressure, in: *Posture and gait: control mechanisms, Volume II*, Woollacoot M., Horak F., eds, University of Oregon Books, Eugene, Oregon, pp. 151-154.

Edin, B., 2001, Cutaneous afferents provide information about knee joint movements in humans, *Journal of Physiology*, 531, 289-297.

Fitzpatrick, R., Rogers, D. K., and McCloskey, D. I., 1994, Stable human standing with lower-limb muscle afferents providing the only sensory input, *Journal of Physiology*, 480, 395-403.

Inglis, J. T., Horak, F. B., Shupert, C. L., and Jones-Rycewicz, C., 1994, The importance of somatosensory information in triggering and scaling automatic postural responses in humans, *Experimental Brain Research*, 101, 159-164.

Johansson, R.S., and Vallbo, Å.B., 1983, Tactile sensory coding in the glabrous skin of the human hand, *Trends in Neurosciences*, 6,27-31.

Kavounoudias, A., Roll, R., and Roll, J. P., 1998, The plantar sole is a 'dynamometric map' for human balance control, *Neuroreport*, 9, 3247-3282.

Kavounoudias, A., Roll, R., and Roll, J. P., 1999, Specific whole-body shifts induced by frequency-modulated vibrations of human plantar soles, *Neuroscience Letters*, 181-184.

Kekoni, J., Hämäläinen, H., Rautio, J., and Tuveka, T., 1989, Mechanical sensibility of the sole of the foot determined with vibratory stimuli of varying frequency, *Experimental Brain Research*, 78, 419-424.

Magnusson, M., Endom, H., Johansson, R., and Wiklund, J., 1990b, Significance of pressor input from the human feet in lateral postural control, *Acta Otolaryngologica Scandinavia*, 110, 321-327.

Maurer, C., Mergner, T., Bolha, B., and Hlavacka, F., 2001, Human balance control during cutaneous stimulation of the plantar soles, *Neuroscience Letters*, 302, 45-48.

Nurse, M. A, and Nigg, B. M., 1999, Quantifying a relationship between tactile and vibration sensitivity of the human foot with plantar pressure distributions during gait, *Clinical Biomechanics*, 14, 667-672.

Orma, E. J., 1957, The effects of cooling the feet and closing the eyes on standing equilibrium. Different patterns of standing equilibrium in young adult men and women, *Acta Physiologica Scandinavia*, 38, 288-297.

Perry, S. D., McIlroy, W. E., and Maki, B. E., 2000, The role of plantar cutaneous mechanoreceptors in the control of compensatory stepping reactions evoked by unpredictable, multidirectional perturbation, *Brain Research*, 877, 401-406.

Ribot-Cisar, E., Vedel, J. P. and Roll, J. P., 1989, Vibration sensitivity of slowly and rapidly adapting cutaneous mechanoreceptors in the human foot and leg, *Neuroscience Letters*, 104, 130-135.

Trulsson, M., 2001, Mechanoreceptive afferents in the human sural nerve, *Experimental Brain Research*, 137, 111-116.

Vallbo, Å. B., and Hagbarth, K.-E., 1968, Activity from skin mechanoreceptors recorded percutaneously in awake human subjects, *Experimental Neurology*, 21, 270-289.

Vedel, J. P., and Roll, J. P., 1982, Response to pressure and vibration of slowly adapting cutaneous mechanoreceptors in the human foot, *Neuroscience Letters*, 34, 289-294.

15

WHAT DOES GALVANIC VESTIBULAR STIMULATION STIMULATE?

Daniel L. Wardman and Richard C. Fitzpatrick*

ABSTRACT

The technique of galvanic vestibular stimulation (GVS) has been used for a long time. The stimulus produces stereotyped automatic postural and ocular responses. The mechanisms underlying these responses are not understood although they are commonly attributed to altered otolith output. Based on animal studies, it seems reasonable to assume that vestibular afferents from the otoliths and semicircular canals are affected similarly by GVS. With this assumption, and anatomical knowledge of the vestibular apparatus, a model is developed to describe the expected responses of vestibular afferents to percutaneous GVS and the physiological implications of this altered sensory signal. Bilateral bipolar GVS, the most commonly used technique, should produce a canal signal consistent with a strong ear-down roll towards the cathodal side, a smaller nose-to-cathode yaw, but no pitch signal. Bilateral bipolar GVS should also produce an otolith signal consistent with tilt towards the cathodal side or a translational acceleration towards the anodal side. The expected responses for other configurations of GVS are also described. The model appears consistent with published data on the ocular and postural responses to GVS, and suggests other testable hypotheses concerning postural, ocular and perceptual responses to GVS.

INTRODUCTION

Galvanic vestibular stimulation (GVS) is achieved by applying a small percutaneous electrical stimulus behind one or both ears. The technique has been known for over 100 years but in recent years there has been a surge of interest in it of as a means of investigating many functions of the vestibular system in human subjects. However, it is not clear what is being stimulated by this technique. Here we outline the history of GVS and, based on studies of GVS in animals, develop a model that describes the effects of GVS on the vestibular organs in human subjects. The model has several testable

* Prince of Wales Medical Research Institute, Barker St, Randwick NSW 2031, Sydney, Australia. Email: r.fitzpatrick@unsw.edu.au

hypotheses, some of which are examined with available data that describe the effects of GVS on ocular movements and posture.

HISTORY OF GALVANIC VESTIBULAR STIMULATION

"Galvanic" refers to electric current. The adjective honours the Italian physiologist, Luigi Galvani (1737-1798), Chair of Anatomy at the Science Academy of the Bologna Institute. His diary for November 6, 1780, records the observation that frog leg muscles contract when stimulated with an electric current produced by dissimilar metals. Galvani's experimental results and hypothesis concerning the origin of this "animal electricity" were disputed by Allessandro Volta (1745-1827), an Italian physicist at Pavia. In the course of his own experiments on animals and humans, Volta, after initially describing the particular taste of applying a current to the tongue, became the first to report dizziness when he applied the current across his own head with his voltaic pile, a forerunner of the modern battery. Ultimately the Galvani-Volta controversy, animal electricity *versus* physical electricity, was resolved in favour of Galvani by the experiments of Alexander von Humboldt (1769-1859) and Eusebio Valli (1755-1816). It is interesting how profoundly the discovery of animal electricity influenced medical and physiological opinion at the turn of the eighteenth century. 'Galvanism' was imbued with the hope that the mysteries of organic life would soon be solved. A movement of popular thought had begun, articulated in the writings of poet Johann von Goethe, and inspiring Mary Shelley's creation of *Frankenstein; or, The Modern Prometheuse.*[1]

After Volta's experience of passing a current through his head, Johannes Purkinje (1787-1869), priest, teacher and physician, also showed that a galvanic current flowing through the head produced an upset in equilibrium in his dissertation of 1820. Later in 1871, the eminent psychiatrist Edouard Hitzig (1838-1907) noted that an electric current sent through the head of a subject elicited eye movement, or nystagmus. Josef Breuer (1842-1925), physician and scientist, used galvanic currents in conjunction with labarynthectomies in animals to demonstrate that the vestibular apparatus was the site of origin of these phenomena. The observations of Hitzig and Breuer on the effects of the galvanic current on human posture and perception are best summarised by Camis (1930).

> "They applied, for example, the two electrodes to the two mastoid processes, and experienced a sensation of falling towards the side of the cathode, which was accompanied by flexion of the head and trunk towards the side of the anode as if to avoid falling down. When the application of the current was continued, falling actually occurred to the side of the anode. During the passage of the current, false sensations of rotation of the surroundings towards the cathode, and some form of ocular nystagmus, are aroused."

GVS has increased in popularity in recent decades as a research and sometimes clinical tool. The technique remains unchanged - two electrodes to the two mastoid processes and a voltaic pile across them. In terms of understanding the aetiology of the human GVS response, not much has been added since the time of Hitzig and Breuer. In particular, the origin of GVS-evoked reflexes is still not understood.

---

[1] Further history can be found in: Camis (1930), Sammartino and Angoff (1968), Rothschuh (1973), Sekitani and Tanaka (1975), Pappas (1984), and Pera (1992).

EFFECTS OF GVS IN ANIMALS

Based on animal studies conducted over more than a century, it is now thought that galvanic stimuli affect the vestibular system by modulating the continuous firing level of groups of peripheral vestibular afferents (Goldberg et al., 1984; Baird et al., 1988). The afferents have high resting discharge rates of approximately 50-100 spikes per second. In general, cathodal currents depolarise and thus increase the firing rate of vestibular afferents whereas anodal currents hyperpolarize and decrease their firing rate. In the squirrel monkey, externally applied galvanic stimuli modify discharge by acting on the spike encoder in the axon terminal, rather than on hair cells or the parent axon (Goldberg et al., 1984).

In the vestibular system, unlike other sensory systems, the division between regularly and irregularly discharging afferents is not well defined. Afferents near the neuroepithelium tend to have regular discharge whereas those more centrally located tend to have an irregular discharge, irrespective of the morphological class of the afferent (Goldberg et al., 1990; Baird et al., 1998; Goldberg, 2000). There is a continuous distribution of the discharge-interval coefficient of variation in vestibular nerve axons from both the semicircular canals and the otolith organs (Goldberg et al., 1984; Baird et al., 1998). Vestibular nerve afferents that discharge irregularly are affected more by galvanic stimuli than those with a constant discharge frequency (Goldberg et al., 1984).

Irregularly discharging afferents also have a greater influence on activity in vestibular efferent pathways than regularly firing afferents (Goldberg, 2000). Where do irregularly firing afferents project? This has been studied by Highstein and colleagues (1987) in the squirrel monkey. The vestibular nuclei output that is stimulated by irregularly firing afferents comprised: (i) 34% of neurones in the rostral pathway that projects to the oculomotor nuclei, (ii) 50% of neurones that terminate on cerebellar-projecting cells innervating the flocculus, and (iii) 60% of the neurones that project caudally to the spinal cord. On this basis it appears that GVS activates vestibulo-spinal, vestibulo-cerebellar and vestibulo-ocular pathways.

The otolith organs and the semicircular canals both have regularly and irregularly firing afferents in similar proportions, and both afferent types innervate overlapping regions in mammalian vestibular nuclei. However, the available physiological data suggest that the segregation of otolith and canal afferents in the central nervous system is probably limited (Peterson 1998; Goldberg 2000). How does the CNS interpret and react to GVS, which appears to excite afferents of different types and is diffuse in its targets?

EFFECTS OF GVS ON HUMAN VESTIBULAR AFFERENTS

If we assume that GVS has similar effects on human and animal afferent fibre populations, we can speculate on the effects of GVS on different groups of vestibular afferents, and the natural stimuli that would produce these firing patterns. Bilateral-bipolar GVS is the most common technique, with the anode at one mastoid process and the cathode at the other. Alternative presentations are unilateral GVS, either anodal or cathodal, or bilateral-monopolar GVS, with both sides simultaneously either anode or cathode and the other electrode at an indifferent site.

Otoliths

The otolith organs are the sensors of linear acceleration or changes in gravitational force with static head tilt. The array of hair cells in the macula of each otolith forms an approximately flat surface. The utricles lie in the horizontal plane and the saccules in vertical planes with antero-posterior orientation (Fig. 1). Each hair cell is aligned so that the optimal stimulus for each is acceleration in a specific direction in the plane of the macula. Thus, afferents from the utricles signal lateral and sagittal components of an applied acceleration, whereas the saccules signal vertical and sagittal components. Within each utricle and saccule, the alignment of the hair cells is fanned out on either side of a striola or dividing line. The hair cells on either side of the striola are aligned so that they signal acceleration in opposite directions. This push-pull pairing of otolith receptors probably enhances detection of acceleration.

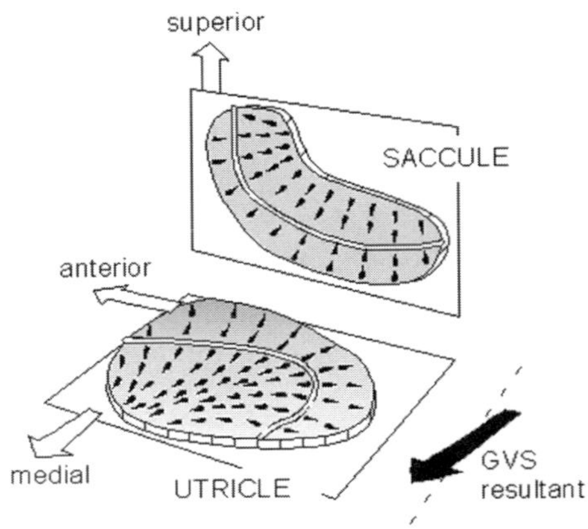

Figure 1. Otoliths. Orientation of the utricle and saccule in head coordinates (open arrows). The saccule is in a vertical plane aligned antero-posterior. The utricle is in a horizontal plane. The alignments of the hair cells are indicated with the kinocilia at the thick ends and sterocilia at the thin ends. Each hair cell responds optimally to acceleration in the direction of its alignment, increasing firing rate when the sterocilia are deflected towards the kinoilium. GVS is assumed to similarly increase (cathode) or decrease (anode) the firing rate of the population of afferents. For most hair cells, the altered discharge is balanced by the same change in a corresponding hair cell on the opposite side of the striola (double lines) signalling the opposite direction. In the utricle there are greater numbers of medial hair cells that increase their firing rate with contra-laterally directed acceleration or ipsilateral tilt. Thus, excitatory GVS should produce a net signal with a resultant indicating this acceleration (black arrow). (Adapted from Spoendlin, 1966.)

Lateral accelerations are detected by the utricles. The striola of the utricle is located towards the lateral edge so that disproportionate numbers of cells are aligned to respond to a laterally directed acceleration than a medially directed acceleration (Fig. 1). Thus, approximately three quarters of the utricular afferents increase their firing with an ipsilateral ear-down tilt or a linear acceleration in the contralateral direction (Gresty et al., 1992). Cathodal GVS will increase the firing rate of afferents from the utricle but, because of the imbalance of hair cell alignment, this will produce a net response that would normally indicate a static tilt towards that side. Decreased firing from the utricle with anodal GVS would produce a net response that signalled tilt away from that side.

Thus, bilateral bipolar GVS should produce a firing pattern consistent with a natural stimulus of static tilt towards the cathodal side or linear acceleration towards the anode whereas bilateral monopolar GVS should produce a "head splitting" firing pattern that would never occur naturally.

Vertical accelerations are detected by the saccules. The striola of the saccule is approximately along the centreline so that there is a closer balance in the number of hair cells that respond to upward and downward accelerations. If GVS increases or decreases the firing rates of these afferents, there would be a relatively minor net response that would indicate vertical acceleration. With bilateral bipolar GVS, any net signal would indicate an upward acceleration on one side of the head and downward acceleration on the other. Together, these should be interpreted as a tilt of the head in roll plane, but no net vertical acceleration. However, with bilateral monopolar GVS, any bias should be reflected equally on both sides, and this could produce a net signal consistent with upward or downward acceleration of the head.

Accelerations in the sagittal plane are signalled by afferents in the utricle and the saccule. Precise information about the sizes of these populations is not available, but the numbers of hair cells that respond to forward and backward accelerations are probably more equally balanced than for lateral accelerations (Spoendlin 1966). Thus, GVS would probably produce a relatively minor net response indicating acceleration in the sagittal plane. If there is a discrepancy in the forward and backward afferent populations, bilateral bipolar GVS should produce firing that indicates anterior acceleration on one side and posterior acceleration on the other. Together, these could be interpreted as a yaw rotation of the head, but no net sagittal acceleration. On the other hand, bilateral monopolar GVS should produce a net signal consistent with forward or backward acceleration of the head.

These considerations suggest that, taken as a whole, the altered firing pattern from the otoliths produced by bilateral bipolar GVS, should correspond to a natural stimulus of a static head-down tilt towards the cathodal side or acceleration towards the anodal side, and perhaps some head rotation in yaw. Bilateral monopolar GVS could produce a firing pattern that corresponds to smaller accelerations or static tilts in the sagittal and vertical planes.

Semicircular Canals

The semicircular canals sense angular accelerations of the head in three dimensions. The three canals on each side of the head are oriented in orthogonal planes, each responding maximally to acceleration in the plane of the canal (Fig. 2A). Thus, the horizontal canal responds maximally to yaw rotation. The two vertical canals, anterior and posterior, are oriented approximately at 45° to the head, so that they respond to both pitch and roll rotations. Bilateral pairing of the semicircular canals enhances detection by acting in a push-pull manner. The horizontal canals work together, but the anterior canal on one side works in tandem with the posterior canal on the opposite side. Hair cells are located at an ampulla in each canal, and the cilia are embedded within a gelatinous cupula that occludes the ampulla. Angular acceleration of the head in the plane of the canal causes the endolymph to exert a differential pressure across the cupula. Depending on the direction of the acceleration, the afferent fibres from the canal increase or decrease their

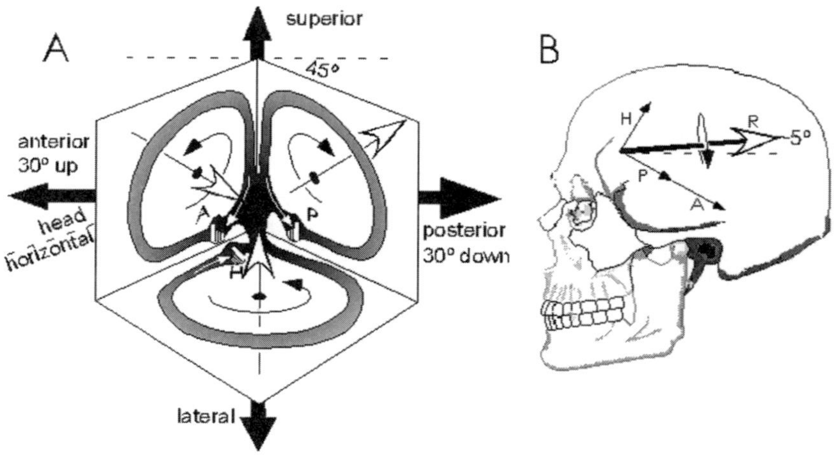

Figure 2. Semicircular canals. *A*, anterior (A) and posterior (P) canals are in "vertical" planes at 90° to each other and 45° to the antero-posterior axis of the head. Orthogonal to both is the "horizontal" (H) canal. In the vertical canals, the kinocilia (thick white lines) are at the canal ends of the ampullae. Thus, they respond in the same way to roll but in opposite ways to pitch rotations. The black arrows show the optimal direction of acceleration to excite each canal, and the central arrow shows the direction of the acceleration vector (RH rule). These are the directions of rotation signalled with excitatory (cathodal) GVS. *B*, Although named "horizontal" and "vertical", the entire canal structure is tilted backward by 30°. Thus, the equivalent rotational vectors (H,A&P) produced by GVS are positioned in head coordinates by tilting backward by 30°. The resultant vector (R) is nearly horizontal (5° elevation) indicating that excitatory GVS should produce a strong roll component to the ipsilateral side, little yaw and no pitch.

spontaneous firing rates. Figure 2A shows the orientation of the kinocilia of the hair cells and, as white arrows, the direction of endolymph flow that would be excitatory and increase the afferent firing rate for each canal. The natural adequate stimuli are angular accelerations in the opposite direction, shown by the black arrows.

The three canals can resolve natural rotational stimuli because they will produce a unique firing pattern for every direction of rotation, but GVS should alter the firing pattern of the canals in a way that has no natural rotational equivalent. When cathodal GVS is applied, all afferents will increase their firing rate. For the horizontal canal, this would be consistent with a natural stimulus of yaw rotation with the nose moving towards the ipsilateral side. This rotation is shown as the black arrow in Figure 2A, and the directional vector (right-hand rule) as the vertical white arrow through the horizontal canal. The increased firing that cathodal GVS would produce from the anterior and posterior canals would be consistent with ipsilateral ear-down roll, and the signals from both canals should be additive (i.e. the antero-posterior components of the directional vectors in Fig. 2A). However, the anterior canal would also indicate nose-down pitch acceleration and the posterior canal a similar nose-up pitch. These opposing signals have no natural equivalent and should cancel because they are approximately equal in size and opposite in direction. These opposing head-relative pitch components from the vertical canals will cancel regardless of head orientation. Thus, the net equivalent rotational vector produced by GVS only has components in yaw and roll relative to the plane of the vestibular apparatus. Furthermore, the roll component should be larger than the yaw component because of the vector addition of signals from both vertical canals.

To complicate matters, the canals are not aligned in the plane of the head. The entire canal structure tilts backward by 30° from the head horizontal (Fig. 2A). This tilt gives the horizontal canals a vector component in the roll plane at the expense of yaw. The vertical canals develop a vector component in the yaw plane at the expense of the roll, while maintaining sensitivity to the pitch component. To predict the GVS effect in head coordinates, we can simply rotate the resultant directional vector produced by GVS backward by 30°. This is shown as the resultant (R) in Figure 2B, directed backward and slightly upward. Thus, most of the yaw component relative to the head should disappear, leaving only a strong roll component. Applying the right-hand rule to the resultant directional vector (R) indicates that when the head is upright, cathodal GVS should be consistent with a strong ipsilateral ear-down roll, and a smaller nose-ipsilateral yaw. Anodal GVS will have opposite effects in each plane.

GVS of the same polarity applied on opposite sides of the head will have opposite effects in yaw and roll but additive effects in pitch because of the mirrored canal system. Thus, unipolar bilateral GVS will produce firing patterns on either side of the head that correspond to roll and yaw rotations in opposite directions and should cancel, leaving a resultant pitch signal. However, this pitch signal should be small because, on each side of the head, the pitch signals from the two vertical canals cancel each other. Bipolar bilateral GVS will produce the same resultant directional vectors on opposite sides of the head for roll and yaw, but cancel for pitch. These should add to produce a signal consistent with a strong roll in the direction of the cathode.

OBSERVED RESPONSES TO GVS

Most of the available experimental data on the effects of GVS in humans examines ocular movements and postural responses during standing. The core findings are outlined below and the adequacy of the above model of GVS to explain them is considered.

Ocular Responses

The vestibular-ocular reflex (VOR) generates counter rotations of the eye that stabilize gaze during head movement. It has been widely believed that the semicircular canals provide the afferent arc for the angular VOR, which stabilizes gaze in the pitch and yaw planes, whereas the otoliths provide for the linear VOR which stabilizes gaze in the roll plane through torsion of the eyes. Thus, examination of eye movements could provide information about the pattern of vestibular afferent firing during GVS.

Static ocular torsion is thought to be a motor consequence of loss of otolithic, probably utricular, function (Halmagyi et al., 1979). The similar imbalance in utricular firing produced by GVS could explain some ocular responses to GVS. Unilateral or bilateral GVS produces ocular torsion, with the upper poles of both eyes torted around the visual axis away from the side of the cathode (Zink et al., 1997; Severac Cauquil et al., 1998; Watson et al., 1998). Ocular torsion is evident at low current levels and the magnitude of ocular torsion is approximately proportional to current strength during bipolar GVS, typical values being 0.5-1° per mA (Zink et al. 1997; Watson et al., 1998; Zink et al. 1998).

Nystagmus, or angular movements of the eyes about the vertical and transverse axes, is thought to arise from stimulation of semicircular canal afferents (Halmagyi et al., 1979). This is the basis of caloric testing. The model of the effects of GVS on the semicircular canals predicts that GVS will evoke VOR responses associated with head roll and, to a lesser extent, head yaw, but not reflexes associated with head pitch. Considering these in reverse order, there have been no reports of vertical nystagmus associated with GVS, consistent with no net pitch signal coming from from the canals. Horizontal nystagmus, associated with head yaw rotations, can be evoked by GVS, but at significantly higher current levels (3-10 mA) than ocular torsion (Watson et al., 1998; Zink et al. 1998). These higher stimulus levels are consistent with the relatively small (~5°) yaw component of the GVS-resultant directional vector. The largest VOR response expected from the semicircular canals with GVS is a torsional response associated with head roll. Recent evidence indicates the semicircular canals can produce ocular torsion (Smith et al., 1995; Kleine et al., 1999). Thus the relatively large torsional response to low-level GVS could arise because it is a combined response to both utricular and canal afferents. The final prediction to come from this analysis is that VOR responses should be seen with bilateral-bipolar GVS or unilateral GVS, whereas they should be minimal or not at all with bilateral-unipolar GVS because the canal responses and lateral utricular responses will cancel. The exception could be a summation of antero-posterior responses from the otoliths to produce some vertical eye movements. All of the studies that have reported VOR reflexes with GVS have used bilateral-bipolar or unilateral GVS.

Postural Responses

We use information from many different sensory systems when we stand. Disturbances of information arising in these sensory systems often lead to clear postural responses. GVS at small stimulus current (0.25-2mA) can evoke a prolonged "galvanic body sway" (Coates, 1973) and transient electromyographic responses in the leg muscles of standing humans (Nashner and Wolfson 1974; Britton et al., 1993). Several hypotheses have been proposed to explain how the CNS interprets and reacts to the altered vestibular firing produced by GVS. The simplest model sees the vestibular system as a sensor that provides an error signal to control vertical alignment (Lund and Broberg, 1983). The onset of the GVS-evoked body sway is rapid and appears mostly complete within one or two seconds of stimulus onset. For bilateral-bipolar GVS, the sway is in the direction of the anode, and is directed to all body segments, with the upper segments tilting further than the lower segments (Day et al., 1997). The posture tends to be maintained for the duration of the stimulus and there is often a reversed sway when the stimulus is stopped.

The sway is sideways to the direction of the head, regardless of the orientation of the body. Thus, the sway directed laterally when the head is facing forward, but is forward and backward when the neck or trunk is rotated so that the head is turned to the side (Nashner and Wolfson 1974; Lund and Broberg, 1983; Britton et al., 1993; Fitzpatrick et al., 1994; Inglis et al., 1995). These results could be consistent with GVS effects on the utricules, the canals or both. The utricle would signal tilt to the cathodal side resulting in a postural correction to the anodal side. Similarly, the canals would signal roll towards the cathodal side also resulting in a correction to the anodal side. When cathodal unilateral GVS is applied, there is a sway of about half the size directed contra-laterally with a smaller anterior component (Severac Cauquil et al., 2000). Again, the lateral component could arise from utricles or canals, and the half-size response indicates an algebraic summing of responses from each side of the head. When cathodal GVS is applied bilaterally, the sway is only anterior and is larger, consistent with the lateral components cancelling and the anterior components summing. The model of canal response to GVS suggests that this anterior sway does not arise because the semicircular canals are signalling a backward pitch. If so, it would have to be explained by an otolith response to GVS. It is possible that there is an imbalance in the sizes of hair-cell populations indicating anterior and posterior sway although precise numbers have not been reported. The drawings of Spoendlin (1966) suggest that a larger number of hair cells in the utricle and the saccule have alignments that respond to posterior acceleration than anterior acceleration. Thus, an otolith response to cathodal GVS could explain the anterior sway response seen in these situations.

CONCLUSIONS

A model has been developed that describes the responses of vestibular afferents to percutaneous GVS. The model appears consistent with published data on the ocular and postural responses to GVS. However, the model suggests several other hypotheses of postural, ocular and perceptual responses that can be tested.

ACKNOWLEDGEMENTS

We are very indebted to Dr Brian Day who was instrumental in developing the ideas behind this work. The work was carried out with support from the Australian National Health and Medical Research Council.

REFERENCES

Baird, R. A., Desmadryl, G., Fernandez, C., and Goldberg, J. M., 1988, The vestibular nerve of the chinchilla. II. Relation between afferent response properties and peripheral innervation patterns in the semicircular canals, *Journal of Neurophysiology*, 60, 182-203.

Britton, T. C., Day, B. L., Brown, P., Rothwell, J. C., Thompson, P. D., and Marsden, C. D., 1993, Postural electromyographic responses in the arm and leg following galvanic vestibular stimulation in man, *Experimental Brain Research*, 94, 143-151.

Camis, M., 1930, *The Physiology of the Vestibular Apparatus*, Oxford University Press, London.

Coats, A. C., 1973, Effect of varying stimulus parameters on the galvanic body-sway response, *Annals of Otology Rhinology and Laryngology*, 82, 96-102.

Day, B. L., Severac Cauquil, A., Bartolomei, L., Pastor, M. A., and Lyon, I. N., 1997, Human body-segment tilts induced by galvanic stimulation: a vestibularly driven balance protection mechanism, *Journal of Physiology*, 500, 661-672.

Fitzpatrick, R. C., Burke, D., and Gandevia, S. C., 1994, Task-dependent reflex responses and movement illusions evoked by galvanic vestibular stimulation in standing humans, *Journal of Physiology*, 478, 363-372.

Goldberg, J. M., 2000, Afferent diversity and the organization of central vestibular pathways, *Experimental Brain Research*, 130, 277-297.

Goldberg, J. M., Desmadryl, G., Baird, R. A., and Fernandez, C., 1990, The vestibular nerve of the chinchilla. IV. Discharge properties of utricular afferents, *Journal of Neurophysiology*, 63, 781-790.

Goldberg, J. M., Smith, C. E., and Fernandez, C., 1984, Relation between discharge regularity and responses to externally applied galvanic currents in vestibular nerve afferents of the squirrel monkey, *Journal of Neurophysiology*, 51, 1236-1256.

Gresty, M. A., Bronstein, A. M., Brandt, T., and Dieterich, M., 1992, Neurology of otolith function. Peripheral and central disorders, *Brain*, 115, 647-673.

Halmagyi, G. M., Gresty, M. A., and Gibson, W. P., 1979, Ocular tilt reaction with peripheral vestibular lesion, *Annals of Neurology*, 6, 80-83.

Highstein, S. M., Goldberg, J. M., Moschovakis, A. K., and Fernandez, C., 1987, Inputs from regularly and irregularly discharging vestibular nerve afferents to secondary neurons in the vestibular nuclei of the squirrel monkey. II. Correlation with output pathways of secondary neurons, *Journal of Neurophysiology*, 58, 719-738.

Inglis, J. T., Shupert, C. L., Hlavacka, F., and Horak, F. B., 1995, Effect of galvanic vestibular stimulation on human postural responses during support surface translations, *Journal of Neurophysiology*, 73, 896-901.

Kleine, J. F., Guldin, W. O., and Clarke, A. H., 1999, Variable otolith contribution to the galvanically induced vestibulo-ocular reflex, *Neuroreport*, 10, 1143-1148.

Lund, S., and Broberg, C., 1983, Effects of different head positions on postural sway in man induced by a reproducible vestibular error signal, *Acta Physiologica Scandinavica*, 117, 307-309.

Nashner, L. M., and Wolfson, P., 1974, Influence of head position and proprioceptive cues on short latency postural reflexes evoked by galvanic stimulation of the human labyrinth, *Brain Research*, 67, 255-268.

Pappas, D. G., 1984, Barany's History of Vestibular Physiology, Translation and commentary, *Annals of Otology, Rhinology and Laryngology*, 110, 1-16.

Pera, M., 1992, *The Ambiguous Frog: The Galvani-Volta Controversy on Animal Electricity*, Translated by Jonathan Mandelbaum, Princeton University Press, Princeton.

Peterson, E. H., 1998, Are there parallel channels in the vestibular nerve? *News in Physiological Sciences*, 13, 194-201.

Rothschuh, K. E., 1973. Physiology of the Enlightenment, in: *Anonymous History of Physiology*, Robert E. Krieger Publishing Company, New York.

Sammartino, P., and Angoff, C., 1968, *The Humanities in the Age of Science. In Honor of Peter Sammartino.*, Farleigh Dickinson University Press, Rutherford, NJ.

Sekitani, T., and Tanaka, M., 1975, Test for galvanic vestibular responses, *Bulletin of the Yamaguchi Medical School*, 22, 439-452.

Severac Cauquil, A., Faldon, M., Popov, K., Bronstein, A., and Day, B. L., 1998, Torsional eye movements induced by galvanic vestibular stimulation in man, *Journal of Physiology*, 506P, 110-111.

Severac Cauquil, A., Martinez, P., Ouaknine, M., and Tardy-Gervet, M-F., 2000, Orientation of the body response to galvanic stimulation as a function of the inter-vestibular balance, *Experimental Brain Research*, 133, 501-505.

Smith, S. T., Curthoys, I. S., and Moore, S. T., 1995, The human ocular torsion position response during yaw angular acceleration, *Vision Research*, 35, 2045-2055.

Spoendlin, H., 1966, Ultrastructure of the Vestibular sense organ, in: *The Vestibular System and its Diseases*, ed. Wolfson, R. J., University of Pennsylvania Press, Philadelphia.

Watson, S. D., Brizuela, A. E., Curthoys, I. S., Colebatch, J. G., Macdougall, H. G., and Halmagyi, G. M. 1998, Maintained ocular torsion produced by bilateral and unilateral galvanic (DC) vestibular stimulation in humans, *Experimental Brain Research*, 122, 453-458.

Zink, R., Bucher, S. F., Weiss, A., Brandt, T., and Dieterich, M., 1998, Effects of galvanic vestibular stimulation on otolithic and semicircular canal eye movements and perceived vertical, *Electromyography and Clinical Neurophysiology*, 107, 200-205.

Zink, R., Steddin, S., Weiss, A., Brandt, T., and Dieterich, M., 1997, Galvanic vestibular stimulation in humans - effects on otolith function in roll, *Neuroscience Letters*, 232, 171-174.

16

SENSORY INTERACTIONS FOR HUMAN BALANCE CONTROL REVEALED BY GALVANIC VESTIBULAR STIMULATION

Brian L. Day[1], Michel Guerraz[2] and Jonathan Cole[3]

ABSTRACT

Many types of sensory information are known to contribute to the human balance control process but little is known about how the different sensory channels interact. Here we consider the postural response to a perturbation delivered to the vestibular channel using galvanic vestibular stimulation. We show that the response is modified by the absence of information in the other sensory channels. Removal of somatosensory information leads to a massive increase in response size. Similarly, removal of visual information augments the response. Furthermore, the response size is graded according to the amount of visual information available. These effects occur through two processes. One that influences the developing response through feedback mechanisms and another that influences the initial response selection through gain changes. The latter is described as a competitive process that can be likened to a proportional representation voting system.

INTRODUCTION

It is well established that the balance process of human upright stance makes use of sensory information from diverse sources. The evidence is found in the degradation of balance performance following the removal of sensory information together with the production of postural responses by stimulation of sensory channels. One surprising fact is that a sensory perturbation does not need to be congruent between sensory channels for a postural response to be evoked. The process can be fooled into action by the stimulation of single channels on their own. Thus, isolated movement of the visual environment (Lestienne et al., 1977; Bronstein and Buckwell, 1997), or excitation of muscle or cutaneous receptors (Eklund, 1972; Hiyashi et al., 1981; Kavounoudias et al., 1998;

[1] MRC Human Movement Group, Sobell Department, Institute of Neurology, Queen Square, London, UK. Email: bday@ion.ucl.ac.uk
[2] Laboratoire "Sport, Performance, Santé" UFR STAPS, Montpellier, France.
[3] Cole, Department of Clinical Neurosciences, University of Southampton and Poole Hospital, UK.

Sensorimotor Control of Movement and Posture
Edited by Gandevia *et al.*, Kluwer Academic/Plenum Publishers, 2002

Kavounoudias et al., 2001) or changes in vestibular afferent input (Njiokiktjien and Folkerts, 1971; Coates, 1973; Nashner and Wolfson, 1974) are each able, on their own, to produce a postural response. Yet it is difficult to think of a physical situation in which a true threat to balance would disturb only a single sensory channel.

Is it that each sensory channel exerts its control independently of the other channels? Such a view is compatible with the finding that the postural responses evoked by separate stimulation of two sensory channels simply sum when the two channels are stimulated together (Hlavacka et al., 1995; Kavounoudias et al., 2001). However, there is evidence that sensory channels do not operate independently to control balance but instead interact with each other at a premotor stage. For example, the postural response to a vestibular perturbation is influenced by non-vestibular information on head and body position (Nashner and Wolfson, 1974; Lund and Broberg, 1983; Pastor et al., 1993), and by visual information (Njiokiktjien and Folkerts, 1971; Smetanin et al., 1990; Britton et al., 1993; Fitzpatrick et al., 1994).

What are the rules that govern such sensory interactions? Here we explore one aspect. We ask, how is the postural response to a perturbation of one sensory channel influenced by the removal of information in the other channels? In this chapter we show that the postural response evoked by a pure vestibular input is modified by the availability of somatosensory and visual information. This interaction occurs in two stages. First by influencing the initial response selection. Secondly, by altering the developing response through feedback.

THE VESTIBULAR STIMULUS AND RESPONSE

To achieve a pure vestibular test stimulus we use the technique of galvanic vestibular stimulation (GVS). It is a simple technique in which a small percutaneous electric current (around 1mA) is passed between electrodes behind the ear. Animal work has shown that GVS acts by modulating the spontaneous firing frequency of vestibular afferents with the firing rate of afferents being increased on the side of the cathode and decreased on the side of the anode (Lowenstein, 1955; Goldberg et al., 1984; Courjon et al., 1987). When applied to standing human subjects this change in vestibular afferent firing evokes a whole-body postural response involving the neck, trunk and legs (Day et al., 1997).

The response to GVS has many interesting features. When viewed electromyographically the response is seen to consist of two sequential components with opposite actions that are thought to arise from different central motor processes (Britton et al., 1993). In leg muscles the first component, which has a latency of around 60 ms, is very small and evokes negligible body movement. The oppositely directed second component occurs at about twice that latency and is strong enough to evoke a characteristic body sway. The body always sways towards the anodal ear irrespective of which direction the head is facing (Nashner and Wolfson, 1974; Lund and Broberg, 1983; Pastor et al., 1993). Thus the pattern of evoked muscle activity depends upon other sources of information that signal head, trunk and leg position. These responses are not restricted to leg muscles. If the arms are being used to help balance the body, for example when standing on a wobbly board and holding on to an earth-fixed handle, then similar EMG responses are found in arm muscles (Britton et al., 1993).

These EMG responses cause the development of forces between the body and supporting surfaces to accelerate the body. In healthy standing subjects this evoked movement continues for a second or two before it is arrested to result in a new posture. As illustrated in Fig. 1, the new posture is characterised by a small tilt and bend of the body in which the head is tilted more than the trunk, and the trunk is tilted more than the pelvis (Day et al., 1997). But again, the response can be readily modified by a change in circumstances. For instance, if the subject stands with feet apart rather than together then the response becomes smaller (Fig. 1). It becomes smaller still if the person changes from a standing to a sitting position (Day et al., 1997).

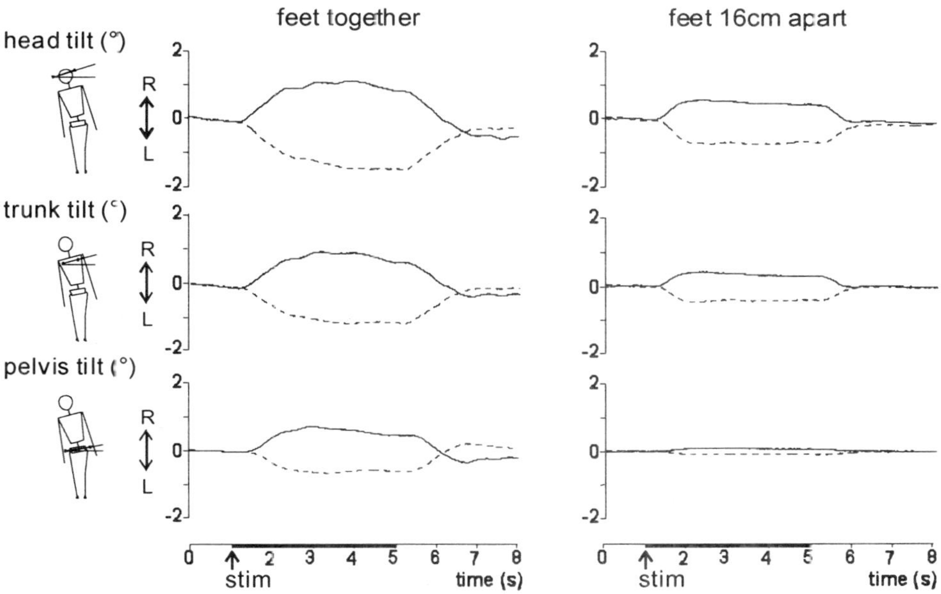

Figure 1. GVS-evoked tilts of body segments. Grand average (10 subjects) tilt responses in the frontal plane of head, trunk and pelvis to a 4 s, 0.7 mA bipolar binaural stimulus. The polarity was such that the stimulating anode was either on the right (solid lines) or the left (dashed lines) mastoid process. Subjects stood with feet either together (left panel) or apart (right panel). (Redrawn from Day et al., 1997.)

It is these rich context-dependant characteristics of the whole-body response to GVS that make it interesting. It does not behave like a simple reflex but at the same time it is highly automatic. It seems to engage sophisticated neural machinery that is devoted to the control of whole-body balance. When used as a probe, therefore, it has the capacity to reveal some of the workings of that machinery.

GVS-EVOKED RESPONSES WITHOUT SOMATOSENSORY INFORMATION

How does the continuous stream of somatosensory information from proprioceptors and cutaneous receptors influence the way in which the balance control process responds to a change in vestibular input? Put another way, what effect would removal of all somatosensory information have on the GVS-evoked response? We have approached this problem by studying the behaviour of a rare subject (IW) with a large-fibre sensory neuronopathy (Sterman et al., 1980). This has left him without the sensations of cutaneous light touch and movement/position sense below the neck (Cole and Sedgwick, 1992). Without large myelinated sensory nerve fibre function IW is unable to feel anything but temperature and pain and without vision has no knowledge of where his limbs and body are in space. Though cutaneous sensation is absent below the collar line he appears to have preserved neck proprioception, either because these afferents are all preserved or, perhaps more likely, because sufficient afferents remain to preserve function. Initially completely incapacitated by his loss of feedback for movement he taught himself to move again, and subsequently to walk and live independently, using visual feedback and cognitive control of movement (Cole, 1995).

Because IW is unable to stand securely with eyes closed we studied his postural responses while he was seated even though only very small responses can be evoked by GVS in seated healthy subjects. This turned out not to be a problem. With eyes closed his head and trunk tilt responses were extremely large being an order of magnitude greater than those of healthy subjects (Fig. 2). The response size was larger than normal right from the beginning of the response and so presumably represents a massive increase in gain of the vestibular channel's influence on balance. With eyes open his responses were smaller than with eyes closed (<50%) but remained considerably larger than normal. Again this effect was observed from the start of the response indicating early gain changes. Although the response was greatly exaggerated the response was normally directed towards the anodal ear and was normally distributed with the head tilting more than the trunk.

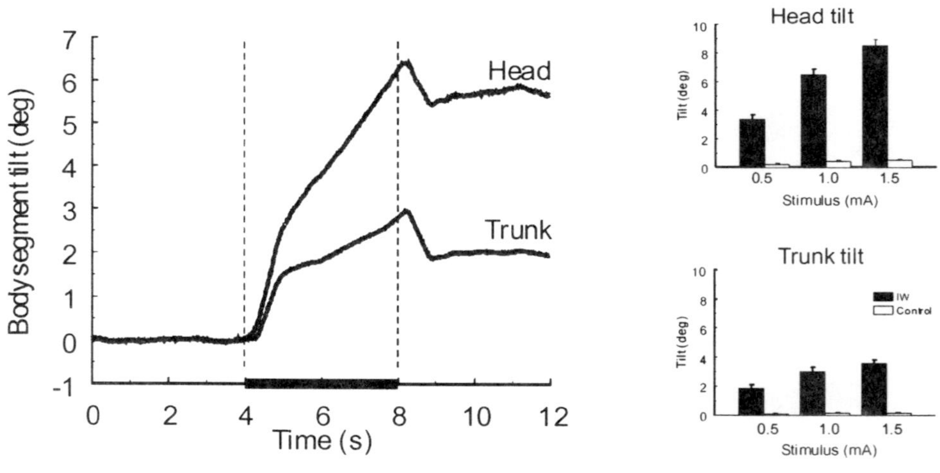

Figure 2. GVS-evoked tilt responses in a deafferented man with eyes closed. Frontal plane body segment tilts in direction of anode (positive direction) in deafferented subject IW. Traces are averaged across 3 stimulus intensities (0.5, 1.0 and 1.5 mA). Stimulus applied for 4 s (thick bar on abscissa). The mean (+ S.E.M.) tilt over the 4-s period is shown for each stimulus intensity and body segment in the bar graphs (right). IW's responses (filled bars) are compared to those of a healthy control subject (empty bars).

IW's response differed from normal in one other important respect. Whereas healthy subjects with intact sensation tend to arrest the developing tilt response to remain fixed in a static tilted posture, IW behaved differently. He continued tilting while the stimulus remained on (Fig. 2). Closer inspection showed that the on-response consisted of two components, an initial relatively fast tilt component followed by a slower continuous tilt component. When the stimulus was turned off he partially tilted back the other way at a third intermediate tilt velocity.

It seems likely that this behaviour is not readily observed in healthy subjects because of the braking effect of somatosensory feedback. As the response develops changes are produced in somatosensory afferents which signal joint motion, muscle state and contact forces. This change is fed back to the balance control process which acts to arrest the response. Without somatosensory feedback, the response in IW (at least for the trunk) therefore represents a true open-loop response to the GVS-evoked vestibular afferent input. Interestingly his two components of response resemble the sum of two independent responses, a position response and a velocity response. It is possible that each of these response types arise from activation of afferents from different vestibular organs. Otolith (utricular) afferents would be a good candidate for the position response and semicircular canals (vertical) for the velocity response.

INFLUENCE OF VISION ON THE GVS-EVOKED RESPONSE

It is considerably easier to manipulate visual than somatosensory information. Simply asking subjects to stand with eyes closed or eyes open during GVS reveals the main effect of vision. With that approach most groups agree that the GVS-evoked postural response is much smaller with the eyes open than with them closed (Njiokiktjien and Folkerts, 1971; Smetanin et al., 1990; Britton et al., 1993; Fitzpatrick et al., 1994). We have investigated this phenomenon in more detail by asking whether the GVS-evoked response is graded according to the amount of visual information available.

The amount of visual information available to the subject was graded in the following way. Subjects looked at the centre of a visual display system which consisted of two parallel planes of small coloured Light Emitting Diodes (LEDs) suspended on fine wires in a grid (Fig. 3). The back plane could be seen through the front plane (which was mainly empty space). This system of leds illuminated in different combinations was used to provide 4 different visual environments. 3-D visual information (*vis3*) was given by illuminating both the back (green) and front (red) planes of LEDs as well as a central fixation (blue) LED. The geometry was such that the green LEDs appeared in the gaps between the red LEDs. This was reduced to a 2-D display (*vis2*), without altering the total amount of light, by turning off the back plane and turning on identical green LEDs located between the red LEDs in the front plane. The visual information was reduced further by turning off all LEDs apart from the blue central fixation LED (*vis1*). Finally, all LEDs were turned off to give the fourth condition of no visual information (*novis*).

As is usual, GVS (0.5 mA) caused the standing subject to sway laterally towards the anodal ear. But the amount of body displacement during the 4s stimulation period depended upon the visual conditions. The maximum displacement occurred in the blackout (*novis*) condition, and the smallest displacement occurred when 3-D vision (*vis3*) was available. Responses during the other two visual conditions lay between these extremes.

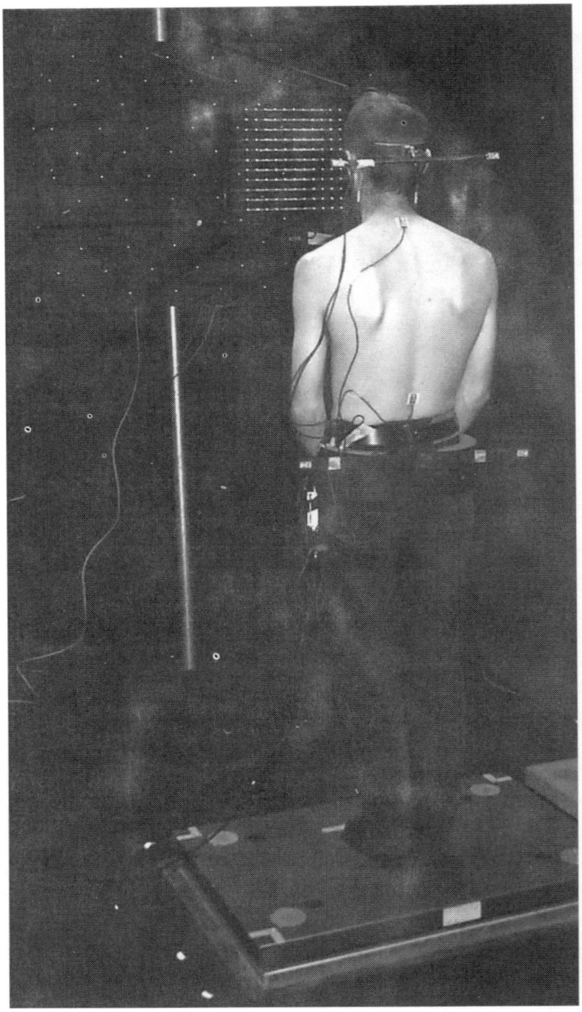

Figure 3. Controlling the visual environment. The subject is shown looking into a 3-D structure consisting of two parallel planes (large back plane and smaller front plane) of LEDs. The visual environment is controlled by switching on the LEDs in different combinations (see text). Infrared emitting diodes are fixed to head gear, hip gear and the subject's back for tracking in three dimensions by a motion analysis system (Selspot II). The subject stands on a force plate (Kistler) which measures the ground reaction forces in three dimensions. GVS is applied to electrodes behind the ears.

Part of this effect is likely to have been due to a visual feedback process. As the body moves in response to GVS the relative motion is detected by the visual system. This feedback information can then act to modify or even arrest the response. Presumably the potency of the visual feedback depends upon the quality or amount of visual information that is available. However, this is not the complete explanation. The very earliest trace of the response, which is given by the initial pulse of force between the feet and the ground, was influenced by the visual conditions. At 400 ms after stimulus onset the lateral ground reaction force, which acted to accelerate the body sideways, was significantly larger with *novis* than with *vis1*. Similarly, the force with *vis1* was greater than with *vis2*. At this early latency the force with *vis2* was not different to that with *vis3*.

It should be borne in mind that visual feedback cannot begin before the body starts to move and that the earliest movement, which occurs as a change in head tilt, has a latency of 150 ms or so. This suggests that these observed changes in force that occur within 400 ms probably are too fast to be produced by visual feedback modification. This was confirmed in separate experiments in which the visual environment was switched from *novis* to *vis3*, or *vice versa*, shortly after the onset of GVS. When this occurred the feedback effect of the new visual environment did not act to change the ground reaction force profile until more than 400 ms had elapsed. These results suggest, therefore, that the gain of the vestibular channel is altered by the amount or quality of visual information that is currently available.

PROPORTIONAL REPRESENTATION MODEL OF SENSORY INTERACTION

Overall, the experimental evidence favours the view that a constant perturbation delivered to the vestibular channel is responded to in the light of information available in the somatosensory and visual channels. The picture is one of a control process in which information from the three sensory channels is dynamically weighted to regulate balance. Each channel has direct access to the balance control process such that a perturbation delivered to any one channel in isolation will always produce a response. However, the response is computed with reference to the current information available from the undisturbed channels. In particular, we suggest that the gain of a particular input-output relationship is constantly updated as a function of the amount of information available in the other channels.

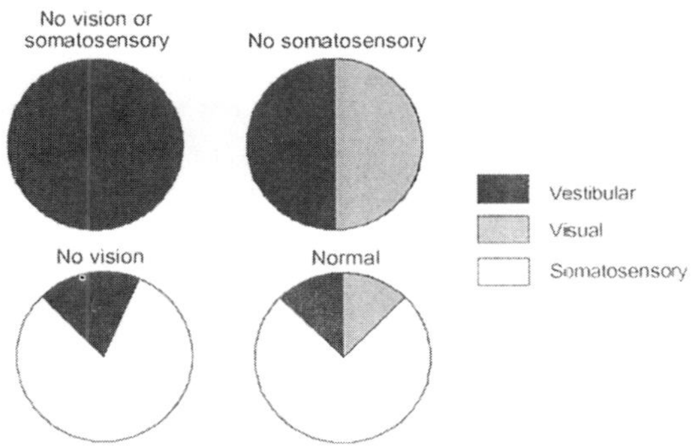

Figure 4. Model of sensory interaction. In this model the weight or gain of each of the three sensory channels is represented by its respective area in the pie charts. Removal of any one channel results in an increase in gain in the remaining channels. Under normal circumstances the somatosensory channel is assumed to have the greatest weight. The size of response to a given galvanic vestibular stimulus is proportional to the size of each 'vestibular' segment.

By analogy, such a system can be likened to a competitive proportional representation voting system in which every single vote has an impact on the outcome but the strength of that impact depends upon the total number of votes (Fig. 4). Thus, without somatosensory or visual information (e.g. subject IW with his eyes closed) those channels have no vote and so vestibular input completely determines the behaviour. The response to GVS is then maximal. When vision is also available (e.g. subject IW with his eyes open) the impact of the vestibular vote is lessened and so the response size is reduced accordingly. Of course, in this model if both sensory systems were perturbed congruently such that they vote for the same outcome then the response size would be restored. Inclusion of somatosensory information weakens the vestibular (and visual) vote considerably more. By the same token, removing or degrading visual information automatically increases the vestibular, and presumably the somatosensory, vote.

Whether or not a process of this sort underlies any part of the intricate machinery of human balance control remains to be seen. But if so, the challenge will be to identify how and where such a process is implemented by central nervous structures.

ACKNOWLEDGEMENTS

This work was supported by the Medical Research Council.

REFERENCES

Britton, T. C., Day, B. L., Brown, P., Rothwell, J. C., Thompson, P. D., and Marsden, C. D., 1993, Postural electromyographic responses in the arm and leg following galvanic vestibular stimulation in man, *Experimental Brain Research,* 94, 143-151.

Bronstein, A. M., and Buckwell, D., 1997, Automatic control of postural sway by visual motion parallax, *Experimental Brain Research,* 113, 243-248.

Coates, A. C., 1973, Effect of varying stimulus parameters on the galvanic body-sway response, *Annals of Otology, Rhinology and Laryngology,* 82, 96-102.

Cole, J., 1995, *Pride and a daily marathon,* The MIT Press, Boston, Mass.

Cole, J. D., and Sedgwick, E. M., 1992, The perceptions of force and of movement in a man without large myelinated sensory afferents below the neck, *Journal of Physiology,* 449, 503-515.

Courjon, J. H., Precht, W., and Sirkin, D. W., 1987, Vestibular nerve and nuclei unit responses and eye movement responses to repetitive galvanic stimulation of the labyrinth in the rat, *Experimental Brain Research,* 66, 41-48.

Day, B. L., Séverac Cauquil, A., Bartolomei, L., Pastor, M. A., and Lyon, I. N., 1997, Human body-segment tilts induced by galvanic stimulation: a vestibularly driven balance protection mechanism, *Journal of Physiology,* 500, 661-672.

Eklund, G., 1972, General features of vibration-induced effects on balance, *Upsala Journal of Medical Science,* 77, 112-124.

Fitzpatrick, R., Burke, D., and Gandevia, S. C., 1994, Task-dependent reflex responses and movement illusions evoked by galvanic vestibular stimulation in standing humans, *Journal of Physiology,* 478, 363-372.

Goldberg, J. M., Smith, C. E., and Fernandez, C., 1984, Relation between discharge regularity and responses to externally applied galvanic currents in vestibular nerve afferents of the squirrel monkey, *Journal of Neurophysiology,* 51, 1236-1256.

Hiyashi, R., Miyake, A., Jijiwa, H., and Watanabe, S., 1981, Postural readjustment to body sway induced by vibration in man, *Experimental Brain Research,* 43, 217-225.

Hlavacka, F., Krizkova, M., and Horak, F. B., 1995, Modification of human postural response to leg muscle vibration by electrical vestibular stimulation, *Neuroscience Letters,* 189, 9-12.

Kavounoudias, A., Roll, R., and Roll, J.-P., 1998, The plantar sole is a 'dynamometric map' for human balance control, *Neuroreport,* 9, 3247-3252.

Kavounoudias, A., Roll, R., and Roll, J.-P., 2001, Foot sole and ankle muscle inputs contribute jointly to human erect posture regulation, *Journal of Physiology,* 532, 869-878.

Lestienne, F., Soechting, J. F., and Berthoz, A., 1977, Postural readjustments induced by linear motion of visual scenes, *Experimental Brain Research,* 28, 363-384.

Lowenstein, O., 1955, The effect of galvanic polarization on the impulse discharge from sense endings in the isolated labyrinth of the thornback ray (raja clavata), *Journal of Physiology,* 127, 104-117.

Lund, S., and Broberg, C., 1983, Effects of different head positions on postural sway in man induced by a reproducible vestibular error signal, *Acta Physiologica Scandinavica,* 117, 307-309.

Nashner, L. M., and Wolfson, P., 1974, Influence of head position and proprioceptive cues on short latency postural reflexes evoked by galvanic stimulation of the human labyrinth, *Brain Research,* 67, 255-268.

Njiokiktjien, C., and Folkerts, J. F., 1971, Displacement of the body's centre of gravity at galvanic stimulation of the labyrinth, *Confinia Neurologica,* 33, 46-54.

Pastor, M. A., Day, B. L., and Marsden, C. D., 1993, Vestibular induced postural responses in Parkinson's disease, *Brain,* 116, 1177-1190.

Smetanin, B. N., Popov, K. E., and Shlykov, V. Yu., 1990, Changes in vestibular postural response determined by information content of visual feedback, *Neirofiziologiya,* 22, 80-87.

Sterman, A. B., Schumberg, H. H., and Asbury, A. K., 1980, The acute sensory neuropathy syndrome: a distinct clinical entity, *Annals of Neurology,* 7, 354-358.

VESTIBULOSPINAL CONTROL OF POSTURE

Fay B. Horak[1], John Buchanan[2], Robert Creath[3], and John Jeka[3]

ABSTRACT

To better understand the role of the vestibular system in postural coordination, we compared the ability of subjects with complete, bilateral loss of vestibular function and age-matched control subjects to maintain equilibrium and postural orientation during sinusoidal displacements of the support surface at a variety of frequencies. We also examined the ability of visual or somatosensory-light touch information to substitute for missing vestibular information in dynamic postural coordination. The results suggest that vestibular information is used as a gravitational reference frame to prevent slow drift of the trunk in space during complex postural tasks. Furthermore, visual information or somatosensory information from light touch of a finger on a stable reference can significantly substitute for loss of vestibular function.

INTRODUCTION

What is the role of vestibular information for control of postural equilibrium and orientation? Our previous studies showed that vestibular information is generally not used to trigger automatic postural responses to surface displacements like a slip or trip. Subjects with complete, bilateral vestibular-loss show normal latencies and spatial-temporal patterns of leg muscle responses to transient surface translations (Horak et. al., 1990). However, vestibular-loss does result in abnormal postural synergies at the neck and trunk suggesting a top-down influence of the vestibulospinal system on postural control (Allum et. al.; 1994; Horak et. al., 1994). In fact, subjects with vestibular-loss fail to control equilibrium over a narrow base of support when a proximal-to-distal hip strategy is required (Horak, et. al., 1990; Runge, et. al., 1998). Disruption of upper body postural control from vestibular-loss is consistent with increased head accelerations and head displacement variability in vestibular patients during gait (Pozzo et al., 1991).

[1] Neurological Sciences Institute of Oregon Health Sciences University, Beaverton, Oregon 97212, USA.
Email: horakf@ohsu.edu
[2] Texas A&M University, College Station, Texas, USA.
[3] Dept. of Kinesiology, Univeristy of Maryland, College Park, MD, USA.

Sensorimotor Control of Movement and Posture
Edited by Gandevia *et al.*, Kluwer Academic/Plenum Publishers, 2002

Previous studies also suggest that the influence of vestibular information on postural equilibrium increases when somatosensory information from the surface and/or visual information is unstable or unavailable (Nashner et. al., 1982).

Our previous study showed that healthy adults gradually change postural strategy from a head strapped-down, ride pattern to a head fixed in space pattern as the frequency of surface translations increased (Buchanan and Horak, 1999). We concluded that vision provides an external reference frame preventing upper body drift, especially at high frequencies of displacement. Other studies showed that somatosensory information associated with light touch of a fingertip on a stable reference was even more effective than vision in stabilizing postural sway during quiet stance in healthy adults and in subjects with vestibular-loss (Jeka and Lackner, 1994; Lackner et al, 1999). In the current studies, we hypothesized that vestibular information, light touch from a fingertip, and vision may play a similar role in preventing upper body drift, especially at higher frequencies of postural displacement requiring head and trunk stabilization in space over the moving feet.

METHODS

The University Institutional Review Board approved experimental protocols in accordance with the Helsinki Declaration. Two separate studies examined responses to sinusoidal surface perturbations. Experiment 1 studied responses to 6 frequencies (0.1 - 1.25 Hz) of 12 cm surface translations with eyes open and closed. Experiment 2 studied responses to 5 frequencies (0.01 - 0.4 Hz) of $\pm$ 1.2° toes-up and toes-down surface *rotations* during eyes closed, with and without light touch. In the touch condition, subjects maintained $<$ 1 Newton of touch contact to a horizontal force plate with their right index fingertip using auditory feedback. In no touch trials, subjects attempted to hold their right index finger slightly above the stationary touch plate.

In both experiments, 6 subjects with bilateral vestibular-loss were paired with age- and sex-matched control subjects. Subjects with vestibular-loss had horizontal vestibulo-ocular reflex sensitivities more than 4 standard deviations below normal. For experiment 1, the vestibular-loss group was subdivided by 3 subjects who were poorly-compensated and 3 subjects who were well-compensated based on their ability to participate in the sinusoidal task across all frequencies. This classification was consistent with differences between subgroups in the i) number of fall trials, ii) variability of the head in space during translations and iii) duration of vestibular-loss. There were no differences between the well-compensated and poorly-compensated groups in the severity of horizontal canal or static or dynamic otolith function based on vestibulo-occular measures.

Here we report anterior-posterior displacement and variability of the trunk, head and index finger in space based on motion of reflective markers placed on the right side of the body (Motion Analysis Corp; Santa Rosa). Displacement was measured as the average peak displacement at the platform driving frequency whereas the standard deviation of displacement at other frequencies was used to quantify response variability. Anterior-posterior displacement and variability of the center of body mass (CoM) was calculated from a 3-segment model (Kane and Levinson, 1985) and morphological measures from each subject. Anterior-posterior excursion of the center of foot pressure (CoP) was calculated from the weighted average of 4 vertical strain gauges under each foot. In Experiment 1, repeated measures 2x2x6 ANOVAs were used to compare trunk, CoM and

CoP displacements and variability between the vestibular-loss and control groups and between eyes open and eyes closed at the 6 frequencies of surface translation. In Experiment 2, repeated measures 2x2x5 ANOVAs compared variability, gain and phase from linear system spectral analysis of CoM and fingertip displacement for the vestibular-loss and control groups, light touch (LT) and no touch (NT) conditions at the 5 frequencies of surface rotation.

RESULTS

Experiment 1: Surface Translations and Effects of Vision

Healthy control subjects and well-compensated vestibular-loss subjects with eyes open gradually changed their postural strategies as sinusoidal surface translations increased in frequency. For slow translation frequencies (< 0.5 Hz), these subjects strapped down their heads to their trunks and to the platform. For fast translation frequencies (> 0.75 Hz), these subjects fixed their upper-trunk and head in space. There was no significant difference in upper trunk displacement between the well-compensated vestibular-loss group and the control group ($p > 0.05$). In contrast, poorly-compensated vestibular-loss subjects produced large swaying motion of the upper trunk and head at all frequencies, even with eyes open ($p < 0.05$). Figure 1 compares the excessive trunk motion of a poorly-compensated vestibular-loss subject with trunk stable-in-space from an age-matched control subject during fast sinusoidal surface translations.

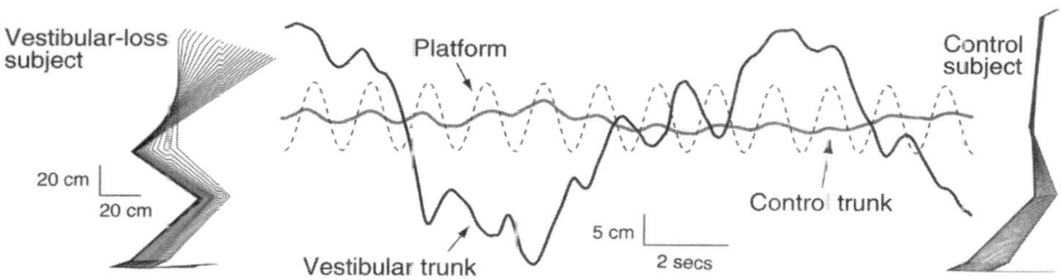

Figure 1. Comparison of stick figures and anterior-posterior displacement of the trunk during 0.75 Hz surface translations in a representative poorly-compensated vestibular-loss subject and age-matched control subject.

All three groups gradually decreased the amplitude of head and upper trunk displacement from nearly equal with the platform's 12 cm displacement at the slowest frequency to only 2 cm at the highest frequencies. The poorly-compensated vestibular-loss subjects, however, usually fell at the three fastest frequencies. All vestibular-loss subjects were less stable than age-matched control subjects based on variability of trunk in space. Whereas the control and well-compensated subjects had consistent variability of upper trunk as well as CoM displacement across all frequencies of perturbation, the poorly-compensated subjects had increasing variability for each increase in frequency of surface translation ($p < 0.05$).

In contrast to the excessively large upper trunk displacements in poorly compensated subjects, CoM displacement was not significantly different among the three groups

($p > 0.05$). All three groups significantly decreased CoM displacement as frequency of translation increased. As anterior-posterior CoM displacement decreased with translation frequency, CoP displacement increased in all three subject groups to counter the increasing surface forces associated with increasing platform velocity. However, both vestibular-loss groups produced larger CoP forces compared to their aged-matched control subjects ($p < 0.05$). Although all groups were able to reduce CoM displacement with increasing platform frequency, both vestibular-loss groups had more variable CoM displacements, especially the poorly compensated group.

Removing vision made it difficult to control the head fixed-in-space pattern and resulted in loss of balance at all frequencies in the vestibular-loss subjects and at the highest frequencies in the control subjects. With eyes closed, a large amplitude, slow-drifting motion of the upper trunk resulted in falls in 80% of attempted trials in poorly compensated subjects, 55% of trials in well-compensated subjects and only 16% of control subjects' trials, always at higher frequencies. The loss of balance resulted from inability to halt and reverse the direction of head and upper trunk slow drift motion. All subjects effectively used vision to stabilize drift of the trunk in space.

Experiment 2: Sinusoidal Rotations and Effects of Light Touch

Control and vestibular-loss subjects used similar postural response strategies in response to sinusoidal surface rotations as with the sinusoidal surface translations. The trunk and CoM were fixed to the surface for low frequencies for both groups with similar displacement amplitudes. Like responses to translations, control subjects fixed the trunk in space for high rotation frequencies ≥ 0.1 Hz (Fig. 2). Although vestibular-loss subjects could stabilize the CoM to the surface at low frequencies, their increasingly high CoM variability at higher frequencies indicated an inability to reference to an earth-centered, gravitational reference frame (Fig. 2). Whereas vestibular-loss subjects showed increased

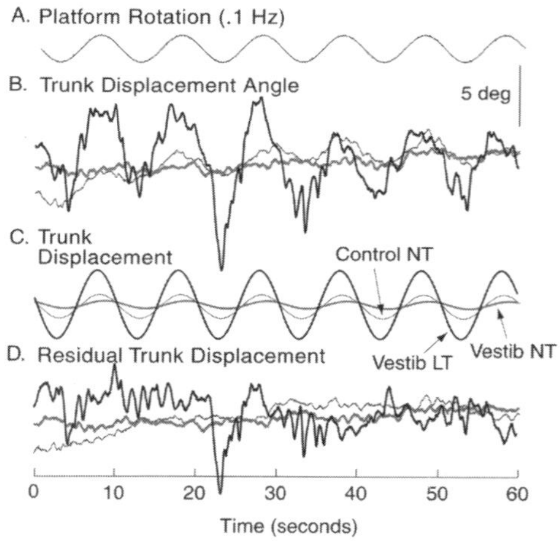

Figure 2. Comparison of control subject (Control NT) and vestibular-loss subject with light touch (Vestib LT) and no touch (Vestib NT) during *A*, sinusoidal rotation of the support surface at 0.1 Hz, *B*, displacement of the trunk in space, *C*, displacement of the trunk at the support surface frequency, and *D*, displacement of the trunk at frequencies other than the platform rotation frequency.

CoM variability as surface frequency increased ($p < 0.02$), control subjects maintained a constant variability of CoM displacement across all frequencies.

When provided with upper extremity somatosensory information from a fingertip in contact with a stable reference, the sway response (CoM variability and gain) of vestibular-loss subjects was equivalent to control subjects at all frequencies, essentially eliminating any deficit in their postural responses (Fig. 2). Both control and vestibular-loss subjects used less than 1 Newton of fingertip touch to significantly reduce CoM motion and variability relative to the rotating surface. The improvement in CoM displacement and variability due to light touch was greater for the vestibular-loss than for the control subjects, especially at the higher frequencies of rotation.

Without light touch contact, vestibular-loss subjects held their fingers fixed relative to a body-centered reference frame, whereas control subjects maintained finger position referenced in space based on the imagined position of the touch surface with eyes closed. The control subjects maintained finger position in space with a level of accuracy approaching that of the light touch condition by decoupling arm motion from platform motion. Like CoM variability, finger variability was higher than normal for the vestibular-loss subjects in the no touch condition with an increasing variability for each increasing rotation frequency.

CONCLUSIONS

The results of these two studies support the hypotheses that vestibular information, vision and light touch play similar roles in providing a reference frame for postural orientation to reduce slow drift in trunk equilibrium position. Vestibular, visual and upper extremity somatosensory information during light touch all significantly decreased the magnitude and variability of trunk displacement during sinusoidal surface displacements. All three sensory systems showed increasing importance as the frequency (and velocity) of body sway increased. For example, whereas healthy subjects showed consistent variability in trunk and CoM displacement regardless of perturbation frequency, vestibular-loss subjects, especially poorly compensated subjects, showed increasing variability as perturbation frequency increased. However, both vision and light touch were able to compensate for loss of vestibular function by reducing variability and slow drift of body sway at these high frequency oscillations.

The vestibular system is not critical for coordinating the legs during dynamic postural motions since well-compensated vestibular-loss subjects with eyes closed and without a touch reference showed a normal, gradual change in postural strategy from a strap-down towards a head/trunk fixed-in-space strategy as the frequency of oscillation increased. However, without sensory substitution with vision or light touch, vestibular-loss subjects were often unsuccessful in controlling equilibrium when the head/trunk fixed-in-space strategy was required, not because of a complete breakdown in postural coordination, but because of increasing slow drift of the upper body combined with increasing variability in upper body position. Although the somatosensory system appears capable of coordinating body segments to control equilibrium during dynamic, continuous surface translations or rotations, it requires an external reference frame for postural tasks involving stabilization of the trunk in space. When the surface translation or rotation was so slow that control of equilibrium involved maintaining the entire body position with respect to the surface, vestibular, visual and light touch information from a fingertip was not required for postural control. However, when the postural equilibrium strategy required stabilization of the trunk with respect to gravity while standing on a surface in motion, either vestibular, visual or light touch somatosensory information was necessary to reduce trunk drift and variability to maintain equilibrium. These studies cannot rule out the possibility that predictive central mechanisms were also used to help coordinate postural movements because after the first oscillation, the sinusoidal perturbations were predictable.

The inability of subjects with vestibular-loss to control equilibrium when the upper body must be decoupled from motion of the lower legs is consistent with the lack of a hip strategy in vestibular-loss subjects in response to discrete surface translations while standing across a narrow beam (Horak et al., 1990), or while attempting quiet stance on one foot or in tandem stance. However, another study demonstrated that vestibular information was not necessary for coordination of the hip strategy for postural correction if sufficient somatosensory information was available (Runge et al., 1998). Although patients with vestibular-loss could not control a hip strategy in conditions that compromised support surface information, such as a narrow base of support, they could generate a hip strategy when required by very fast translations of a firm, flat support surface. That is, vestibular-loss subjects showed a gradual increase in hip torques and hip motion to control CoM as surface translation velocity increased, similar to control subjects (Runge et. al., 1998).

Results from these studies are consistent with a top-down influence of the vestibulospinal system on postural control. Stability of the head and trunk were much more dependent on the availability of vestibular information than was CoM stability or coordination of the leg segments. The vestibular-loss subjects' postural control strategy suggested that all body parts were coupled and driven by movements of the support platform across all frequencies. It was difficult for vestibular-loss subjects to uncouple motion of the head and trunk from the legs or to uncouple motion of the hand in space from the CoM, suggesting difficulty uncoupling degrees of freedom to simplify control.

The ability to substitute visual or somatosensory information for missing vestibular information in stabilization of the trunk in space varied among subjects with similar vestibular-loss. It is not clear why some subjects with vestibular-loss compensated better than others, but all vestibular-loss subject benefited from sensory substitution. Somatosensory information from the arm associated with light touch contact of a single fingertip was even more effective than vision in stabilizing the trunk in space both during the dynamic postural task in these studies and in previous studies of quiet stance in vestibular-loss subjects and in healthy control subjects (Lackner et.al., 1999).

ACKNOWLEDGEMENTS

Supported by NIH grants DC 01849, AG 06457 and NRSA DC 00372. We thank Dr. Robert Peterka for performing tests of vestibulo-occular reflex function.

REFERENCES

Allum, J. H. J., Honegger, F., and Schicks, H., 1994, The influence of a bilateral peripheral vestibular deficit on postural synergies, *Journal of Vestibular Research,* 4, 49-70.

Pozzo, T., Berthoz, A., and Lefort, L., 1991, Head stabilization during various locomotor tasks in humans II. Patients with bilateral peripheral vestibular deficits, *Experimental Brain Research,* 85, 208-217.

Buchanan J. J., and Horak F. B., 1998, Role of vestibular and visual systems in controlling head and trunk position in space, *Society for Neuroscience,* 24, 153.

Buchanan, J. J., and Horak, F. B., 1999, Emergence of postural patterns as a function of vision and translation frequency, *Journal of Neurophysiology,* 81, 2325-2339.

Horak, F. B., Shupert, C. L., Dietz, V., and Horstmann, G., 1994, Vestibular and somatosensory contributions to responses to head and body displacements, *Experimental Brain Research,* 100, 93-106.

Horak, F. B., and Shupert, C. L., 1994, Role of the vestibular system in postural control, in: *Vestibular Rehabilitation*, Herdman, S. J., Whitney, S. L., and Borello-France, D. F. eds., Publisher, F. A. Davis, Philadelphia.

Horak, F. B., Nashner, L. M., and Diener, H. C. 1990, Postural strategies associated with somatosensory and vestibular loss, *Experimental Brain Research,* 82, 167-177.

Jeka, J. J., and Lackner J. R., 1994, Fingertip contact influences human postural control, *Experimental Brain Research,* 100, 485-502.

Kane, T. R., and Levinson, D. A., 1985, *Dynamics: Theory and Applications,* McGraw Hill, New York.

Lackner, J. R., DiZio, P., Jeka, J. J., Horak, F. B., Krebs, D., and Rabin, E., 1999, Precision contact of the fingertip reduces postural sway of individuals with bilateral vestibular loss, *Experimental Brain Research,* 126, 459-466.

Nashner, L. M., Black, F. O., and Wall, III, C., 1982, Adaptation to altered support and visual conditions during stance: Patients with vestibular deficits, *Journal of Neuroscience,* 2, 536-544.

Runge, C. F., Shupert C. L., Horak, F. B., and Zajac, F. E., 1998, Role of vestibular information in initiation of rapid postural responses, *Experimental Brain Research,* 122, 403-412.

18

SENSORY CONTRIBUTIONS TO THE CONTROL OF STANCE

A posture control model

Thomas Mergner[1], Christoph Maurer[1], and Robert J. Peterka[2]

ABSTRACT

We present the outline of a dual kinetic-kinematic postural control model. It is based on concepts of inter-sensory interaction (sensor fusion) which we consider instrumental for sensorimotor integration. Separation into kinetic and kinematic control signals begins at the level of the sensors (e.g., vestibular system - otoliths: force field meters, canals: head angular speedometers). Sensor fusion mechanisms are used to yield separate internal representations for foot support kinematics, force fields such as gravity, and contact forces such as pull or push having impact on the body. These representations are fed as global set point signals into local proprioceptive control loops of the joints. Fed into an ankle joint proprioceptive loop for body-on-support stabilization, they yield compensation of support tilt, gravity and contact forces, even when these stimuli are combined and, furthermore, voluntary lean is superimposed. Model simulations parallel our experimental findings so far obtained.

INTRODUCTION

It is still an enigma how humans control their upright stance and how they embed voluntary actions into postural mechanisms. The problems faced by research in this field are the multi-segment dynamics and biomechanics of the human body, the fact that the external force fields (such as gravity) and contact forces (reaction forces) tend to change during the interaction with the environment, and the many sensor systems involved (vestibular, visual, proprioceptive, somatosensory). This complexity led researchers in the past to resort to a "reduced system", by studying postural reflexes in decerebrate cats. These studies suggested the existence of a vestibulo-spinal reflex (VSR) and a cervico-spinal reflex (CSR; see Mergner et al., 1997). It is generally believed that these reflexes

[1] Neurological University Clinic, Freiburg, Germany. Email: mergner@uni-freiburg.de
[2] Neurological Sciences Institute, Oregon Health & Science University, Portland, Oregon, USA.

also exist in humans, but come under the control of higher CNS centers shortly after birth, in order to allow for cognitive mechanisms and volition. The latter are often believed to be responsible for our dilemma, in that they obscure in an almost insurmountable way the reflex mechanisms.

We hold, in contrast, that the problems with the classical reflex concept mainly arise from the fact that it is confined to the otolith antigravity function and does not deal with other relevant aspects of posture control such as compensation for external stimuli. Furthermore, implementation of several control functions into one mechanism requires a sensorimotor integration that goes beyond the classical reflex concept. Along this line, we present an extended and modified postural control concept in the form of a dynamic model and mention preliminary experimental evidence that support this model.

THE CLASSICAL REFLEX CONCEPT

When dealing with the classical concept of a tonic antigravity function of the otolith system, von Holst and Mittelstaedt (1950) postulated an interaction between VSR and CSR, by which "the brain is acting as if the vestibular organ was located in the trunk" (where the center of mass, COM, is located). Both the VSR and the CSR lead to activation of forelimb extensors on the side ipsilateral to the stimulus, which for the VSR is a head tilt in space and for the CSR is a trunk tilt with respect to the head. During head rotation on stationary trunk the stimuli combine with opposite signs, such that the reflexes cancel each other and trunk posture is not affected (for literature, see Mergner et al., 1997). Textbook schemes of this concept usually show a cat that is somehow free floating and they illustrate the reflex responses in terms of kinematics (forelimb extension/flexion). In our view a more realistic representation requires consideration of kinetics (forces acting on the support surface and other external forces).

In our scheme we consider, accordingly, a cat on a stationary support and assume that the cat has adopted the illustrated head and trunk postures during spontaneous behavior (Fig. 1A-C). During tilt of the trunk towards the left side, the cat compensates for the leftward shift of the COM's gravitational force vector by increasing/decreasing the extensor muscle tone of its left/right leg (for didactic reasons: the COM force vector is given by downward arrows and the compensating ground reaction forces, supported by extensor muscle tone, by upward arrows). This applies whether, during the trunk lean, the cat keeps the head aligned with respect to the trunk (VSR; A) or vertical in space (CSR; B). In contrast, head tilt on a stationary trunk requires essentially no reaction (C). This VSR-CSR interaction in Fig. 1A-C can be viewed as a neck proprioceptive coordinate transformation that uses the otolith signal arising in the head to reference the COM in the trunk to absolute space. Since this control concerns forces, we call it a kinetic one and consider in this respect the otolith system as a force field meter which mainly indicates the direction of the gravitational vector.

This otolith antigravity mechanism, however, does not allow the cat to cope optimally with external perturbations that may add at any moment to its spontaneous behavior. We consider two perturbations as most relevant. One would be an external contact force having impact on the body (Fig. 1D). Conceivably, the otolith feedback control mechanism that is appropriate for the conditions A-C would no longer provide optimal body stabilization in space with the additional force. Specifically, an otolith mechanism would sense the perturbation only to the extent that body orientation has

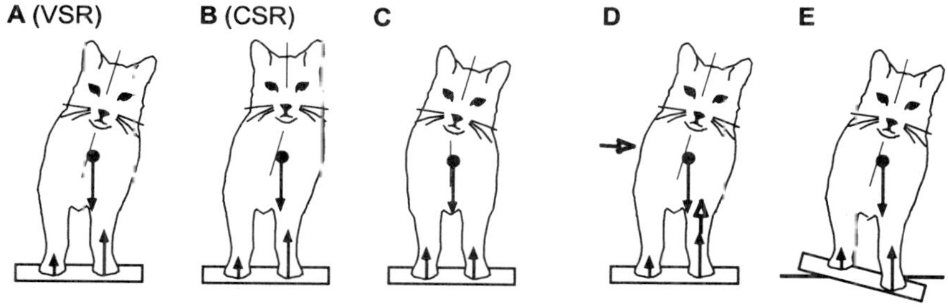

Figure 1. Modified version of the classical postural reflex scheme which is used to illustrate the antagonistic interaction between the vestibulo-spinal reflex (VSR; *A*) and the cervico-spinal reflex (CSR; *B*) during head tilt on stationary trunk (*C*). Perturbations by an external force (push; *D*) or a tilt of the support surface (*E*), superimposed on spontaneous body tilt (A), are compensated for by an extra counter-force (see hollow arrows) and a change in the length of the legs, respectively.

changed. The same applies to the second perturbation considered, i.e. a tilt of the support surface (Fig. 1E), which requires a change in length of the legs for maintenance of body posture. A control system that can cope with all three external stimuli (gravity, external contact force, support tilt) and is solely built upon the antigravity mechanism appears unrealistic (e.g., it would require an extremely high internal gain and almost ideal dynamics), a notion that is counter to current evidence.

An alternative would be to include additional sensors in the control mechanism that would allow the individual to distinguish among the external stimuli and to perform the appropriate weighting of the sensory inputs. Next we consider evidence for such an alternative solution, where combinations of different sensor signals yield separate internal representations of gravity, external forces, and support tilt, and then feed into a single postural control mechanism with appropriate weighting.

A VESTIBULAR MECHANISM SPECIFYING SUPPORT KINEMATICS

In psychophysical studies of human self-motion perception (see Mergner et al., 1997) we showed that there exists, in addition to the above described coordinate transformation of the otolith signal, a chain of axial proprioceptive transformations of vestibular *canal* information downwards to the feet, which yield an internal representation of foot support kinematics. To appreciate this finding intuitively, consider a subject standing with eyes closed on a platform that starts to rotate; the subject experiences primarily the platform as rotating in space and rotation of his own body as the consequence of the platform rotation. The underlying mechanism could be described in the form of a dynamic model in which proprioceptive body-on-support signals are combined with a support-in-space signal derived from the proprioceptive transformation of the canal signal to the feet. We have suggested that this kinematic mechanism may also be part of the human postural control system (Mergner and Rosemeier, 1998; see also below).

Noticeably, it was this concept which led us to assume that perception and sensory control of action is based on internal representations of the external physical events, as reconstructed from fusion of two or more sensory cues - rather than on a single sensory input (as reflected in the concept of reflexes, like the VSR, which alone would be equivocal as to its functional meaning).

INTERNAL REPRESENTATION OF CONTACT FORCES

It is commonly held that somatosensory cues, arising from the ground reaction forces in the feet, participate in the control of human upright stance. This view is supported by investigations of patients with chronic bilateral loss of vestibular function, who balanced on a sinusoidally tilting platform with eyes closed (Maurer et al., 2000). In the mid-frequency range they tended to maintain body uprightness with respect to the platform rather than in space, compatible with loss of the vestibular cues. At low frequencies (≤0.1 Hz), however, they shifted to a body stabilization in space. The conclusion was that humans make use of a low frequency somatosensory cue which yields a measure of the shift of the center of pressure (COP) under their feet.

We deem it likely that the COP-like cue also is used by intact individuals. One function would be to keep the COM's gravitational vector within the limits of the foot base. Another likely function would be to identify external forces such as a horizontal push or pull on the body. As mentioned before, the otolith system would detect the stimulus only to the extent that it led to a change in body orientation. However, the stimulus is also transmitted, more directly, to the ground reaction forces where it adds to the gravity-related force component. Given that subjects dispose of an internal estimate of the gravity component, they would be able to infer from the COP-like cue the external contact force.

IMPLEMENTATION INTO A POSTURAL CONTROL MODEL

The sensory information of joint angle, arising mainly from muscle spindle input, is required for the above mentioned coordinate transformations of the vestibular signals. Furthermore, the proprioceptive input locally controls joint position and movement. The question is how to modify this local proprioceptive joint control to account for external stimuli (gravity, pull, platform tilt). When trying to implement the above mentioned psychophysical findings into a postural control scheme, we suggested previously that the internal representations of the external stimuli are used as set point signals for the local proprioceptive loops (Mergner et al., 1997; Mergner and Rosemeier, 1998). Formally, combining the internal support-in-space representation with the local proprioceptive body-on-foot(support) control transforms the latter into a body-in-space control. As a simplified example consider a subject in upright stance who uses ankle proprioceptive input to control spontaneous body sway, having set the desired body-to-foot angle to 90°. Upon a toe-down tilt of the support surface by 8°, for instance, the internal representation of the stimulus would shift the set point towards 98°, while proprioceptive sway compensation continues.

We assume that this set point control principle also applies to the internal representation of gravity, having in mind Sherington's observation that the increased vestibular tone in forelimb extensors of the decerebrate cat collapses after cutting the dorsal spinal roots. This may appear surprising, since there exist direct (monosynaptic) projections from vestibular nuclear neurons to spinal motoneurons. However, it can be explained by assuming that the vestibulo-(reticulo)-spinal network is providing a set point signal to the local proprioceptive loops, rather than providing a direct motor input. In line with this notion is the finding that lesions of direct vestibulo-collic reflex pathways in the cat leave no major functional deficits (see Wilson and Schor, 1999). An analogous set point mechanism is assumed for the internal representation of the contact force.

A KINETIC-KINEMATIC MODEL OF HUMAN POSTURAL CONTROL

A simplified view on our human postural control model is given in Fig. 2. The key features are (i) the internal representations of the external stimuli, i.e. of support kinematics, of the gravity force field and of external contact forces, and (ii) the use of these internal representations to provide set point signals to a local (ankle joint) proprioceptive feedback loop that tries to stabilize the body on its support. Omitted are details of the sensor fusion mechanisms by which the internal representations are created. The set of sensors (ankle proprioception, canal and otolith vestibular inputs, and a somatosensory cue in the feet) are considered the minimum required to establish these representations (other receptors, such as Golgi tendon organs, are considered as redundant as to this aim, at least). Also omitted are the physics of the body and its support (implemented as two mutually coupled segments, one being the head-trunk-leg segment, the other a foot-on-support segment), a dead time (100 ms), and the controler (which generates corrective torque based on sensory inputs).

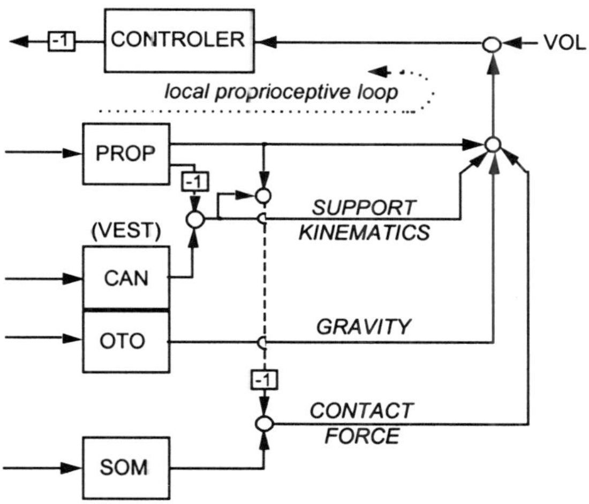

Figure 2. Dual kinetic-kinematic model of human posture control (simplified representation, physics of body and stimuli omitted). The internal stimulus representations of gravity, external contact force, and support kinematics are derived from sensor fusion mechanisms (details not shown) which receive inputs from ankle proprioception (PROP), the canal (CAN) and otolith (OTO) parts of the vestibular system (VEST), and from somatosensory cues in the feet (SOM). The representation of the external force is derived by subtracting an estimate of the COM's contribution to the ground reaction force (dashed arrow) from the SOM cue. Signals derived from the internal representations are used, together with a voluntary control signal (VOL), to alter the set point of the local (ankle joint) proprioceptive loop that stabilizes the body on its support. The sign reversal of the controler output (box -1) indicates the negative feedback character of the control loop.

We refer to the internal representations of the external stimuli as global variables, in so far that they may act on different joints depending on momentary body geometry and functional demands. The distinction between global and local variables in the control of a multi-segment body is advantageous as concerns the computational effort (Mergner and Rosemeier, 1998). However, the mechanism which assigns the global variables to the appropriate joint effectors is still an enigma (possibly it involves load receptors).

COMPARISON BETWEEN EXPERIMENTAL FINDINGS AND MODEL

For an experimental test of our model (Mergner et al., 2000), we presented normal subjects and vestibular loss patients with sinusoidal tilts on a motion platform and body pull stimuli (f= 0.05-0.8 Hz). Visual orientation cues were excluded. We applied the pull stimuli via a body harness by means of two force controlled cable winches, thereby varying peak ankle torque from 1-16 Nm. In another experiment the platform serving as body support during the tilt and pull stimuli was coupled to body position in terms of a "body sway referencing" (which kept the ankle angle essentially constant). The COP and COM responses obtained in normal subjects could well be matched by model simulations in terms of gain and phase characteristics with one and the same model settings for all stimulus conditions and parameters. This also applied to the data of the patients after removing the model's vestibular components and increasing the gain of the somatosensory input. In line with the model's predictions, patients were unable to balance (compensate) the stimuli in the two conditions with body sway referencing of the platform.

We hold that our model is biologically plausible, in that it not only delivers simulation results which closely parallel the experimental results, but also fulfils the criteria of parsimony, robustness, and computational speed. Furthermore, it can cope with complex stimulus conditions (without the need for explicit weighting or reweighting of sensory signals) and even voluntary lean can be superimposed. Because of its modular character, it can be applied to the movement control of single body segments (e.g. the head or the arm) as well as to a multi-segment body during posture control.

ACKNOWLEDGEMENTS

DFG Me 715/4-3, NIH AG17960

REFERENCES

Maurer, C., Mergner, T., Bolha, B., and Hlavacka, F., 2000, Vestibular, visual, and somatosenory contributions to human control of upright stance, *Neuroscience Letters,* 281, 99-102.

Mergner, T., and Rosemeier, T., 1998, Interaction of vestibular, somatosensory and visual signals for posture control and motion perception under terrestrial and microgravity conditions, *Brain Research, Brain Research Reviews,* 28, 118-135.

Mergner, T., Maurer, C., and Peterka, R. J., 2000, Human postural control: New dynamic model and experimental evidence, *Society for Neuroscience,* 26, 785.12.

Mergner, T., Huber, W., and Becker, W., 1997, Vestibular-neck interaction and transformation of sensory coordinates, *Journal of Vestibular Research,* 7, 347-367.

von Holst, E., and Mittelstaedt, H., 1950, Das Reafferenzprinzip (Wechselwirkungen zwischen Zentral-nervensystem und Peripherie), *Naturwissenschaften,* 37, 464-476.

Wilson, V. J., and Schor, R. H., 1999, The neural substrate of the vestibulocollic reflex: what needs to be learned, *Experimental Brain Research,* 129, 483-493.

SECTION IV

Motoneurones and Motor Units

Sherrington's dictum that the motoneurone is the final common pathway for movement generation is a physiological truism still emphasised in most textbooks, but the concept is increasingly questioned by spinal cord neurobiologists (e.g., Burke, 1985; see also Chapter 31). The familiarity of this phrase masks the imprecision of our understanding of how motoneurones transduce their myriad of inputs. This Section combines contributions from two sessions from the Symposium and hence the chapters cover a range of perspectives on motoneurones and the ways in which they are driven by peripheral inputs - from commentaries on what is still to be determined through to reviews of recent research findings.

This Section begins with a comprehensive review of the operation of classical presynaptic inhibitory mechanisms in animals by P. Rudomin (Chapter 19). These mechanisms can focally alter transmission along a particular reflex pathway such that individual collaterals of the same afferent can be affected in different ways. The challenge here is to determine the overall behavioural effect of such a "fine-grained" capacity to modulate motoneurone discharge, how it is integrated with postsynaptic effects, and how it helps in the selection of motoneurones in a task-related way. R. Burke (Chapter 20) considers the importance of the particular motor task in the control of the firing and selection of motoneurones. While the utility of the size principle of orderly recruitment is accepted, and exceptions to it proposed (although not necessarily agreed upon), the underlying mechanisms for this principle are, in his words "still unclear". He argues for a functionally useful selection of the "right" motoneurones for each task. Such a selection process is critically dependent on the activity of spinal interneurones (see also Chapter 31).

Given that motoneurones must adapt throughout life to changes in length, strength and endurance of the muscle fibres they innervate and the tasks they must perform, it is likely that their own properties change on times scales much longer than those usually analysed in acute experimental studies. D. Kernell (Chapter 22) argues that these properties include the "baseline" ones (defining threshold, and resistance) and those controlling repetitive firing, superimposed on which are not only the ionic synaptic currents, but also the modulating influences (see below).

Once a motoneurone is firing steadily then it is possible to analyse the pattern of firing to deduce characteristics of the transformation between the input driving it, the trajectory of the action potentials' after-hyperpolarization and the level of synaptic noise. The methods rely on the analysis of time interval histograms and the interval "death rate" (when a firing "terminates" the interspike period) generated by long trains of impulses. This approach was greatly refined by P. Matthews (Matthews, 1996; see also Powers and Binder, 2000) and he amplifies this in Chapter 23, as do R. Powers, M. Binder and colleagues in Chapter 24. The latter group has tested experimentally some of Matthews' predictions and deduced that the death rate of interspike intervals reflects the instantaneous distance (time) to threshold during the interval. They propose that the death rate measure may provide a functionally relevant index of changes in motoneurone excitability during repetitive discharge. Their method is applicable during periods of steady firing and it takes account of changes in spike threshold during the preceding spike's after-hyperpolarization. The challenging question of how to assess "excitability" during phasic changes in motoneuronal output remains for the future.

The contribution by M. Binder (Chapter 25) draws together several of his and Powers' group's studies on the distribution of effective synaptic currents to motoneurones of different size and threshold in the lumbar cord of the cat. Virtually all segmental, sensory, and supraspinal inputs exert their effects on cat motoneurones via interneurones (see also Chapter 31). Now Binder reports near-linear summation between the currents produced by various combination of two inputs activated together, a finding which implies that the spatially distributed inputs act independently on the motoneurone (see also Prather et al., 2001). An implication of these studies is the ability of a particular input to alter the range of recruitment thresholds within the motoneurone pool and thereby alter the order of recruitment. Nonetheless, the actual order of recruitment remains quite robust (for recent views, see Haftel et al., 2001). For example, A. Schmied and colleagues (Chapter 21) show that presynaptic inhibition of human wrist extensor motoneurones evoked by flexor Ia afferents affects the motoneurones of higher recruitment threshold more than those of lower threshold. The homonymous Ia effect itself (produced by radial nerve stimulation) is larger in low-threshold motoneurones, however. In their work, frank reversals in the order of recruitment of simultaneously recorded pairs of motor units do not occur.

The mechanisms and function of persistent inward currents ("plateau currents") in maintaining repetitive motoneurone discharge are currently under intense investigation (for historical aspects, see Hornby et al., 2002). J. Hounsgaard and colleagues (Chapter 27). review recent evidence that plateau currents are critically dependent on a particular subunit expressed by L-type Ca^{++} channels. At least in the turtle, they argue that the intracellular Ca^{++} level plays a critical role in the physiological actions of these channels. They further emphasise the powerful modulation of plateau currents via several agents that operate on both G-protein coupled metabotropic receptors and ligand (ion)-gated receptors.

H. Hultborn (Chapter 26) and D. Collins (Chapter 28) and their colleagues explore the impact of plateau currents on the motoneurone's firing rate behaviour. Hultborn describes the acceleration in firing frequency as the threshold for the non-inactivating (persistent) inward current is exceeded and points out that this acceleration (at least in some experimental preparations) has a relatively long delay and slow onset. To bring out this behaviour, his group uses intracellularly injected triangular current pulses (e.g., Bennett et al., 1998), which exposes the counter-clockwise hysteresis in the voltage-current relation and can be used to describe the stimulus current-spike-frequency relation. Collins and colleagues describe in Chapter 28 how a triangular pattern of sensory input also appears to trigger a plateau-like response in the motoneurones innervating ankle extensors of conscious human subjects. This response appears capable of recruiting many motor units, accelerating their discharge and generating large forces under selected experimental conditions in human subjects (see also Collins et al., 2001). It remains to reach agreement on what role these responses play during more natural conditions and during pathophysiological circumstances (see, for example, Brownstone et al., 1992; Baldissera et al., 1994; Hornby et al., 2001, 2002). The major difficulty here is that in human studies, the manifestation and action of motoneuronal plateau currents must be inferred, rather than measured directly as in animal studies.

Pathological changes in motoneurone behaviour can be examined by analysing the firing characteristics of motor units during voluntary tasks. C. Thomas and colleagues (Chapter 29) collate much useful information about the increased variability of motor unit discharge when descending inputs to spinal motoneurones are completely or partially disrupted, as in spinal cord injury. They propose that the increase in discharge variability, and the presence of doublet discharges, may optimise force production in weakened muscles.

REFERENCES

Burke, R.E., 1985, Integration of sensory information and motor commands in the spinal cord, in: *Motor Control: From Movement Trajectories to Neural mechanisms. Short Course Syllabus*, P.S.G. Stein, ed., Society for Neuroscience, Bethesda, pp. 44-66.

Baldissera F., Cavallari P., and Dworzak F., 1994. Motor neuron 'bistability'. A pathogenetic mechanism for cramps and myokymia, *Brain*, 117, 929-939.

Bennett D.J., Hultborn H., Fedirchuk B., and Gorassini M., 1998, Synaptic activation of plateaus in hindlimb motoneurons of decerebrate cats, *Journal of Neurophysiology*, 80, 2023-2037.

Brownstone R.M., Jordan L.M., Kriellaars D.J., Noga B.R., and Shefchyk S.J., 1992, On the regulation of repetitive firing in lumbar motoneurones during fictive locomotion in the cat, *Experimental Brain Research*, 90, 441-455.

Collins D.F., Burke D., and Gandevia S.C., 2001, Large involuntary forces consistent with plateau-like behavior of human motoneurons, *Journal of Neuroscience*, 21, 4059-4065.

Haftel V.K., Prather J.F., Heckman C.J., and Cope T.C., 2001, Recruitment of cat motoneurons in the absence of homonymous afferent feedback, *Journal of Neurophysiology*, 86, 616-628.

Hornby, T.G., McDonagh, J.C., Reinking, R.M., and Stuart, D.G., 2001, Open issues on the functional role of plateau potentials in the repetitive discharge of motoneurons in experimental animals and humans, in: Sensorimotor Control, R.Dengler and A. Kossev, eds., IOS Press, Amsterdam, pp. 65-74.

Hornby, T.G., McDonagh, J.C., Reinking, R.M., and Stuart, D.G., 2002, Motoneurons: A preferred firing range across vertebrate species? *Muscle & Nerve*, in press.

Matthews P.B.C., 1996, Relationship of firing intervals of human motor units to the trajectory of post-spike after-hyperpolarization and synaptic noise, *Journal of Physiology*, 492, 597-628.

Powers R.K., and Binder M.D., 2000, Relationship between the time course of the afterhyperpolarization and discharge variability in cat spinal motoneurones, *Journal of Physiology*, 528, 131-150.

Prather J.F., Powers R.K., and Cope T.C., 2001, Amplification and linear summation of synaptic effects on motoneuron firing rate, *Journal of Neurophysiology*, 85, 43-53.

19

SELECTIVITY OF THE CENTRAL CONTROL OF SENSORY INFORMATION IN THE MAMMALIAN SPINAL CORD

Pablo Rudomin*

ABSTRACT

Afferent feedback from muscle proprioceptors, as well as movement-induced activation of skin receptors plays an important role in the patterning of motor activity for stepping and postural control. An important component in this control is the presynaptic GABAergic modulation of the synaptic effectiveness of muscle and cutaneous afferents, known to change in phase with the locomotor cycle, during the execution of voluntary movements, or after a peripheral nerve injury. Recent electrophysiological studies, together with ultrastructural observations, indicate that the distribution of GABAa synapses in the intraspinal arborizations of muscle spindle and tendon organ afferents is not homogeneous. Namely, that some collaterals are the targets of one, or more, GABAergic interneurones, while other collaterals of the same fibre receive no GABAergic connections. In addition, both PAD and inhibition of PAD have a local character. This allows, at least in principle, decoupling the information arising from common sensory inputs. A spatially restricted modulation of PAD could play a significant role in the adjustment of the synaptic effectiveness of Ia afferents at the onset of voluntary contractions in humans, during movement-induced stimulation of the skin, or during the compensation of motor activity following partial denervation of muscles. Changes in the synchronization of the PAD-mediating interneurones can also have a profound effect on the information transmitted by a given set of afferent fibres. Data are presented that in the anesthetized cat, variation in the spontaneous activity of a population of dorsal horn neurones in laminae III-VI, that respond to stimulation of low-threshold cutaneous afferents, produce correlated fluctuations of monosynaptic reflexes by means of pre- and postsynaptic mechanisms. It is suggested that correlated changes in the level of PAD can also play a significant role in the presynaptic adjustment of the synaptic effectiveness of the afferent fibres during specific motor tasks.

* Department of Physiology, Biophysics and Neurosciences, Centro de Investigación y de Estudios Avanzados del Instituto Politécnico Nacional, México D.F. 07000, Mexico. Email: rudomin@fisio.cinvestav.mx

Intraspinal Collaterals of Sensory Fibres

Using the Golgi methods to stain myelinated fibres in longitudinal sections of the mouse spinal cord, Ramón y Cajal found about 100 years ago a series of transversal fibres merging into the gray matter and proposed that most of them were right angle collaterals of the longitudinal fibres in the white matter that ended in the gray matter at various segmental levels (Ramón y Cajal, 1904). He also noted that when collaterals arrive to their area of distribution in the gray matter, they loose their myelin sheet and contact neuronal cell bodies and dendrites, including motoneurones. In his book, Ramón y Cajal commented that Kölliker considered the finding of the intraspinal collaterals of the sensory fibres as one of the "most important advances in the knowledge of the structure of the spinal cord". The nerve impulses generated in the periphery would arrive to the dorsal columns where they would brake down in one ascending and one descending current, of the same or different intensity. The nervous wave would then propagate through the intraspinal collaterals to be transmitted to motor and funicular neurones. The excitatory current would be further propagated within the intraespinal arborizations of the axons like "the blood current in the vascular tree, where the energy of the circulating waves would be proportional to the diameter of the conductors". According to Ramón y Cajal, the largest axon branches would "absorb" most of the nervous impulse that represented the ordinary pathway for spinal reflexes.

Ramón y Cajal considered that the neurones mediating the inborn reflexes such as the rotulian reflex had secure connections that were "fatally" established during the foetal stage or during the first months after birth. Their "invariability" and hereditary character would be the consequence of a long history of plastic adaptations of the nervous system to the most urgent defensive needs of the organism. Early in the phylogenetic development, these reflexes could have shown some variation, but with time and selection of the most secure arrangements, they would become invariable.

The view that impulse transmission between neurones was of electrical nature prevailed in this, or in different forms, for about 50 years. Electrical transmission implied, to some extent, little variability in synaptic transmission. Yet, there were data suggesting that the conduction of impulses within the intraespinal arborizations of the afferent fibres could be curtailed either by spontaneous variations (Barron and Matthews, 1939), or by stimulation of a neighboring dorsal root (Howland et al., 1955).

Presynaptic Inhibition and Primary Afferent Depolarization

When the quantal nature of chemical transmission in peripheral synapses was more firmly established (Del Castillo and Katz, 1954), it became clear that variability of transmission in central synapses could also arise from fluctuations in the amount of chemical transmitter released from the presynaptic terminals by the action potential (see Redman, 1990).

But the real conceptual change came following the proposal that the synaptic effectiveness of the sensory fibres in the spinal cord could be controlled by extrinsic mechanisms (Frank and Fourtes, 1957). Yet, it was not until the early sixties when Eccles and collaborators suggested that the extrinsic control of the synaptic effectiveness was mediated by specific sets of GABAergic interneurones via axo-axonic contacts with the terminal arborizations of the afferent fibres (for review see Rudomin and Schmidt, 1999).

It is now fairly well established that the terminal arborizations of muscle and cutaneous afferents have GABAa as well as GABAb receptors. Activation of the GABAa receptors increases the permeability to chloride ions, which move according their electrochemical gradient and produce primary afferent depolarization (PAD). Reduction of transmitter release (presynaptic inhibition) may occur either because of the depolarization of the terminal arborizations of the afferent fibres, or because the associated shunt that may prevent conduction of action potentials. Activation of GABAb receptors appears to reduce the calcium currents associated with the action potential and transmitter release (for references see Rudomin and Schmidt, 1999).

It has been suggested that GABAa and GABAb receptors in muscle spindle afferents can be spatially separated (Stuart and Redman, 1992). However, it is not known if these two classes of receptors are activated by the same set or by independent sets of GABAergic interneurones. As pointed out some time ago (Rudomin, 1993), a separate activation of GABAa and GABAb receptors would allow an independent control of impulse conduction in the intraspinal branch points (via GABAa receptors) and of the frequency behavior of synaptic transmission (via GABAb receptors; see Lev-Tov et al., 1988; Peshori et al., 1998). Yet, there is still no compelling evidence that this can occur.

Descending fibre terminals in the spinal cord appear not to receive GABAergic axo-axonic synapses and have few, if any, GABAa and GABAb receptors (for review see Rudomin and Schmidt, 1999). Hence, segmental stimuli that produce PAD and presynaptic inhibition of Ia afferents, have no effects on the monosynaptic EPSPs produced in spinal motoneurones by descending volleys (see Fig. 1).

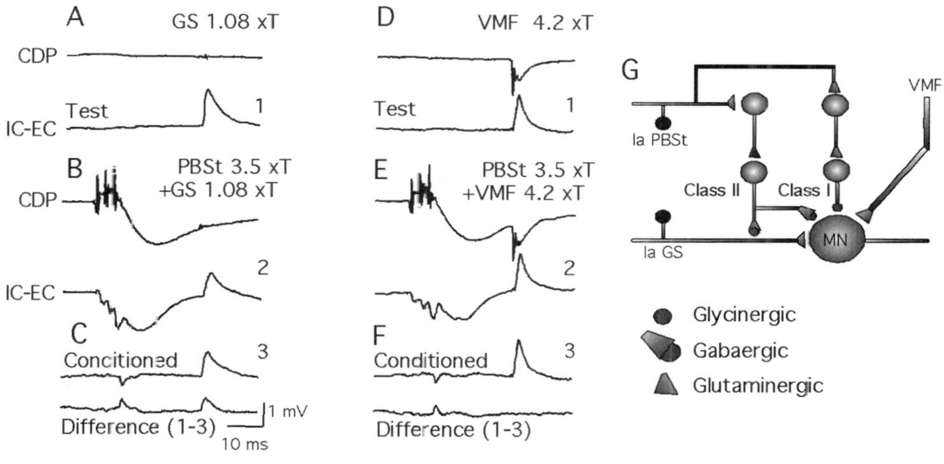

Figure 1. Differential inhibition of Ia and descending EPSPs. *A-C*, effects of posterior biceps semitendinosus nerve (PBSt) conditioning stimulation on Ia monosynaptic EPSPs elicited in a spinal motoneurone by stimulation of the gastrocnemius (GS) nerve. CDP, Cord dorsum potential. IC-EC, intracellular minus extracellular potential. *A*, Test Ia EPSP. *B*, effects produced by PBSt plus GS nerve stimulation. *C*, upper trace (3), Conditioned Ia EPSP obtained by subtracting, from trace 2, the potential changes produced by PBSt stimulation alone (not illustrated). Lower trace, difference between test and conditioned Ia EPSP (1-3). *D-F*, effects of same PBSt conditioning stimulation on monosynaptic EPSPs produced by stimulation of the ventro-medial fasciculus (VMF). Same format an in *A-C*. Note that PBSt conditioning depresses the Ia EPSP without affecting the VMF EPSP. Stimulus strengths are indicated in nthe figure and are expressed as times threshold (xT) of most excitable fibres. *G*, diagram of some of the neuronal pathways mediating pre- and postsynaptic inhibition. Presynaptic inhibition is mediated by last-order GABAergic synapsing with Ia GS terminals as well as with motoneurons (Class II interneurons). Postsynaptic inhibition is mediated by glycinergic (class I) interneurons. (Modified from Rudomin et al., 1991 and Rudomin and Schmidt, 1999.)

Patterns of PAD in Muscle Afferents

Not all sensory stimuli were equally effective in producing PAD (see Rudomin and Schmidt, 1999). It appeared that muscle spindle and tendon afferents had completely different PAD patterns. Ia fibres were depolarized by stimulation of group I fibres, mostly from flexors, as well as by stimulation of vestibulospinal fibres, but not by stimulation of the motor cortex, bulbar reticular formation, raphe nuclei and the red nucleus, that instead inhibited the PAD elicited by group I and vestibulospinal fibres (type A PAD pattern). In contrast, Ib fibres appeared to be depolarized by group I muscle afferents, as well as by vestibulospinal, reticulospinal, raphe-spinal, rubrospinal and corticospinal fibres. Stimulation of cutaneous nerves had dual actions. It produced PAD in some Ib fibres (type B PAD pattern) and inhibited PAD in other Ib fibres (type C PAD pattern).

To explain these differences in PAD patterns it was assumed that PAD of Ia and Ib fibres was mediated by separate sets of last-order GABAergic interneurones (Rudomin et al., 1983). An independent control of the synaptic effectiveness of Ia and Ib fibres, could have functional relevance in those cases where both inputs converge onto the same spinal interneurones. The non-linear interactions between DRPs produced by stimulation of Ia and Ib afferents reported by Brink et al. (1983) could result from convergence of these inputs on the first-order interneurones mediating PAD (Rudomin et al., 1983).

The view that muscle spindles were not depolarized by cortico-spinal, rubro-spinal and reticulo-spinal pathways that instead inhibited the PAD elicited by group I muscle and vestibulo-spinal fibres persisted for some time (Jiménez et al., 1988; Jankowska et al., 1993). However, in a more recent study, also made on functionally identified muscle afferents, Enríquez et al. (1996a) found that only 52% of the afferents had a type A PAD pattern, while 26% had a type B and 13% a type C PAD pattern. That is, there was a substantial number of muscle spindle afferents in which descending inputs produced PAD. Afferents from tendon organs also displayed the three types of PAD patterns, but in this case only the 11% of the analyzed fibres had a type A PAD pattern, 35% a type B and 54% a type C PAD pattern.

In this study, about 9% of the examined muscle spindle afferents had PAD patterns that could not be included within the type A, B or C categories. For example, some fibres were depolarized by group I muscle afferents, by stimulation of the motor cortex and the reticular formation, or by stimulation of the sural nerve, but not by stimulation of the superficial peroneus nerve nor by the raphe nuclei. This suggests that the division of the PAD patterns of muscle afferents in three categories is probably an oversimplification. In fact, we have found pairs of collaterals of the same afferent with different PAD patterns (see below).

Changes in PAD Patterns after a Peripheral Nerve Crush

The percentage of Ia and Ib gastrocnemius afferents with type A, B or C PAD patterns was changed by crushing the nerve and allowing regeneration and reconnection of the damaged afferents with the receptors (Enríquez et al., 1996b). Two to twelve weeks after the nerve crush, the proportion of fibres reconnected to functionally identified muscle spindles with a type A PAD pattern was reduced to 35%, while the proportion of afferent fibres in which stimulation of the bulbar reticular formation but not of cutaneous afferents produced PAD (type C PAD pattern) was increased to 65%. On the other hand, all afferents reconnected with tendon organs were depolarized by stimulation of cutaneous nerves and also by stimulation of the bulbar reticular formation. That is, they expressed a type B PAD pattern.

The alterations in the PAD patterns observed after a peripheral nerve crush have been explained by changes in the spinal pathways leading to PAD of muscle spindles and of tendon organs. It is not clear, however, if these changes result from reorganization of the spinal pathways, so that muscle spindles with a former type A PAD pattern acquired a type C PAD pattern and all tendon organs acquired a type B PAD pattern, or if reconnected individual afferents retained their original PAD patterns, and the observed alterations resulted from changes in the number of afferents with different PAD patterns.

Although the functional implications of the changes in the profiles of PAD patterns following a peripheral nerve crush have not been yet established, we have recently examined whether or not the PAD produced in the regenerated fibres is associated with presynaptic inhibition. Two to twelve weeks after a peripheral nerve crush, when the PAD patterns of the Ia afferents are still different from those observed before the peripheral nerve lesion, the presynaptic control of their synaptic efficacy is already in operation. (Enríquez, Manjarrez and Rudomin, unpublished observations).

This is in contrast with what has been observed in cutaneous afferents where PAD cannot be elicited in many fibres one month after a peripheral nerve crush (Horch and Lisney, 1981). Although there is no clear cut explanation for the higher susceptibility of the pathways mediating the PAD of cutaneous than of muscle afferents after a peripheral nerve crush, it is interesting to note that the depression of PAD following cutaneous nerve injury can be reduced by continuous application of nerve growth factor to the lesioned nerve (Fitzgerald et al., 1985). Furthermore, damage to a cutaneous nerve produces a significant reduction in the amount of substance P, somatostatin and calcitonin-related peptide in the dorsal horn of the spinal cord (Jessell et al., 1979; Barbut et al., 1981; Wall et al., 1981; Tessler et al., 1984). It thus seems that changes in PAD of cutaneous fibres following a peripheral nerve injury are associated with changes in the availability of trophic factors transported from the periphery to the spinal arborizations of afferent fibres (see also Devor, 1983). It is therefore possible that the expression of PAD in cutaneous afferents, unlike expression of PAD in group I muscle afferents, requires the presence of these peptides, but this remains to be investigated.

Differential Control of PAD

The disclosure of a differential control of presynaptic inhibition in muscle spindle afferents during voluntary contractions in humans (Hultborn et al., 1987), together with the finding that muscle spindle afferents have at least three different PAD patterns (Rudomin et al., 1983; Jiménez et al., 1988) raised the question on the extent to which all intraspinal collaterals of the same muscle spindle afferent had the same or different PAD patterns, and how these patterns could be modified by central actions. Eguibar et al. (1997) found that although stimulation of the PBSt nerve produced PAD of about the same magnitude in pairs of collaterals of individual muscle afferents ending in the intermediate zone at the L6-L7 segmental level, conditioning stimulation of cutaneous nerves and of the motor cortex could produce differential inhibition of the PAD. The relative magnitude of the inhibition of the PAD elicited in each collateral could be changed by varying the strength of the segmental inputs used to produce the background PAD, the strength and source of the inhibitory conditioning stimuli, or the site of cortical stimulation. In fact, it was possible in several fibres to completely suppress the background PAD in one collateral while the PAD elicited in the other collateral remained practically the same.

Differential inhibition of PAD has been found also in pairs of collaterals of single muscle afferents with a type A PAD pattern, one ending in the L6 segmental level, within the intermediate nucleus region, and the other in the L3 level, around the Clarke's column (Lomelí et al., 1998). In most of the examined fibres the inhibition of the PAD was stronger in the L3 than in the L6 collaterals (Fig. 2), while in some fibres the inhibition of PAD was larger at L6 than at L3. Yet, reversible spinalization could change the rostral or caudal dominance in the inhibition of the PAD, suggesting that the differential inhibition observed in these fibres, and therefore the information flowing through individual collaterals, can be subjected to a descending control.

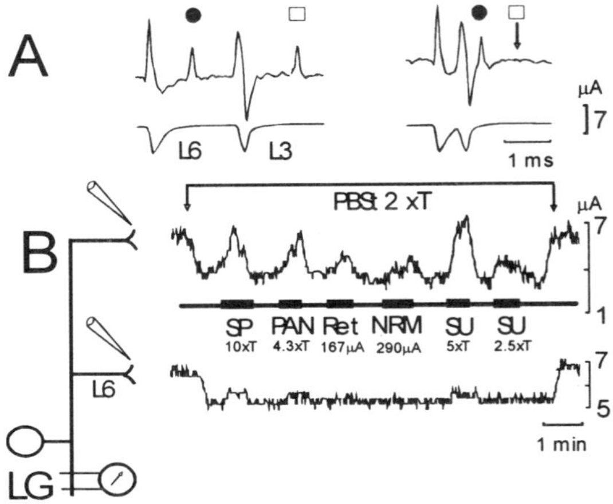

Figure 2. Selective inhibition of primary afferent depolarization (PAD) in segmental and ascending collaterals of a single group I muscle afferent with a type A PAD pattern. *A*, two separated stimulating micropipettes were placed at the L3 and L6 segmental levels within Clarke's column and intermediate zone, respectively. Antidromic action potentials produced through each micropipette were recorded from a fine lateral gastrocnemius nerve filament (LG). Interaction, because of refractoriness, between antidromic action potential produced by stimulation of both micropipettes was taken as evidence for activation of two collaterals of the same afferent fibre. *B*, continuous recordings of the intraspinal threshold changes produced in the L3 and L6 collaterals of the same fibre following a variety of sensory and supraspinal stimuli. Threshold reduction indicates PAD. Conditioning stimulation of sensory nerves [superficial peroneus (SP), sural (SU) and posterior articular (PAN) nerves] and of supraspinal structures [reticulospinal (Ret) and raphespinal (NRM) fibres] inhibited the PAD produced by stimulation of the PBSt nerve. However, the inhibition of PAD was stronger in the L3 than in the L6 collateral. The PBSt stimulus was a train of four pulses, 400 Hz, 2 xT applied once per second 25 ms before the threshold testing pulse. SU and SP nerves were stimulated with one pulse applied 50 ms before the testing pulse; PAN, Ret and NRM with a train of eight pulses at 700 Hz, preceding the threshold testing pulse by 75 ms. Stimulus strengths are indicated. (Modified from Rudomin and Schmidt, 1999.)

The PAD of afferent fibres with a type B PAD pattern (usually ascribed to tendon organ afferents) is also subjected to a differential segmental and supraspinal control. Stimulation of cutaneous nerves and/or the motor cortex or other supraspinal structures produced PAD of different magnitude in each of the two collaterals that were examined, regardless of whether both ended within the intermediate nucleus at the L6 level, or one at the L6 and the other at the L3 segment. In some of these fibres, the asymmetry in PAD magnitude was so large, that stimulation of cutaneous nerves or of the motor cortex could produce PAD in one collateral, practically without producing PAD in the other collateral (see Fig. 3 and Eguibar et al., 1997).

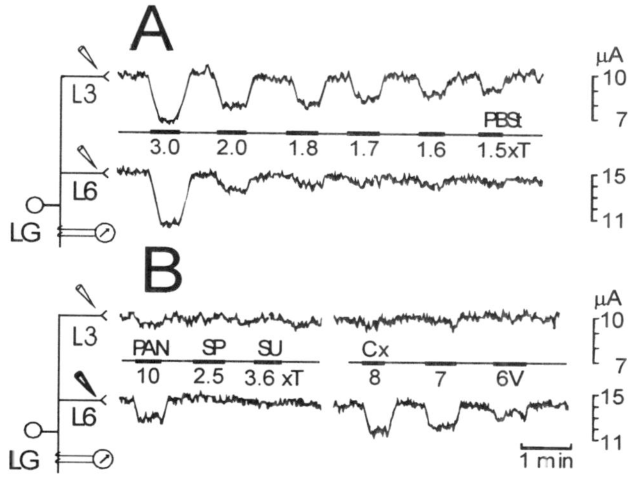

Figure 3. Selective PAD produced in segmental and ascending collaterals of a single group I muscle afferent with a type B pattern in the L6 collateral. *A-B*, continuous recordings of the intraspinal threshold changes produced in the L3 and L6 collaterals of the same fibre following stimulation of muscle, cutaneous and joint afferents and of the motor cortex. Stimulation paradigms as in Fig. 2. *A*, effects of graded stimulation of the PBSt nerve. Note that stimuli between the group I range (< 2 xT) produced a larger PAD in the L3 than in the L6 collateral, while stimuli in the group II range (3 xT) had marked effects on both collaterals. *B*, PAN stimulation also produced PAD in both collaterals, but SU and SP were ineffective. On the other hand, stimulation of the motor cortex produced strong PAD in the L6 collateral but had no effect on the L3 collateral. Lomelí and Rudomin, unpublished observations.

All these findings indicate that the intraspinal branches of sensory fibres are not hard-wired routes that diverge excitation to spinal neurones in a invariable manner, but rather dynamic pathways where information flow can be centrally addressed to reach specific neuronal targets (Lomelí et al., 1998). This central control of information flow is achieved, at least in part, by means of GABAergic interneurones connected, through axo-axonic synapses, with the intraspinal terminals of the afferent fibres (see Rudomin and Schmidt, 1999), and could play an important role in the selection of information flow in muscle spindles that occurs at the onset of voluntary contractions in humans (Hultborn et al., 1987; Iles, 1996). A selective control of presynaptic inhibition in axonal terminals has been also observed in reticulospinal neurones in the lamprey and in the nervous system of invertebrates (for review see Nusbaum et al., 1997).

One question that remains to be analyzed is the extent to which this differential control of PAD in individual collaterals of the same afferent fibres leads to a effective decoupling of the information transmitted by the afferent fibres to their target neurones. There are some indications that this may be so in the case of group II afferents (see Riddell et al., 2001). Yet, it must be pointed out that the postsynaptic expression of the differential control of PAD at the population level will depend on other factors as well, such as the degree of synchronization in the activity of all of the interneuronal sets that mediate the PAD of the intraspinal arborizations of particular sets of afferent fibres. Earlier studies (Rudomin et al., 1969; Rudomin et al., 1975) have indicated that variability of monosynaptic reflexes and of Ia monosynaptic EPSPs in single motoneurones is introduced, at least in part, by synchronized fluctuations of interneuronal activity acting on the Ia fibre terminals. They have also shown that the level of the synchronization in the interneuronal activity determines the extent to which the monosynaptic responses of individual motoneurones are correlated with the monosynaptic responses of other members of the population (Rudomin and Madrid, 1972).

Local Character of PAD

The magnitude of the differential control of the synaptic effectiveness of different collaterals of individual muscle afferent will depend on the extent to which the PAD elicited in one set of collaterals remains confined within that set and does not spread to other intraspinal arborizations of the same fibre. It also requires that PAD elicited in single collaterals or in restricted sets of collaterals is produced by separate populations of last-order GABAergic interneurones (see Harrison and Jankowska, 1984 and Jankowska and Padel, 1984).

Studies made using intraspinal threshold changes as indicators of PAD (Eguibar et al., 1997; Quevedo et al., 1997; Lomelí et al., 1998), together with ultrastructural observations (Lamotte et al., 1998) have indicated that the distribution of GABAa synapses within the intraspinal arborizations of muscle spindle and tendon organ afferents is not homogeneous. Some collaterals appear to be the targets of one or more GABAergic interneurones, while other collaterals receive no axo-axonic connections from these interneurones. Moreover, direct activation (by means of intraspinal microstimulation) of single, or small groups of GABAergic interneurones, has indicated that the monosynaptic PAD produced by this procedure can remain spatially confined within a reduced set of intraspinal arborizations of the afferent fibres, without spreading to nearby collaterals (Quevedo et al., 1997).

Intrafibre recordings of PAD of single fibres in the dorsal columns (Eccles et al., 1961), or recordings of dorsal root potentials (DRPs; see Lloyd and McIntyre, 1949) most likely represent the weighted average of the PAD's elicited in individual collaterals, and may not be the best method to disclose the local character of PAD. Measurement of intraspinal threshold changes of single fibres to localize the sites of origin of PAD appears to be more suitable, but has its own restrictions. As shown by Quevedo et al. (1997), the percentage threshold reduction produced in one collateral by a fixed PBSt conditioning stimulus, did not vary significantly by displacing the threshold testing micropipette 100 to 200 μm away from the fibre terminals. This imposes a limit on the spatial resolution of the threshold testing method to detect local changes in PAD, but this will certainly depend on the geometry of the intraspinal collaterals and the position of the threshold testing micropipette relative to the collaterals.

Tonic PAD of Muscle Afferents

Most of the information regarding the existence of a tonic PAD has been derived from studies made on cutaneous fibres projecting to the dorsal horn (see Willis, 1999 for review). Wall and McMahon (1994) and Wall (1994), have provided evidence suggesting that the long range caudal projections of cutaneous afferents in the rat spinal cord do not normally conduct action potentials, because of a tonic GABAergic influence, and that removal of the GABAergic actions would resume conduction. According to Wall (1995), this presynaptic mechanism could serve to spatially shape the input from afferents in the normal operation of the dorsal horn, and could be the means by which somatotopically inappropriate connections are prevented from influencing dorsal horn neurones in the mature system. Along the same line, Biella and Sotgiu (1995) reported that after blocking conduction of action potentials in the intact sciatic nerve of the rat, stimulation of the saphenous nerve activates neurones in the sciatic projection area. That is, there is an unmasking of the synaptic activity of inappropriate connections. This effect has been attributed, at least in part, to a reduction of a tonic presynaptic inhibition exerted on the saphenous intraspinal terminals by impulses conveyed by sensory fibres in the sciatic nerve.

Evidence of a tonic PAD in muscle spindle afferents is more limited. Dorsal root potentials produced by stimulation of Ia afferents appear not to be significantly changed by spinalization in anesthetized and decerebrate preparations, in contrast with the facilitation of DRPs produced by Ib and cutaneous afferents (see Lundberg, 1964). In the anesthetized cat, stimulation of cutaneous afferents reduces the amplitude of the antidromic responses of Ia fibres ending within the motor nucleus and increases the amplitude of the Ia monosynaptic EPSPs recorded from motoneurones. These effects were attributed to inhibition of a tonic PAD (Rudomin et al., 1974). Similar conclusions have been reached by Lomelí et al. (1998) who have shown that stimulation of cutaneous nerves as well as of the bulbar reticular formation, raphe nucleus and contralateral motor cortex may increase the intraspinal threshold of the L3 and L6 collaterals of single muscle afferents with a type A PAD pattern.

To disclose the role of descending influences on the tonic PAD, we have examined the effects of spinalization on the resting threshold of L3 and L6 collaterals of 18 individual fibres with a type A PAD pattern (Lomelí and Rudomin, unpublished observations). We found that in 6/18 fibres the resting threshold was increased in both collaterals between 22-23% during spinal block. In 3 fibres the threshold of the L3 collateral was clearly increased during spinalization, while the threshold of the L6 collateral was unchanged, and in 4 fibres the threshold of the L6 collaterals increased while that of the L3 collaterals was slightly reduced. That is, in 13/18 fibres reversible spinalization increased the intraspinal threshold in at least one of the two collaterals of the examined afferent fibres. These effects can be attributed to the removal of a tonic descending activity that increases transmission along the spinal pathways that mediate PAD and presynaptic inhibition. The available data suggest in addition that there is a rostro-caudal gradient in the magnitude of the tonic PAD, but this requires more solid evidence.

Autogenetic PAD and Presynaptic Inhibition of Muscle Afferents

Although individual muscle spindle and tendon organ afferents have type A, B or C patterns of PAD, there are some basic differences in the behavior of PAD and presynaptic inhibition of the two fibre systems that could have interesting functional implications. Repetitive activation of Ib fibres, either by electrical stimulation or during muscle contraction, generates a steady level of PAD (Lafleur et al., 1992). This leads to a transient inhibition of motoneurones (Zytnicki et al., 1990). It has been suggested that in this case presynaptic inhibition acts as a filter of input from tendon organs during muscle contraction (Zytnicki and Jami, 1998). Yet, the extent to which post-activation depression (Hultborn and Nielsen, 1998) contributes to the decay of the responses of the inhibitory interneurones activated by the tendon organ afferents has not been determined.

Single fibre Ia monosynaptic EPSPs of small amplitude do not appear to decline with repetitive stimulation, but rather summate and produce a sustained depolarization, while large EPSPs grow initially and are subsequently depressed (Peshori et al., 1998). After (-)-baclofen, a GABAb agonist, the large EPSPs are reduced in amplitude, but now summate during high frequency stimulation and lead to a sustained depolarization.

Peshori et al. (1998) suggested that, in addition to postsynaptic mechanisms, differences in the tonic levels of presynaptic inhibition in different terminals could play a relevant role in the frequency behavior of the synaptic actions mediated by the muscle afferents. It is also possible that, unlike Ib afferents, Ia fibres display a low autogenetic PAD, but this remains to be investigated (however see Lev-Tov et al., 1983).

Post-activation depression of transmitter release (Hultborn and Nielsen, 1998) could also play some role in the reduction of Ia EPSP amplitude when these are generated by trains of stimuli. Another question pending to be answered is the extent to which the magnitude of autogenetic PAD relates to the PAD patterns of the afferent fibre. Also, it is not clear if different intraspinal collaterals of the same fibre express the same or different degrees of autogenetic PAD and of the extent to which this feature can be centrally controlled.

Spontaneous Activity of Dorsal Horn Neurons and Variability of Monosynaptic Reflexes

Rudomin et al. (1987) used spike triggered averaging to examine the connections of intermediate nucleus interneurones with afferent fibres and motoneurones. This procedure disclosed the presence of negative dorsal root potentials and inhibitory ventral root potentials that followed the interneuronal activity with a rather short, presumably monosynaptic latency. Recordings from the cord dorsum revealed in addition the existence of a negative wave that preceded the interneuronal activity by 40-50 ms and lasted about 100 ms. It was suggested that this wave was generated by a set of dorsal horn neurones that had connections with the pathways mediating PAD of muscle afferents.

More recently, Manjarrez et al. (2000) provided evidence supporting this proposal. They showed in addition that during the generation of the spontaneous negative potentials in the cord dorsum (nCDPs), the monosynaptic responses of dorsal horn neurones, as well as the DRPs produced by stimulation of low threshold cutaneous and Ib muscle afferents, are facilitated, while the DRPs produced by stimulation of Ia afferents are inhibited. They also found that during the occurrence of the spontaneous nCDPs, there is a strong facilitation of the Ia-monosynaptic reflexes, both of flexors and extensors, possibly through both pre- and postsynaptic mechanisms. It is now clear that the intermittent activity of this set of dorsal horn neurones is one of the causes of the correlated fluctuations of monosynaptic reflexes that were analyzed with some detail about 30 years ago (Rudomin and Dutton, 1969); Rudomin and Madrid, 1972; Rudomin et al.,1975; Gossard et al., 1994).

We have now evidence (Piña et al., 2001) showing that during the generation of the spontaneous nCDPs, there is a synchronous activation of a population of dorsal horn neurones along several spinal segments (S1 to L5); (Eblen-Zajjur and Sandkulher, 1997). These neurones respond to stimulation of low-threshold cutaneous afferents with mono- and oligosynaptic latencies. A spinal dorsal hemisection, or a lesion comprising the ipsilateral dorsal horn and part of the intermediate zone between L6 and L7, clearly decouples the spontaneous neuronal activity generated rostrally from that generated caudally to the section (Piña et al., 2001). These findings support the existence of an ensemble of highly interconnected interneurones in the dorsal horn that extends rostro-caudally along several spinal segments that are spontaneously active and also respond to a variety of inputs besides those from low-threshold cutaneous afferents. When activated, this set of neurones, among its many actions on reflex pathways, inhibits transmission along the pathways mediating PAD of Ia afferents in a highly correlated manner and facilitates PAD of Ib afferents. It is tempting to suggest that this ensemble of dorsal horn neurones is involved in the reduction of presynaptic inhibition of muscle spindle afferents at the onset of a voluntary contraction (Hultborn et al., 1987), as well as in the resetting of presynaptic inhibition of muscle spindle afferents at the end of a step cycle, or when finding unexpected obstacles (Iles, 1996)

CONCLUSIONS

Ramón y Cajal´s finding, 100 years ago, that sensory fibres in the dorsal columns divide in ascending and descending branches, providing collaterals that enter the gray matter and made synaptic contacts with a variety of spinal neurones, was a turning point for the understanding of the functional organization of the spinal cord. It is now clear that the synaptic effectiveness of these collaterals can be modulated by extrinsic mechanisms, among them those mediated by GABAergic interneurones. This modulation transforms the intraspinal arborizations of the sensory fibres into complex and dynamic ensembles that allow addressing of information flow to specific neuronal targets. Since the synaptic effectiveness of each particular group of sensory fibres seems to be modulated by different sets of spinal neurones, each responding in a characteristic manner to supraspinal and sensory inputs, the possible outcomes of these "distributed" systems are enormous, and there is no question that their functional relevance in a given motor task will greatly depend on the magnitude of the synchronization of the actions occurring in individual collaterals. Our studies on the neuronal elements contributing to the spontaneous cord dorsum potentials have already revealed a system of highly interconnected dorsal horn neurones that modulate, in a correlated manner along several spinal segments, impulse transmission through many reflex pathways, including those controlling the synaptic effectiveness of sensory afferents. There are still many questions pertaining the functional role of this correlating system. A more complete appraisal of its functional significance will require studies in behaving mammals during the execution of specific motor tasks as well as during sensory discrimination.

ACKNOWLEDGEMENTS

Partly supported by grant NS 09196 and CONACyT grant s 41739 and 3908N.

REFERENCES

Barbut, D., Pollak, J. M., and Wall, P. D., 1981, Substance-P in spinal cord dorsal horn decreases following peripheral nerve injury, *Brain Research,* 205, 289-298.

Barron, D. H., and Matthews, B. 1939, Intermittent conduction in spinal cord, *Journal of Physiology,* 85, 73-103.

Biella, G., and Sotgiu, M. L. 1995, Evidence that inhibitory mechanisms mask inappropriate somatotopic connections in the spinal cord of normal rat, *Journal of Neurophysiology,* 74, 495-505.

Brink, E., Harrison, P. J., Jankowska, E., McCrea, D. A., and Skoog, B., 1983, Post-synaptic potentials in a population of motoneurones following activity of single interneurones in the cat, *Journal of Physiology,* 343, 341-359.

Burke, R. E., and Rudomin, P., 1977, Spinal neurones and synapses, in: *Handbook of Physiology, The Nervous System, Section 1, Vol. 1, Part 2.*, Brookhart, J. M. and Mountcastle, V. B., ed., American Physiological Society, Bethesda, MD, pp. 877-944.

Del Castillo, J., and Katz, B., 1954, Quantal components of the end-plate potential, *Journal of Physiology,* 124, 566-573.

Devor, M., 1983, Plasticity of spinal cord somatotopy in adult mammals: involvement of relatively ineffective synapses, *Dimes Birth Defects Foundation,* 19, 287-314.

Eblen-Zajjur, A. A., and Sandkulher, J., 1997, Synchronicity of nociceptive and non-nociceptive adjacent neurons in the spinal dorsal horn of the rat: stimulus-induced plasticity, *Neuroscience,* 76, 39-54.

Eccles, J. C., Eccles, R. M., and Magni, F., 1961, Central inhibitory action attributable to presynaptic depolarization produced by muscle afferent volleys, *Journal of Physiology,* 159, 147-166.

Eguibar, J. R., Quevedo, J., and Rudomin, P., 1997, Selective cortical and segmental control of primary afferent depolarization of single muscle afferents in the cat spinal cord, *Experimental Brain Research,* 113, 411-430.

Enríquez, M., Jiménez, I., and Rudomin, P., 1996a, Segmental and supraspinal control of synaptic effectiveness of functionally identified muscle afferents in the cat, *Experimental Brain Research,* 107, 391-404.

Enríquez, M., Jiménez, I., and Rudomin, P., 1996b, Changes in PAD patterns of group I muscle afferents after a peripheral nerve crush, *Experimental Brain Research,* 107, 405-420.

Fitzgerald, M., Wall, P. D., Goedert, M., and Emson, P. C., 1985, Nerve growth factor counteracts the neurophysiological and neurochemical effects of chronic sciatic nerve section, *Brain Research,* 232, 131-141.

Frank, K., and Fourtes, M. G. F., 1957, Presynaptic and postsynaptic inhibition of monosynaptic reflex, *Federation Proceedings,* 16, 39-40.

Gossard, J. P., Floeter, M. K., Kawai, Y., Burke, R. E., Chang, T., and Schiff, S. J., 1994, Fluctuations of excitability in the monosynaptic reflex pathway to lumbar motoneurons in the cat, *Journal of Neurophysiology,* 72, 1227-1239.

Harrison, P. J., and Jankowska, E., 1984, Do interneurones in lower lumbar segments contribute to the presynaptic depolarization of group I muscle afferents in Clarke's column? *Brain Research,* 295, 203-210.

Horch, K. W., and Lisney, J. W., 1981, Changes in primary afferent depolarization of sensory neurones during peripheral nerve regeneration in the cat, *Journal of Physiology,* 313, 287-299.

Howland, J., Lettvin, J. T., McCulloch, W. S., Pitts, W., and Wall, P. D., 1955, Reflex inhibition by dorsal root interaction, *Journal of Neurophysiology,* 18, 1-17.

Hultborn, H., Meunier, S., Morin, C., and Pierrot-Deseilligny, E., 1987, Assessing changes in presynaptic inhibition of I a fibres: a study in man and the cat, *Journal of Physiology,* 389, 729-756.

Hultborn, H., and Nielsen, J. B., 1998, Modulation of transmitter release from Ia afferents by their preceding activity - a "postactivation depression", in: *Presynaptic Inhibition and Neural Control,* Rudomin, P., Romo, R. and Mendell, L., ed., Oxford University Press, London, pp. 178-191.

Iles, J. F., 1996, Evidence for cutaneous and corticospinal modulation of presynaptic inhibition of Ia afferents from the human lower limb, *Journal of Physiology,* 491, 197-207.

Jankowska, E., and Padel, Y., 1984, On the origin of presynaptic depolarization of group I muscle afferents in Clarke's column in the cat, *Brain Research,* 295, 195-201.

Jankowska, E., Riddell, J. S., and McCrea, D. A., 1993, Primary afferent depolarization of myelinated fibres in the joint and interosseous nerves of the cat, *Journal of Physiology,* 466, 115-131.

Jessell, T., Tsunoo, A., Kanazawa, I., and Otsuka, M., 1979, Substance P: Depletion in the dorsal horn of rat spinal cord after section of the peripheral processes of primary sensory neurons, *Brain Research,* 168, 247-259.

Jiménez, I., Rudomin, P., and Solodkin, M., 1988, PAD patterns of physiologically identified afferent fibers from the medial gastrocnemius muscle, *Experimental Brain Research,* 71, 643-657.

Lafleur, J., Zytnicki, D., Horcholle-Bossavit, G., and Jami, L., 1992, Depolarization of Ib afferent axons in the cat spinal cord during homonymous muscle contraction, *Journal of Physiology,* 445, 345-354.

Lamotte, d. B., Destombes, J., Thiesson, D., Hellio, R., Lasserre, X., Kouchtir-Devanne, N., Jami, L., and Zytnicki, D., 1998, Indications for GABA-immunoreactive axo-axonic contacts on the intraspinal arborization of a Ib fiber in cat: a confocal microscope study, *Journal of Neuroscience,* 18, 10030-10036.

Lev-Tov, A., Fleshman, J. W., and Burke, R. E., 1983, Primary afferent depolarization and presynaptic inhibition of monosynaptic group Ia EPSPs during posttetanic potentiation, *Journal of Neurophysiology,* 50, 413-427.

Lev-Tov, A., Meyers, D. E. R., and Burke, R. E., 1988, Modification of primary afferent depolarization in cat group Ia afferents following high frequency intra-axonal tetanization of individual afferents, *Brain Research* 438, 328-330.

Lloyd, D. P. C., and McIntyre, A. K., 1949, On the origins of dorsal root potentials, *Journal of General Physiology,* 32, 409-443.

Lomelí, J., Quevedo, J., Linares, P., and Rudomin, P., 1998, Local control of information flow in segmental and ascending collaterals of single afferents, *Nature,* 395, 600-604.

Lundberg, A., 1964, Supraspinal control of transmission in reflex paths to motoneurones and primary afferents, *Progress in Brain Research,* 12, 197-221.

Manjarrez, E., Rojas-Piloni, J. G., Jiménez, I., and Rudomin, P., 2000, Modulation of synaptic transmission from segmental afferents by spontaneous activity of dorsal horn spinal neurones in the cat, *Journal of Physiology,* 529, 445-460.

Nusbaum, M. P., El Manira, A., and Gossard, J. P., 1997, Presynaptic mechanisms during activity in vertebrates and invertebrates, MIT Press, Cambridge, Massachusetts, pp. 237-253.

Peshori, K. R., Collins, W. F., and Mendell, L. M., 1998, EPSP amplitude modulation at the rat Ia-alpha motoneuron synapse: effects of GABAB receptor agonists and antagonists, *Journal of Neurophysiology,* 79, 181-189.

Piña, S., Manjarrez, E., Jiménez, I., and Rudomin, P., 2001, Segmental Distribution of Dorsal Horn Neurons whose Activity is Synchronized with Spontaneous Cord Dorsum Potentials, *Society for Neuroscience Abstracts,* pp. 402.2.

Quevedo, J., Eguibar, J. R., Lomelí, J., and Rudomin, P., 1997, Patterns of connectivity of spinal interneurons with single muscle afferents, *Experimental Brain Research,* 115, 387-402.

Ramón y Cajal, S., 1904, Textura del Sistema Nervioso del Hombre y de los Vertebrados. Madrid, Nicolás Moya.

Redman, S., 1990, Quantal analysis of synaptic potentials in neurons of the central nervous system, *Physiological Reviews,* 70, 165-198.

Riddell, J. S., Stecina, K., Gosgnach, S., Chakrabarty, S., Sefchyk, S. J., and McCrea, D., ,2001, Selective Supression of Group II Sensory input to Spinal Interneurons during Fictive Locomotion in the Cat, *Society for Neuroscience Abstracts,* pp. 402.6.

Rudomin, P., 1993, Central Control of Sensory Information, in: *Neuroscience: From Neural Networks to Artificial Intelligence,* Rudomin, P., Arbib, M., Cervantes-Pérez F., and Romo, R., ed., Springer Verlag. Heidelberg, pp. 116-135.

Rudomin, P., Burke, R. E., Núñez, R., Madrid, J., and Dutton, H., 1975, Control by Presynaptic correlation: a mechanism affecting information transmission from Ia fibers to motoneurons, *Journal of Neurophysiology,* 38, 267-284.

Rudomin, P., and Dutton, H., 1969, Effects of conditioning afferent volleys on variability of monosynaptic responses of extensor motoneurons, *Journal of Neurophysiology,* 32, 140-157.

Rudomin, P., Dutton, H., and Muñoz-Martinez, E. J., 1969, Changes in correlation between monosynaptic reflexes produced by conditioning afferent volleys, *Journal of Neurophysiology,* 32, 759-772.

Rudomin, P., Jiménez, I., Solodkin, M., and Dueñas, S., 1983, Sites of action of segmental and descending control of transmission on pathways mediating PAD of Ia- and Ib-afferent fibers in cat spinal cord, *Journal of Neurophysiology,* 50, 743-769.

Rudomin, P., and Madrid, J., 1972, Changes in correlation between monosynaptic responses of single motoneurons and in information transmission produced by conditioning volleys to cutaneous nerve, *Journal of Neurophysiology,* 35, 44-64.

Rudomin, P., Nunez, R., Madrid, J., and Burke, R. E., 1974, Primary afferent hyperpolarization and presynaptic facilitation of Ia afferent terminals induced by large cutaneous fibers, *Journal of Neurophysiology,* 37, 413-429.

Rudomin, P., and Schmidt, R. F., 1999, Presynaptic inhibition in the vertebrate spinal cord revisited, *Experimental Brain Research,* 129, 1-37.

Rudomin, P., Solodkin, M., and Jiménez, I., 1987, Synaptic potentials of primary afferent fibers and motoneurons evoked by single intermediate nucleus interneurons in the cat spinal cord, *Journal of Neurophysiology,* 57, 1288-1313.

Stuart, G. J., and Redman, S. J., 1992, The role of GABAA and GABAB receptors in presynaptic inhibition of Ia EPSPs in cat spinal motoneurones, *Journal of Physiology,* 447, 675-692.

Tessler, A., Himes, B. T., Soper, K., Murray, M., Goldberger, M. E., and Reichlin, S., 1984, Recovery of substance P but not somatostatin in the cat spinal cord after unilateral lumbosacral dorsal rhizotomy: a quantitative study, *Brain Research,* 305, 95-102.

Wall, P. D., 1994, Control of impulse conduction in long range branches of afferents by increases and decreases of primary afferent depolarization in the rat, *European Journal of Neuroscience,* 6, 1136-1142.

Wall, P. D., 1995, Do nerve impulses penetrate terminal arborizations? A pre-presynaptic control mechanism, *Trends in Neuroscience,* 18, 99-103.

Wall, P. D., Fitzgerald, M., and Gibson, S. J., 1981, The response of rat spinal cord cells to unmyelinated afferents after peripheral nerve section and after changes in substance P levels, *Neuroscience,* 6, 2205-2215.

Wall, P. D., and McMahon, S. B., 1994, Long range afferents in rat spinal cord. III. Failure of impulse transmission in axons and relief of the failure after rhizotomy of dorsal roots, *Philosophical Transactions of the Royal Society of London. Series B: Biological Sciences,* 343, 211-223.

Willis, W. D., 1999, Dorsal root potentials and dorsal root reflexes: a double-edged sword, *Experimental Brain Research,* 124, 395-421.

Zytnicki, D., and Jami, L., 1998, Presynaptic inhibition can act as a filter of input from tendon organs during muscle contraction, i*n: Presynaptic Inhibition and Neural Control.* Rudomin P., Romo R. and Mendell L., ed., Oxford University Press, London, pp. 303-314.

Zytnicki, D., Lafleur, J., Horcholle-Bossavit, G., Lamy, F., and Jami, L., 1990, Reduction of Ib autogenetic inhibition in motoneurons during contractions of an ankle extensor muscle in the cat, *Journal of Neurophysiology,* 64, 1380-1389.

20

SOME UNRESOLVED ISSUES IN MOTOR UNIT RESEARCH

Robert E. Burke*

ABSTRACT

The intrinsic properties of motoneurones, muscle units, and synaptic inputs exhibit correlated variations that subserve a wide range of functional demands. In large limb muscles, these correlations suggest distinct "types" of motor units, while in smaller, distal muscles their distributions are more continuous. The CNS mechanisms that control recruitment patterns are still unclear, particularly the organization of spinal interneurone circuits. We need new approaches to identify segmental interneurones by their inputs and output targets. However, *functional* circuitry is changeable, depending on the "state" of the system. Shifting alliances of interneurone groups can in principle produce virtually unlimited permutations of motor unit coactivation and suppression. Although such state-dependence plasticity is a challenge, it can also be a useful tool in unraveling interneurone organization.

INTRODUCTION

Sherrington's concept of motoneurones as "the final common path" (Sherrington, 1904), inextricably linked with the muscle fibers (the "motor unit") that are innervated by them, and recognition that the aggregate force output of a muscle is regulated by the numbers of motor units that are recruited during a given action (Liddell and Sherrington, 1925) are now part of the everyday language of neuroscience. The introduction by Eccles and others of intracellular recording from motoneurones (Brock et al., 1952; Frank and Fuortes, 1955) led to a flood of new information about motoneurones and motor units. Anders Lundberg extended the work to the organization of specific segmental interneurone circuits that operate during reflex action, as well as their interactions with each other and with descending control of spinal pathways (Lundberg, 1975; see also Baldissera et al., 1981; Burke, 1999b; Jankowska, 1992). The inter-relations between central systems for controlling motoneurones and the properties of the muscle units that

* Laboratory of Neural Control, National Institute of Neurological Disorders and Stroke, National Institutes of Health, Bethesda, MD 20092-4455, USA. Email: reburke@helix.nih.gov

Sensorimotor Control of Movement and Posture
Edited by Gandevia *et al.*, Kluwer Academic/Plenum Publishers, 2002

they innervate developed in parallel since the mid-1960's (Burke, 1981; Henneman and Mendell, 1981).

Thus it seems that the systematic study of motoneurones and motor units is a mature discipline. This does not mean that the big questions have been answered satisfactorily but rather that we spend most of our time wrestling with issues that have been with us for decades. Among them are: Are there really distinct motor unit "types" and, if so, how many are there? How are motor units with different properties used in real world movements with widely varying demands? How stable are recruitment sequences and how are they produced? And finally, how are populations of motoneurones and motor units linked together to produce coordinated action? In a mature discipline, it seems useful to re-examine such basic issues periodically, lest we fall into systems of belief that are too rigid (Stuart, 1999).

MOTOR UNIT TYPES

The development of the notion of distinct 'types' of motor units on the basis of muscle unit properties has been reviewed recently (Burke, 1999a). The properties of motoneurones themselves and their synaptic inputs also vary systematically in relation to muscle unit properties[8], although these characteristics alone make less compelling clusters than do the combinations of muscle unit properties (Zengel et al., 1985).

So how many motor unit types are there? Not surprisingly, the answer is "It depends". For large limb muscles in the cat and rat, and probably for humans as well, three basic types, usually called FF, FR, and S which predict the histochemical makeup of muscle units, seem adequate to organize much of the existing evidence (Burke, 1999a; Burke, 1981). However, there is evidence that the fourth unit type intermediate between FF and FR unit groups, referred to as F(int) (McDonagh et al., 1980), may also exhibit a distinct myosin isoform (type IIX myosin; Larsson et al., 1991).

Although this scheme is convenient, it does encompass the wide variations found among the highly specialized motor units in extra-ocular muscles (Goldberg and Shall, 1999). In addition, the mechanical properties of motor units in small, distal limb muscles also appear to be distributed as continua (Appelberg and Emonet-Denand, 1967; Kernell et al., 1975), without the clustering that suggests distinct 'types' (but cf. Gates et al., 1991). A vexing problem is the considerable diversity among the myosin isoforms that identify muscle fiber types (Pette and Staron, 1990). Perhaps some of the permutations of these structural proteins might explain some of the mechanical variability between motor units within a given 'type', or between unit categories in different muscles or species.

RECRUITMENT ORDER

Doug Stuart recently reviewed the development of ideas about orderly motor unit recruitment, with emphasis on the progression from small to large force units during increasing force output (Stuart, 1999). Because of the coordination of motor unit properties, this 'size-principle ordering can be viewed as consistent with any of the coordinated properties that are subsumed under the rubric of motor unit types (Cope and Clark, 1991; Zajac, 1990). Systematic work on the mechanisms that underlie recruitment sequences was catalyzed by a seminal series of papers from Elwood Henneman's group in 1965 (reviewed in Burke, 1999a; Burke, 1981; Henneman and Mendell, 1981). Henneman proposed a 'size principle' in which an invariant recruitment sequence is dictated by the physical size of motoneurones or intrinsic motoneurone properties closely related to it (Henneman et al., 1965). Despite contemporary evidence for considerable stochastic variation in the recruitment of single motor units with respect to population discharge (Rall and Hunt, 1956), the invariant size principle hypothesis has had a powerful heuristic appeal.

Intracellular recording methods confirmed many of Henneman's predictions about inter-relations between motoneurone input resistance, monosynaptic group Ia excitatory synaptic potential (EPSP) amplitudes, muscle unit properties, and relative susceptibility to recruitment in stretch reflexes within a motor nucleus (Burke, 1968a, b; Burke et al., 1976; Fleshman et al., 1981a). The intrinsic excitability of motoneurones, measured by rheobase current, is also biassed to produce activation of small, slow twitch (type S) motor units before the larger force, faster type FR, which in turn are more excitable than the large force, fast-twitch, fatigable type FF units (Fleshman et al., 1981b; Zengel et al., 1985). More recently, variations in the ability of motoneurones to sustain plateau potentials (Hultborn, 1999), which can be adaptively controlled by neuromodulators (Delgado-Lezama and Hounsgaard, 1999), have been added to the list of mechanisms that are linked to orderly recruitment (Heckman and Lee, 1999).

On the other hand, intracellular studies did not support the original size principle assumption that all synaptic systems impinging on motoneurones within a given motor pool should have the same qualitative organization. In particular, some polysynaptic systems produce excitatory drive directed to the larger, fast twitch end of the motor unit population spectrum (Binder et al., 1996; Burke et al., 1970). In addition, the existence of stochastic variations around a central tendency for orderly recruitment (Gossard et al., 1994; Rall and Hunt, 1956; Rudomin and Dutton, 1969), argues rather forcefully against an invariant organization dominated by intrinsic motoneurone properties (Cope and Sokoloff, 1999a; Henneman et al., 1974). Over time the term "size principle" has come to be used as a rubric to describe a general but not completely deterministic recruitment sequence in which small force, slow twitch, fatigue resistant motor units are activated before larger force, faster contracting, and less fatigue-resistant units, rather than an explanation of mechanisms responsible for it (Cope and Sokoloff, 1999a; Henneman et al., 1974). This ordering makes excellent biomechanical and metabolic sense under many conditions.

SELECTIVE RECRUITMENT?

Despite the advantages of size-ordered recruitment, there is evidence that under some conditions large force, fast-contracting motor units may be selectively recruited (Creed et al., 1932; Gossard et al., 1994; Kanda et al., 1977). This may be related to predominant polysynaptic cutaneous and descending excitation found among some fast twitch motor units (Binder et al., 1996; Burke et al., 1970). Analogous reversals of voluntary recruitment thresholds have been found in the human first dorsal interosseous motor units during stimulation of skin afferent from the index finger (Datta et al., 1991; Garnett and Stephens, 1981). Nardone and coworkers (Nardone et al., 1989) observed apparent selective recruitment of high threshold motor units in the human tibialis anterior during isotonic lengthening of the ankle. Howell and colleagues (Howell et al., 1995) made similar observations in the human first dorsal interosseous muscle, although Bawa and Jones (Bawa and Jones, 1999) failed to confirm selective recruitment during eccentric contractions in the human flexor carpi radialis. These and other studies of selective recruitment have been reviewed elsewhere (Burke, 1981, 1990; Cope and Pinter, 1995).

It is fair to say that the idea of selective recruitment of normally high threshold, presumably fast twitch motor units has aroused some controversy. A number of studies have reported inability to produce clear recruitment reversals (Bawa and Jones, 1999; Cope and Pinter, 1995; Cope and Sokoloff, 1999a). The evident difficulty in demonstrating differential recruitment of normally high-threshold motor units probably reflects the strength of the factors that produce size-ordered recruitment (Burke, 1981; Henneman and Mendell, 1981). It has been argued that examples of "violations" of the size principle (a rather remarkable phrase) are not functionally significant (Cope and Pinter, 1995; Cope and Sokoloff, 1999a). Although functional rationales have been offered for selective recruitment in movements that require simultaneous recruitment of an entire motor unit population (Burke, 1981, 1990), or in actions that require rapid relaxation of force (Nardone et al., 1989; Nardone and Schieppati, 1988; Smith et al., 1980), it must be acknowledged there is still no definitive evidence on this point. However, it seems to explain the situation in specialized muscles of the turtle neck (Callister et al., 1999). Differential recruitment has one compelling importance, which is that synaptic organization cannot be left out of the recruitment control equation.

One factor that complicates demonstration of differential recruitment is that the synaptic input systems that are organized to foster recruitment reversals are largely polysynaptic. This makes them dependent on the 'state' of the spinal circuitry at the moment of testing (Burke, 1999b), which may require conditions that are difficult to attain in controlled laboratory situations. Indeed, group Ia EPSPs had been recorded in alpha-motoneurones for a quarter-century before the existence of a strong disynaptic EPSP component was documented by Schomburg and Behrends during fictive locomotion (Schomburg and Behrends, 1978; see also Angel et al., 1996; Degtyarenko et al., 1998). There is increasing evidence that segmental interneurones in other di- and polysynaptic reflex pathways also receive powerful control from central pattern generators (CPGs) for rhythmic movements like locomotion and scratching (Burke, 1999b) , as well as from descending motor command systems (Baldissera et al., 1981; Lundberg, 1975). It seems likely that selective recruitment of normally high-threshold motor units through polysynaptic segmental pathways is similarly state-dependent.

HOW CAN WE DEFINE MOTONEURONE 'POOLS'?

The idea of recruitment order is inextricably bound to the notion of a motor unit 'pool' within which recruitment takes place but such pools are not easily defined (Burke, 1990; Cope and Sokoloff, 1999b). A functional pool is not necessarily the motor nucleus defined by innervation of an anatomical muscle. It has been suggested that differential recruitment simply represents the existence of distinct, task-related subpopulations of motor units rather than "violations" of the size principle (Cope and Pinter, 1995; Cope and Sokoloff, 1999a). There are indeed examples in which two or more functional motor unit pools share a single muscle compartment. Perhaps the most striking is the bi-articular cat sartorius, in which the anterior portion (antSart) exhibits two bursts during walking, one during the stance phase and the other during the swing phase (Pratt and Loeb, 1991). Using chronic recording methods in intact, walking cats, Hoffer and coworkers (1987) showed that the two antSart bursts per step cycle involve independent groups of muscle units that share the same antSart territory, while units in medial sartorius have yet another activity pattern. Apparently the motoneurones of these three functional pools are intermixed within the same spinal motor nucleus (Gordon et al., 1991).

Sokoloff and Cope (1996) have recently described another example of two functional motor unit pools within the cat soleus, which is made up of mostly type S motor units and has a single functional compartment. In this case, motor units with different ranges of axonal conduction velocity (CV) in the cat soleus exhibited differential recruitment to reflex stimuli. Some units with relatively fast CVs were excited by stimuli that simultaneously inhibited the majority population. Again, differences in synaptic organization are necessary to explain such differential control.

Over 50 years ago, David Lloyd (1946) suggested that group Ia input defines a 'myotatic unit', which is the set of muscles that are excited or inhibited during the stretch reflex. One can extrapolate this idea to include motor unit pools that exhibit consistent patterns of linkage that imply common synaptic drive. Such alliances are not hard-wired; they can shift depending on the motor task at hand. In fact, group Ia excitatory linkage does not necessarily mean that muscles always act synergistically. The long flexor muscle of the cat toes, flexor hallucis longus (FHL) and flexor digitorum longus (FDL), are strongly linked by heteronymous group Ia excitation (Fleshman et al., 1984) but they nevertheless exhibit strikingly different activity patterns during walking in intact cats (O'Donovan et al., 1982; Trank and Smith, 1996). Such disparate activity patterns depend on different distributions of the other synaptic systems that project to FHL and FDL motoneurones (Fleshman et al., 1984).

Co-activation of motor units in a given motor act implies the existence of common synaptic drive. Because most sources of such common drive are segmental interneurones acting in combinations that are state-dependent, it is not easy to provide direct evidence for such alliances. Recent work on modulation of disynaptic cutaneous excitation during fictive locomotion suggests that one group of excitatory last-order interneurones projects to multiple motor nuclei of cat hindlimb flexor muscles (Burke, 1999b). Such common effects are necessary but not sufficient to establish that the same interneurones have axons that diverge to contact multiple motor nuclei. On the other hand, demonstrations of statistically synchronous firing among motor units in different muscle provides strong evidence for divergent common interneurones (Datta and Stephens, 1990; Stephens et al., 1999). Although such studies, mostly from human subjects, do not allow clear definition of the input systems involved, they do show that synaptic linkages change with the tasks under study. Both approaches have considerable merit for further analysis of this question.

CONCLUSION

It should be clear from the above discussion that there are important unresolved issues in this field. The most important of them have to do with the organization of synaptic input systems that coordinate the activity of motor unit populations, whether within a single motor nucleus or between multiple nuclei. Although we have reasonable information about the few afferent and long descending systems that project directly to motoneurones, much remains to be done to unravel the organization of segmental interneurones that are the major source of synaptic drive to motoneurones.

REFERENCES

Angel, M. J., Guertin, P., Jiminez, I., and McCrea, D. A., 1996, Group I extensor afferents evoke disynaptic EPSPs in cat hindlimb extensor motoneurones during fictive locomotion, *Journal of Physiology,* 494, 851-861.

Appelberg, B., and Emonet-Denand, F., 1967, Motor units of the first superficial lumbrical muscle of the cat, *Journal of Neurophysiology,* 30, 154-160.

Baldissera, F., Hultborn, H., and Illert, M., 1981, Integration in spinal neuronal systems, in: *Handbook of Physiology. Section 1: The Nervous System. Vol. III. Motor Control, Part 1*, V. B. Brooks, ed., American Physiological Society, Washington, DC, pp. 509-595.

Bawa, P., and Jones, K. E., 1999, Do lengthening contractions represent a case of reversal in recruitment order? in: *Peripheral and Spinal Mechanisms in the Neural Control of Movement, Progress in Brain Research, Vol. 123.*, M. D. Binder, ed., Elsevier, Amsterdam, pp. 215-220.

Binder, M. D., Heckman, C. J., and Powers, R. K., 1996, The physiological control of motoneuron activity. In: *Handbook of Physiology, Sect. 12: Exercise: Regulation and Integration of Multiple Systems*, L. B. Rowell and J. T. Shepherd, ed., Oxford University Press, New York, pp. 3-53.

Brock, L. G., Coombs, J. S., and Eccles, J. C., 1952, The recording of potentials from motoneurones with an intracellular electrode, *Journal of Physiology,* 117, 431-460.

Burke, R., 1999a, Revisiting the notion of 'motor unit types', in: *Peripheral and Spinal Mechanisms in the Neural Control of Movement. Progress in Brain Research, Vol. 123*, M. Binder, ed., Elsevier, Amsterdam, pp. 167-175.

Burke, R., 1999b, The use of state-dependent modulation of spinal reflexes as a tool to investigate the organization of spinal interneurons, *Experimental Brain Research*, 128, 263-277.

Burke, R. E., 1968a, Firing patterns of gastrocnemius motor units in the decerebrate cat, *Journal of Physiology,* 196, 631-645.

Burke, R. E., 1968b, Group Ia synaptic input to fast and slow twitch motor units of cat triceps surae, *Journal of Physiology,* 196, 605-630.

Burke, R. E., 1981, Motor Units: Anatomy, Physiology and Functional Organization, in: *Handbook of Physiology, Sect. 1: The Nervous System, Vol. II. Motor Control, Part 1*, V. B. Brooks, ed., American Physiological Society, Washington, DC, pp. 345-422.

Burke, R. E., 1990, Selective recruitment of motor units, in: *Freedom to Move: Dissolving Boundaries in Motor Control,* D. R. a. F. Humphrey, H.-J., ed.. John Wiley & Sons, Ltd., Chichester.

Burke, R. E., Jankowska, E., and ten Bruggencate, G., 1970, A comparison of peripheral and rubrospinal synaptic input to slow and fast twitch motor units of triceps surae, *Journal of Physiology,* 207, 709-732.

Burke, R. E., Rymer, W. Z., and Walsh, J. V., 1976, Relative strength of synaptic input from short latency pathways to motor units of defined type in cat medial gastrocnemius, *Journal of Neurophysiology,* 39, 447-458.

Callister, R. J., Peterson, E. H., and Brichta, A. M., 1999, Neuromuscular stretegies underlying ballistic moements, in: *Peripheral and Spinal Mechanisms in the Neural Control of Movement. Progress in Brain Research, Vol. 123.*, M. D. Binder, ed., Elsevier, Amsterdam, pp. 233-243.

Cope, T. C., and Clark, B. D., 1991, Motor units recruitment in the decerebrate cat: several unit properties are equally good predictors of order, *Journal of Neurophysiology,* 66, 1127-1138.

Cope, T. C., and Pinter, M. J., 1995, The size principle: still working after all these years, *News in Physiological Sciences,* 10, 280-286.

Cope, T. C., and Sokoloff, A. J., 1999a, Orderly recruitment among motoneurons supplying different muscles, *Journal de Physiologie,* 93, 81-85.

Cope, T. C., and Sokoloff, A. J., 1999b, Orderly recruitment tested across muscle boundaries, in: *Peripheral and Spinal Mechanisms in the Neural Control of Movement. Progress in Brain Research, Vol. 123.*, M. D. Binder, ed., Elsevier, Amsterdam, pp. 177-190.

Creed, R. S., Denny-Brown, D., Eccles, J. C., Liddell, E. G. T., and Sherrington, C. S., 1932, *Reflex Activity of the Spinal Cord,* London: Oxford University Press.

Datta, A. K., Farmer, S. F., and Stephens, J. A., 1991, Central Nervous Pathways Underlying Synchronization of Human Motor Unit Firing Studied During Voluntary Contractions, *Journal of Physiology,* 432, 401-425.

Datta, A. K., and Stephens, J. A., 1990, Synchronization of motor unit activity during voluntary contraction in man, *Journal of Physiology*, 422, 397-419.

Degtyarenko, A. M., Simon, E. S., Norden-Krichmar, T., and Burke, R. E., 1998, Modulation of oligosynaptic cutaneous and muscle afferent reflex pathways during fictive locomotion and scratching in the Cat [In Process Citation], *Journal of Neurophysiology*, 79, 447-63.

Delgado-Lezama, R., and Hounsgaard, J., 1999, Adapting motoneurons for motor behavior, in: *Peripheral and Spinal Mechanisms in the Neural Control of Movement, Progress in Brain Research, Vol. 123.*, M. D. Binder, ed., Elsevier, Amsterdam, pp. 57-63.

Fleshman, J. W., Lev-Tov, A., and Burke, R. E., 1984, Peripheral and central control of flexor digitorum longus and flexor hallucis longus motoneurons: The synaptic basis of functional diversity, *Experimental Brain Research*, 54, 133-149.

Fleshman, J. W., Munson, J. B., and Sypert, G. W., 1981a, Homonymous projection of individual group Ia-fibers to physiologically characterized medial gastrocnemius motoneurons in the cat, *Journal of Neurophysiology*, 46, 1339-1348.

Fleshman, J. W., Munson, J. B., Sypert, G. W., and Friedman, W. A., 1981b, Rheobase, input resistance, and motor-unit type in medial gastrocnemius motoneurons in the cat, *Journal of Neurophysiology*, 46, 1326-1338.

Frank, K., and Fuortes, M. G. F., 1955, Potentials recorded from the spinal cord with microelectrodes, *Journal of Physiology*, 130, 625-654.

Garnett, R. and Stephens, J. A., 1981, Changes in the recruitment threshold of motor units produced by cutaneous stimulation in man, *Journal of Physiology*, 311, 463-473.

Gates, H. J., Ridge, R. M. A. P., and Rowlerson, A., 1991, Motor units of the fourth deep lumbrical muscle of the adult rat - isometric contractions and fibre type compositions, *Journal of Physiology*, 443, 193-215.

Goldberg, S. J., and Shall, M. S., 1999, Motor units of extraocular muscles: recent findings, in: *Peripheral and Spinal Mechanisms in the Neural Control of Movement. Progress in Brain Research, Vol. 123.*, M. D. Binder, ed. Elsevier, Amsterdam, pp. 221-232.

Gordon, D. C., Loeb, G. E., and Richmond, F. J. R., 1991, Distribution of motoneurons supplying cat sartorius and tensor fasciae latae, demonstrated by retrograde multiple-labelling methods, *Journal of Comparative Neurology*, 304, 357-372.

Gossard, J.-P., Floeter, M. K., Kawai, Y., Burke, R. E., Chang, T., and Schiff, S. J., 1994, Fluctuations of excitability in the monosynaptic reflex pathway to lumbar motoneurons in the cat, *Journal of Neurophysiology*, 72, 1227 - 1239.

Heckman, C. J., and Lee, R. H., 1999, Synaptic integration in bistable motoneurons, in: *Peripheral and Spinal Mechanisms in the Neural Control of Movement. Progress in Brain Research, Vol. 123.*, M. D. Binder, ed., Elsevier, Amsterdam, pp. 49-56.

Henneman, E., Clamann, H. P., Gillies, J. D., and Skinner, R. D., 1974, Rank order of motoneurons within a pool: Law of combination, *Journal of Neurophysiology*, 37, 1338-1349.

Henneman, E., and Mendell, L. M., 1981, Functional organization of motoneuron pool and its inputs, in: *Handbook of Physiology. Sect. I. The Nervous System. Vol. II, Part 1*, V. B. Brooks, ed., American Physiological Society, Bethesda, MD, pp. 423-507.

Henneman, E., Somjen, G. G., and Carpenter, D. O., 1965, Functional significance of cell size in spinal motoneurons, *Journal of Neurophysiology*, 28, 560-580.

Hoffer, J. A., Loeb, G. E., Sugano, N., Marks, W. B., O'Donovan, M. J., and Pratt, C. A., 1987, Cat hindlimb motoneurons during locomotion. III. Functional segregation in sartorius, *Journal of Neurophysiology*, 57, 554-562.

Howell, J. N., Fugelvand, A. J., Walsh, M. L., and Bigland-Ritchie, B., 1995, Motor unit activity during isometric and concentric-eccentric contractions of the human first dorsal interosseus muscle, *Journal of Neurophysiology*, 74, 901-904.

Hultborn, H., 1999, Plateau potentials and their role in regulating motoneuronal firing, in: *Peripheral and Spinal Mechanisms in the Neural Control of Movement. Progress in Brain Research, Vol. 123.*, M. D. Binder, ed., Elsevier, Amsterdam, pp. 39-56.

Jankowska, E., 1992, Interneuronal relay in spinal pathways from proprioceptors, *Progress in Neurobiology*, 38, 335-378.

Kanda, K., Burke, R. E., and Walmsley, B., 1977, Differential control of fast and slow twitch motor units in the decerebrate cat, *Experimental Brain Research*, 29, 57-74.

Kernell, D., Ducati, A., and Sjöholm, H., 1975, Properties of motor units in the first deep lumbrical muscle of the cat's foot, *Brain Research*, 98, 37-55.

Larsson, L., Edstrom, L., Lindegren, B., Gorza, L., and Schiaffino, S., 1991, MHC Composition and enzyme-histochemical and physiological properties of a novel fast-twitch motor unit type, *American Journal of Physiology*, 261, C93-C101.

Liddell, E. G. T., and Sherrington, C. S., 1925, Recruitment and some other features of reflex inhibition, *Proceedings of the Royal Society London,* 97B, 488 - 518.

Lloyd, D. P. C., 1946, Integration pattern of excitation and inhibition in two-neuron reflex arcs, *Journal of Neurophysiology,* 9, 439-444.

Lundberg, A., 1975, Control of spinal mechanisms from the brain, in: *The Nervous System. Vol 1: The Basic Neurosciences*, R. Brady, ed., Raven Press, New York, pp. 253-265.

McDonagh, J. C., Binder, M. D., Reinking, R. M., and Stuart, D. G., 1980, Tetrapartite classification of motor units of cat tibialis anterior, *Journal of Neurophysiology,* 44, 696 - 712.

Nardone, A., Romano, C., and Schieppati, M., 1989, Selective recruitment of high-threshold human motor units during voluntary isotonic lengthening of active muscles, *Journal of Physiology,* 409, 451 - 471.

Nardone, A., and Schieppati, M., 1988, Shift of activity from slow to fast muscle during voluntary lengthening contractions of the triceps surae muscles in humans, *Journal of Physiology,* 395, 363 - 381.

O'Donovan, M. J., Pinter, M. J., Dum, R. P., and Burke, R. E., 1982, The actions of FDL and FHL muscles in intact cats: Functional dissociation between anatomical synergists, *Journal of Neurophysiology,* 47, 1126-1143.

Pette, D., and Staron, R. S., 1990, Cellular and molecular diversities of mammalian skeletal muscle fibers, *Reviews of Physiology, Biochemistry and Pharmacology,* 116, 2 - 76.

Pratt, C. A., and Loeb, G. E., 1991, Functionally complex muscle of the cat hindlimb. I. Patterns of activation across sartorius, *Experimental Brain Rèsearch*, 85, 243-256.

Rall, W., and Hunt, C. C., 1956, Analysis of reflex variability in terms of partially correlated excitability fluctuations in a population of motoneurons, *Journal of General Physiology,* 39, 397-422.

Rudomin, P., and Dutton, H., 1969, Effects of conditioning afferent volleys on variability of monosynaptic responses of extensor motoneurons, *Journal of Neurophysiology,* 32, 140-157.

Schomburg, E. D., and Behrends, H. B., 1978, The possibility of phase-dependent monosynaptic and polysynaptic is excitation to homonymous motoneurones during fictive locomotion, *Brain Research,* 143, 533-7.

Sherrington, C. S., 1904, The correlation of reflexes and the principle of the common path, *British Association Reports,* 74, 728-741.

Smith, J. L., Betts, B., Edgerton, V. R., and Zernicke, R. F., 1980, Rapid ankle extension during paw shakes: selective recruitment of fast ankle extensors, *Journal of Neurophysiology,* 43, 612-620.

Sokoloff, A. J., and Cope, T. C., 1996, Recruitment of triceps surae motor units in the decerebrate cat II. Heterogeneity among soleus motor units, *Journal of Neurophysiology,* 75, 2005-2016.

Stephens, J. A., Harrison, L. M., Mayston, M. J., Carr, L. J., and Gibbs, J., 1999, The sharing principle, in: *Peripheral and Spinal Mechanisms in the Neural Control of Movement, Progress in Brain Research, Vol. 123.*, M. D. Binder, ed., Elsevier, Amsterdam, pp. 419-426.

Stuart, D. G., 1999, The segmental motor system - advances, issues, and possibilities, in: *Peripheral and Spinal Mechanisms in the Neural Control of Movement, Progress in Brain Research, Vol. 123,* M. D. Binder, ed., Elsevier, Amsterdam, pp. 3-28.

Trank, T. V., and Smith, J. L., 1996, Adaptive control for backward quadrupedal walking VI. Metatarsophalageal joint dynamics and motor patterns of digit muscles, *Journal of Neurophysiology,* 75, 678-694.

Zajac, F. E., 1990, Coupling of recruitment order to the force produced by motor units: The "size principle hypothesis" revisited, in: *The Segmental Motor System*, M. Binder and L. Mendell, ed., Oxford University Press, New York, pp. 96 - 111.

Zengel, J. E., Reid, S. A., Sypert, G. W., and Munson, J. B., 1985, Membrane electrical properties and prediction of motor-unit type of cat medial gastrocnemius motoneurons in the cat, *Journal of Neurophysiology,* 53, 1323-1344.

21

PRESYNAPTIC AND DISYNAPTIC INHIBITION INDUCED BY GROUP I MUSCLE AFFERENTS

Annie Schmied, Jean-Marc Aimonetti, and Jean-Pierre Vedel*

ABSTRACT

The task related changes in the Gp I inputs were investigated in type-identified motor units in the wrist extensor muscles. During wrist extension, the monosynaptic inputs generated by applying radial nerve stimulation were distributed among the motoneurone pool in line with the size principle. Their effectiveness was enhanced in the same way during hand clenching and during wrist extension combined with stimulation of the palm and finger cutaneous receptors. The orderly distribution of the monosynaptic Gp I inputs was reversed by the presynaptic inhibition induced by stimulating the Gp I flexor afferents. The effects of the presynaptic inhibition were partially released by applying cutaneous stimulation. During wrist extension, the Gp I flexor afferents generated disynaptic excitatory inputs acting specifically on high-threshold motor units together with disynaptic inhibitory inputs distributed in line with the size principle among the wrist extensor motor nucleus. During hand clenching, their effectiveness was differentially modulated depending on the motor unit type.

INTRODUCTION

Intensive investigations on animals as well as humans have firmly established the validity of the size principle (Henneman and Mendell, 1981; Burke, 1981) according to which the motor units consisting of small-sized motoneurones (type S MNs) which innervate muscle fibres developing small, slow contraction forces are recruited before the units consisting of larger (type F) MNs which innervate muscle fibres developing larger, faster contraction forces. How flexible this orderly recruitment may be in response to the multiple biomechanical demands of the motor repertoire still remains, however, to be established (Kernell and Hultborn, 1990; Burke, 1991; Cope and Pinter, 1995).

In addition to the MN membrane properties (Powers and Binder, 2001), the recruitment and firing patterns of the MNs depend on the distribution of the synaptic inputs reaching the MN pool (Burke, 1991; Cope and Pinter, 1995). Orderly recruitment

* DPM-CNRS 31 chemin Joseph Aiguier, 13402 Marseille cedex 20 France. Email: schmied@dpm.cnrs-mrs.fr

takes place when the net excitatory drive generated by the concurrent excitatory (EPSPs) and inhibitory (IPSPs) post-synaptic potentials is distributed with a decreasing gradient from the type S to F MNs. This is the case with the monosynaptic primary afferents (Ia) originating from the muscles spindles in cats (see Munson, 1990; Powers and Binder, 2001). Some discrepancies have been reported, however, in human studies (Schmied et al., 1997a). The question arises as to whether the presynaptic inhibition which controls the transmission of the Ia monosynaptic projections (Rudomin, 1990) might contribute to modulating their effectiveness differentially, depending on the motor unit type. Here we summarize data indicating that the presynaptic inhibition might act differentially on the S and F MNs during voluntary movements in humans (Aimonetti et al., 2000a, b, c).

When applied to inhibitory inputs, Hennemann's size principle stated that less inhibitory drive was required to make the least excitable MNs (type F) stop firing (Henneman and Mendell, 1981). In cats, the disynaptic IPSP of the reciprocal inhibition generated by Ia afferents in antagonist MN pools actually showed an increasing gradient towards type S MNs (Burke, 1981; Powers and Binder, 2001). Here we observed that the short-latency responses generated by the Gp I wrist flexor afferents on wrist extensor MNs show a similar task-dependent gradient during voluntary contraction in humans (Aimonetti et al., 2001).

Other inputs transmitted by polysynaptic pathways such as the cutaneous ones mainly elicit EPSPs in the type F and IPSPs in the type S MNs (Munson, 1990; Burke, 1991). These inputs therefore play a key role in controlling the MN pool output (Kernell and Hultborn, 1990; Burke, 1991; Cope and Pinter, 1995). Moreover, cutaneous inputs have been found to take part in the down-regulation of the presynaptic inhibition of the Gp I inputs in cats (Rudomin, 1990) and humans (Nakashima et al., 1990). Here we show how hand cutaneous inputs may modulate the short-latency proprioceptive assistance of wrist extensor MNs during voluntary contraction in humans (Aimonetti et al., 2000b).

These issues were addressed by stimulating the radial and/or median nerves in order to activate synergistic and antagonistic Gp I muscle afferents. The changes in the firing probability of extensor carpi radialis MNs were assessed by computing peristimulus time histograms (PSTHs) during isometric wrist extension, during hand clenching while the antagonist wrist flexor muscles were coactivated and during wrist extension while the palm and fingers were being steadily brushed in order to activate the numerous cutaneous receptors liable to be turned on during hand clenching. As previously reported (Schmied et al., 1997a), a consistent correlation was observed between the motor unit recruitment thresholds and twitch rise times. On this basis, fatigue-resistant motor units presumably of the S and F types could be distinguished.

MONOSYNAPTIC-LIKE EXCITATION GENERATED BY GROUP I HOMONYMOUS MUSCLE AFFERENTS: TASK-DEPENDENCE AND EFFECTS OF CUTANEOUS STIMULATION

Early increases in firing probability appeared in the PSTHs in the form of narrow peaks (mean latency: 20 ms) which reflected the effects of monosynaptic Ia EPSPs on all the MNs tested. The first 0.5 ms of the response was taken to be of monosynaptic origin (Hultborn et al., 1987). The strength of this monosynaptic component was plotted with respect to the recruitment threshold (Fig. 1A) and the twitch rise time (Fig. 1B). During wrist extension (circles), hand clenching (squares) and wrist extension combined with

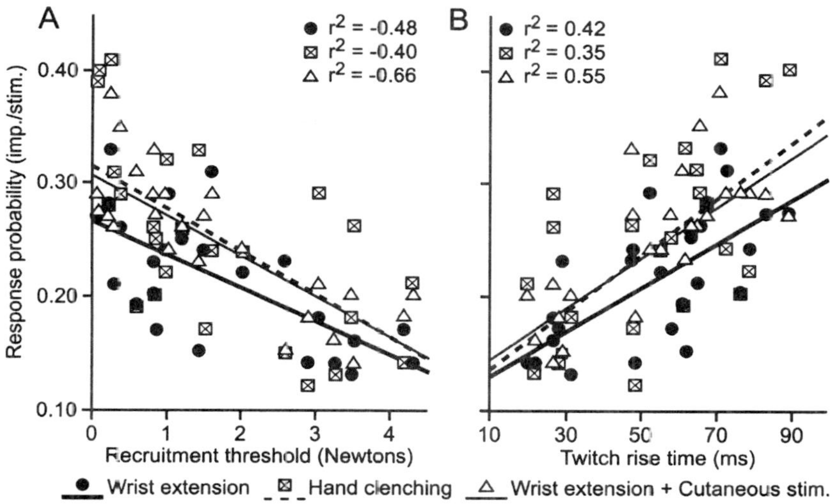

Figure 1. The monosynaptic component (first 0.5 ms) of the increase in the motor unit firing probability induced by the radial nerve stimulation is plotted against the recruitment threshold (A) and the twitch rise time (B). During wrist extension (●), the monosynaptic group I excitatory inputs are distributed following the size-principle in the wrist extensor motoneurone pool. Their effectiveness is enhanced similarly during hand clenching (⊠) and during wrist extension with hand cutaneous stimulation (△).

cutaneous stimulation (triangles), the amplitude of the monosynaptic-like component was correlated with the motor unit parameters in keeping with the size principle.

In the case of hand clenching, however, the response was larger than during wrist extension, as previously observed (Schmied et al., 1997b). This was more marked in the case of the low threshold (Fig. 1A) and slowly contracting units (Fig. 1B).

Applying cutaneous stimulation to the hand during wrist extension enhanced the monosynaptic component of the response as during hand clenching. This supported the hypothesis that cutaneous afferents activated during hand clenching might release presynaptic inhibition.[9]

PRESYNAPTIC INHIBITION OF GROUP I HOMONYMOUS MUSCLE INPUTS DEPENDING ON THE MOTOR UNIT TYPE

Stimulation of Gp I afferents from the wrist flexor muscles induces a presynaptic inhibition of the monosynaptic response of wrist extensor MNs to homonymous Gp I inputs (Berardelli et al., 1987). The amplitude of the monosynaptic part of PSTH peaks (first 0.5 ms) during wrist extension was plotted (ordinate) with respect to the recruitment threshold and the twitch rise time (Fig. 2A, B). The responses to radial nerve stimulation (black circles) are shown together with the responses to combined median and radial nerve stimulation (delay: 20 ms) with and without cutaneous stimulation (crosses and squares, respectively) The size-principle gradient was reversed by the presynaptic inhibition induced by median nerve stimulation, so that low threshold and slowly

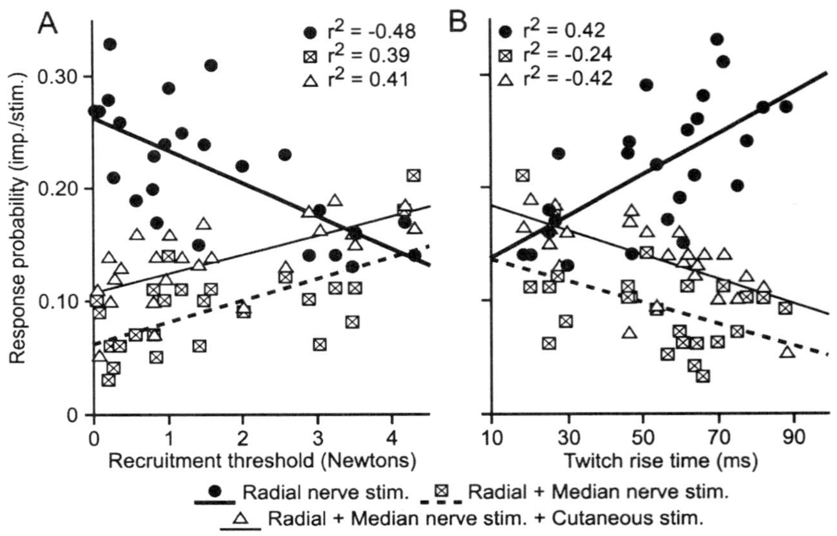

Figure 2. The monosynaptic component (first 0.5 ms) of the increase in firing probability of wrist extensor motoneurone induced by the radial nerve stimulation (●), the radial and median nerve stimulation with and without hand cutaneous stimulation (△ ⊠, respectively) during wrist extension is plotted against the recruitment threshold (A) and the twitch rise time (B). The size-principle ordered distribution of the monosynaptic group I input (●) is reverted by the presynaptic inhibition induced by the stimulation of the group I flexor afferents 20 ms before (⊠). The effect of the presynaptic inhibition is partially released by sweeping the inside of the hand (△).

contracting motor units produced the smallest responses (Fig. 2A, B). As previously reported (Nakashima et al., 1990), stimulating the hand cutaneous receptors released presynaptic inhibition, which affected the motor units quite evenly (Fig. 2A, B). The presynaptic inhibition decreased in a similar way during hand clenching (not illustrated here, Aimonetti et al., 2000c).

The presynaptic inhibition, expressed as a percentage of the monosynaptic response, affected the type S motor units more strongly than the type F ones. This differs from the homogenous pattern of distribution of the presynaptic inhibition to all motor unit types observed in cats (Zengel et al., 1983).

SHORT-LATENCY RESPONSES GENERATED BY GROUP I ANTAGONIST MUSCLE AFFERENTS: DEPENDENCE ON THE MOTOR UNIT TYPE AND ON THE TASK

Gp I flexor afferent stimulation decreased the firing probability of all the extensor motor units tested, at latencies compatible with the disynaptic delay expected to occur with reciprocal inhibition (Day et al., 1984). In type F units, however, the decrease in firing probability was preceded by a small increase. The latency of the excitatory component (whenever present) that of the purely inhibitory component observed in other units (Aimonetti et al., 2001) and were compatible with a disynaptic delay.

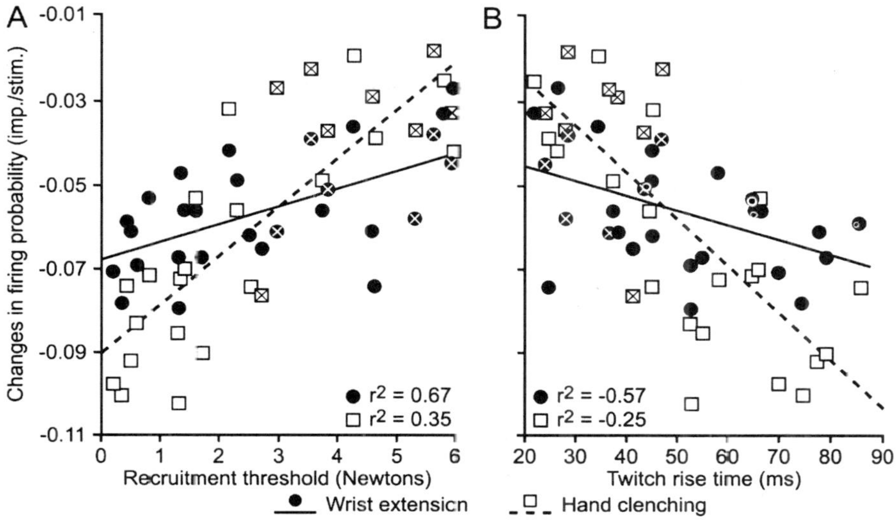

Figure 3. The decrease in the motor unit firing probability induced by the median nerve stimulation is plotted against the recruitment threshold (A) and the twitch rise time (B). During wrist extension (●), the disynaptic inhibitory inputs generated by group I flexor afferents are distributed following the size-principle in the wrist extensor motoneurone pool. During hand clenching (□) their effectiveness is differentially modulated depending on the motor unit properties.

The strength of the inhibitory response was plotted versus the recruitment threshold (Fig. 3A) and the twitch rise times (Fig. 3B) during wrist extension (circles) and hand clenching (squares). The inhibitory responses were preceded by an excitatory component more often during hand clenching (crosses in squares) than during wrist extension (crosses in circles). In both conditions, the low threshold, slowly contracting motor units produced the largest inhibitory responses (Fig. 3A, B). During hand clenching, however, the slope of the regression line between the response strength and the motor unit parameters became steeper, since the inhibitory response was stronger in type S units and weaker in the type F ones (Fig. 3A, B).

Given the similar latencies of the short-latency excitatory and inhibitory responses generated in response to Gp I antagonist nerve afferent stimulation, the simplest hypothesis would be that the wrist extensor MNs may be receiving a mixture of EPSPs and IPSPs from interneurones innervated by Ib and Ia afferents (Jankowska and McCrea, 1983). The excitatory interneurones appear to preferentially affect the type F MNs. A similar finding has been made in the case of cat Gp I afferents originating from synergistic muscles (Powers and Binder, 1985). The wrist flexor and extensor MN pools might therefore interact more like synergistic than antagonistic muscles. In keeping with this hypothesis, the Gp I disynaptic inhibition observed in these muscles was not affected by recurrent inhibition (Aymard et al., 1995), which suggests that interneurones not behaving like Ia interneurones may be involved.

CONCLUSION

The Gp I pathways provided only part of the excitatory and inhibitory drive to the wrist extensor MNs during wrist extension and hand clenching. Despite the potential ability of the flexor-induced presynaptic inhibition to reverse the distribution of the Gp I monosynaptic inputs in favour of the type F MNs, and despite the differential modulation of the flexor-induced excitatory inhibition again in favour of the type F MNs, no changes in the order of recruitment were ever observed in the paired recordings of type S and F units. It can therefore be concluded in agreement with Cope and Pinter (1995) that the size principle is still working.

REFERENCES

Aimonetti, J.-M., Vedel, J.-P., Schmied, A., and Pagni, S., 2000a, Distribution of presynaptic inhibition on type identified motoneurones in the extensor carpi radialis pool in man, *Journal of Physiology,* 522, 125-135.

Aimonetti, J.-M., Vedel, J.-P., Schmied, A., and Pagni, S., 2000b,Mechanical cutaneous stimulation alters Ia presynaptic inhibition in human wrist extensor muscles: a single motor unit study, *Journal of Physiology,* 522, 137-145.

Aimonetti, J.-M., Vedel, J.-P., Schmied, A., and Pagni, S., 2000c,Task dependence of Ia presynaptic inhibition in human wrist extensor muscles: a single motor unit study, *Clinical Neurophysiology,* 111, 1165-1174.

Aimonetti, J.-M., Vedel, J.-P., Schmied, A., and Pagni, S., 2001,Changes in the tonic activity of wrist extensor motor units induced by stimulating antagonistic group I afferents in humans, *Experimental Brain Research,* 141, 21-32.

Aymard, C., Chia, L., Katz, R., Lafitte, C., and Pénicaud, A., 1995, Reciprocal inhibition between wrist flexors and extensors in man: a new set of interneurones? *Journal of Physiology,* 437, 221-235.

Berardelli, A., Day, B. L., Marsden, C. D., and Rothwell, J. C., 1987, Evidence favouring presynaptic inhibition between antagonist muscle afferents in human forearm, *Journal of Physiology,* 391, 71-83.

Burke, R. E., 1981, *Handbook of Physiology, sect. I, The Nervous System, vol. II, Motor Control,* Brooks, V. B.,ed., American Physiological Society, Bethesda, pp. 345-421.

Burke, R. E., 1991, *Motor Control: Concepts and Issues*, D. R. Humphrey, D. R. and Freund, H.-J., eds., John Wiley and Sons, Chichester, pp. 5-21.

Cope, T. C., and Pinter, M. J., 1995, The size principle: still working after all these years, *International Union for Physiological Sciences,* 10, 280-286.

Day, B. L., Marsden, C. D., Obeso, J. A., and Rothwell, J. C., 1984, Reciprocal inhibition between the muscles of the human forearm, *Journal of Physiology,* 349, 519-534.

Henneman, E., and Mendell, L. M., 1981, *Handbook of Physiology, sect. I, The Nervous System, vol. II, Motor Control,* edited by Brooks, V. B., American Physiological Society, Bethesda, pp. 423-507.

Hultborn, H., Meunier, S., Morin, C., and Pierrot-Deseilligny, E., 1987, Assessing changes in presynaptic inhibition of Ia fibres: a study in man and the cat, *Journal of Physiology,* 389, 729-756.

Jankowska, E., and McCrea, D. A., 1983, Shared reflex pathways from Ib tendon organ afferents and Ia muscle spindle afferents in the cat, *Journal of Physiology*, 338, 99-111.

Kernell, D., and Hultborn, H., 1990, Synaptic effects on recruitment gain: a mechanism of importance for the input-output relations of motoneurone pools? *Brain Research,* 507, 176-179.

Munson, J. B., 1990, *The Segmental Motor System*, edited by Binder, M. D. and Mendell, L. M., Oxford University Press, Oxford, pp. 291-307.

Nakashima, K., Rothwell, J. C., Day, B. L., Thompson, P. D., and Marsden, C.D., 1990, Cutaneous effects on presynaptic inhibition of flexor Ia afferents in the human forearm, *Journal of Physiology,* 426, 369-380.

Powers, R. K., and Binder, M. D., 1985, Distribution of oligosynaptic group I input to the cat medial gastrocnemius motoneurone pool, *Journal of Neurophysiology*, 53, 497-517.

Powers, R. K., and Binder, M. D., 2001, Input-output functions of mammalian motoneurons, *Review of Physiology, Biochemistry and Pharmacology,* 143, 137-263.

Rudomin, P., 1990, *The Segmental Motor System*, Binder, M. D. and Mendell, L. M., eds., Oxford University Press, Oxford, pp. 349-380.

Schmied, A., Morin, D., Vedel, J.-P., and Pagni, S., 1997a, The "size principle" and synaptic effectiveness of muscle afferent projections to human extensor carpi radialis motoneurones during wrist extension, *Experimental Brain Research*, 113, 214-229.

Schmied, A., Vedel, J.-P., Calvin-Figuière, S., Rossi-Durand, C., and Pagni, S., 1997b, Task-dependence of muscle afferent monosynaptic inputs to human extensor carpi radialis motoneurones, *EEG and Clinical Neurophysiology*, 105, 220-234.

Zengel, J. E., Reid, S. A., Sypert, G. W., and Munson, J. B., 1983, Presynaptic inhibition, EPSP amplitude, and motor-unit type in triceps surae motoneurons in the cat, *Journal of Neurophysiology*, 49, 922-931.

THINGS WE KNOW AND DO NOT KNOW ABOUT MOTONEURONES

Daniel Kernell*

ABSTRACT

An introductory survey is given of the cellular physiology of motoneurones (MNs). Steady driving currents, applied to individual cells through microelectrodes, may be used for determining such key parameters as the range of possible discharge rates and the shape and steepness of the curve relating discharge frequency to current intensity (f-I relation). Quantitatively, MN properties may vary considerably between animal species and between cells innervating different types of muscle fibres. Central synapses impinging upon MNs often simply provide "driving" currents, altering MN discharge rate largely in accordance with the f-I relation. In addition, metabotropic synapses may have "MN-modifying" effects, altering MN membrane and activation properties in various ways. Studies of MN firing and response patterns in normal and pathological motor behaviour is essential for evaluating the functional role of short- and long-term modifications of MN properties.

ELECTROPHYSIOLOGICAL "ACTIVATION PROPERTIES"

This introductory survey provides some general background information for the following chapters of this Section concerning "somatic" alpha motoneurones (MNs), i.e. cells of the spinal cord or brain stem which innervate skeletal muscle fibres. In this context, I will mainly deal with the "activation properties" of the MNs, i.e. the cell (membrane) properties which determine how the MNs will respond with action potentials (APs) when activated. Under normal conditions, most MN responses consist of trains of repetitive APs and those trains are elicited by relatively maintained currents, which represent the sum of brief postsynaptic currents from many soma-dendritic synapses. The relation between activating current intensity and MN response (e.g. discharge rate) is typically investigated using currents which are artificially applied to a single cell via a microelectrode (intracellular sharp electrode or whole-cell patch-clamp electrode). Many

* Department of Medical Physiology, University of Groningen, PO Box 196, 9700 AD Groningen, The Netherlands. Email: dhkernell@hotmail.com

such measurements have been performed during the last 30-40 years (reviews: Kernell, 1992; Binder et al., 1996; Powers and Binder, 2001).

Each skeletal muscle is innervated and controlled by a group of MNs (MN pool). The force of the muscle is modulated using two different strategies more or less in paralell by changing the number of active MNs (recruitment gradation), and by altering the discharge rate of recruited MNs (rate gradation).

The manner in which the recruitment gradation takes place depends on the distribution of the relevant synapses across the MN pool and also, to an important extent, on the presence of systematic differences in electrical excitability between the various MNs. In a given cell, the net excitatory current needed for AP generation will depend on the voltage threshold level, the "resting" membrane potential at that moment and the electrical input resistance of the cell. The latter entity depends on factors such as the cell morphology and size and the mean specific resistance of the passive membrane (i.e. its "leak conductance"). Measurements of input resistance and resting membrane potential may, of course, be influenced to various extents by ongoing synaptic activity. Still, there are large and systematic differences between MNs with regard to several of these key excitability parameters (e.g. for threshold current, input resistance, membrane resistivity, neuronal size; see reviews cited above).

In addition to the set of "static" properties mentioned above, the intensity of activating current required for single or repetitive AP discharges will depend on the intensity and kinetics of several kinds of voltage-gated ion currents which are influenced already in the subthreshold range of membrane potentials (review: Binder et al., 1996). Such influences are probably responsible for the fact, long known, that the activating current needed for eliciting maintained repetitive discharges is often higher than that required for a single AP. Furthermore, the presence of a subthreshold-activated fast persistent inward current is essential for the ability of MNs to generate maintained repetitive impulse firing (Lee and Heckman, 2001).

With regard to the rate gradation of MN activity, steady injected currents may be used for determining key parameters such as the possible range of steady discharge rates (Fmin, Fmax), the shape and slope of the curve relating impulse frequency to activating current intensity (f-I relation), and the manner in which discharge rate changes with time during steady activation ("frequency-adaptation"). Adaptation is typically shown as a decline of rate, taking place in one or two early rapid phases and a late slower phase.

There are many types of ion channels which together determine the characteristics of motoneuronal repetitive firing. In this context it is important to remember that not only the properties of these various channels are important, but also how they are spatially distributed across the soma and the huge dendritic trees of a MN. This distribution may be markedly non-homogeneous (see below).

Some of the membrane properties involved in repetitive impulse firing give rise to easily measureable phenomena also with single APs, the main one being the period of hyperpolarization which follows a single spike during a few tens to hundreds of ms. In modern terminology this afterpotential is often referred to as the "medium-duration" afterhyperpolarization (AHP), and it largely reflects the time course and intensity of a calcium-activated potassium current (Powers and Binder, 2001). The size and time course of the currents underlying the AHP are of importance for several (but not all) of the repetitive firing properties: they contribute to the setting of the Fmin (not Fmax), they will affect the shape and steepness of the f-I curve, and their properties are important for the earliest phase of spike-frequency adaptation, but probably not for the late phase.

Important AHP characteristics can be indirectly deduced from recordings of AP discharges during normal motor behaviour (e.g. Matthews, 1996; Powers and Binder, 2000a; see also Powers et al., Chapter 24).

TYPES OF SYNAPTIC INFLUENCE: DRIVE AND MODIFICATION

There are two major classes of synaptic influence on motoneuronal discharges:

1. Postsynaptic effects of a conventional kind, providing excitatory (or inhibitory) "driving" currents, causing cells to change their discharge rates in accordance with their intrinsic f-I relations (Granit et al., 1966; Kernell, 1969). In anaesthetized animals, activated synapses provide mainly "driving currents", i.e. there is then often a good agreement between the effects of postsynaptic currents and corresponding intensities of injected currents (Powers and Binder, 1995; Powers and Binder, 2000b; Powers and Binder, 2001).

2. Postsynaptic effects of a "MN-modifying" (also called "MN-modulating") kind, altering the way in which the MNs respond to driving currents, i.e. modifying the activation properties of the MNs (Hultborn, 1999; Delgado-Lezama and Hounsgaard, 1999; Heckman and Lee, 1999). The best known effect of this kind concerns a modification of MN membrane properties such that, when depolarized above a certain threshold, the MN will itself become more likely to generate a depolarizing persistent inward current (cf. Schwindt and Crill, 1980), adding to any externally supplied excitation. This current is predominantly mediated by L-type Ca^{++} chanels (see Alaburda et al., Chapter 27). Once started, such a "plateau current" may sometimes suffice to let the MN continue firing repetitively even if the external synaptic excitation becomes interrupted ("plateau firing"). In addition to facilitating the emergence of plateau currents, MN-modifying synapses may also cause a decrease in the size of the AHP and steepen the f-I relation of the MN, thus changing its responses to driving currents from elsewhere (e.g. Hounsgaard and Kiehn, 1989; cf. also Van Dongen et al., 1986). Also other effects on electrical excitability and f-I relation may occur (Powers and Binder, 2001). It is still unclear to which extent modifications of excitability and f-I relation take place in combination with the facilitation of plateau currents.

With regard to MNs, examples of "plateau-facilitating" synapses were published by Hultborn, Hounsgaaard, Kiehn and their colleagues (Conway et al., 1988; Hounsgaard et al., 1988; Hounsgaard and Kiehn, 1989), and those cases provided evidence for MN-modifying synapses using noradrenaline or serotonin as transmitter. Since then several other kinds of metabotropic receptors have been found whose activation may facilitate the emergence of plateau currents (Delgado-Lezama and Hounsgaard, 1999). The synapses involved in such actions typically have a preferentially dendritic localization (e.g. Bennett et al., 1998; Delgado-Lezama et al., 1999; Carlin et al., 2000; Lee and Heckman, 2000). The effects and mechanisms of plateau currents and MN-modifying synapses are extensively dealt with by contributions in Chapters 26 and 27.

MN-modifying postsynaptic effects, such as those described in this section, act in the course of seconds to minutes. It must not be forgotten that there are also MN-modifying processes happening on much longer time scales.

MOTONEURONAL LONG-TERM PLASTICITY

Marked and long-term gradual changes in MN activation properties do, of course, take place during ontogenetic development (pre- and postnatally). There is an increasing recent literature relating developmental changes in membrane properties to those of the MN activation characteristics (e.g. Berger et al., 1996; Gao and Ziskind-Conhaim, 1998; Jiang et al., 1999; Martin-Caraballo and Greer, 1999).

There are many observations demonstrating that, also in the adult state, MN activation properties may become affected by various long-term influences such as, for instance: axotomy (Heyer and Llinas, 1977); transection of spinal cord (Czeh et al., 1978); change of innervation target (Foehring et al., 1987); long-term increase of activity, as produced by peripheral electrical stimulation (Czeh et al., 1978; Munson et al., 1997); tissue culture conditions (Perrier et al., 2000), brain-derived neurotrophic factor (BDNF; Gonzalez and Collins, 1997). The (molecular) mechanisms generating such long-term changes are still to a great extent uncertain. It would, in this context, be interesting to know what the long-term effects might be of activity of MN-modifying synapses, such as those discussed above.

RELATION TO MUSCLE AND ANIMAL SPECIES

When comparing MN data from different investigations, it should be realized that there are normallly large and systematic quantitative variations in activation properties among motoneurones. Even within the pool of MNs innervating a single muscle, large differences exist in excitability as well as in repetitive firing properties (review: Kernell, 1992), and in the characteristics of "plateau" discharges (Lee and Heckman, 1998). These various differences are generally well adapted to differences in the contractile properties of the muscle fibres innervated by the respective individual MNs (e.g. higher excitability and slower range of discharge rates for MNs of "slow" vs. those of "fast" muscle fibres; Kernell, 1992). The mechanisms underlying the "muscle-adapted" differentiation of MN properties are largely unknown. Some of these MN differences may be genetically specified and/or determined during early developmental phases. Mechanisms for various kinds of long-term plasticity might be of importance as well.

Different animal species may show marked differences in the activation properties of their skeletal muscle fibres and MNs. Thus, for instance, the contractile speed of muscles of rats is generally faster than that for homologous muscles of cats. Correspondingly, in the MNs concerned, the AHP duration (important for minimum rates) is briefer for rats than for cats (Gardiner and Kernell, 1990).

So far, motoneuronal subspecializations have often been described and analysed in terms of being matched to muscle fibres along a "fast vs. slow" axis of differentiation. Cytochemical investigations suggest that other, still enigmatic specializations exist as well (Copray and Kernell, 2000).

MOTONEURONAL USAGE IN MOTOR BEHAVIOUR

We do not yet understand much about the functional role of acute or long-term changes of MN activation properties. It is essential to be able to relate these properties to the manner in which MNs are used in motor behaviour. Such problems are at present being tackled by several groups of investigators, typically using electromyographic (EMG) techniques for the recording of MN discharges in awake animals and humans (see also Chapters 26, 28 and 29). Various indirect methods are then used for making deductions about MN activation properties from the recorded discharge patterns and possibly other experimental data (e.g. various stimulation procedures).

In this context animal studies remain important, partly because of the possibilities for changing experimental conditions (e.g. recordings after depletion of transmitter in descending monoaminergic fibres, Kiehn et al., 1996). Furthermore, in animal preparations, predictions concerning cellular MN physiology from recorded AP discharges may be directly tested and validated (Powers et al., Chapter 24). Human studies are, of course, of interest in their own right. They also offer the methodological advantage that verbal instructions may be used for eliciting a wide range of motor behaviours. In addition, human disease states (which primarily have to be studied for the development of diagnostic and/or therapeutic procedures) may offer unique experimental opportunities for understanding how MNs are affected by various types of long-term processes (e.g. Piotrkiewicz et al., 1999; Thomas et al., Chapter 29).

REFERENCES

Bennett, D. J., Hultborn, H., Fedirchuk, B., and Gorassini, M., 1998, Synaptic activation of plateaus in hindlimb motoneurons of decerebrate cats, *Journal of Neurophysiology,* 80, 2023-2037.

Berger, A. J., Bayliss, D. A., and Viana, F., 1996, Development of hypoglossal motoneurons, *Journal of Applied Physiology,* 81, 1039-1048.

Binder, M. D., Heckman, C. J., and Powers, R. K., 1996, The physiological control of motoneuron activity, in: *Handbook of Physiology. Exercise. Regulation and Integration of Multiple Systems,* American Physiological Society, New York, pp. 3-53.

Carlin, K. P., Jones, K. E., Jiang, Z., Jordan, L. M., and Brownstone, R. M., 2000, Dendritic L-type calcium currents in mouse spinal motoneurons: implications for bistability, *European Journal of Neuroscience,* 12, 1635-1646

Conway, B. A., Hultborn, H., Kiehn, O., and Mintz, I., 1988, Plateau potentials in alpha-motoneurones induced by intravenous injection of L-dopa and clonidine in the spinal cat, *Journal of Physiology,* 405, 369-384.

Copray, S., and Kernell, D., 2000, Neurotrophins and trk-receptors in adult rat spinal motoneurons: differences related to cell size but not to 'slow/fast' specialization, *Neuroscience Letters,* 289, 217-220.

Czeh, G., Gallego, R., Kudo, N., and Kuno, M., 1978, Evidence for the maintenance of motoneurone properties by muscle activity, *Journal of Physiology,* 281, 239-252.

Delgado-Lezama, R., and Hounsgaard, J., 1999, Adapting motoneurons for motor behavior, *Progress in Brain Research,* 123, 57-63.

Delgado-Lezama, R., Perrier, J. F., and Hounsgaard, J., 1999, Local facilitation of plateau potentials in dendrites of turtle motoneurones by synaptic activation of metabotropic receptors, *Journal of Physiology,* 515, 203-207.

Foehring, R. C., Sypert, G. W., and Munson, J. B., 1987, Motor-unit properties following cross-reinnervation of cat lateral gastrocnemius and soleus muscles with medial gastrocnemius nerve. II. Influence of muscle on motoneurons, *Journal of Neurophysiology,* 57, 1227-1245.

Gao, B. X., and Ziskind-Conhaim, L., 1998, Development of ionic currents underlying changes in action potential waveforms in rat spinal motoneurons, *Journal of Neurophysiology,* 80, 3047-3061.

Gardiner, P. F., and Kernell, D., 1990, The fastness of rat motoneurones: time course of afterhyperpolarization in relation to axonal conduction velocity and muscle unit contractile speed, *Pflügers Archives,* 415, 762-766.

Gonzalez, M., and Collins, W. F., 3rd, 1997, Modulation of motoneuron excitability by brain-derived neurotrophic factor, *Journal of Neurophysiology,* 77, 502-506.

Granit, R., Kernell, D., and Lamarre, Y., 1966, Algebraical summation in synaptic activation of motoneurones firing within the 'primary range' to injected currents, *Journal of Physiology,* 187, 379-399.

Heckman, C. J., and Lee, R. H., 1999, Synaptic integration in bistable motoneurons, *Progress in Brain Research,* 123, 49-56.
Heyer, C. B., and Llinas, R., 1977, Control of rhythmic firing in normal and axotomized cat spinal motoneurons, *Journal of Neurophysiology,* 40, 480-488.
Hounsgaard, J., Hultborn, H., Jespersen, B., and Kiehn, O., 1988, Bistability of 'alpha'-motoneurones in the decerebrate cat and in the acute spinal cat after intravenous 5-hydroxytryptophan, *Journal of Physiology,* 405, 345-367.
Hounsgaard, J., and Kiehn, O., 1989, Serotonin-induced bistability of turtle motoneurones caused by a nifedipine-sensitive calcium plateau potential, *Journal of Physiology,* 414, 265-282.
Hultborn, H., 1999, Plateau potentials and their role in regulating motoneuronal firing, *Progress in Brain Research,* 123, 39-48.
Jiang, Z., Rempel, J., Li, J., Sawchuk, M. A., Carlin, K. P., and Brownstone, R. M., 1999, Development of L-type calcium channels and a nifedipine-sensitive motor activity in the postnatal mouse spinal cord, *European Journal of Neuroscience,* 11, 3481-3487.
Kernell, D., 1969, Synaptic conductance changes and the repetitive impulse discharge of spinal motoneurones, *Brain Research,* 15, 291-294.
Kernell, D., 1992, Organized variability in the neuromuscular system: A survey of task-related adaptations, *Archives Italiennes de Biologie,* 130, 19-66.
Kiehn, O., Erdal, J., Eken, T., and Bruhn, T., 1996, Selective depletion of spinal monoamines changes the rat soleus EMG from a tonic to a more phasic pattern, *Journal of Physiology,* 492, 173-184.
Lee, R. H., and Heckman, C. J., 1998, Bistability in spinal motoneurons in vivo: systematic variations in rhythmic firing patterns, *Journal of Neurophysiology,* 80, 572-582.
Lee, R. H., and Heckman, C. J., 2000, Adjustable amplification of synaptic input in the dendrites of spinal motoneurons in vivo, *Journal of Neuroscience,* 20, 6734-6740.
Lee, R. H., and Heckman, C. J., 2001, Essential role of a fast persistent inward current in action potential initiation and control of rhythmic firing, *Journal of Neurophysiology,* 85, 472-475.
Martin-Caraballo, M., and Greer, J. J., 1999, Electrophysiological properties of rat phrenic motoneurons during perinatal development, *Journal of Neurophysiology,* 81, 1365-1378.
Matthews, P. B., 1996, Relationship of firing intervals of human motor units to the trajectory of post-spike after-hyperpolarization and synaptic noise, *Journal of Physiology,* 492, 597-628.
Munson, J. B., Foehring, R. C., Mendell, L. M., and Gordon, T., 1997, Fast-to-slow conversion following chronic low-frequency activation of medial gastrocnemius muscle in cats. II. Motoneuron properties, *Journal of Neurophysiology,* 77, 2605-2615.
Perrier, J. F., Noraberg, J., Simon, M., and Hounsgaard, J., 2000, Dedifferentiation of intrinsic response properties of motoneurons in organotypic cultures of the spinal cord of the adult turtle, *European Journal of Neuroscience,* 12, 2397-2404.
Piotrkiewicz, M., Hausmanowa-Petrusewicz, I., and Mierzejewska, J., 1999, Motoneurons are altered in muscular dystrophy, *Journal de Physiologie,* 93, 167-173.
Powers, R. K., and Binder, M. D., 1995, Effective synaptic current and motoneuron firing rate modulation, *Journal of Neurophysiology,* 74, 793-801.
Powers, R. K., and Binder, M. D., 2000a, Relationship between the time course of the afterhyperpolarization and discharge variability in cat spinal motoneurones, *Journal of Physiology,* 528, 131-150.
Powers, R. K., and Binder, M. D., 2000b, Summation of effective synaptic currents and firing rate modulation in cat spinal motoneurons, *Journal of Neurophysiology,* 83, 483-500.
Powers, R. K., and Binder, M. D., 2001, Input-output functions of mammalian motoneurons, *Reviews in Physiology, Biochemistry and Pharmacology,* 143, 137-263.
Schwindt, P. C., and Crill, W. E., 1980, Properties of a persistent inward current in normal and TEA-injected motoneurons, *Journal of Neurophysiology,* 43, 1700-1724.
Van Dongen, P. A., Grillner, S., and Hokfelt, T., 1986, 5-Hydroxytryptamine (serotonin) causes a reduction in the afterhyperpolarization following the action potential in lamprey motoneurons and premotor interneurons, *Brain Research,* 366, 320-325.

23

A NEW WAY OF USING MODELLING TO ESTIMATE THE SIZE OF A MOTONEURONE'S EPSP

Peter B.C. Matthews*

ABSTRACT

Earlier modelling of a noisy motoneurone has been extended to provide a new measure of excitability. The distance-to-threshold estimate of an MN's AHP, derived from its interval histogram, is used to create a simplified daughter model to mimic the behaviour of its parent and determine a new measure of an MN's response to a stimulus. This Estimated Potential (EP) provides a linear measure of the size of the parent's underlying EPSP, irrespective of its firing rate, and thereby improves on the classic firing index from which it is derived. The EP is applicable with both random and spike-triggered stimulation. It is emphasized that in the presence of noise a tonically firing MN's average responsiveness at a given time during its AHP depends upon what may be termed the "survivor's mean trajectory", rather than upon the "distance to threshold" AHP found in the absence of noise; these differ because noise-induced spiking eliminates individual trajectories when they reach threshold.

INTRODUCTION

The following Chapter 24 by Powers, Turker and Binder introduces the essential ideas and emphasizes that a motoneurone's response to a given EPSP varies with its initial firing rate, depending upon how closely the trajectory of its AHP approaches threshold. Thus the simple Firing Index (no of spikes per stimulus) fails to provide a reliable measure of the magnitude of the underlying EPSP; moreover, summation of two EPSPs rarely produces an equivalent, proportional, increase in the Firing Index (see Matthews, 1999). However, as explained below, when an MN's Interval Histogram can be used to estimate its "distance-to-threshold" AHP the problem can be resolved by combining this with the Firing Index to derive a new and better measure, termed the EP (Estimated Potential). This awaits

---

* University Laboratory of Physiology, Parks Road, Oxford, OX1 3PT, UK.
Email: peter.matthews@physiol.ox.ac.uk

physiological validation, but seems securely based on Powers and Binder's (2000) comparison of estimated and measured AHPs in real MNs.

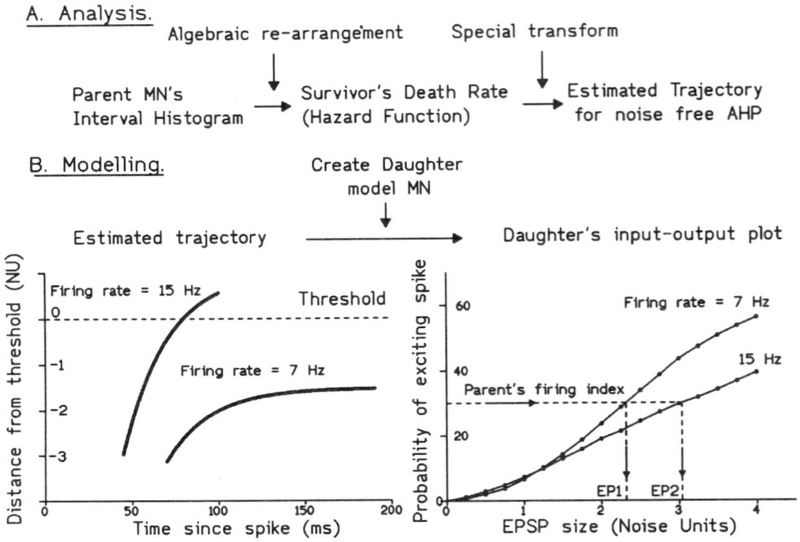

Figure 1. Schema to illustrate the determination of the Estimated Potential (EP) from an MN's firing index by using its estimated AHP to create a fixed-threshold "daughter" model MN whose response to a range of inputs can be determined. Voltages are scaled in Noise Units, corresponding to the standard deviation of the synaptic noise. With low firing rates, the AHP comes to equilibrium well below threshold, with excitation entirely due to the noise. The model MN is then less responsive to small inputs and more responsive to large inputs than with high firing rates; the transition occurs when the input EPSP approximates to the noise. The EP is defined as the input that equalizes the daughter's firing index with the parent's; as shown above, the EP for a given firing index varies with the MN's firing rate (i.e. EP2 > EP1). Provided the daughter faithfully mirrors its parent, the EP gives a linear, frequency invariant, measure of the parent's EPSP. However, as noted below, the EP continued to provide a useful measure when the parent had a variable threshold even though the daughter had a constant threshold. (Daughter MN modelled purely in terms of voltage as in Matthews, 1996, with an AHP time constant of 25 ms and a membrane time constant of 4 ms; final equilibria at +1 and −1.5 NU. The 15 Hz interval histogram was symmetrical while the 7 Hz one was skewed; their coefficients of variation were 0.13 and 0.27).

RANDOM STIMULATION OF FIXED THRESHOLD MODEL

The method has been developed using a one-compartment model MN with a constant threshold, but is equally applicable to more complex models and to real MNs. Figure 1 in Chapter 24 confirms for a two compartment model that its distance-to-threshold AHP is reliably recovered by first converting its interval histogram to an "interval death rate function" and then using a special transform to transmute this to the AHP, measured relative to the threshold (Chapter 24). A basically similar one-compartment model with AHP and leak conductances (Matthews, 2000, with injected current) has been presently tested with a brief EPSP consisting of a 1 ms voltage pulse and both the firing index and the distance-to-threshold AHP determined. The estimated AHP of this parent model is then fitted with an exponential and used to create a simpler daughter model, operating solely in terms of voltage (Matthews, 1996) which mimics the behaviour of the parent at its particular firing rate. The daughter is then tested with a range of sizes of EPSP, randomly timed with regard to the ongoing spike train, to produce a plot of firing index against EPSP amplitude (see Fig. 3B in Chapter 24). Using this calibration curve, the parent's EPSP can be immediately estimated from its firing index, albeit with the value now scaled in Noise Units corresponding to the SD of the parent's voltage noise. With a single-compartment constant-threshold parent the estimate is rather good; the parent and daughter models then largely correspond. The same procedures can be applied to more complex parent models (see later) so for greater generality the estimated EPSP is better termed the Estimated Potential or EP. Figure1 illustrates the manipulations involved in determining the EP.

SPIKE-TRIGGERED STIMULATION AND THE SURVIVORS' AHP

The same procedure can be used to determine the EP from the fixed-threshold parent's firing index when the stimuli are locked to a particular phase of the AHP. In this case it might be thought that there would be no need to actually run the daughter model and that the parent's EPSP could be directly determined from its distance-to-threshold AHP, as estimated from its interval histogram. This, however, does not work because of the statistical complexities resulting from the "absorption barrier" produced by the spike threshold, as illustrated in Figure 2. Due to the nature of its derivation, the transform applied to the death rate estimates the AHP in the absence of noise. If there were no spiking this would be the same as the mean value found in the presence of noise by averaging the values for the individual trajectories at each time. But spiking preferentially eliminates those individual trajectories with an excess of noise-induced depolarisation, without affecting those with an excess of hyperpolarisation. Thus what may be termed "the survivors' mean trajectory" will be more hyperpolarised than the value of the AHP expected from the noise-free underlying conductances. Figure 2 shows that the effect becomes surprisingly large as threshold is approached. The firing index obtained by averaging the individual responses to an EPSP given at a fixed time during a noisy AHP depends upon the survivors' mean and the distribution from which it is derived, rather than directly upon the underlying AHP and the noise amplitude; only those intervals which have survived up to a given time can contribute to the ongoing average excitability. It bears emphasis that expressing the response either as a firing index or as the more informative change in death rate provides a non-linear measure of the underlying EPSP; morever, both these indices vary on testing a given EPSP at different stages of the AHP and on varying the model's mean firing rate. The EP behaves reliably throughout because the daughter model has the same "survivors' mean AHP" as its constant-threshold parent.

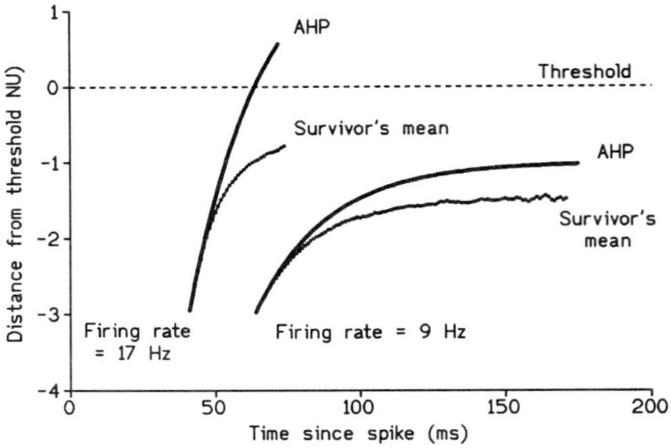

Figure2. Two examples of the progressive deviation of the "survivor's mean trajectory" from the AHP as hreshold is approached; the response to a spike-triggered stimulus depends upon the value of the survivor's mean rather than upon the AHP per se. The AHP represents the membrane potential due to the underlying conductances without noise (equivalent to the mean membrane potential in the presence of noise with spiking inactivated); this is the value obtained by applying the standard transform to the interval histogram. The "survivor's mean" was obtained from a noisy discharge by averaging the value of the membrane potential of those intervals which had "survived" up to the time in question. Each successive point is based on a progressively smaller subset of remaining individual trajectories; these survived because they had less depolarising noise than those which had been eliminated by reaching threshold, through noise-induced excitation. (Same fixed-threshold model as in Fig.1, but using different firing rates; the coefficients of variation of the underlying histograms were 0.11 and 0.22.)

CONCLUSION

Using a constant-threshold parent model MN the EP provides a linear measure of the parent's EPSP. However, the threshold of many real MNs varies during the course of the AHP (Powers & Binder, 1996). Expanding the modelling to allow for this complication suggests that EP remains an effective measure for small to medium sized EPSPs, up to about three times the noise level, eliciting firing indices of up to about 60% (Matthews in preparation). The EP's limitation is that it is scaled in units of voltage whose absolute value will in general be unknown. First, it depends on the amount of voltage noise of the parent MN or model, since the EP is expressed in Noise Units corresponding to the SD of the noise. Second, on testing the variable threshold model the EP gave the size of the EPSP multiplied by a constant scaling factor whose value was set by the detailed properties of the parent. The EP is thus best used for making comparative measurements in the same MN. The temporal composition of the parent's noise can also affect the scaling, but this does not have to be known to create an effective daughter model. The EP remained a linear measure of the EPSP even when the daughter's noise spectrum was made to differ from that of its parent (the parent's distance-to-threshold AHP was then estimated using the transform appropriate to the daughter's noise). Testing the utility of the EP for real MN's should present no difficulties; the point at issue is whether it provides a more linear and less frequency-dependent measure of response than the classic firing index. Finally, it should be noted that the present computations were performed with a step size of 1 ms and an EPSP restricted to a single bin; this is equivalent to determining a Cusum over the rising phase of the EPSP. Recovery of the full time course of an EPSP would require an iterative calculation performed bin by bin.

REFERENCES

Matthews, P. B. C., 1996, Relationship of firing intervals of human motor units to the trajectory of post-spike after-hyperpolarisation and synaptic noise, *Journal of Physiology*, 492, 597-628.

Matthews, P. B. C., 1999, The effect of firing on a model motoneurone and its implications for cortical stimulation, *Journal of Physiology*, 518, 867-882.

Matthews, P. B. C., 2000, Modelling the tonic firing elicited by noisy inputs, Appendix pp. 145-149 in Powers and Binder, op. cit

Powers, R. K., and Binder, M. D., 1996, Experimental evaluation of input-output models of motoneuron discharge, *Journal of Neurophysiology*, 79, 367-379.

Powers, R. K., and Binder, M. D., 2000, Relationship between the time course of the afterhyperpolarisation and discharge variability in cat spinal motoneurones, *Journal of Physiology*, 528, 131-150.

WHAT CAN BE LEARNED ABOUT MOTONEURONE PROPERTIES FROM STUDYING FIRING PATTERNS?

Randall K. Powers[1], Kemal S. Türker[2], and Marc D. Binder[1]

ABSTRACT

The discharge patterns of tonically-firing neurones are influenced by both the characteristics of their presynaptic input and their intrinsic properties. The regularity of the discharge of motoneurones is thought to reflect their prominent post-spike afterhyperpolarization (AHP). When a motoneurone fires at steady mean rate, the distribution of interspike intervals is determined by the amplitude and frequency content of the synaptic noise together with the decrease in excitability following a spike due to AHP. This paper considers how motoneurone discharge statistics can be used to estimate AHP trajectories as well as a motoneurone's sensitivity to excitatory input.

INTRODUCTION

Electrical recordings made from motoneurones in experimental animals display a prominent post-spike afterhyperpolarization (AHP), which is an important determinant of their repetitive discharge behaviour (e.g., Kernell, 1965). This finding led to the expectation that the characteristics of the discharge of a human motoneurone can be used to infer characteristics of its AHP (Person and Kudina, 1972; Matthews, 1996; Piotrkiewicz, 1999). When cat motoneurones are excited with steady injected current, they discharge with remarkable regularity, and their minimum steady discharge rate is approximately equal to the reciprocal of their AHP duration (Kernell, 1965). The discharge of voluntarily-activated human motoneurones is less regular, presumably reflecting the presence of appreciable synaptic noise (cf. Matthews, 1996). The relative variance of interspike intervals is generally low at high firing rates and increases at lower firing rates (Tokizane and Shimazu, 1964; Clamann, 1969; Person and Kudina, 1972; Piotrkiewicz, 1999). A plot of the standard deviation of the interspike intervals (ISIs) versus the mean ISI often exhibits an upward bend, and the mean ISI at which this bend

[1] Dept of Physiology and Biophysics, Univ. of Washington School of Medicine, Seattle WA 98195, USA. Email: rkpowers@u washington.edu [2] Dept of Physiology, Univ. of Adelaide, Adelaide SA 5005, Australia.

occurs is thought to reflect the AHP duration (Person and Kudina, 1972; Piotrkiewicz, 1999). Alternatively, Matthews (1996) proposed that the interval death rate or hazard function derived from the ISIs of human motoneurones may be used to reveal part of the time course of the AHP.

We have recently tested the predictions of these statistical methods by eliciting repetitive discharge in cat motoneurones with a combination of steady injected current and superimposed noise (Powers and Binder, 2000). The discharge statistics of these noise-driven cat motoneurones were similar to those of voluntarily-activated human motoneurones. Plots of the ISI standard deviation versus ISI mean exhibited an upward bend, but the mean ISI at which the bend occurred was generally shorter than the measured AHP duration. AHP trajectories were estimated from a function derived from the ISI histogram (the interval death rate) as described by Matthews (1996). Although the estimated trajectories were smaller in amplitude than the measured AHP trajectories, it was argued that the interval death rate still provides a functionally relevant measure of fluctuations in motoneurone excitability during repetitive discharge (see Matthews, Chapter 23).

The present report extends these findings using a two-compartment motoneurone model to simulate discharge elicited by noisy synaptic drive. AHP trajectories predicted from interval death rates were very close to the actual interspike trajectories in the model. The model was also used to test whether alternative measures of discharge statistics could be used to obtain a rough estimate of relative excitability at different mean discharge rates. Finally, low-frequency, simulated EPSPs of different amplitude were superimposed on different levels of excitatory drive to examine the relation between their spike-triggering efficacy and the background discharge rate of the motoneurone.

METHODS

A two-compartment threshold-crossing model was used to simulate a synaptically-driven motoneurone. The passive behaviour of the model was determined by six parameters: the relative area of the two compartments, the total resistance and capacitance of each compartment and the coupling conductance between compartments (see Pinsky and Rinzel, 1994). The proximal compartment, which represented the soma and proximal dendrites, was given 30% of the total surface area. The rest of the parameter values were chosen by trial and error to match the passive behaviour of a particularly well-studied cat motoneurone. When the membrane potential in the proximal compartment exceeded a fixed threshold (15 mV above the resting potential), a 'spike' was generated along with an exponentially-decaying potassium conductance, whose time constant and peak amplitude were chosen to match the AHP of the same motoneneurone. Repetitive discharge was produced by applying low-pass filtered (250 Hz cutoff) Gaussian conductance noise to both compartments. The equilibrium potential for the conductance noise was set at 40 mV above the resting potential to simulate a mixture of excitatory and inhibitory input, and the standard deviation of the noise was set to produce voltage noise in the proximal compartment with a standard deviation of about 1 mV.

Interspike interval (ISI) histograms were compiled from 10 minutes of simulated discharge at different mean rates. The ISI histogram represents an estimate of the underlying probability density function, f(t). The interval death rate is defined as: $\phi(t)$

$= f(t)/(1 - \int f(t))$, and for finite bin widths can be approximated as $\phi(i) = (\ln(N0/N1)/\text{bin width}$, where N1 is the sum of all the spikes occurring in the bins greater than i and N0 is the same value plus the number of spikes in bin i (cf. Matthews, 1996). In some of the simulations, the excitability of the model motoneurone was tested by superimposing a low-frequency train of conductance transients whose time course and amplitude were chosen to mimic composite EPSPs of different sizes (0.25 – 4 mV).

RESULTS

Measured and Predicted AHP Trajectories

The discharge behaviour of the model motoneurone was similar to that seen in voluntarily-activated human motoneurones (e.g., Matthews, 1996) and noise-driven cat spinal motoneurones (Powers and Binder, 2000). Figure 1A and B show ISI histograms (bell-shaped curves) and interval death rates at two different mean discharge rates. At the lower rate (8.1 imp/s, panel A), the ISI histogram shows a positive skew and the interval death rate reaches a constant level at about 120 ms. In contrast, at the higher rate (14.1 imp/s, panel B), the ISI histogram is more symmetric and the interval death rate does not reach a plateau. As argued by Matthews (1996) and later confirmed by us (Powers and Binder, 2000), a plateau in the interval death rate indicates that many interspike intervals are longer than the AHP duration, and that after the AHP is completed, the mean

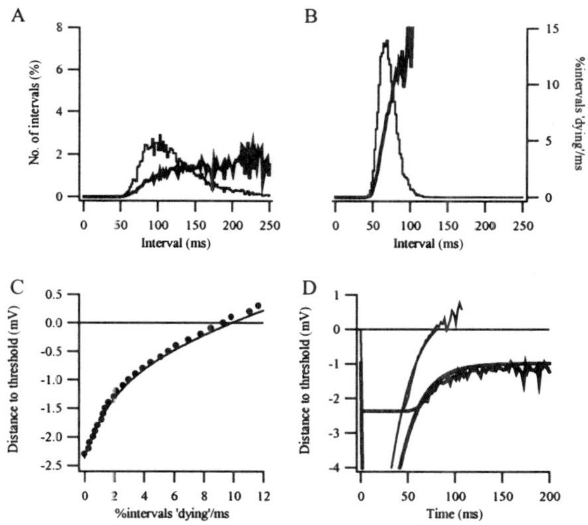

Figure 1. Predicting AHPs from discharge statistics. *A* and *B*, ISI histograms and interval death rates derived from model discharge at a low (A) and high (B) firing rate. *C*, Relation between mean distance to threshold and interval death rate, derived by applying membrane noise of different mean levels to a threshold-crossing detector. D. Comparison of actual AHP trajectories (smooth traces) and predicted AHP trajectories at low (bold lines) and high firing rates.

membrane potential is below threshold. Since the interval death rate provides an estimate of the instantaneous probability of spike occurrence as a function of time since the previous spike, this function should reflect the distance to spike threshold during the interspike interval. If spike threshold is fixed (as in the model), the distance to threshold represents the AHP trajectory.

In order to convert the interval death rate to the distance to threshold, the voltage noise produced by the synaptic input is used as the input to a threshold detector with the threshold set to different levels relative to the mean value of the voltage noise. The output of the detector provides estimates of interval death rate at different distances to threshold (filled circles in Figure 1C). A double exponential fit to this function is used to convert death rates obtained from the two-compartment model into estimates of distance to threshold as a function of time since the previous spike. Figure 1D shows that these estimates (jagged traces) are quite close to the 'measured' AHP trajectories (smooth lines). Moreover, as predicted, at the low firing rate (bold traces), the mean membrane trajectory reaches a plateau that is below spike threshold (see Chapter 23).

Predictions from Other Discharge Statistics

There are two main factors that affect the accuracy of the predicted AHP trajectories: (i) accurate estimates of the interval death rates, and (ii) precise knowledge of the characteristics of the synaptic noise. Accurate estimates of interval death rates require very long (> 10 min) recordings of discharge at a constant mean rate, which are difficult to obtain from human motoneurones. Moreover, the characteristics of synaptic noise in these motoneurones are unknown. Additional simulations were therefore performed in order to determine if it is possible to obtain some information about interspike membrane trajectories without depending upon long discharge records or assumptions about synaptic noise.

Several different types of synaptic noise were simulated to determine if certain characteristics of the ISI distributions could be used to infer underlying interspike membrane trajectories independent of the type of noise used. In addition to the voltage noise produced by our standard conductance noise (see Methods), we used conductance noise with one-half the normal standard deviation, and also produced voltage noise with a different frequency content by halving or doubling the membrane time constants. Regardless of the type of noise used, ISI distributions showed characteristic changes in shape as the mean discharge rate was varied. The ISI standard deviation increased with increasing mean ISI, as reported for human motoneurones and noise-driven cat motoneurones (see Introduction), but there was no distinct bend in the relation at the mean ISI corresponding to the AHP duration (not shown). At longer mean ISIs, the ISI distribution was broader and the amount of positive skew increased.

Figure 2 shows the relations between two different ISI statistics and the distance to threshold at the time of the mean ISI for different mean firing rates and different types of synaptic noise. As firing rate increased, the distance to threshold at the mean ISI and both the values of skew (A) and coefficient of variation (B) decreased in parallel. These simulations suggest that values of skew greater than 0.5 and coefficients of variation above about 0.15 can be taken to indicate that most motoneurone spikes are triggered by positive noise excursions superimposed upon a mean membrane potential that is below the threshold for spike initiation.

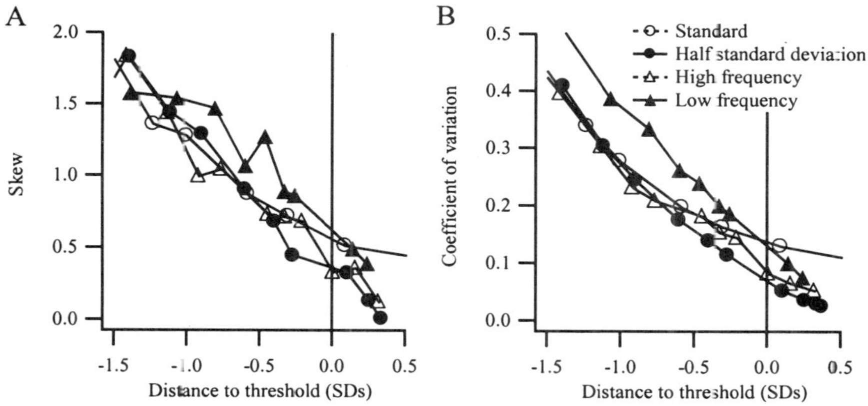

Figure 2. ISI statistics. Data for different mean rates and their relation to the distance to threshold at the end of the mean interspike interval. *A*, Relation between skew and the distance to threshold at the time of the mean ISI. *B*, Relation between the coefficient of variation of the mean ISI and distance to threshold.

Motoneurone Excitability at Different Mean Firing Rates

The spike-triggering efficacy of an EPSP depends upon the proportion of the ISI over which it is within range of threshold. At the lowest firing rates, the membrane potential reaches a constant level below threshold after the AHP is completed (Türker, 1995; Matthews, 1996; Powers and Binder, 2000; and see above). EPSPs that are larger than the distance to threshold at the end of the AHP will thus be able to trigger spikes over a large proportion of the AHP. This model of interspike trajectories predicts a complex relationship between background excitatory drive and the spike-triggering efficacy of EPSPs that depends upon EPSP size (Matthews, 1999). For relatively large EPSPs, spike-triggering efficacy is predicted to decrease with increasing discharge rate over the range of discharge rates typically studied in voluntarily-activated human motor units (see Fig. 4 in Matthews, 1999). We have recently confirmed this relation in rat hypoglossal motoneurons using injected current to mimic background synaptic noise and superimposed EPSPs (Türker and Powers, 2001). In contrast, for small EPSPs, spike-triggering efficacy increases with increasing mean rate (Poliakov et al., 1997; Türker and Powers, 2001).

Simulations with the two-compartment model also showed that the relation between discharge rate and the spike-triggering efficacy of EPSPs depended on EPSP size. A train of conductance transients was applied to the distal compartment while the model was discharging in response to a steady conductance input with superimposed noise. Post-stimulus time histograms (PSTHs) were compiled between the times of EPSP occurrence and those of motoneurone spikes, and spike-triggering efficacy was measured from the area above the baseline in the PSTH peak divided by the number of EPSPs applied. Figure 3A illustrates the spike-triggering effect of a conductance transient that produced a 1 mV EPSP (upper dotted trace) in a model motoneurone discharging at 8.1 imp/s. The

PSTH (lower trace) exhibits a peak during the rising phase of the EPSP and the area of the peak (measured from the rising phase of the CUSUM, upper solid line) indicates that the probability of a given EPSP evoking an extra spike was about 0.15. Figure 3B illustrates the relationship between EPSP amplitude and spike-triggering efficacy for a low (8.1 imp/s, open circles) and a high (14.1 imp/s, filled circles) background firing rate. For EPSPs with peak amplitudes less than 1 mV, spike-triggering efficacy was higher at the higher background firing rate, whereas the opposite relation held for EPSPs with peak amplitudes greater than 1 mV.

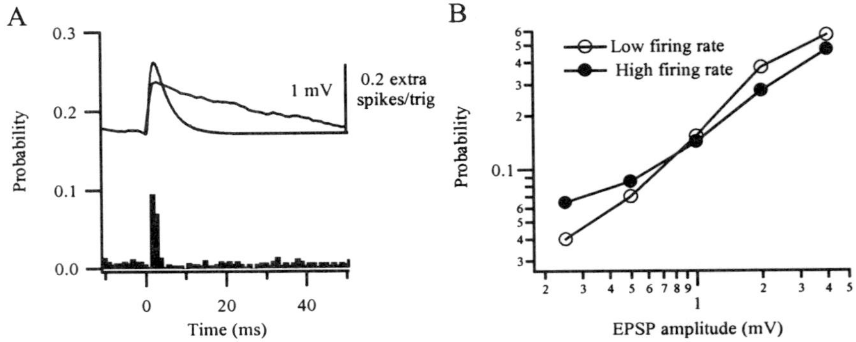

Figure 3. Spike-triggering efficacy of different size EPSPs at low and high firing rates. *A,* Post-stimulus time histogram (PSTH, lower trace) and CUSUM (upper solid line) between the times of occurrence of an EPSP (upper dashed line) and motoneurone spikes. Both the PSTH and CUSUM are normalized by the number of EPSPs applied. *B,* Relation between EPSP size and firing probability.

CONCLUSIONS

Motoneurone interspike interval (ISI) statistics can be used to infer characteristics of the underlying membrane potential trajectories between spikes. Specifically, the interval death rate is proportional to the distance from threshold at different times during the ISI. This distance-to-threshold function can in turn be used to predict how transient excitatory inputs will affect firing probability. Unfortunately, accurate estimates of this function require long discharge records together with knowledge of the characteristics of synaptic noise. Nonetheless, other characteristics of the ISI distribution, such as skew, can be used to determine whether or not a significant proportion of the spikes are triggered at intervals longer than the AHP duration. If this is the case, the interspike membrane trajectory will be characterised by a final plateau, over which the mean membrane potential is below threshold. Under these conditions, large EPSPs have a higher probability of triggering a spike than they do at higher background discharge rates where the interspike membrane trajectory rises to threshold in a nearly linear fashion.

ACKNOWLEDGEMENTS

This work was supported by NIH grants NS 31925 and NS 26840, NSF grant IBN-9986167, NH&MRC of Australia and a Human Frontiers Science Project Short Term Fellowship.

REFERENCES

Clamann, P. H., 1969, Statistical analysis of motor unit firing patterns in a human skeletal muscle, *Biophysical Journal*, 9, 1233-1251.

Kernell, D., 1965, The limits of firing frequency in cat lumbosacral motoneurones possessing different time course of afterhyperpolarization, *Acta Physiologica Scandinavica*, 65, 87-100.

Matthews, P. B. C , 1996, Relationship of firing intervals of human motor units to the trajectory of post-spike after-hyperpolarization and synaptic noise, *Journal of Physiology*, 492, 597-628.

Matthews, P. B. C , 1999, The effect of firing on the excitability of a model motoneurone and its implications for cortical stimulation, *Journal of Physiology*, 518, 867-882.

Person, R. S., and Kudina, L. P, 1972, Discharge frequency and discharge pattern of human motor units during voluntary contraction of muscle, *Electroencephalography & Clinical Neurophysiology*, 32, 471-483.

Piotrkiewicz, M., 1999, An influence of afterhyperpolarization on the pattern of motoneuronal rhythmic activity, *Journal de Physiologie*, 93, 125-133.

Pinsky, P. F., and Rinzel, J., 1994, Intrinsic and network rhythmogenesis in a reduced Traub model for CA3 neurons, *Journal of Computer Neuroscience*, 1, 39-60.

Poliakov, A. V., Powers, R. K., and Binder, M. D., 1997, Functional identification of the input-output transforms of motoneurones in the rat and cat, *Journal of Physiology*, 504, 401-424.

Powers, R. K., and Binder, M. D., 2000, Relationship between the time course of the afterhyperpolarization and discharge variability in cat spinal motoneurones, *Journal of Physiology*, 528, 131-150.

Tokizane, T., and Shimazu, H., 1964, *Functional Differentiation of Human Skeletal Muscle*, University of Tokyo Press, Tokyo.

Türker, K. S., 1995, The shape of the membrane potential trajectory in tonically-active human motoneurons, *Journal of Electromyography & Kinesiology*, 5, 3-14.

Türker, K. S., and Powers, R. K., 2001, The effect of motoneuron discharge rate on synchronization, *Society for Neuroscience, Abstract*, 27, 625.9.

25

RELATIVE STRENGTHS AND DISTRIBUTIONS OF DIFFERENT SOURCES OF SYNAPTIC INPUT TO THE MOTONEURONE POOL

Implications for motor unit recruitment

Marc D. Binder[1], C.J. Heckman[1] and Randall K. Powers[2]

ABSTRACT

Understanding how synaptic inputs from segmental and descending systems shape motor output from the spinal cord requires information on the relative magnitudes of the synaptic currents produced by the different systems and their patterns of distribution within a motoneurone pool. Equally important are quantitative descriptions of how different synaptic inputs are integrated when they are concurrently active and of how voltage- and ligand-gated conductances on the dendrites of motoneurones affect the transfer of synaptic currents to the soma. We have carried out a number of experimental studies of these inter-related problems on motoneurones in the cat spinal cord and have explored the implications of our findings with computer simulations utilizing a synthetic model of the cat medial gastrocnemius motoneurone pool. This paper provides a brief review of the principal results of our studies.

DISTRIBUTION OF EFFECTIVE SYNAPTIC CURRENTS

We have measured the effective synaptic currents produced by six different input systems to lumbar motoneurones in barbiturate-anesthetized cats. The effective synaptic current is defined as the amount of current that reaches the soma of a neurone during high-frequency repetitive activation of a population of presynaptic cells (Heckman and Binder, 1988). Effective synaptic currents can be measured with a single-electrode voltage clamp recording in the soma of a motoneurone (Heckman and Binder, 1988; Lindsay and Binder, 1991) or inferred from the change in the steady-state firing rate that the synaptic activation

---

[1] Dept of Physiology & Biophysics, University of Washington School of Medicine, Seattle WA 98195, USA. Email: mdbinder@u.washington.edu
[2] Department of Physiology, Northwestern University Medical School, Chicago, IL 60611, USA.

produces in the motoneurone (Powers et al., 1993; Powers and Binder, 1995). The average magnitudes of these effective synaptic currents varied widely from an average of < 1 nA for recurrent inhibition to about 4 nA for the contralateral pyramidal tract. Further, we noted several distinct patterns in the distribution of these currents within a motoneurone pool. Both recurrent inhibitory and reciprocal Ia-inhibitory effective synaptic currents appeared to be distributed uniformly within a pool of motoneurones. The Ia-afferent excitatory input was larger to low-threshold than to high-threshold motoneurones, whereas the ipsilateral lateral vestibulospinal input was larger to the high-threshold motoneurones. Stimulating either the contralateral red nucleus or pyramidal tract produced predominantly depolarizing effective synaptic currents in high-threshold motoneurones and predominantly hyperpolarizing currents in low-threshold motoneurones. These results are summarized in Figure 1.

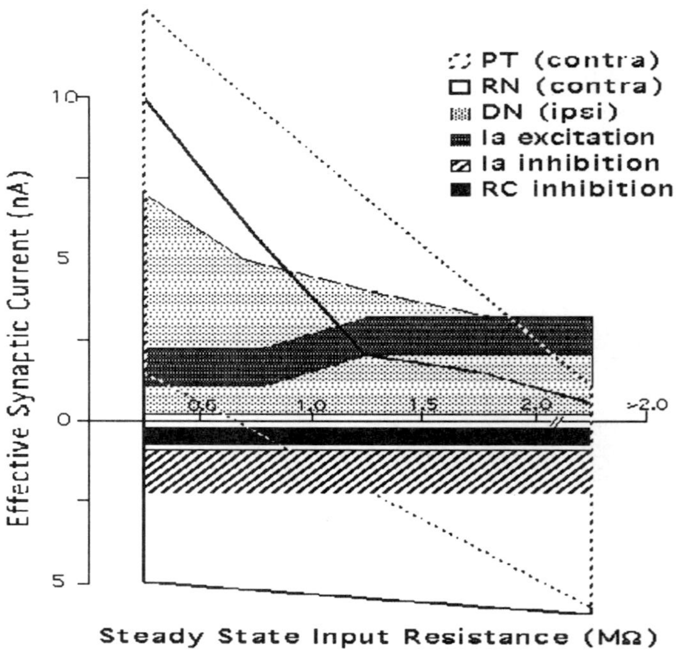

Figure 1. Graphical representation of the magnitude and distribution of the effective synaptic currents (I_N) from six different input systems measured in cat lumbar motoneurons at rest. The stripped line represents I_N from the contralateral pyramidal tract (Binder et al. 1998); the dark, stippled band represents I_N from homonymous Ia afferent fibres (Heckman and Binder, 1988); the stripped band represents I_N from Ia-inhibitory interneurons (Heckman and Binder, 1991a); the black band represents the I_N from Renshaw interneurones (Lindsay and Binder, 1991); the thick lines outline the I_N from contralateral rubrospinal neuronees (Powers et al., 1993); and the light stippled band represents I_N from ipsilateral Deiter's nucleus (Westcott et al., 1995). (Modified from Figure 6 of Binder et al., 1998.)

SUMMATION OF EFFECTIVE SYNAPTIC CURRENTS

We also studied how spinal motoneurones integrate the synaptic currents generated by the concurrent activation of different groups of presynaptic neurones (Powers and Binder, 2000). We made intracellular recordings from triceps surae motoneurones in barbiturate-anaesthetised cats and measured their responses to repetitive activity in different sets of presynaptic neurones produced by electrical stimulation of descending fibres or peripheral nerves and by longitudinal muscle vibration. We studied eight different pairs of synaptic inputs and found that when any two synaptic inputs were activated concurrently, both the effective synaptic currents and the synaptically-evoked changes in firing rate were generally equal to or slightly less than the linear sum of the effects produced by activating each input alone (Fig. 2). Overall, our results indicate that the spatial distribution of synaptic boutons on motoneurones acts to minimize electrical interactions between synaptic sites permitting near linear summation of synaptic currents. However, as discussed in the following section, modulation of voltage-gated conductances on the soma and dendrites of the motoneurone can lead to marked non-linearities in synaptic integration.

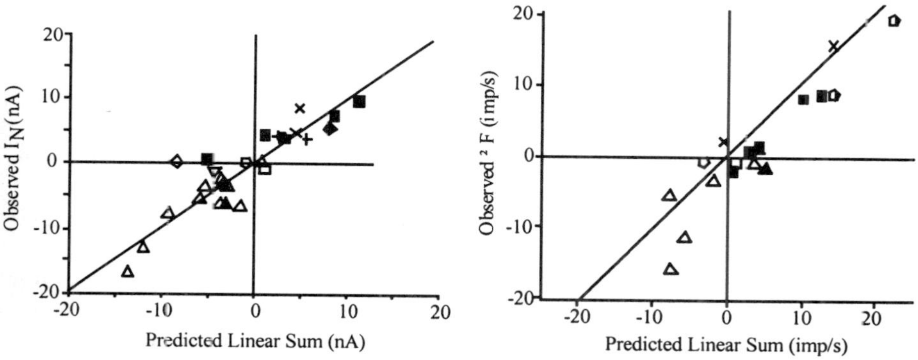

Figure 2. Summation of effective synaptic currents and firing rate modulation in cat triceps surae motoneurones. The left graph plots effective synaptic current (I_N) during concurrent activation of two different inputs versus the linear sum of the currents produced by each input alone. The right graph shows the changes in discharge rate produced by concurrent activation of two inputs versus the linear sum of the changes produced by each input alone. Open symbols refer to cases in which synaptic activation produced a greater than 20% decrease in the input resistance of the cell, whereas filled symbols refer to cases in which the synaptic input produced less than a 20% change in input resistance. Diagonal lines are lines of identity. Circles designate Ia afferents (Ia) + descending fibres from the contralateral red nucleus (RN); upright triangles designate Ia + low and high threshold afferents in the mixed common peroneal nerve (CP); inverted triangles designate CP + RN; squares designate Ia + descending fibres from the ipsilateral pyramidal tract (PT); plus symbols designate Ia + descending fibres from the ipsilateral Dieter's nucleus (DN); stacked triangles designate CP + DN; crosses designate Ia + low and high threshold cutaneous afferents from the sural nerve (SN); and diamonds designate DN + SN. The half-filled circles in the right graph indicate two motoneurones in which the summation of Ia and RN effects on firing rate were measured, but their effects on membrane conductance were not. (Modified from Figs. 2 and 7 in Powers and Binder, 2000.)

AMPLIFICATION OF SYNAPTIC CURRENTS BY DENDRITIC CONDUCTANCES

Effective synaptic currents can be substantially amplified (up to 6-fold) in cat preparations in which persistent inward currents are expressed in motoneurone dendrites (e.g. Lee and Heckman, 2000). In our studies of summation of effective synaptic currents (Powers and Binder, 2000), we induced a partial blockade of potassium channels in some motoneurones by adding tetraethylammonium or cesium to the electrolyte solution in the intracellular pipette. Persistent inward currents could be evoked in these motoneurones with small, depolarizing current pulses which led to instances of substantially greater-than linear summation of injected and synaptic currents. Greater-than linear summation of effective synaptic currents is also observed when individual inputs do not produce sufficient depolarization in a motoneurone to activate persistent inward currents alone, but do so when activated concurrently. However, in cases in which two different sources of synaptic input are sufficiently large to activate these currents, the amplified effective synaptic currents may show linear summation (Prather et al, 2001). This latter finding is particularly important in that it demonstrates that active dendritic currents can amplify synaptic inputs in a graded, rather than in an all-or-none fashion.

COMPUTER SIMULATION OF MOTOR UNIT RECRUITMENT

We have examined the potential effects of different synaptic input systems on recruitment order within a mammalian motoneurone pool using computer simulations (Heckman and Binder, 1991b; 1993). The synaptic inputs and motor unit properties in the model were based as closely as possible on the available experimental data for the cat medial gastrocnemius (MG) pool and muscle. Monte Carlo techniques were employed to add random variance to the motor unit thresholds and forces and to sample the resulting recruitment orders. We found that the effects of the different synaptic inputs on recruitment order depended on how they modified the range of recruitment thresholds established by differences in the intrinsic current thresholds of the motoneurones (Fig. 3). Application of a uniform synaptic input to the pool (i.e., one distributed equally to all motoneurones) resulted in a recruitment sequence that was quite stable: even with 50% added random variance, the recruitment reversals did not exceed 8%.

The simulated Ia afferent input generated a 2-fold expansion of the range of recruitment thresholds beyond that attributed to the differences in the intrinsic current thresholds. The Ia input generated a small reduction in the number of recruitment reversals due to random variance (6% reversals at 50% random variance). The simulated ipsilateral vestibulospinal input generated a 2-fold compression of the range of recruitment thresholds that exerted a small increase in the number of recruitment reversals (12% reversals at 50% random variance).

More dramatic effects on recruitment order were obtained when the model motoneurone pool was driven by the simulated oligosynaptic rubrospinal excitatory input. This input imparted a 9-fold compression in the recruitment threshold range resulting in a recruitment that became nearly random (40% reversals at the standard 50% added variance level).

Reciprocal Ia inhibition was simulated by a uniform distribution within the pool, but its effects on recruitment order were highly dependent on the distribution of the excitatory input. Reciprocal inhibition exerted only minor effects on recruitment order when combined with the Ia or vestibulospinal inputs. However, when the excitatory drive was supplied by the rubrospinal input, even small amounts of reciprocal inhibition were sufficient to completely reverse the normal recruitment sequence.

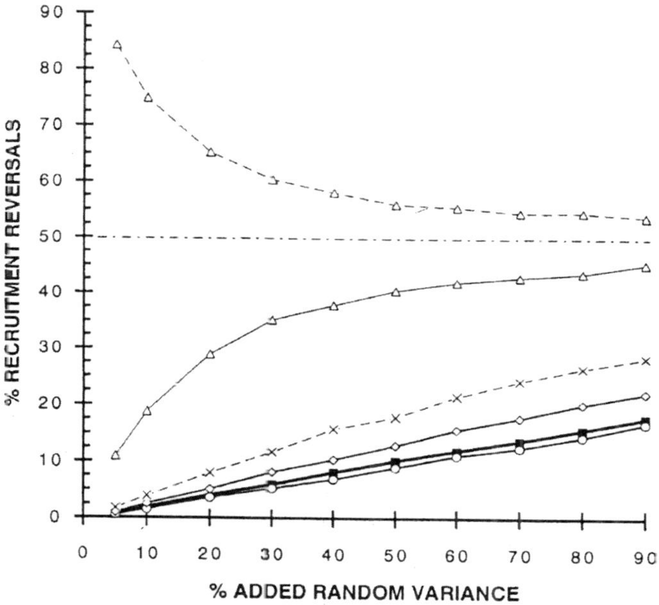

Figure 3. Effects of synaptic inputs on percentage of recruitment reversals in the presence of varying amounts of random variance. Filled squares: uniform input. Open circles: Ia input. Open diamonds: vestibulospinal input. Open triangles: Excitatory rubrospinal input. Open triangles with dashed line: Combined rubrospinal excitation and inhibition. X's: 50/50 combination of Ia and rubrospinal excitatory inputs. (Modified from Fig. 5A in Heckman and Binder, 1993.)

More recent simulations have included the effects of active motoneurone dendrites on the distributions of synaptic inputs within the MG pool (Haftel et al., 2001). Persistent inward currents tend to be larger in motoneurones with high input conductances than in those with low input conductances, resulting, for example, in a distribution of *amplified* Ia effective synaptic currents that is the reverse of that observed in motoneurones with passive dendrites (cf. Lee and Heckman, 2000). Nonetheless, how the specific synaptic input modified the range of recruitment thresholds established by differences in the intrinsic current thresholds of the motoneurones remained the critical factor with respect to recruitment order.

ACKNOWLEDGEMENTS

Our work is supported by grants NS-26840 (MDB), NS-31925 (RKP) and NS-34382 (CJH) from the National Institute of Neurological Diseases and Stroke and grant IBN-9986167 (MDB) from the National Science Foundation.

REFERENCES

Binder, M. D., Heckman, C. J., and Powers, R. K., 1996, The physiological control of motoneuron activity, in: *Handbook of Physiology. Section 12, Exercise: Regulation and Integration of Multiple Systems* L. B. Rowell and J. T. Shepherd, eds., American Physiological Society, Oxford, New York, pp. 3-53.

Binder, M. D., Robinson, F. R., and Powers, R.K., 1998, Distribution of effective synaptic currents in triceps surae motoneurons. VI. Contralateral pyramidal tract, *Journal of Neurophysiology,* 80, 241-298.

Heckman, C. J., and Binder, M. D., 1988, Analysis of effective synaptic currents generated by homonymous Ia afferent fibers in motoneurons of the cat, *Journal of Neurophysiology,* 60, 1946-1966.

Heckman, C. J., and Binder, M. D., 1991a, Analysis of Ia-inhibitory synaptic input to cat spinal motoneurons evoked by vibration of antagonist muscles, *Journal of Neurophysiology,* 66, 1888-1893.

Heckman, C. J., and Binder, M. D., 1991b, Computer simulation of the steady-state input-output function of the cat medial gastrocnemius motoneuron pool, *Journal of Neurophysiology,* 65, 952-967.

Heckman, C. J., and Binder, M. D., 1993, Computer simulations of the effects of different synaptic input systems on motor unit recruitment, *Journal of Neurophysiology,* 70, 1827-1840.

Lee, R. H., and Heckman, C.J., 2000, Adjustable amplification of synaptic input in the dendrites of spinal motoneurons *in vivo*, *Journal of Neuroscience,* 20, 6734-6740.

Lindsay, A. D., and Binder, M. D., 1991, Distribution of effective synaptic currents underlying recurrent inhibition in cat triceps surae motoneurons, *Journal of Neurophysiology,* 65, 168-177.

Powers, R. K., and Binder, M. D., 1995, Effective synaptic current and motoneuron firing rate modulation, *Journal of Neurophysiology,* 74, 793-801

Powers, R. K., and Binder, M. D., 2000, Summation of effective synaptic currents and firing rate modulation in cat spinal motoneurons, *Journal of Neurophysiology,* 83, 483-500.

Powers, R. K., Robinson, F. R., Konodi, M. A., and Binder, M. D., 1992, Effective synaptic current can be estimated from measurements of neuronal discharge, *Journal of Neurophysiology,* 68, 964-968.

Powers, R. K., Robinson, F. R., Konodi, M. A., and Binder, M. D., 1993, Distribution of rubrospinal synaptic input to cat triceps surae motoneurons, *Journal of Neurophysiology,* 70, 1460-1468.

Powers R. K., and Binder, M.D., 2001, Input-output functions of mammalian motoneurons, *Reviews of Physiology, Biochemistry and Pharmacology,* 143, 137-263.

Prather, J. F., Powers R. K., and Cope T. C., 2001, Amplification and linear summation of synaptic effects on motoneuron firing rate, *Journal of Neurophysiology,* 85, 43-53.

Westcott, S. L., Powers, R. K., Robinson, F. R., and Binder, M. D., 1995, Distribution of vestibulospinal input to cat triceps surae motoneurons, *Experimental Brain Research,* 107, 1-8.

26

PLATEAU POTENTIALS AND THEIR ROLE IN REGULATING MOTONEURONAL FIRING

Hans Hultborn*

ABSTRACT

The classical view of the mammalian spinal motoneurone, which emerged from the laboratories of Eccles and Granit in the 1950s and 1960s, held that the cell membrane was essentially passive in areas of synaptic contact (largely the dendrites). The relation between the synaptic excitation and firing frequency was then determined by the post-spike afterhyperpolarisation (Granit and Kernell in the 1960s). During the last 15 years, it has been known that several active membrane properties, including voltage-dependent, non-inactivating inward currents, further shape the motoneuronal output. These inward currents may produce prolonged depolarizations (plateau potentials). It has been demonstrated that neurotransmitters can modify both the plateau properties and the afterhyperpolarisation, thus effectively controlling the input-output relation for the motoneurones.

INTRODUCTION

When Schwindt and Crill (1984) discovered that motoneurones can display prolonged plateau potentials and self-sustained firing it certainly opened new horizons in the exploration of motoneuronal function. They demonstrated that the plateau potentials were due to voltage-dependent non-inactivating Ca^{2+} currents. It now appears obvious that this persistant inward current has the capability to boast and amplify the "classical" synaptic excitation also in the case when full-blown *self-sustained* plateau potentials are not triggered. The degree to which this amplification of normal synaptic excitation take place may to a large extent depend on the spatial (co-)localization of the excitatory synapses and the voltage-sensitive, non-inactivating Ca^{2+} channels, as well as the (synaptic) control of these Ca^{2+} channels.

In the experiments of Schwindt and Crill, penicillin was used to block outward K^+ currents, thereby "uncovering" the inward non-inactivating Ca^{2+} current. They analyzed these mechanisms as a model of 'spinal epilepsia'. During the same period, Hultborn et al

* Department of Medical Physiology, Faculty of Health Sciences, University of Copenhagen, Blegdamsvej 3, Copenhagen, DK-2200, Denmark. Email: h.hultborn@mfi.ku.dk

(1976) discovered a phenomenon of prolonged muscle contraction resulting from short lasting excitatory synaptic inputs to the spinal cord. The further analysis of this phenomenon proved plateau potentials in motoneurones to be responsible for this long-lasting motoneuronal activity (Crone et al., 1988, Hounsgaard et al., 1988). In these experiments, in decerebrate unaneasthetized cats, no penicillin was used. It turned out that endogenously present amounts of monoamines could enable the plateau properties in spinal motoneurones. This Chapter will give a short acount of the phenomenology of the plateau properties in spinal motoneurones together with a discussion on the possible functional consequences. The following Chapter by Alaburda et al. (Chapter 27) focuses on the underlying mechanisms and the control of the responsible L-Ca^{2+} channels.

PLATEAU POTENTIALS IN THE CAT SPINAL MOTONEURONES – PHENOMENOLOGY AND TERMINOLOGY

Figure 1 illustrates how the voltage-dependent persistant inward current has been rescorded in individual motoneurones as a result of intracellular current injections, either as short lasting rectagular current pulses, or triangular current pulses. Figure 1A shows self-sustained firing that is initiated and terminated by depolarising and hyperpolarising current pulses respectively. In many cases, when the motoneurones started with a ('spontaneous') steady firing, these pulses could switch the firing frequency between two stable levels ('bistable firing'). The presence of plateau potentials is best shown following inactivation of the fast sodium spikes (Fig. 1B).

When the plateau potentials were present, there was a peculiar 'frequency acceleration' *during* the current pulse (Fig. 1C). Reflecting the activation of the persistant inward current, this firing acceleration had a long delay and slow onset. The initiation of the plateau could also be seen in the response to long-lasting 'triangular' current pulses (Fig. 1D). At the pulse onset, the membrane potential increased linearly with the current until firing began, at which point the mean potential transiently dropped due to the afterhyperpolarisation. Then, the membrane potential (and the firing rate) again increased linearly with the current until a critical transition frequency was reached, where the frequency and (potential) increased steeply. This step rise, or jump, in firing frequency indicated the initiation of the plateau (or rather the associated non-inactivating inward currents), since a reduction in injected current after the jump did not bring the firing rate back to that before the jump at matched current levels. This can be seen most clearly by plotting the firing rate against the current amplitude and observing the rate difference at a given current level (Fig. 1D, bottom graph; referred to as 'counter-clockwise hysteresis'). Again, the plateau properties during triangular current pulses are even easier to visualize when the Na^+ spikes are inactivated (Fig. 1E).

To summarize, when using classical 'current-clamp' recording conditions the persistant inward current is visualized most directly as a plateau potential after inactivation of the Na^+ spikes. The 'plateau current' is also reflected in the firing pattern as i) self-sustained firing, ii) frequency acceleration during constant current injection and iii) a counter-clockwise f/I, or V/I, hysteresis in response to triangular current pulses.

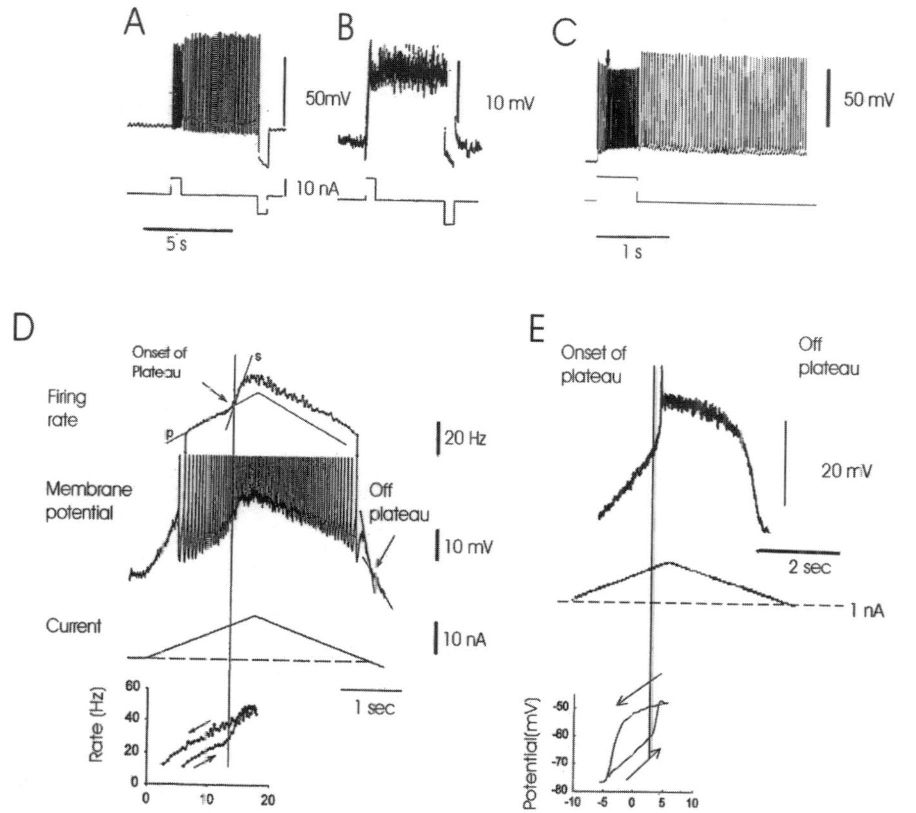

Figure 1. Signs of plateau potentials recorded in spinal motoneurones in the decerebrate cat. *A-C,* Sustained shifts in excitability triggered by depolarizing and hyperpolarizing currents injected intracellularly. Upper traces are intracellular recordings, lower traces monitor the amount and timing of injected current. A. Sustained repetitive firing initiated by a short depolarizing current pulse and terminated by a short hyperpolarizing current pulse. B. In this record the spike-generating mechanism was inactivated. Current pulses now evoked and terminated plateau potentials. C. Record from another motoneurone demonstrates the frequency acceleration during the rectangular current pulse as a sign of induction of the plateau current. Decerebrate, unanaestetized cat preparation (Rearranged from Hounsgaard et al., 1988). *D-E,* Firing pattern and membrane potentials during injection of triangular current pulses. *D,* the firing rate (first trace); intracellular recording of the membrane potential and spike activity (second trace); injected current (third trace); graph at bottom shows the frequency plotted against current for the data shown above. A linearly increasing current (ramp) was injected into the cell, while measuring the membrane potential and instantaneous firing frequency. The plateau is initiated at the point where the frequency and potential jumped steeply (plateau threshold). A subsequent decrease in current did not reverse this steep jump. In this case the inward current persisted even after the cell ceased to fire, and was inactivated at the point marked with an arrow. Tops of spikes were clipped. *E.* the membrane potential shifts during injection of triangular current pulses are more directly visualised when action potentials were inactivated by QX314 injected in the motoneurones by the recording microelectrode. Note the sudden depolarizing jump as the plateau potential is triggered, and that the depolarisation remains for a considerable period although the injected current is decreased. The graph at the bottom shows the membrane potential plotted against current for data shown above. Decerebrate, unanaestetized cat preparation. (Modified from Bennett et al., 1998.)

MONOAMINES ENABLES THE PLATEAU POTENTIALS

The ability to generate plateau properties is dependent on the activity in descending monoaminergic projections in the unanaesthetized cat. Thus the plateau properties illustrated in Figure 1 disappears following a spinal transection, but returns after intravenous injections of serotonin or noradrenaline precursors (Hounsgaard et al., 1988, Conway et al., 1988). At this period we argued that the monoamines enabled the plateau properties primarily by reducing the outward K^+ currents, but subsequent studies have shown that several other transmitters may be involved, and that part of the effect is due to a facilitation of the inward L-Ca^{2+} channels (see Chapter 27).

THRESHOLDS FOR ACTION POTENTIALS AND PLATEAU POTENTIALS

When the persistent plateau current was induced by intracellular current injection alone (Hounsgaard et al., 1988), its threshold was typically reached when the firing rate was in the range of 20 - 50 Hz. The localised depolarisation of the soma by currents through the recording microelectrode, could make a difference in relation to the more natural distributed input by *synaptic* excitation, as the plateau current mainly originates from the distal dendrites (Hounsgaard and Kiehn, 1993). This issue obviously required further attention, and it was on this background that Bennett and colleagues (1998) systematically investigated the threshold of plateau potentials to intracellular current injections during tonic (subthreshold) synaptic excitation and inhibition. In this study the synaptic excitation was evoked by maintained stretch of the triceps surae muscle, and the inhibition by maintained trains of stimulation of the 'antagonist' nerve (reciprocal inhibition). The main finding was that the plateau threshold decreased substantially during synaptic excitation. Obvioulsy it is expected that less current was needed (through the microelectrode) as this would be added on the top of the synaptic excitation. The novel finding was that the intracellular current ramps initiated the plateau at significantly less depolarized membrane potential (as recorded from the soma), or at lower frequencies of firing (Fig. 2) than without tonic synaptic excitation. The reduction in plateau threshold was graded with the amount of excitation; with larger stretches, the plateau threshold was often lowered to near the initial recruitment level. Figure 2 also shows that synaptic inhibition had the opposite effect and increased the plateau threshold.

The important conclusion from that study was that with synaptic excitation of a motoneruone the thresholds for initiating the plateau potential and recruitment of the motoneurone is almost the same. At several occasions the plateau potential was initiated at a lower threshold, and therefore directly contributed to the recruitment of the motoneurone (Bennett et al., 1998). The results of these experiments rather suggest that the persistent inward current has the function to secure a stable recruitment, and to support a steady firing (and thus contraction of the muscle) even with a weak and unsteady synaptic excitation of the motoneurones.

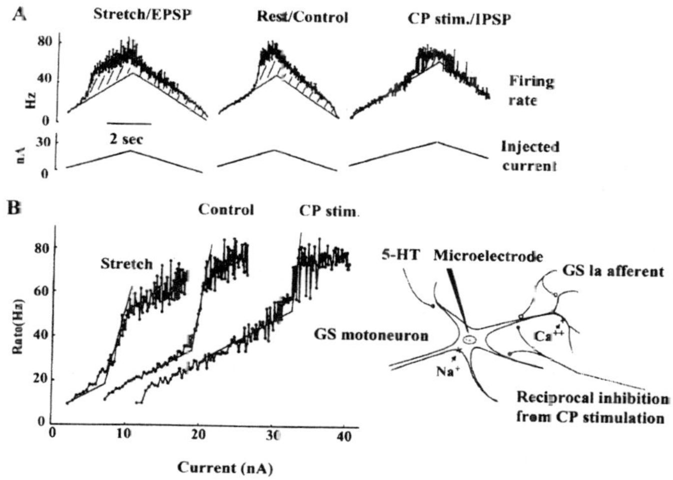

Figure 2. Synaptic input changes the threshold at which intracellular current injections recruit plateau potentials. Plateau activation by intracellular current injection was studied with and without a steady (tonic) peripheral synaptic input (Ia afferent excitation, and reciprocal Ia inhibition, cf. inset drawing). *A*, Middle plot, control situation without added tonic synaptic input. Left plot, effect of tonic muscle stretch (10mm) applied throughout the current ramp shown. This tonic synaptic excitation was adjusted to be below the recruitment level (subthreshold) before the current ramp. Right plot, effect of tonic inhibitory nerve stimulation (IPSPs from common peroneal nerve (CP) stim. 2T, 100 Hz). Hashed regions show contributions of plateau in each case. *B*, Composite of results from *A*, showing responses during ascending phase of ramp plotted against current. Note that the firing frequency at which the plateau current is triggered is lower during tonic synaptic excitation and raised by tonic synaptic inhibition. (Modified from Bennett et al., 1998.)

PLATEAU POTENTIALS UNDER NORMAL CONDITIONS – IN ANIMALS AND HUMANS

It is obviously of great importance to determine whether plateau potentials are present in intact animals, and humans, under normal motor behaviour. It could be argued that the plateau properties studied in decerebrate preparations (or under *in vitro* conditions) represent an unphysiological extreme, never encountered under normal circumstances. In order to evaluate this problem, Eken and Kiehn (1989) used single motor unit recording from freely moving rats. The study focused on the firing behaviour of soleus motor units during quiet standing. In this case it was possible to 'reproduce' the *bistable* firing behaviour seen previously in the cat motoneurones in decerebrate preparations (Hounsgaard et al., 1988). Abrupt maintained increases in firing frequency were evoked by trains of low threshold afferent stimulation causing a burst of excitation, while lasting decreases in firing rate was triggered from other nerves. Furthermore, they occurred in individual motor units without any change in firing frequency of other simultaneously active units. Therefore, it was suggested that the frequency shifts were related to activation of intrinsic properties of individual motoneurones rather than by long lasting changes in the descending synaptic drive.

The firing pattern in human motoneurones has also been investigated for bistable firing during tonic voluntary contractions in combination with short-lasting bursts of vibration (Kiehn and Eken, 1997; Gorassini et al.,, 1998) and electrical stimulation (Collins et al., 2001, 2002, see also Chapter 28). Although a bistable firing pattern was never demonstrated, short periods of vibration often recruited new motor units into a 'self-sustained' maintained activity. As this occurred without an increase in firing frequency of other active motor units (simultaneously recorded), it was assumed that the descending drive had remained constant. Subsequent studies reinvestigating the intact rat (Gorassini et al., 1999), with motor units from different muscles and different recruitment thresholds, produced results more in line with the human motor unit studies than the initial studies on the rat soleus motor units.

These results, both from humans and the intact rat, are thus compatible with the notion that activation of the plateau potentials occurs at a low threshold, and actually is part and parcel of the normal recruitment process.

It should finally be admitted that the experimental paradigms used in these studies have emphasized the ‘all-or-none’ character of the voltage-dependent non-inactivating inward ‘plateau’ current. Under physiological circumstances it is likely that the most significant aspect is the powerful ‘amplification’ it is supplying to the ‘classical’ synaptic excitation. How this ‘variable gain control’ at motoneuronal level is used in the normal control of movements is a major challenge in the future research in this field.

REFERENCES

Bennett, D. J. Hultborn, H., Fedirchuk, B., and Gorassini, M., 1998, Synaptic activation of plateaus in hindlimb motoneurons of decerebrate cats, *Journal of Neurophysiology,* 80, 2023-2037.

Collins, D. F., Burke, D., and Gandevia, S.C., 2001, Large involuntary forces consistent with plateau-like behavior of human motoneurons, *Journal of Neuroscience,* 21, 4059-4065.

Collins, D. F., Burke, D., and Gandevia, S. C., 2002, Sustained contractions produced by plateau-like behaviour in human motoneurones, *Journal of Physiology,* 289-301.

Conway, B. A., Hultborn. H., Kiehn, O., and Mintz, I., 1988, Plateau potential in α-motoneurons induced by intravenous injection of L-DOPA and clonidine in the spinal cat, *Journal of Physiology,* 405, 369-384.

Crone. C., Hultborn, H., Kiehn, O., Mazieres, L., and Wigström, H., 1988, Maintained changes in motoneuronal excitability by short-lasting synaptic inputs in the decerebrate cat, *Journal of Physiology,* 405, 321-343.

Eken, T., and Kiehn, O., 1989, Bistable firing properties of soleus motor units in unrestrained rats, *Acta Physiologica Scandinavica,* 136, 383-394.

Gorassini, M., Bennett, D. J., Kiehn, O., Eken, T., and Hultborn, H., 1999, Activation patterns of hindlimb motor units in the awake rat and their relation to motoneuron intrinsic properties, *Journal of Neurophysiology,* 82, 709-717.

Gorassini, M. A., Bennett, D. J., and Yang, J. F., 1998, Self-sustained firing of human motor units, *Neuroscience Letters,* 247, 13-16.

Hounsgaard, J., Hultborn, H., Jespersen, B., and Kiehn, O., 1988, Bistability of α-motoneurones in the decerebrate cat and in the acute spinal cat after intravenous 5-hydroxy-tryptophan, *Journal of Physiology,* 405, 345-367.

Hounsgaard, J., and Kiehn, O., 1993, Calcium spikes and calcium plateaux evoked by differential polarization in dendrites of turtle motoneurones *in vitro, Journal of Physiology,* 468, 245-259.

Hultborn, H., Wigström, H., and Wängberg, B., 1976, Prolonged activation of soleus motoneurones following a conditioning train in soleus Ia afferents - a case for a reverberating loop? *Neuroscience Letters,* 1, 147-152.

Kiehn, O., and Eken, T., 1997, Prolonged firing in motor units: evidence of plateau potentials in human motoneurons? *Journal of Neurophysiology,* 78, 3061-3068.

Schwindt, P. C., and Crill, W. E., 1984, Membrane properties of cat spinal motoneurons, in: *Handbook of the Spinal Cord, Vols. 2 and 3.* Davidoff, R. A., ed., Marcel Dekker, New York, pp. 199-242.

MECHANISMS CAUSING PLATEAU POTENTIALS IN SPINAL MOTONEURONES

Aidas Alaburda, Jean-François Perrier and Jørn Hounsgaard[1]

ABSTRACT

Plateau potentials are generated by a voltage sensitive persistent inward current. In spinal motoneurones this current is predominantly mediated by influx of Ca^{2+} through L-type Ca^{2+} channels of the $Ca_V1.3$ subtype. Depolarisation-induced facilitation of L-type Ca^{2+} channels is thought to be the mechanism for delayed activation (wind-up and warm-up) of the plateau potential and for the hysteresis in firing frequency and I-V relation during triangular depolarisation. L-type Ca^{2+} channels and plateau potentials in spinal motoneurones are facilitated by activation of metabotropic receptors for glutamate, acetylcholine, noradrenaline and serotonin and down regulated by activation of $GABA_B$ receptors. The facilitation has been shown to depend on activated calmodulin.

PLATEAU POTENTIALS IN SPINAL MOTONEURONES

Plateau potentials were originally observed in motoneurones during experimentally induced spinal seizures (Kao and Crill, 1972) and later shown to be an intrinsic property mediated by a persistent inward current, I_i (Schwindt and Crill, 1977; Schwindt and Crill, 1980a; Schwindt and Crill, 1984). The potential for a physiological role emerged from two observations. First it was shown that plateau properties were induced in spinal motoneurones by activation of a range of metabotropic transmitter receptors (Hounsgaard et al., 1984; Hounsgaard and Kiehn, 1985; Hounsgaard et al., 1988; Hounsgaard and Kiehn, 1989; Conway et al., 1988; Lee and Heckman, 1996; Svirskis and Hounsgaard, 1998). Secondly, although plateau potentials were only directly observable in reduced preparations, recordings of unit activity and force development from muscles in the intact organism provide strong evidence that plateau potentials are part of the normal physiological repertoire of spinal motoneurones (Eken and Kiehn, 1989; Kiehn and Eken, 1997; Kiehn and Eken, 1998; Gorassini et al., 1998; Gorassini et al., 1999; Collins et al., 2001, 2002). The fact that the ability to generate plateau potentials is a highly conserved intrinsic property of spinal motoneurones in mature terrestrial vertebrates (Perrier and

[1] MFI 12.5.9. The Panum Institute, Copenhagen University, Blegdamsvej 3, DK-2200N, Denmark.
Email: j.hounsgaard@mfi.ku.dk

Hounsgaard, 2000) suggests a fundamental role in motor behaviour. This is supported by the finding that maturation of motor behaviour and plateau potentials evolves in parallel during development (Jiang et al., 1999; Perrier and Hounsgaard, 2000).

ION CHANNELS MEDIATING PLATEAU POTENTIALS IN SPINAL MOTONEURONES

Plateau potentials in spinal motoneurones (Fig. 1A) are mediated by a persistent inward current, I_i (Schwindt and Crill, 1984; Svirskis and Hounsgaard, 1997) and associated with a conductance increase (Hounsgaard and Kiehn, 1989). The nature of I_i was first explored by Schwindt and Crill (1977, 1980a, 1984). In the cat *in vivo* they found that I_i was sensitive to iontophoretically applied Ba^{2+} and therefore concluded that at least part of I_i was mediated by Ca^{2+} channels. They also found that I_i was insensitive to QX314, excluding a contribution from voltage sensitive Na^+ channels. This was confirmed more directly in a slice preparation of the spinal cord of the turtle. In this preparation plateau potentials in motoneurones persisted in the presence of TTX and were blocked by Co^{2+} (Hounsgaard and Kiehn, 1985). Moreover, removal of Na^+ ions from the extracellular medium did not affect plateau potentials (Perrier and Hounsgaard, 1999).

Dihydropyridine sensitivity being the hallmark of L-type Ca^{2+} channels (Bean, 1985) established this channel as the main contributor since both plateau potentials and the underlying I_i were blocked by nifedipine (Hounsgaard and Mintz, 1988; Svirskis and Hounsgaard, 1997; Svirskis and Hounsgaard, 1998). Key properties of plateau potentials can be related directly to the properties of I_i and L-type Ca^{2+} channels.

Activation Range. In spinal motoneurones, plateau potentials are activated near the threshold for action potentials (Fig. 1A). Activation of I_i and plateau potentials occur in the same voltage range, 10-30 mV depolarised from the resting membrane potential, in the spinal and decerebrate cat *in vivo* (Schwindt and Crill, 1984; Hounsgaard et al., 1988; Lee and Heckman, 1998) and in the turtle spinal cord *in vitro* (Hounsgaard and Mintz, 1988; Svirskis and Hounsgaard, 1998). $Ca_V1.3$, the subtype of L-channels expressing the α1D subunit, has a similar activation range and is expressed in dendrites of adult motoneurones in mouse (Carlin et al., 2000) and turtle (Simon, Perrier and Hounsgaard, article in preparation). Also in favour of $Ca_V1.3$ being the channel mediating plateau potentials in spinal motoneurones, is the shared 10 fold lower sensitivity to dihydropyridines compared to other L-channel subtypes (Koschak et al., 2001; Xu and Lipscombe, 2001).

Delayed activation. The slow or delayed activation of plateau potentials and I_i can be related to the properties of a population of L-type Ca^{2+} channels (Perrier et al., 2001). The opening of L-type Ca^{2+} channels in response to depolarisation is facilitated by preceding depolarisation (Dolphin, 1996). In motoneurones and plateau generating interneurons in the spinal cord, depolarisation induced facilitation of L-type Ca^{2+} channels is the mechanism for wind-up of the response to repeated depolarisations (Fig. 1B) and for the gradually reduced threshold for activating plateau potentials during a maintained, initially subthreshold depolarisation (Russo and Hounsgaard, 1994; Russo and Hounsgaard, 1996; Delgado-Lezama and Hounsgaard, 1999). It has also been found that an increased intracellular calcium concentration facilitates opening of L-type Ca^{2+} channels via calmodulin activation (Zuhlke et al., 1999; Zuhlke et al., 2000) and activation of plateau potentials in motoneurones (Perrier et al., 2000). For this reason it was proposed that the depolarisation induced transition of closed Ca^{2+} channels from a reluctant to a willing state, suggested to account for the warm-up phenomenon associated with windup (Russo and Hounsgaard, 1994; Russo and Hounsgaard, 1996; Svirskis and Hounsgaard, 1997; Delgado-Lezama and Hounsgaard, 1999), could be mediated by gradually increasing calcium accumulation during successive depolarisations (Perrier et al., 2000).

One of the indicators for plateau potentials in motoneurones is the counter clockwise hysteresis in firing rate during activation by slow triangular current ramps (Hounsgaard et al., 1988; Bennett et al., 1998) and in motor unit activity and force generation in response to synaptic activation (Gorassini et al., 1998; Gorassini et al., 1999; Collins et al., 2001). These properties are direct reflections of the hysteresis in I_i during a triangular depolarising ramp in voltage clamp (Svirskis and Hounsgaard, 1997).

BIOPHYSICAL MECHANISMS FOR PLATEAU POTENTIALS IN SPINAL MOTONEURONES

A neuron represents a system of distributed parameters. From a biophysical point of view the three following characteristics are essential for plateau generation and bistable behavior in spinal motoneurones: 1. The shape of I-V characteristics of the neuron; 2. the electrotonic structure of the dendritic tree and 3. delayed activation of a persistent inward current.

The I-V characteristics from three different motoneurones in Figure 1C-E illustrate the requirements for plateau potentials (Fig. 1D) and bistability (Fig. 1E) (Schwindt and Crill, 1977). Without inward currents the I-V relation consists of two linear segments - an increased slope at depolarized levels due to outward rectification. The intersection with the potential axis is the resting membrane potential (RMP) (Fig. 1C). A net inward current generated by an active conductance can formally be represented by a negative conductance added to the I-V relation. If the depolarization activated conductance for the inward current exceeds the leak conductance, the I-V relation becomes N- shaped with a region of negative resistance (Fig. 1D). This leads to hysteresis in current clamp mode: one value of current corresponds to more than one value of voltage i.e. generation of a plateau potential. If the inward conductance is even bigger, the N-shaped I-V may cross the voltage axis more than once (Fig. 1E). In this case the I-V relation will have two stable zero current states: RMP and stable depolarization (SD). SD means that at this depolarized membrane potential the inward current equals the outward leak current. Transitions between the branches of positive resistance of the N-shape I-V are observed experimentally as activation and deactivation of plateau potentials. The steeper positive slope of the I-V close to SD explains the conductance increase associated with plateau potentials (Hounsgaard and Kiehn, 1989).

Two factors contribute to the hysteresis in firing rate in current clamp and in the I-V relation in voltage clamp during triangular activation. First to consider is the electrotonic distribution of the generator for the persistent inward current. If the generator is distributed and dendrites are long – i.e. the electrotonic length exceeds a certain critical value, then the distal parts of the dendritic tree can be at two different stable membrane potentials, one near RMP and one near SD at the same fixed potential at the soma (Baginskas et al., 1999). Each state of the distal dendrites corresponds to a different value of the potential fixation current, or in other words, such a neuron can be bistable even in voltage clamp conditions at the soma (Baginskas et al., 1999). Secondly, hysteresis is produced by the unusually slow rates of activation and deactivation due to depolarisation induced facilitation (Russo and Hounsgaard, 1994; Russo and Hounsgaard, 1996;

Delgado-Lezama and Hounsgaard, 1999). Together these electrotonic and kinetic properties generate complex spatio-temporal dynamics to the activation/deactivation of the persistent inward current in spinal motoneurones (Svirskis and Hounsgaard, 1997; Baginskas et al., 1999; Svirskis et al., 2001).

MODULATION

In spinal motoneurones the plateau potential is a latent property revealed in the presence of promoting agents (Hounsgaard and Mintz, 1988). Brief trains of stimuli applied in the dorsolateral or medial funiculus induced facilitation of the response to a depolarising current pulse, even in presence of ionotropic receptor blockers, as illustrated in fig. 1F (Delgado-Lezama et al., 1997; Delgado-Lezama et al., 1999). This facilitation was mediated by L-type Ca^{2+} channels. Plateau properties are promoted by penicillin and blockers of certain voltage sensitive K^+ channels (Schwindt and Crill, 1980a; Schwindt and Crill, 1984; Hounsgaard and Mintz, 1988) and by a range of G protein coupled metabotropic receptors for serotonin, noradrenaline, acetylcholine and glutamate (Hounsgaard et al., 1984; Hounsgaard and Kiehn, 1985; Hounsgaard and Kiehn, 1989; Delgado-Lezama et al., 1997; Conway et al., 1988; Lee and Heckman, 1996). This led to the idea that the voltage sensitive persistent inward current in motoneurones was normally present but was masked by outward currents. Only in the absence of the outward current, was the inward current capable of generating plateau potentials (Hounsgaard and Kiehn, 1985; Hounsgaard and Mintz, 1988). Indeed, the facilitation of plateau potentials by serotonin and muscarine is associated with a reduction of several potassium conductances (Hounsgaard and Kiehn, 1989; Svirskis and Hounsgaard, 1998). On the other hand experiments also show that the persistent inward current mediated by L-type Ca^{2+} channels is reduced by activation of $GABA_B$ receptors and enhanced by activation of muscarine receptors, serotonin receptors and group I metabotropic glutamate receptors (mGluR I) (Delgado-Lezama et al., 1997; Svirskis and Hounsgaard, 1998). It seems, therefore, that the modulating neurotransmitters activate intracellular pathways that target several types of ion channels and in this way regulate outward and inward current generators in parallel. The modulating neurotransmitters that facilitate L-type Ca^{2+} channels may all converge on receptor subtypes coupled to the phospholipase C, diacyl glycerol, IP3 pathway. This is supported by the recent finding that plateau potentials, pre-existing or promoted by activation of mGlu I receptors, depend on activated calmodulin (Perrier et al., 2000). The role of elevated intracellular Ca^{2+} in facilitation of L-channels seems central. Not only could plateau potentials be promoted by Ca^{2+} influx through N-type Ca^{2+} channels rather than via activation of metabotropic receptors, but these plateau potentials, mediated by L-type Ca^{2+} channels, were also calmodulin dependent.

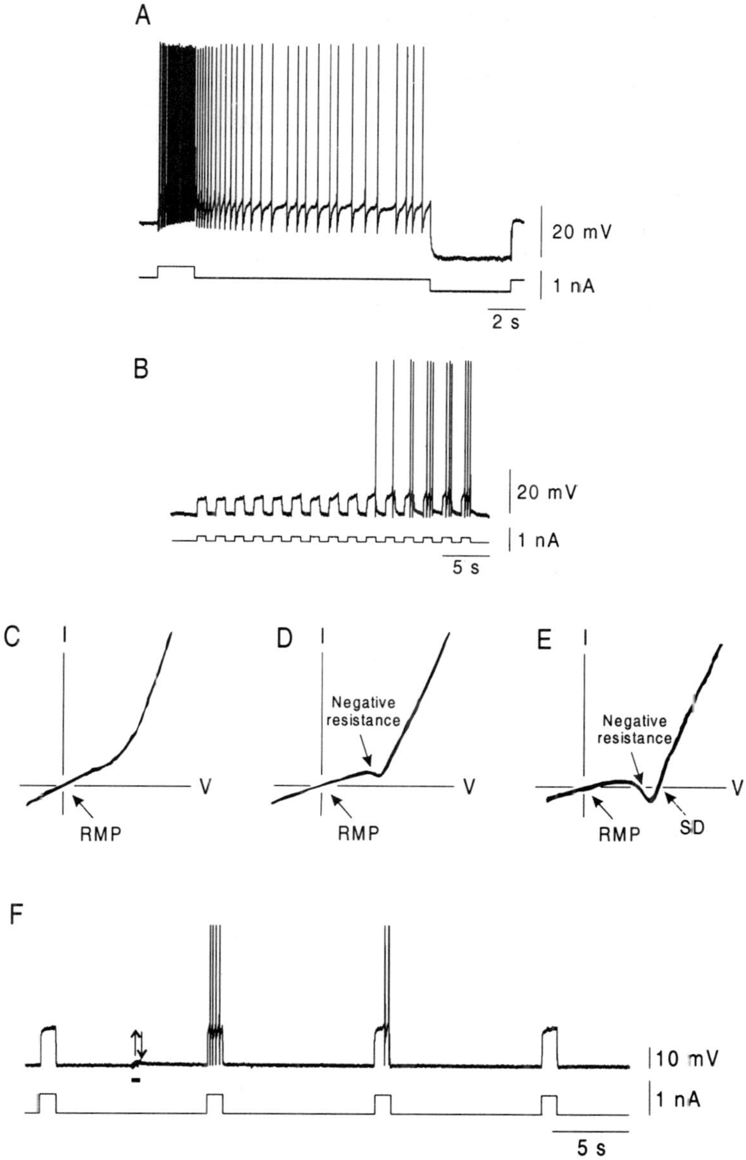

Figure 1. *A*, Plateau potential in a turtle spinal motoneurone recorded in the presence of CNQX (20 μM), AP5 (50 μM), strychnine (10 μM) in order to block ionotropic receptors and of the mGluRI agonist ACPD (40 μM). Note the delayed spiking acceleration during depolarizing current pulse and the afterdischarge following the pulse, which was "reset" by hyperpolarizing current pulse. *B*, Windup of the response to a repeated current pulse (same motoneurone as in A). *C-E*, The I-V relations from three different motoneurones. Arrows indicates resting membrane potential (RMP), stable depolarisation (SD) and negative resistance. (Modified from Schwindt and Crill, 1977). *F*, Metabotropic synaptic facilitation of intrinsic response properties of turtle spinal motoneuronees. Subthreshold for spike initiation current pulse became suprathreshold after brief dorsolateral funiculus stimulus train (↑↓). Excitability remained increased during more than 10 s. Ionotropic receptors were blocked by application of CNQX, AP5 and strychnine in the bath. (From Delgado-Lezama et al., 1997.)

OTHER PERSISTENT INWARD CURRENTS IN SPINAL MOTONEURONES

In addition to the plateau generating, persistent inward current mediated by $Ca_V1.3$ L-type Ca^{2+} channels, two other persistent inward currents have been identified in spinal motoneurones: a calcium-activated nonselective cationic current (I_{CAN}) and a persistent component of the TTX sensitive Na^+ current.

In the rat, interneurons in the deep dorsal horn express plateau potentials that depend on L-type Ca^{2+} channels but also involve I_{CAN} (Morisset and Nagy, 1999). However, although this current is present in spinal motoneurones it does not generate inward current during plateau potentials and no normal function has so far been attributed to I_{CAN} in motoneurones (Perrier and Hounsgaard, 1999). The voltage insensitivity of I_{CAN} is also incompatible with the warm-up phenomenon in motoneurones and dorsal horn neurons in the turtle (Svirskis and Hounsgaard, 1997; Russo and Hounsgaard, 1994).

A persistent component of the TTX sensitive Na^+ current present in many cell types and previously shown to support plateau potential in Purkinje cells (Llinas and Sugimori, 1980; Hounsgaard and Midtgaard, 1988) was recently suggested to be present in cat spinal motoneurones (Lee and Heckman, 2001). There is also a persistent TTX sensitive component of inward current in adult turtle spinal motoneurones (unpublished data). Although this current may serve an essential role in spike initiation and control of rhythmic firing it seems not to be significant for generation of plateau potentials. First, plateau potentials are absent when Ca^{2+} channels are blocked (Hounsgaard and Kiehn, 1985; Hounsgaard and Mintz, 1988) secondly, the generation and the properties of plateau potentials and wind-up are unaffected by block of Na^+ channels externally with TTX or internally with QX314 (Schwindt and Crill, 1980b; Hounsgaard and Mintz, 1988; Svirskis and Hounsgaard, 1997; Delgado-Lezama et al., 1997) and remain unaffected in the absence of external Na^+ (Perrier and Hounsgaard, 1999).

REFERENCES

Baginskas, A., Gutman, A., Hounsgaard, J., Svirskiene, N., and Svirskis, G., 1999, Semi-quantitative theory of bistable dendrites with wind-up, in: *Modeling in the Neurosciences. From Ionic Channels to Neural Networks,* R. R. Poznanski, ed., Harwood Academic Publishers, Australia, pp. 417-437.

Bean, B. P., 1985, Two kinds of calcium channels in canine atrial cells, Differences in kinetics, selectivity, and pharmacology, *Journal of General Physiology,* 86, 1-30.

Bennett, D. J., Hultborn, H., Fedirchuk, B., and Gorassini, M., 1998, Synaptic activation of plateaus in hindlimb motoneurons of decerebrate cats, *Journal of Neurophysiology,* 80, 2023-2037.

Carlin, K. P., Jones, K. E., Jiang, Z., Jordan, L. M., and Brownstone, R. M., 2000, Dendritic L-type calcium currents in mouse spinal motoneurons: implications for bistability, *European Journal of Neuroscience,* 12, 1635-1646.

Collins, D. F., Burke, D., and Gandevia, S. C., 2001, Large involuntary forces consistent with plateau-like behavior of human motoneurons, *Journal of Neuroscience,* 21, 4059-4065.

Collins, D. F., Burke, D., and Gandevia, S. C., 2002, Sustained contractions produced by plateau-like behaviour in human motoneurones, *Journal of Physiology,* 289-301.

Conway, B. A., Hultborn, H., Kiehn, O., and Mintz, I., 1988, Plateau potentials in alpha-motoneurones induced by intravenous injection of L-dopa and clonidine in the spinal cat, *Journal of Physiology,* 405, 369-384.

Delgado-Lezama, R., and Hounsgaard, J., 1999, Adapting motoneurons for motor behavior, *Progress in Brain Research,* 123, 57-63.

Delgado-Lezama, R., Perrier, J. F., and Hounsgaard, J., 1999, Local facilitation of plateau potentials in dendrites of turtle motoneurones by synaptic activation of metabotropic receptors, *Journal of Physiology,* 515, 203-207.

Delgado-Lezama, R., Perrie,r J. F., Nedergaard, S., Svirskis, G., and Hounsgaard, J., 1997, Metabotropic synaptic regulation of intrinsic response properties of turtle spinal motoneurones, *Journal of Physiology,* 504, 97-102.

Dolphin, A. C., 1996, Facilitation of Ca2+ current in excitable cells, *Trends in Neuroscience,* 19, 35-43.

Eken, T., and Kiehn, O., 1989, Bistable firing properties of soleus motor units in unrestrained rats, *Acta Physiologica Scandinavica,* 136, 383-394.

Gorassini, M., Bennett, D. J., Kiehn, O., Eken, T., and Hultborn, H., 1999, Activation patterns of hindlimb motor units in the awake rat and their relation to motoneuron intrinsic properties, *Journal of Neurophysiology,* 82, 709-717.

Gorassini, M. A., Bennett, D. J., and Yang, J. F., 1998, Self-sustained firing of human motor units, *Neuroscience Letters,* 247, 13-16.

Hounsgaard, J., Hultborn, H., Jespersen, B., and Kiehn, O., 1984, Intrinsic membrane properties causing a bistable behaviour of alpha- motoneurones, *Experimental Brain Research,* 55, 391-394.

Hounsgaard, J., Hultborn, H., Jespersen, B., and Kiehn, O., 1988, Bistability of alpha-motoneurones in the decerebrate cat and in the acute spinal cat after intravenous 5-hydroxytryptophan, *Journal of Physiology,* 405, 345-367.

Hounsgaard, J., and Kiehn, O., 1985, Ca++ dependent bistability induced by serotonin in spinal motoneurons, *Experimental Brain Research,* 57, 422-425.

Hounsgaard, J., and Kiehn, O., 1989, Serotonin-induced bistability of turtle motoneurones caused by a nifedipine-sensitive calcium plateau potential, *Journal of Physiology,* 414, 265-282.

Hounsgaard, J., and Midtgaard, J., 1988, Intrinsic determinants of firing pattern in Purkinje cells of the turtle cerebellum in vitro, *Journal of Physiology,* 402, 731-749.

Hounsgaard, J., and Mintz, I., 1988, Calcium conductance and firing properties of spinal motoneurones in the turtle, *Journal of Physiology,* 398, 591-603.

Jiang, Z., Rempel, J., Li, J., Sawchuk, M. A., Carlin, K. P., and Brownstone, R. M., 1999, Development of L-type calcium channels and a nifedipine-sensitive motor activity in the postnatal mouse spinal cord, *European Journal of Neuroscience,* 11, 3481-3487.

Kao, L.I., and Crill, W. E., 1972, Penicillin-induced segmental myoclonus. I. Motor responses and intracellular recording from motoneurons, *Archives of Neurology,* 26, 156-161.

Kiehn, O., and Eken, T., 1997, Prolonged firing in motor units: evidence of plateau potentials in human motoneurons?, *Journal of Neurophysiology,* 78, 3061-3068.

Kiehn, O., and Eken, T., 1998, Functional role of plateau potentials in vertebrate motor neurons, *Current Opinion in Neurobiology,* 8, 746-752.

Koschak, A., Reimer, D., Huber, I., Grabner, M., Glossmann, H., Engel, J., and Striessnig, J., 2001, alpha 1D (Cav1.3) subunits can form l-type Ca2+ channels activating at negative voltages, *Journal of Biological Chemistry,* 276, 22100-22106.

Lee, R. H., and Heckman, C. J., 1996, Influence of voltage-sensitive dendritic conductances on bistable firing and effective synaptic current in cat spinal motoneurons in vivo, *Journal of Neurophysiology,* 76, 2107-2110.

Lee, R. H. and Heckman, C. J., 1998, Bistability in spinal motoneurons in vivo: systematic variations in persistent inward currents, *Journal of Neurophysiology,* 80, 583-593.

Lee, R. H., and Heckman, C. J., 2001, Essential role of a fast persistent inward current in action potential initiation and control of rhythmic firing, *Journal of Neurophysiology,* 85, 472-475.

Llinas, R., and Sugimori, M., 1980, Electrophysiological properties of in vitro Purkinje cell dendrites in mammalian cerebellar slices, *Journal of Physiology,* 305, 197-213.

Morisset, V., and Nagy, F., 1999, Ionic basis for plateau potentials in deep dorsal horn neurons of the rat spinal cord, *J Neurosci.,* 19, 7309-7316.

Perrier, J. F., Alaburda, A, and Hounsgaard, J, 2002, Spinal platicity mediated by L-Type Ca++ channels, in: Proceedings from The 1st Segerfalk Symposium 'Spinal Cord Function, Plasticity and Repair', *Brain Research Reviews,* in press.

Perrier, J. F., and Hounsgaard, J., 1999, Ca(2+)-activated nonselective cationic current (I(CAN)) in turtle motoneurons, *Journal of Neurophysiology,* 82, 730-735.

Perrier, J. F., and Hounsgaard, J., 2000, Development and regulation of response properties in spinal cord motoneurons, *Brain Research Bulletin,* 53, 529-535.

Perrier, J. F., Mejia-Gervacio, S., and Hounsgaard, J., 2000, Facilitation of plateau potentials in turtle motoneurones by a pathway dependent on calcium and calmodulin, *Journal of Physiology,* 528, 107-113.

Russo, R. E., and Hounsgaard, J., 1994, Short-term plasticity in turtle dorsal horn neurons mediated by L-type Ca2+ channels, *Neuroscience,* 61, 191-197.

Russo, R. E., and Hounsgaard, J., 1996, Plateau-generating neurones in the dorsal horn in an in vitro preparation of the turtle spinal cord, *Journal of Physiology,* 493, 39-54.

Schwindt, P. C., and Crill, W. E., 1977, A persistent negative resistance in cat lumbar motoneurons, *Brain Research,* 120, 173-178.

Schwindt, P. C., and Crill, W. E., 1980a, Properties of a persistent inward current in normal and TEA-injected motoneurons, *Journal of Neurophysiology,* 43, 1700-1724.

Schwindt, P. C., and Crill, W. E., 1980b, Effects of barium on cat spinal motoneurons studied by voltage clamp, *Journal of Neurophysiology,* 44, 827-846.

Schwindt, P. C., and Crill, W. E., 1984, Membrane properties of cat motoneurons, in: *Handbook of the Spinal Cord,* Davidoff R. A., ed., Marcel Dekker Inc., New York, Basel, pp. 199-242.

Svirskis, G., Gutman, A., and Hounsgaard, J., 2001, Electrotonic structure of motoneurons in the spinal cord of the turtle: inferences for the mechanisms of bistability, *Journal of Neurophysiology,* 85, 391-398.

Svirskis, G., and Hounsgaard, J., 1997, Depolarization-induced facilitation of a plateau-generating current in ventral horn neurons in the turtle spinal cord, *Journal of Neurophysiology,* 78, 1740-1742.

Svirskis, G., and Hounsgaard, J., 1998, Transmitter regulation of plateau properties in turtle motoneurons, *Journal of Neurophysiology,* 79, 45-50.

Xu, W., and Lipscombe, D., 2001, Neuronal Ca(V)1.3alpha(1) L-type channels activate at relatively hyperpolarized membrane potentials and are incompletely inhibited by dihydropyridines, *Journal of Neuroscience,* 21, 5944-5951.

Zuhlke, R. D., Pitt, G. S., Deisseroth, K., Tsien, R. W., and Reuter, H., 1999, Calmodulin supports both inactivation and facilitation of L-type calcium channels, *Nature,* 399, 159-162.

Zuhlke, R. D., Pitt, G. S., Tsien, R. W., and Reuter, H., 2000, Ca2+-sensitive inactivation and facilitation of L-type Ca2+ channels both depend on specific amino acid residues in a consensus calmodulin- binding motif in the(alpha)1C subunit, *Journal of Biological Chemistry,* 275, 21121-21129.

RECENT EVIDENCE FOR PLATEAU POTENTIALS IN HUMAN MOTONEURONES

David F. Collins[1], Monica Gorassini[2], David Bennett[2], David Burke[3] and Simon C. Gandevia[3]

ABSTRACT

Motoneurones in reduced animal preparations can exhibit plateau potentials that amplify their response to synaptic inputs and can persist for prolonged periods in the absence of synaptic drive. There is mounting evidence that a similar mechanism may be an integral part of the normal activation of motoneurones. Some of the work describing plateau potentials in reduced animal preparations is reviewed and then evidence that similar properties contribute to the normal activation of motoneurones in rats and humans is presented. Recent data have shown that during high-frequency electrical stimulation over human muscle, large contractions can develop which originate within the central nervous system and are present in addition to the contraction due to the direct activation of motor axons. These "extra" contractions may in part be due to plateau potentials in spinal motoneurones. It is becoming clear that intrinsic properties of human motoneurones may make a large contribution to muscle contractions during normal movements. The extent to which the nervous system uses this as a gain control mechanism to tailor motor output for a given task needs to be further explored.

INTRODUCTION

Motoneurones in reduced animal preparations can be induced to develop plateau potentials which are sustained membrane depolarisations produced by the activation of persistent inward currents (e.g. Schwindt and Crill, 1980; Hounsgaard et al., 1988; Lee and Heckman, 1996). Such potentials markedly amplify the cells' response to synaptic inputs and can lead to periods of self-sustained activity in the absence of any synaptic drive. Since motoneurones comprise the final link between the central nervous system and skeletal muscle, how they respond to synaptic inputs could dramatically affect the

[1] Faculty of Physical Education and Recreation, University of Alberta, Edmonton, Alberta, Canada. Email: dave.collins@ualberta.ca
[2] University Centre for Neuroscience, University of Alberta, Edmonton, Alberta, Canada.
[3] Prince of Wales Medical Research Institute and University of New South Wales, Sydney 2031, Australia.

Sensorimotor Control of Movement and Posture
Edited by Gandevia *et al.*, Kluwer Academic/Plenum Publishers, 2002

resulting motor output. Such an active role for motoneurones in motor control contrasts sharply with classical ideas. Early experiments on motoneurones in spinal or deeply anaesthetised preparations showed a nearly linear relationship between synaptic drive and cell discharge (Eccles, 1957; Granit et al., 1966). Thus it was thought that motoneurone discharge faithfully reflected the net synaptic drive and motoneurones were passive participants shaping motor output. The finding that motoneurones have intrinsic properties which could allow them to play a much more active role in synaptic integration has led to a re-evaluation about how motoneurones contribute to motor control.

Even before intrinsic motoneurone properties were identified, there was evidence of motoneurone behaviour that was inconsistent with traditional ideas. These early data generally involved sustained motor output that outlasted the stimulus that evoked it. Sherrington (1906) described "afterdischarge" in motoneurones activated by the crossed extension reflex. Forbes (1929) proposed that such sustained motor output may be due to reverberating activity in spinal circuits. Granit and colleagues (1957; 1959) described prolonged discharges that could be potentiated by repeated short trains of synaptic drive. Similar prolonged discharges have been reported periodically throughout the literature (e.g. Hultborn et al., 1975, 1980; Hounsgaard et al., 1984; Wada et al., 1989).

During the 1970s and 1980s there was much interest in contractions that developed during prolonged periods of vibration applied over muscle tendons. These "tonic vibration reflexes" (TVRs) were often associated with sustained motor activity which outlasted the vibratory stimulus (Hagbarth and Eklund, 1966; Marsden et al., 1969). It was also shown that similar contractions could be evoked by electrical stimulation of the nerve to the muscle (De Gail et al., 1966). However this result has remained largely unexplored. Two pieces of evidence have led to the idea that TVRs were due to a buildup of activity in polysynaptic, supraspinal (possibly brainstem), reflex pathways, (Kanda, 1972). First, TVRs in animals were abolished after a spinal transection. Secondly, the short-latency reflex excitation of motoneurones (at presumed monosynaptic latency) was greatly attenuated at stimulus frequencies used for the TVR (e.g. Lang and Vallbo, 1967; Burke and Schiller, 1976) suggesting the effect is not due to a build-up of activity in such "simple" reflex pathways. In hindsight, these "reflex" contractions may be due to properties intrinsic to the motoneurones themselves.

PLATEAU POTENTIALS IN REDUCED ANIMAL PREPARATONS

Schwindt and Crill (1980) described persistent inward currents in motoneurones that could account for the sustained motor output observed previously in response to transient synaptic input. These inward currents result in sustained depolarisations of the cell membrane (plateau potentials) which can be turned-on and turned-off by transient excitatory and inhibitory inputs, respectively (Hounsgaard et al., 1984, 1986; Crone et al., 1988). In this way the cell can be switched between two stable levels of depolarisation. This "bistable" behaviour depends on the presence of specific neuromodulators and neurotransmitters in the spinal cord (Hounsgaard et al., 1988; Conway et al., 1988). Plateau activity was absent after a transection of the spinal cord but was restored by i.v. injection of serotonergic and noradrenergic precursors. Hence supraspinal centres may regulate the activation of plateau potentials in motoneurones via descending serotonergic and noradrenergic pathways.

The persistent inward currents are due primarily to activity in dendritic calcium channels (Hounsgaard and Kiehn, 1993; Lee and Heckman, 1998, 2000) suggesting that plateau activation may be particularly sensitive to changes in synaptic drive. How such changes alter the threshold for plateau activation has been explored by Bennett and colleagues (1988a,b). During triangular current injection the cells typically showed a sudden jump in firing frequency as the plateau was activated during increasing injected current, after which the cells discharged at significantly higher rates during the decreasing current injection compared to similar current on the ascending limb (hysteresis). Synaptic drive provided by stretch of the homonymous muscle (soleus) or stimulation of its' nerve reduced the threshold for plateau activation. In some cells, the plateau could be activated below the threshold for spike initiation.

Intracellular stimulation and recording techniques in reduced vertebrate preparations have made it possible to demonstrate persistent inward currents in motoneurones which underlie the activation of plateau potentials. The extent to which similar currents are activated in conscious animals and humans and how they contribute to the normal activation of motoneurones is now being explored.

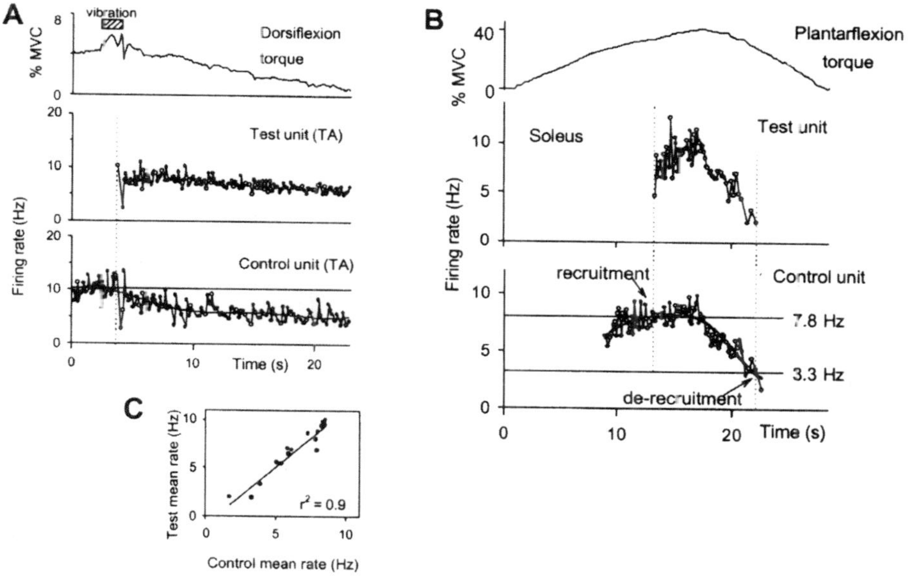

Figure 1. *A*, sustained activation of a test tibialis anterior (TA) motor unit (second panel) after removal of muscle vibration (marked by dashed bar) and after subject reduces contraction effort (synaptic drive) as marked by reduction in firing rate of lower-threshold control motor unit (bottom panel) from 10 to 5 Hz. *B*, during a triangular isometric torque contraction, a soleus test unit was de-recruited when a control unit was firing at a rate 5.5 Hz lower than when the test unit was initially recruited (7.8 vs. 3.3 Hz). This represents a possible 40% reduction in estimated synaptic drive required to maintain firing of a motor unit compared to the amount needed to recruit a unit. C, mean firing rate (calculated every 500 ms) of control unit plotted against the mean rate of test unit during the contraction in *B*. The highly linear relationship in firing rate modulation of the two units suggests that they received common synaptic drives and that the firing rate of the control unit is good indicator of estimated synaptic drive to the test motor unit. (From Gorassini et al. 2002a.)

PLATEAU POTENTIALS INFERRED FROM MOTOR UNIT ACTIVITY

The strongest evidence suggesting that intrinsic motoneurone activation may contribute to the firing behaviour of motor units in freely moving animals and in humans has come from experiments using paired motor unit recording techniques (rat: Eken and Kiehn, 1989; Eken, 1998; Gorassini et al., 1999; human: Kiehn and Eken, 1997; Gorassini et al., 1998; Zijdewind and Thomas, 2001). In these experiments, the firing frequency of a tonically active unit of the pair (control unit) is often used to estimate indirectly the synaptic drive to the other motor unit in the pair (test unit). This is done so that any discharge of the test unit that cannot be accounted for by synaptic inputs alone can be attributed to intrinsic activation of the parent motoneurone (see Gorassini et al., 2002a for Discussion). For example, a test motor unit that is recruited by a brief muscle vibration will continue to discharge after the vibration is removed, even though the subject maintains a constant (Kiehn and Eken, 1997), or even lowered (Gorassini et al., 1998; 2002a), contraction effort after the muscle vibration (see Fig. 1A). Thus, with respect to the estimated profile of the synaptic drive, the test unit is recruited when the synaptic drive is transiently increased by vibration-sensitive afferent inputs, as indicated by an increase in the firing rate of the control unit. Further, the test unit continues to fire when the synaptic input that initially recruited it is removed and when the background synaptic drive is lower than pre-vibration levels. Note that in Figure 1A the firing rate of the control unit can be reduced to below pre-vibration levels and yet the test unit continues to fire. Only control and test unit pairs that have tightly correlated firing patterns are used as this indicates that the unit pairs likely share common synaptic inputs and that the firing rate of the control unit is a good estimator of synaptic drive to the test unit under study (Gorassini et al., 2002a, Fig. 1C). The continued discharge of the test unit at levels of synaptic drive that are too low to recruit the unit initially may be explained by an added depolarization of the parent motoneurone that comes from intrinsic sources (plateaus).

Recently, the contribution of this intrinsic activation has been estimated through use of paired motor unit recordings during triangular, isometric torque contractions (Gorassini et al., 2002a,b). The firing rate of a control motor unit was compared during the recruitment and de-recruitment of a relatively higher-threshold test motor unit to quantify the difference between the synaptic drive needed to recruit a test unit and the level required to sustain its firing at its minimal discharge rate, i.e., just before de-recruitment during the descending phase of a triangular contraction (Fig. 1B). The decrease in extrinsic synaptic drive required to maintain a minimal discharge from recruitment is proposed to reflect the amount of depolarization produced from the sustained activation of intrinsic currents in the parent motoneurone. On average, the firing rate of the control unit was reduced by 3.6 Hz from an initial recruitment rate of 9.8 Hz and this represents a possible 40% reduction in estimated synaptic drive during the moderate, isometric contractions used in this study (< 30% maximum voluntary force). In addition, during repeated contractions, the level of estimated synaptic drive (control unit firing rate) required to re-recruit a test motor unit was reduced by 30 to 60% when the test units were re-recruited within 4-6 seconds or less. Thus, the estimated size of the intrinsic contribution to motor unit firing (4 Hz or about 30 to 40% contribution to a motoneurone's firing rate) and the time course of activation-dependent increases in intrinsic excitability of the motor units or "warm-up" (4-6 seconds or less) are similar to data for plateau potentials in reduced animal preparations (Svirskis and Hounsgaard,

1997; Bennett et al. 1998a,b; Lee and Heckman, 1998). This similarity suggests that the intrinsic activation of motor units in the earlier human studies (previous paragraph) is due to the activation of plateau potentials in the parent motoneurones.

THE CENTRAL CONTRIBUTION TO THE FORCES EVOKED BY ELECTRICAL STIMULATION OVER HUMAN MUSCLE

Percutaneous electrical stimulation of human muscle initiates contractions by activating directly the motor axons beneath the stimulating electrodes. However, recently it has been shown that when using slightly wider pulse widths and higher stimulation frequencies (1 ms pulse width, 100 Hz) than used conventionally, "extra" contractions can develop which originate within the central nervous system (Collins et al., 2001a, 2002; see also De Gail et al., 1966; Lang and Vallbo, 1967). Contractions evoked by direct motor axon stimulation have been well-studied and the resulting forces are consistent and predictable. For example, brief stimulus trains at constant-frequency evoke contractions of constant force. Longer trains (>~2s) can evoke contractions which may fatigue somewhat due to properties of the muscle fibres and/or a decrease in the number of motor axons recruited. Ramp changes in stimulus frequency evoke contractions which produce a peak force between ~30-40 Hz and a given stimulation frequency evokes a force which changes only slightly depending on whether the frequency is increasing or decreasing (see Binder-Macleod and Clamann, 1989). In contrast are the "extra" contractions which can develop in relaxed subjects in the muscles which plantarflex or dorsiflex the ankle using modified stimulation parameters (Collins et al. 2001a, 2002). Constant frequency and intensity stimulation resulted in contractions which often progressively increased (Fig. 2A, thick line) and generated forces of up to ~40% of that during a maximal voluntary contraction. Triangular patterns of stimulation (~4 Hz to 100 Hz and back to ~4 Hz in 6 s) often produced an abrupt increase in the contraction force despite the linearly increasing stimulus frequency and the contraction often continued to grow even as the stimulus frequency declined (Fig. 2B, thick line). These extra contractions were absent when the nerve was completely blocked proximal to the stimulation site (Fig. 2, thin lines). Therefore, they originate within the central nervous system and develop in addition to the effects from direct stimulation of the motor axons (force remaining during the nerve block, Fig. 2, thin lines). This central mechanism could be triggered at innocuous stimulation intensities, in sleeping subjects and even in those with a complete spinal cord injury, suggesting that these extra contractions are not due to inadvertent volitional drive to the motoneurones. This behaviour could also be observed at stimulation intensities below the threshold for activating motor axons, suggesting that activation of the largest diameter afferents provides sufficient input to produce the extra contractions. The modified parameters for the stimulation (e.g. use of a wide pulse width, 1 ms) probably ensured that stimulation evokes a relatively larger afferent volley than that during more conventional stimulation. This may be why these "central" contractions have not been reported more frequently (cf. De Gail et al., 1966; Lang and Vallbo, 1967).

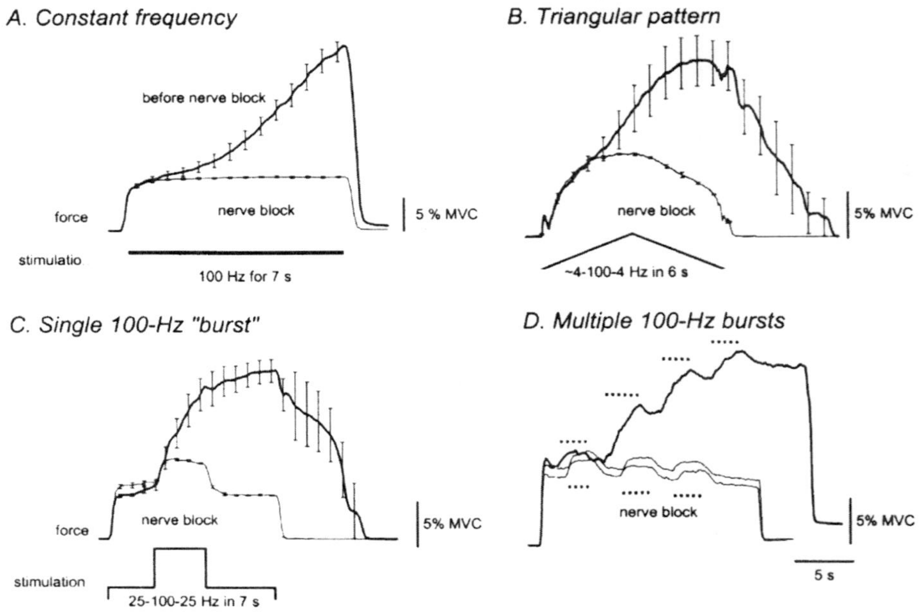

Figure 2. Forces evoked in single subjects by stimulation over triceps surae (*A*) and tibialis anterior (*B-D*). The stimuli were delivered before (thick lines) and during (thin lines) a complete nerve block proximal to the stimulation site. Data in panels *A-C* shown the mean and SEM of the force evoked by 5 stimulus trains. Panel *D* shows the force evoked by single stimulus trains which alternated between 25 and 100 Hz. The periods of 100 Hz stimulation are shown by the horizontal dotted lines. (Panel *A* from Collins et al. 2001a, panels *C-D* adapted from Collins et al. 2002).

Many characteristics of these "extra contractions" are consistent with the activation of plateau potentials in spinal neurones (Collins et al., 2001a, 2002). For example, they are triggered by high frequency stimulation of large diameter afferents (Collins et al., 2001a) and can be turned off by inhibitory inputs associated with cutaneous nerve stimulation or contraction of the antagonist muscle (unpublished observations, cf. Collins et al., 2002). Triangular stimulation profiles resulted in marked hysteresis in the relationship between force and stimulation frequency (Fig. 2B). In many instances a short stimulus train could trigger a sustained motor output which decayed over seconds (Fig. 2B-D) or could persist for several minutes. Similarly, a "burst" of 100 Hz stimulation during a 25 Hz stimulus train (Fig. 2C) resulted in a sustained increase in force remaining after the 100 Hz burst. Often the magnitude of the response to such 100 Hz bursts became progressively larger with repeated stimulation (Fig. 2D), similar to the wind-up or warm-up phenomena shown for plateaus in motoneurones of reduced preparations. All of these phenomenon were absent during the nerve block (Fig. 2, thin lines). The contractions also share similarities with those described during the TVR (Hagbarth and Eklund, 1966, Marsden et al., 1969) and its electrical analogue (De Gail et al., 1966; Lang and Vallbo, 1967) and probably share a similar mechanism. Recordings of motor unit activity during the stimulation showed that new units were recruited during the developing contractions and their discharge was not time-locked to the stimulus pulses

(Collins et al., 2001a). Thus, these units were not recruited by direct motor axon stimulation or by activity in short-latency (monosynaptic) reflex pathways. Single motor unit recordings obtained during the sustained discharge which often persisted between stimulus trains has shown that many of the active units discharged at a constant frequency despite increases in ongoing contraction torque (Collins et al., 2001b). The same units activated by volition increased their discharge frequency with increases in contraction force (Collins et al., 2001b). This is strong evidence that, for these units, the sustained discharges are not due to are not due to changes in synaptic drive but are due to mechanisms intrinsic to the motoneurones themselves. Some of the increased torque recorded after successive stimulus trains is probably due to the recruitment of additional motor units whose discharge is maintained by plateau potentials. However, some units that remained active after the electrical stimulation did show increases in discharge rate with increases in background force. Thus, the extra contractions produced by the high frequency stimulus likely results from a combined effect of intrinsic activation (e.g., plateau potentials) in motoneurones and increases in background synaptic drive. The extent to which these two mechanisms contribute to the observed force increments remains to be determined.

It has been suggested that plateau potential activation makes an important contribution to motoneurone depolarisation during normal contractions (Kiehn and Eken, 1997; Gorassini et al., 1998, 2002a,b). If a large component of the presently described electrically-evoked contractions depend on a similar mechanism, the ability for human motoneurones to express this behaviour may be similar to that shown previously in reduced animal preparations. Interestingly, weak and brief voluntary contractions (~5% MVC for 2 s) performed during longer trains of stimulation (25 Hz for 15 s) could also trigger the central mechanism, resulting in an elevated or even increasing force after the voluntary contraction ended (Collins et al., 2002). Thus, neural activity associated with even small voluntary contractions is sufficient to trigger this behaviour. If the central nervous system can regulate the strength of this mechanism by altering the levels of neuronmodulators and neurotransmitters in the spinal cord, it would provide a mechanism whereby the gain of motor output can be regulated over a wide range.

CONCLUSIONS

Motoneurones in reduced animal preparations can develop plateau potentials which amplify their response to synaptic drive and can also lead to periods of sustained motor output. The strength of the underlying persistent inward currents is dependent on the concentrations of specific neurotransmitters and neuromodulators present in the spinal cord. Such regulation could potentially provide a powerful "gain control" mechanism over all motor output. Until recently it was unclear whether plateau potentials were unique to the laboratory or if they represented a fundamental part of the normal activation of motoneurones. There is now mounting evidence that plateau potentials can be induced experimentally in motoneurones in vivo and the presumed underlying currents may contribute significantly to those responsible for maintaining motoneurone discharge during natural motor behaviour. It can no longer be assumed that a motoneurone discharge simply reflects net synaptic drive but rather motoneurones have the capacity to actively shape their output. The extent to which the nervous system adjusts the gain of this mechanism according to task demands remains to be determined.

ACKNOWLEDGEMENTS

This work was funded by the Alberta Heritage Foundation for Medical Research, Canadian Institute for Health Research and the National Health and Medical Research Council of Australia (#3206).

REFERENCES

Bennett, D. J., Hultborn, H., Fedirchuk, B., and Gorassini, M., 1998a, Short-term plasticity in hindlimb motoneurons of decerebrate cats, *Journal of Neurophysiology*, 80, 2038-2045.

Bennett, D. J., Hultborn, H., Fedirchuk, B., and Gorassini, M., 1998b, Synaptic activation of plateaus in hindlimb motoneurons of decerebrate cats, *Journal of Neurophysiology*, 80, 2023-2037.

Binder-Macleod, S. A., and Clamann, H. P., 1989, Force output of cat motor units stimulated with trains of linearly varying frequency, *Journal of Neurophysiology*, 61, 208-217.

Burke, D., and Schiller, H. H., 1976, Discharge pattern of single motor units in the tonic vibration reflex of human triceps surae, *Journal of Neurology, Neurosurgery and Psychiatry*, 39, 729-741.

Collins, D. F., Burke, D., and Gandevia, S. C., 2001a, Large involuntary forces consistent with plateau-like behavior of human motoneurons, *Journal of Neuroscience*, 21, 4059-4065.

Collins, D. F., Stein, R. B., and Gorassini, M., 2001b, Possible contribution of motoneuron plateau potentials to sustained torque generation by high frequency stimulation over human muscle, *Society for Neuroscience Abstracts*, 625.8.

Collins, D. F., Burke, D., and Gandevia, S. C., 2002, Sustained contractions produced by plateau-like behaviour in human motoneurones, *Journal of Physiology*, 538, 289-301.

Conway, B. A., Hultborn, H., Kiehn, O., and Mintz, I., 1988, Plateau potentials in alpha-motoneurones induced by intravenous injection of L-dopa and clonidine in the spinal cat, *Journal of Physiology*, 405, 369-384.

Crone, C., Hultborn, H., Kiehn, O., Mazieres, L., and Wigstrom, H., 1988, Maintained changes in motoneuronal excitability by short-lasting synaptic inputs in the decerebrate cat, *Journal of Physiology*, 405, 321-343.

De Gail, P., Lance, J. W., and Neilson, P. D., 1966, Differential effects on tonic and phasic reflex mechanisms produced by vibration of muscles in man, *Journal of Neurology, Neurosurgery and Psychiatry*, 29, 1-11.

Eccles, J. C., 1957, The Physiology of Nerve Cells, Johns Hopkins Press, Baltimore.

Eken, T., 1998, Spontaneous electromyographic activity in adult rat soleus muscle, *Journal of Neurophysiology*, 80, 365-376.

Eken, T., and Kiehn, O., 1989, Bistable firing properties of soleus motor units in unrestrained rats, *Acta Physiologica Scandinavica*, 136, 383-394.

Forbes, A., 1929, The Foundations of Experimental Psychology, Clark University Press, Worcester.

Gorassini, M., Bennett, D. J., Kiehn, O., Eken, T., and Hultborn, H., 1999, Activation patterns of hindlimb motor units in the awake rat and their relation to motoneuron intrinsic properties, *Journal of Neurophysiology*, 82, 709-717.

Gorassini, M., Bennett, D. J., and Yang, J. F., 1998, Self-sustained firing of human motor units, *Neuroscience Letters*, 247, 13-16.

Gorassini, M., Yang, J. F., Siu, M., and Bennett, D. J., 2002a, Intrinsic activation of human motoneurons: possible contribution to motor unit excitation, *Journal of Neurophysiology*, 87, in press.

Gorassini, M., Yang, J. F., Siu, M., and Bennett, D. J., 2002b, Intrinsic activation of human motoneurons: reduction of motor unit recruitment thresholds by repeated contractions, *Journal of Neurophysiology*, 87, in press.

Granit, R., Phillips, C.G., Skogland ,S., and Steg, G., 1957, Differentiation of tonic from phasic alpha ventral horn cells by stretch, pinna and crossed extension reflexes, *Journal of Neurophysiology*, 20, 470-481.

Granit, R., Homma, S., and Matthews, P. B. C., 1959, Prolonged changes in the discharge of mammalian muscle spindles following tendon taps or muscle twitches, *Acta Physiologica Scandinavica*, 46, 185-193.

Granit, R., Kernell, D., and Lamarre, Y., 1966, Algebraical summation in synaptic activation of motoneurones firing within the 'primary range' to injected currents, *Journal of Physiology*, 187, 379-399.

Hagbarth, K. E., and Eklund, G., 1966, Motor Effects of Vibratory Muscle Stimuli in Man, in: *Muscular Afferents and Motor Control, Proceedings of the First Noble Symposium*. Stockholm, ed. Granit, R, Almqvist and Wiksell, Stockholm, pp. 177-186

Hounsgaard, J., Hultborn, H., Jespersen, B., and Kiehn, O., 1984, Intrinsic membrane properties causing a bistable behaviour of alpha-motoneurones, *Experimental Brain Research*, 55, 391-394.

Hounsgaard, J., Hultborn, H. and Kiehn, O., 1986, Transmitter-controlled properties of alpha-motoneurones causing long-lasting motor discharge to brief excitatory inputs, *Progress in Brain Research*, 64, 39-49.
Hounsgaard, J., Hultborn, H., Jespersen, B., and Kiehn, O., 1988, Bistability of alpha-motoneurones in the decerebrate cat and in the acute spinal cat after intravenous 5-hydroxytryptophan, *Journal of Physiology*, 405, 345-367.
Hounsgaard, J., and Kiehn, O., 1993, Calcium spikes and calcium plateaux evoked by differential polarization in dendrites of turtle motoneurones in vitro, *Journal of Physiology*, 468, 245-259.
Hultborn, H., and Wigstrom, H., 1980, Motor response with long latency and maintained duration evoked by activity in Ia afferents, In *Progress in Clinical Neurophysiology*, ed. Desmedt, J. E., Karger, Basel, pp. 99-116.
Hultborn, H., Wigstrom, H., and Wangberg, B., 1975, Prolonged activation of soleus motoneurones following a conditioning train in soleus Ia afferents - a case for a reverberating loop?, *Neuroscience Letters*, 1, 147-152.
Kanda, K., 1972, Contribution of polysynaptic pathways to the tonic vibration reflex, *Japanese Journal of Physiology*, 22, 367-377.
Kiehn, O., and Eken, T., 1997, Prolonged firing in motor units: evidence of plateau potentials in human motoneurons?, *Journal of Neurophysiology*, 78, 3061-3068.
Lang, A. H., and Vallbo, Å. B., 1967, Motoneuron activation by low intensity tetanic stimulation of muscle afferents in man, *Experimental Neurology*, 18, 383-391.
Lee, R. H., and Heckman, C. J., 1996, Influence of voltage-sensitive dendritic conductances on bistable firing and effective synaptic current in cat spinal motoneurons in vivo, *Journal of Neurophysiology*, 76, 2107-2110.
Lee, R. H., and Heckman, C. J., 1998, Bistability in spinal motoneurons in vivo: systematic variations in persistent inward currents, *Journal of Neurophysiology*, 80, 583-593.
Lee, R. H., and Heckman, C. J., 2000, Adjustable amplification of synaptic input in the dendrites of spinal motoneurons in vivo, *Journal of Neuroscience*, 20, 6734-6740.
Marsden, C. D., Meadows, J. C., and Hodgson, H. J. F., 1969, Observations on the reflex response to muscle vibration and its voluntary control, *Brain*, 92, 829-846.
Schwindt, P., and Crill, W., 1980, Role of a persistent inward current in motoneuron bursting during spinal seizures, *Journal of Neurophysiology*, 43, 1296-1318.
Sherrington, C.S., 1906, The Integrative Action of the Nervous System, Yale University Press, New Haven.
Svirskis, G., and Hounsgaard, J., 1997, Depolarization-induced facilitation of plateau generating currents in ventral horn neurons in the turtle spinal cord, *Journal of Neurophysiology*, 78: 1740-1742.
Wada, N., Nakajima, Y., and Homma, S., 1989, Long-lasting afterdischarge of alpha-motoneurons after muscle vibration or electrical stimulation of Group I afferent fibers in the anemically decerebrated cat, *Neuroscience Research*, 6, 234-247.
Zijdewind, I., and Thomas, C. K., 2001, Spontaneous motor unit behaviour in human thenar muscles after spinal cord injury, *Muscle and Nerve*, 24, 952-962.

29

PATTERNS OF PATHOLOGICAL FIRING IN HUMAN MOTOR UNITS

Christine K. Thomas,[1] Jane E. Butler,[1] and Inge Zijdewind[2]

ABSTRACT

Modulation of motor unit firing rates can change muscle force production. Motor unit firing rates are often reduced during voluntary contractions of muscles influenced by disorders such as stroke or multiple sclerosis, while higher firing rates are typical of muscles innervated by a reduced number of motoneurones. An expanded range of motor unit firing rates is characteristic of disorders in which damage to various systems and neurons occur. Most neuromuscular disorders result in an increase in motor unit discharge variability, in part due to a higher incidence of doublets. In spinal cord injured subjects, long lasting involuntary contractions are common. This activity may reflect persistent inward currents that are revealed due to a lack of (voluntary) inhibition. Some of these changes in motor unit behaviour may actually work to enhance muscle force rather than to reduce it.

INTRODUCTION

Muscle force is usually graded by the recruitment and rate modulation of motoneurones. This chapter focuses on the motor unit rate modulation that occurs during brief voluntary and involuntary contractions of muscles influenced by various disorders of the nervous system. Variability in motor unit firing, including the occurrence of doublets and repetitive discharges, are also described. Finally, some of the functional implications of these data are discussed.

[1] The Miami Project to Cure Paralysis and Department of Neurological Surgery, University of Miami School of Medicine, Miami, FL 33101, USA. Email: cthomas@miami.edu
[2] Department of Medical Physiology, University of Groningen, Groningen, The Netherlands

VOLUNTARY CONTRACTIONS

Motor Unit Firing Rate Modulation

During voluntary contractions, motor units usually begin to fire at 5-12 Hz (Person and Kudina, 1972; Tanji and Kato, 1973). These rates typically increase during maximal voluntary contractions (to 5-92 Hz, Enoka and Fugelvand, 2001), rate modulation that arises from descending drive, intraspinal and afferent sources. Some of these inputs can be interrupted by a neuromuscular disorder or trauma. Comparisons between motor unit behaviour in these situations and control data from healthy muscles can thus help us to understand the role that various input sources play in motor unit rate modulation.

Table 1 shows the onset and maximum motor unit firing rates typically recorded during voluntary contractions of muscles influenced by various neuromuscular disorders. Data from the uninvolved limb or control muscles are also given (and see Person and Kudina, 1972; Tanji and Kato, 1973; Enoka and Fuglevand, 2001). Onset and maximal motor unit firing rates are usually reduced when there is some disruption of descending voluntary drive due to stroke for example, or as a result of the demyelination that characterizes multiple sclerosis. In contrast, disorders such as amyotrophic lateral sclerosis and poliomyelitis that involve motoneurone death usually result in increased motor unit firing rates. Force is presumably generated by rate modulating units that survive. After spinal cord injury however, the range of motor unit firing rate modulation increases in muscles like triceps brachii that are innervated from spinal levels near the injury site. In many of these disorders, motor units within the same muscle are influenced to various extents. There is also variation in unit behaviour across muscles. Presumably this reflects differences in the severity of damage, effects on different systems and/or neuron death. Another factor that is often overlooked is damage to sensory systems. Temporary interruption of afferent input to motoneurones by anesthetic block of a peripheral nerve reduces motor unit firing rates (Macefield et al., 1993).

Alterations in motor unit rate modulation also occur for other reasons. For example, inspiratory muscles of people with chronic obstructive pulmonary disease (COPD) are constantly subjected to increased loads. These muscles are thus required to generate larger negative pressures to inflate the lungs. They are also forced to work at short muscle lengths due to chronic hyperinflation. To overcome the increased loads, even when breathing at rest, the neural drive to the major obligatory inspiratory muscles is larger. This drive is reflected by increased unit firing rates. Greater rate differences occur for diaphragm units (COPD: 17.9 Hz vs. control: 10.5 Hz) than for scalenes (11.4 Hz vs. 8.5 Hz) or parasternal intercostal muscles (13.4 Hz vs. 10.1 Hz) (Gandevia et al., 1996; De Troyer et al., 1997). This disproportionate change in firing rates for different motor pools suggests that frequency modulation is more prominent than recruitment for diaphragm motoneurones compared to parasternal intercostal and scalene muscles.

Other factors that may contribute to changes in motor unit firing rate modulation with disease or trauma may include neuromuscular adaptation, injury to certain motor unit types, muscle-specific features, different recruitment and rate modulation strategies across motor units and muscles, inadequate voluntary drive, poor coordination, coactivation of muscles and/or changes in muscle activation patterns and use. Controls for medication, data pooling across different disorders and use of submaximal contractions that are not referred to maximum also need consideration.

Table 1. Voluntary motor unit firing frequencies and variability

		Frequency		Firing variability			
Condition	**Muscle**	**Onset Hz**	**MVC Hz**	**On-set**	**MVC**	**Doub-lets**	**Authors**
Stroke	Deltoid ADM	7±1 (8±1) 7±1 (7±1)		=	↑		Frascarelli et al. 1998
Stroke	Biceps brachii	9-15 (11-18)					Gemperline et al. 1995
Stroke	EIP			↑			Young & Shahani 1980
Stroke	TA	5-13 (6-20)		↑	↑	Yes	Rosenfalck & Andreassen 1980
Spastic hemiparesis	FDI			↑			Freund et al. 1973
Pyramidal lesions	FDI			↑			Freund et al. 1973
Multiple sclerosis	TA	5-13 (6-20)		↑	↑	Yes	Rosenfalck & Andreassen 1980
Multiple sclerosis	Quadriceps	4-9 (7-14)	13±2 (24±6)				Rice et al. 1992
Multiple sclerosis	Biceps, triceps, TA			↑	↑		Dorfman et al. 1989
Chorea	FDI	11±3 (11±1)		↑			Dengler et al. 1986
Parkinson	FDI	10±3 (11±1)		↑			Dengler et al. 1986
Parkinson	Biceps FDI	6-8 (8±1) 6-8 (7±2)		↑		Yes	Petajan & Jarcho 1975
Parkinson	FDI, LFE	3-5		↑		Yes	Dietz et al. 1974
Parkinson				↑			Das Gupta 1963
Akinetic Parkinson	FDI			↑			Freund et al. 1973
Cerebellar ataxia	EIP			↑			Young & Shahani 1980
Cerebellar conditions	FDI			↓			Freund et al. 1973
Spinal cord injury	Biceps Gluteus maximus		5-8 10-13				Little et al. 1996
Spinal cord injury	Triceps brachii TA	18±2 (10±1) 7±2 (9±1)		↑		Yes	Wiegner et al. 1993
Spinal cord injury	Triceps brachii Thenar	2-25 (6-15) 5±3	4-67 (10-45) 13±5 (34±10)	↑	↑	Yes	Thomas & Kozhina 1999 Thomas & Zijdewind 2002 Thomas 1997b
Post polio	Quadriceps	9±2 (8±1)					Rodriquez et al. 1991
Post polio	TA	<10-20					Borg et al. 1995
ALS		16±5		↑			Petajan 1974
Motoneuron disease	Biceps, triceps, TA			↑	↑		Dorfman et al. 1989
Lower motoneuron				↑		Yes	Rowińska-Marcińska et al. 1999

(*Continued*)

Table 1 (Continued)

Lower motoneuron	FDI weak, weaker	9±2 11±3 (9±2)	20±6				Miller & Sherratt 1978
Neuropathy				↑		Yes	Partanen 1978
Neuropathy		21±5 (8±2)		=			Petajan 1974
Neuropathy	Biceps brachii	11±2 (11±1)		↑			Fuglsang-Frederiksen et al. 1987
Neuropathy	TA	9±2 (8±2)		↑			Halonen et al. 1981
Poly-neuropathy	FDI			↑			Freund et al. 1973
Polymyositis						Yes	Partanen 1978
Myopathy	Biceps, triceps, TA			=	=		Dorfman et al. 1989
Myopathy						No	Rowinska-Marcinska et al. 1999
Myopathy		15±2 (8±2)		=			Petajan 1974
Myopathy	Biceps brachii	11±2 (11±1)		↑			Fuglsang-Frederiksen et al. 1987
Myopathy	TA	11±2 (8±2)		=			Halonen et al. 1981
Myotonic dystrophy						Yes	Partanen 1978
Muscular dystrophy	Biceps brachii	8-16	12-53	↓	↓		Piotrkiewicz et al. 1999

Control or uninvolved limb data are in brackets; Onset, the frequency at which units fire steadily; MVC, maximum voluntary contraction; ↑, increased; ↓, decreased; =, no change; ADM, abductor digiti minimi; EIP, extensor indicis proprius; FDI, first dorsal interosseous; LFE, long finger extensors; TA, tibialis anterior

Variability in Motor Unit Firing Behaviour

In healthy muscles, motor unit discharge variability is reduced as voluntary muscle force increases (Person and Kudina, 1972). Most studies involving muscles influenced by stroke, multiple sclerosis or various neuropathies report an increased discharge variability of motor units during voluntary contractions. Only in cerebellar conditions, various myopathies or in children with muscular dystrophy have reductions in discharge variability been reported (Table 1). These studies also emphasize that motor units vary in the extent to which their discharge variability is altered. For some units it is unchanged. Thus, the means reported for any condition probably underestimate the actual changes in some units.

Motor unit firing variability takes various forms. For example, in pathological tremor (e.g. Parkinson's disease), units can fire at tremor frequency, as paired responses (20-80 ms apart) or in synchronized groups during sustained isometric contractions (Das Gupta, 1963; Mori 1975; Elble and Randall, 1976; Dengler et al., 1986; Baker et al., 1992;). Repetitive discharges also often occur in muscles influenced by motoneurone disorders like amyotrophic lateral sclerosis but are less frequent or absent in various myopathic conditions (Rowinska-Marcinska et al., 1999). The incidence of doublets (two potentials <20 ms apart) also increases in muscles influenced by stroke, multiple sclerosis, various neuropathies and spinal cord injury (Table 1). Doublets can also exacerbate tremor (Dengler et al., 1989) and spasms of muscles paralyzed by spinal cord injury (Thomas and Ross, 1997).

INVOLUNTARY CONTRACTIONS

Involuntary activity is motor unit firing that is not intended by the subject. Short lasting, bursting activity such as clonus and spasms can be observed after spinal cord injury for instance. Long-lasting, tonic involuntary activity is seen in both control muscles (Bawa et al., 2000; Søgaard et al., 2001) and those influenced by pathology (Stein et al., 1990; Thomas, 1997a; Zijdewind and Thomas, 2001). Detailed investigation of unit firing patterns during these involuntary contractions may give clues as to the mechanisms underlying these different types of unintended activity.

Short Lasting Activity

During spasms of medial gastrocnemius muscles paralyzed completely by spinal cord injury, motor unit firing rates either increased and then decreased with spasm intensity or were relatively constant. Mean (± SD) peak spasm firing rates were 18 ± 9 Hz for rate modulated units and 11 ± 10 Hz for units with little rate modulation (Thomas and Ross, 1997). These variations in firing rate modulation may reflect adaptation and/or that units depend differently on descending (voluntary) input for rate modulation.

In thenar units influenced by spinal cord injury, the firing frequencies were higher during a spasm than during maximal voluntary contractions. The higher unit firing rates during spasms denotes the importance of afferent input to motoneurones and/or reflects the decline in the efficiency of the voluntary input to these motoneurones (Zijdewind and Thomas, 2001; also see Gandevia et al., 1990). Short-lasting, low frequency spontaneous unit activity is also seen in thenar muscles of spinal cord injured subjects at rest (2.2 ± 3.7 Hz). This activity could result from brief bursts of afferent input (Zijdewind and Thomas, 2001).

Long Lasting Activity

Long-lasting involuntary motor unit activity occurs in muscles paralyzed by spinal cord injury (Stein et al., 1990; Thomas 1997a; Gorassini et al., 1999; Collins et al., 2001; Zijdewind and Thomas, 2001). At rest, motor units in both paralyzed and partially paralyzed muscles were spontaneously active at low frequencies (6.1 ± 1.1 Hz; Zijdewind and Thomas, 2001). Long lasting involuntary activity also occurs in control muscles. For example, motor unit activity sometimes outlasts inputs such as tendon vibration (Kiehn and Eken, 1997; Gorassini et al., 1998), electrical nerve stimulation (Collins et al., 2001) and voluntary activation (Bawa et al., 2000; Søgaard et al., 2001; Collins et al., 2002). These findings in control and spinal cord injured subjects could represent the expression of persisting inward currents in human motoneurones (see Hounsgaard et al., 1988; Hultborn, 1999).

Activation of persistent inward currents in motoneurones may be more common than expected. 'Unintended' prolonged unit firing may reflect a lack of active inhibition rather than deviant firing behaviour. For example, a 'relax' command initiated thenar motor unit activity in spinal cord injured subjects (Zijdewind and Thomas, 2001). Thus, active inhibition may be necessary to prevent motoneurones from firing. Lack of this inhibition may thus underlie the relatively high incidence of involuntary contractions after spinal cord injury.

FUNCTIONAL CONSIDERATIONS

While onset and maximum motor unit firing rates are often reduced in muscles influenced by neuromuscular disorders or trauma, motor unit firing variability is often increased. In part this relates to a higher incidence of doublets. These changes in motor unit behaviour may not all be detrimental. For example, a doublet at contraction onset markedly enhances motor unit (Macefield et al., 1996; Thomas et al., 1999) or whole muscle force (Cooper and Eccles, 1930). After an initial doublet, synchronized stimulation at low rates optimizes muscle and motor unit force-time integral in different muscles, across species and after chronic paralysis (reviewed by Griffin et al., 2001). Thus, if these firing patterns occur during voluntary contractions, even weak drive to a muscle may activate it near optimally. Moreover, an increase in motor unit firing variability may work to increase asynchronous motor unit firing. This may act to smooth the force (Rack and Westbury, 1969), even when motor unit numbers are reduced.

ACKNOWLEDGEMENTS

The authors thank Bette Mas for assisting in the preparation of this review. This work was funded by USPHS grant 30226, the NHMRC Neil Hamilton Fairley Fellowship 007148, the University of Groningen and The Miami Project to Cure Paralysis.

REFERENCES

Baker, J. R., Davey, N. J., Ellaway, P. H., and Friedland, C. L., 1992, Short-term synchrony of motor unit discharge during weak isometric contraction in Parkinson's disease, *Brain,* 115, 137-154.

Bawa, P., Chalmers, G. R., Jones, K. E., Søgaard, K., and Walsh, M. L., 2000, Control of the wrist joint in humans, *European Journal of Applied Physiology*, 83, 116-127.

Borg, J., Borg, K., Edstrom, L., Grimby, L., Henriksson, J., Larsson, L., and Tollback, A., 1995, Motoneuron and muscle fiber properties of remaining motor units in weak tibialis anterior muscles in prior polio, *Annals of the New York Academy of Sciences,* 753, 335-342.

Collins, D. F., Burke, D., and Gandevia, S. C., 2001, Large involuntary forces consistent with plateau-like behavior of human motoneurons, *Journal of Neuroscience*, 21, 4059-4065.

Collins, D. F., Burke, D., and Gandevia, S. C., 2002, Sustained contractions produced by plateau-like behaviour in human motoneurones, *Journal of Physiology,* 289-301.

Cooper, S., and Eccles, J. C., 1930, The isometric responses of mammalian muscles, *Journal of Physiology*, 69, 377-385.

Das Gupta, A., 1963, Paired response of motor units during voluntary contraction in Parkinsonism, *Journal of Neurology, Neurosurgery and Psychiatry*, 26, 265-268.

De Troyer, A., Leeper, J. B., McKenzie, D. K., and Gandevia, S. C., 1997, Neural drive to the diaphragm in patients with severe COPD, *American Journal of Respiratory and Critical Care Medicine*, 155, 1335-1340.

Dengler, R., Gillespie, J., Argenta, M., Elek, J., Wolf, W., and Struppler, A., 1989, The impact of paired motor unit discharges on tremor, *Electroencephalography and Clinical Neurophysiology*, 29, 113-117.

Dengler, R., Wolf, W., Schubert, M., and Struppler, A., 1986, Discharge pattern of single motor units in basal ganglia disorders, *Neurology*, 36, 1061-1066.

Dietz, V., Hillesheimer, W., and Freund, H. -J., 1974, Correlation between tremor, voluntary contraction, and firing pattern of motor units in Parkinson's disease, *Journal of Neurology, Neurosurgery and Psychiatry*, 37, 927-937.

Dorfman, L. J., Howard, J. E., and McGill, K. C., 1989, Motor unit firing rates and firing rate variability in the detection of neuromuscular disorders, *Electroencephalography and Clinical Neurophysiology*, 73, 215-224.

Elble, R. J., and Randall, J. E., 1976, Motor-unit activity responsible for 8- to 12-Hz component of human physiological finger tremor, *Journal of Neurophysiology*, 39, 370-380.

Enoka, R. M., and Fuglevand, A. J., 2001, Motor unit physiology: some unresolved issues, *Muscle and Nerve*, 24, 4-17.

Frascarelli, M., Mastrogregori, L., and Conforti, L., 1998, Initial motor unit recruitment in patients with spastic hemiplegia, *Electromyography and Clinical Neurophysiology*, 38, 267-271.

Freund, H. -J., Dietz, V., Wita, C. W., and Kapp, H., 1973, Discharge characteristics of single motor units in normal subjects and patients with supraspinal motor disturbances, in: *New Developments in Electromyography and Clinical Neurophysiology*, J. E. Desmedt, ed., Karger, Basel, pp. 242-250.

Fuglsang-Frederiksen, A., Smith, T., and Hogenhaven, H., 1987, Motor unit firing intervals and other parameters of electrical activity in normal and pathological muscle, *Journal of Neurological Science*, 78, 51-62.

Gandevia, S. C., Leeper, J. B., McKenzie, D. K., and De Troyer, A., 1996, Discharge frequencies of parasternal intercostal and scalene motor units during breathing in normal and COPD subjects, *American Journal of Respiratory Critical Care in Medicine*, 153, 622-628.

Gandevia, S. C., Macefield, G., Burke, D., and McKenzie, D. K., 1990, Voluntary activation of human motor axons in the absence of muscle afferent feedback. The control of the deafferented hand, *Brain*, 113, 1563-1581.

Gemperline, J. J., Allen, S., Walk, D., and Rymer, W. Z., 1995, Characteristics of motor unit discharge in subjects with hemiparesis, *Muscle and Nerve*, 18, 1101-1114.

Gorassini, M., Bennett, D. J., and Yang, J.F., 1999, Excitability of motor units in persons with spasticity from spinal cord injury, *Neuroscience Abstracts*, 25, 120.

Gorassini, M. A., Bennett, D. J., and Yang, J. F., 1998, Self-sustained firing of human motor units, *Neuroscience Letters*, 247, 13-16.

Griffin, L., Godfrey, S., and Thomas, C. K., 2001, Stimulation pattern that maximizes force in paralyzed and control whole thenar muscles, *Journal of Neurophysiology*, submitted.

Halonen, J. -P., Falck, B., and Kalimo, H., 1981, The firing rate of motor units in neuromuscular disorders, *Journal of Neurology*, 225, 269-276.

Hounsgaard, J., Hultborn, H., Jespersen, B., and Kiehn, O., 1988, Bistability of α-motoneurones in the decerebrate cat and in the acute spinal cat after intravenous 5-hydroxytryptophan, *Journal of Physiology*, 405, 345-367.

Hultborn, H., 1999, Plateau potentials and their role in regulating motoneuronal firing, *Progress in Brain Research*, 123, 39-48.

Kiehn, O., and Eken, T., 1997, Prolonged firing in motor units: evidence of plateau potentials in human motoneurons? *Journal of Neurophysiology*, 78, 3061-3068.

Little, J. W., Goldstein, B., Gitter, A., and Haselkorn, J. K., 1996, Spinal cord infarction: varying degrees of upper and lower motoneuron dysfunction, *Journal of Spinal Cord Medicine*, 19, 242-248.

Macefield, V. G., Fuglevand, A. J., and Bigland-Ritchie, B., 1996, Contractile properties of single motor units in human toe extensors assessed by intraneural motor axon stimulation, *Journal of Neurophysiology*, 75, 2509-2519.

Macefield, V. G., Gandevia, S. C., Bigland-Ritchie, B., Gorman, R. B., and Burke, D., 1993, The firing rates of human motoneurones voluntarily activated in the absence of muscle afferent feedback, *Journal of Physiology*, 471, 429-443.

Miller, R. G., and Sherratt, M., 1978, Firing rates of human motor units in partially denervated muscle, *Neurology*, 28, 1241-1248.

Mori, S., 1975, Entrainment of motor-unit discharges as a neuronal mechanism of synchronization, *Journal of Neurophysiology*, 38, 859-870.

Partanen, V. S. J., 1978, Double discharges in neuromuscular diseases, *Journal of Neurological Science*, 36, 377-382.

Person, R. S., and Kudina, L. P., 1972, Discharge frequency and discharge pattern of human motor units during voluntary contraction of muscle, *Electroencephalography and Clinical Neurophysiology*, 32, 471-483.

Petajan, J. H., 1974, Clinical electromyographic studies of diseases of the motor unit, *Electroencephalography and Clinical Neurophysiology*, 36, 395-401.

Petajan, J. H., and Jarcho, L. W., 1975, Motor unit control in Parkinson's disease and the influence of levodopa, *Neurology*, 25, 866-869.

Piotrkiewicz, M., Hausmanowa-Petrusewicz, I., and Mierzejewska, J., 1999, Are motoneurons involved in muscular dystrophy? *Clinical Neurophysiology*, 110, 1111-1122.

Rack, P. M. H., and Westbury, D. R., 1969, The effects of length and stimulus rate on tension in the isometric cat soleus muscle, *Journal of Physiology*, 204, 443-460.

Rice, C. L., Vollmer, T. L., and Bigland-Ritchie, B., 1992, Neuromuscular responses of patients with multiple sclerosis, *Muscle And Nerve*, 15, 1123-1132.

Rodriquez, A. A., Agre, J. C., Black, P. O., and Franke, T. M., 1991, Motor unit firing rates in postpolio and control subjects during submaximal contraction, *American Journal of Physical Medicine and Rehabilitation*, 70, 191-194.

Rosenfalck, A., and Andreassen, S., 1980, Impaired regulation of force and firing pattern of single motor units in patients with spasticity, *Journal of Neurology, Neurosurgery and Psychiatry*, 43, 907-916.

Rowińska-Marcińska, K., Zalewska, E., and Hausmanowa-Petrusewicz, I., 1999, Double discharges of motor units in neuromuscular disorders, *Journal de Physiologie*, 93, 175-182.

Søgaard, K., Sjogaard, G., Finsen, L., Olsen, H. B., and Christensen, H., 2001, Motor unit activity during stereotyped finger tasks and computer mouse work, *Journal of Electromyography and Kinesiology*, 11, 197-206.

Stein, R. B., Brucker, B. S., and Ayyar, D. R., 1990, Motor units in incomplete spinal cord injury: electrical activity, contractile properties and the effects of biofeedback, *Journal of Neurology, Neurosurgery and Psychiatry*, 53, 880-885.

Tanji, J., and Kato, M., 1973, Firing rate of individual motor units in voluntary contraction of abductor digiti minimi muscle in man, *Experimental Neurology*, 40, 771-783.

Thomas, C. K., 1997a, Contractile properties of human thenar muscles paralyzed by spinal cord injury, *Muscle and Nerve*, 20, 788-799a.

Thomas, C. K., 1997b, Fatigue in human thenar muscles paralysed by spinal cord injury, *Journal of Electromyography and Kinesiology*, 7, 15-26.

Thomas, C. K., Johansson, R. S., and Bigland-Ritchie, B., 1999, Pattern of pulses that maximize force output from single human thenar motor units, *Journal of Neurophysiology*, 82, 3188-3195.

Thomas, C. K., and Kozhina, G., 1999, Motor unit firing rates during voluntary contractions, XVIIth Congress of the International Society of Biomechanics, Abstract: 242.

Thomas, C. K., and Ross, B. H., 1997, Distinct patterns of motor unit behavior during muscle spasms in spinal cord injured subjects, *Journal of Neurophysiology*, 77, 2847-2850.

Thomas, C. K., and Zijdewind, I., 2002, Force gradation during voluntary contractions of human thenar muscles after spinal cord injury, Motor Control & Proprioception: Physiology, pathophysiology & recovery, Abstract.

Wiegner, A. W., Wierzbicka, M. M., Davies, L., and Young, R. R., 1993, Discharge properties of single motor units in patients with spinal cord injuries, *Muscle and Nerve*, 16, 661-671.

Young, R. R., and Shahani, B. T., 1980, A clinical neurophysiological analysis of single motor unit discharge patterns in spasticity, in: *Spasticity: Disordered Motor Control*, R. G. Feldman, R. R. Young, and W. P. Koella, eds., Year Book, Chicago, pp. 219-231.

Zijdewind, I., and Thomas, C. K., 2001, Spontaneous motor unit behavior in human thenar muscles after spinal cord injury, *Muscle and Nerve*, 24, 952-962.

SECTION V

Propriospinal Neurones and Spinal Reflexes

This Section provides three chapters on spinal reflexes, which are also addressed in other sections of the volume, a chapter on a new opening on the properties of spinal interneurones, and three chapters which cover a lively debate at the Symposium on the recent phylogeny of cortico-propriospinal connections with cervical interneurones and motoneurones. All of this work comes under the rubric of the convergence of descending command signals and sensory feedback onto propriospinal and segmental spinal interneurones, with the latter often engaged in pattern generation for various movements.

Electrophysiological approaches, often combined with modern anatomical methods, have provided an encyclopaedic amount of data on spinal interneuronal connectivity, particularly in the cat and non-mammalian vertebrates. Far less is known about how this circuitry is actually used during the movement of most vertebrates species, however. For technical reasons, this problem is exacerbated in the primate, where in both non-human and human subjects, the focus has been on the motor effects of descending command signals rather than their integration with sensory input and ongoing spinal interneuronal activity. Many who study humans now seek insights from the increasingly complex connectivity of the spinal cord and the properties of its neural elements: the task is formidable, but progress is nonetheless evident!

To set the scene, D. Stuart (Chapter 30) reviews the current state-of-the-play on the phase- and context-dependency of low-threshold, spinal proprioceptive reflexes. Such effects were shown in the late 19th C, but their advancement and general acceptance has been an iterative process, which began in the late 1950s and is still far from complete. It is argued that such reflexes, which are fundamental to the elaboration of smoothly co-ordinated movement, are best appreciated by movement neuroscience trainees if they first gain appreciation of the comparative neurobiology of central pattern generation, with which descending command signals and sensory input interact in both invertebrate and vertebrate species, including the human.

J. Rothwell (Chapter 31) briefly introduces topics on spinal interneurones and propriospinal neurones with the reminder that knowledge about the properties of spinal interneurones active in a reflex circuit must be integrated with what is known about descending motor commands and motoneuronal output. Important recent studies described by E. Fetz provide a new opening on the behaviour of interneurones in the cervical spinal cord of the awake, actively moving monkey (Chapter 32). Interneuronal activity and electromyographic activity recorded for many upper limb muscles, while the monkey makes flexion and extension movements at the wrist, have revealed functional linkages between motoneurone's first-order interneurones (premotoneurones) and various cervical motoneurone pools (nuclei). The patterns of such linkages have been compared with those that were previously established for motor cortical and rubrospinal cells. Conceptually, this seminal work is beginning to reveal how a motor command might be represented in the discharge of various interneuronal inputs to motoneurones. A de novo finding is that some interneurones become active during the preparatory period before the movement begins, just as do many motor cortical neurones. These changes also fit with changes in spinal reflexes during the preparatory period, this being another example of the integration of descending command signals, ongoing interneuronal activity and sensory input at the spinal level. Additional modifications of the Fetz techniques promise to provide information which not only further advances understanding of spinal interneuronal activity during movement, but also facilitates estimation of changes in the central actions of sensory afferents in a particular nerve, including, for example, the level of presynaptic inhibition to which they are subjected.

While the interneurones studied by Fetz and colleagues have not been classified according to their afferent input and role in specific reflex paths, there is one group of interneurones in the cervical spinal cord that has received recent attention. This is the class of C3-C4 propriospinal neurones projecting to forelimb motoneurones. The properties of this group of neurones have been well characterised in the cat and a role for them has been proposed for movements which involve reaching to a target. The point of current dispute is the extent to which such a system operates in primates. First, there is a growing body of indirect evidence from reflex studies in the human upper limb that is quite consistent with such a system acting on forearm and upper arm muscles. This evidence has been strengthened by recent studies from E. Pierrot-Deseilligny and colleagues, which are described here (Chapter 33). Small increases in the intensity of transcranial stimulation of the motor cortex change peripheral nerve-evoked facilitation of the motoneurone responses into a depression. This is consistent with the convergence of descending and peripheral inputs onto neurones which inhibit propriospinal neurones.

B. Alstermark and T. Isa review their recent studies in the monkey in which they found the “missing” disynaptic excitation of motoneurones to pyramidal tract stimulation (Chapter 34). However, to unmask it, it was necessary to reduce descending inhibition of the pathway Their studies show unequivocally the existence of propriospinal neurones with some similar projections to those of the C3-C4 system the cat. It remains uncertain how this affects the evolutionary scaling of the “system”, and indeed if it should even be regarded as one system. Further points for consideration are the relative strengths and sources of cortical projections to the propriospinal neurones. An important question is the extent to which it is valid to compare the size of excitatory post-synaptic potentials recorded in different species of anaesthetised animals following, for example, stimulation of descending tracts with inputs to motoneurones. To resolve this. it may be necessary to measure the size of mono-, di- and oligo-synaptic EPSPs to stimulation of different systems in the same motoneurone. This task is not simple, however, because it is difficult to assess the size of disynaptic EPSPs in the presence of earlier monosynaptic activity.

By way of reply, P. Kirkwood and colleagues refer to their initial attempts to establish the existence of a C3-C4 system transmitting corticospinal excitation to upper limb motoneurones in two non-human primate species using similar methods to those used in the cat (Chapter 35). These attempts proved rather disappointing, and so they argued that the C3-C4 propriospinal path to motoneurones had been progressively downgraded or even replaced in the course of evolution of upper limb control by the more direct cortico-motoneuronal connection (see also Olivier et al., 2001). Furthermore, they point to the many assumptions that are necessary for acceptance of the view that a C3-C4 propriospinal system has been retained in human subjects. In terms of comparison of results obtained in anaesthetised cats and primates, they reiterate their fundamental conclusion that great caution is needed in generating arguments by analogy, when, for example, a particular propriospinal system has been studied in minute detail in one species (the cat) but much less studied in other species (the non-human primate). Thus, it is not safe to link the system as it exists in the cat to humans, if there is evidence that the system behaves differently in intermediate species. They conclude with proposals for further studies to examine this intriguing issue. In such future studies, there is need for consideration of both new and long-standing neuroethological and neurophylogentic approaches to augment the current essentially electrophysiological and neuroanatomical information (e.g., Iwaniuk et al., 1999).

The remaining two chapters in the Section are concerned with pathological changes in segmental reflexes. The contribution by S. Meunier and colleagues proposes a role for non-dopaminergic systems in the induction of changes in segmental pathways in Parkinson's disease (Chapter 36). Their studies reveal pathological changes in disynaptic reciprocal inhibition of wrist muscles and also in presumed group II muscle afferent paths to upper-leg muscles. The results are interpreted in the light of growing evidence for damage to non-dopaminergic paths in Parkinson's disease (Hellstrom-Lindahl and Court, 2000). Z. Rymer reports that extensor spasms in patients with partial spinal cord injuries can resemble the pattern of electromyographic activity and joint torques normally present at the end of the stance phase of gait (Chapter 37). The flexor spasm in these patients appears as a classical withdrawal reflex and the proposition now is that the extensor spasm may represent a fragmentary locomotor response from the damaged spinal cord. This contribution emphasises the view that spasticity, one of the positive symptoms in upper motoneurone lesions and an index of hyperreflexia, need not be prominent following spinal cord injury, and that the presence of spasms may reflect a different element of the segmental response to injury.

REFERENCES

Iwaniuk, A.N., Pellis, S.M., and Whishaw, I.Q., 1999, Is digital dexterity really related to corticospinal projections? A re-analysis of the Heffner and Masterton data set using modern comparative statistics, *Behavioural Brain Research*, 101, 173-87.

Hellstrom-Lindahl E., and Court J.A., 2000, Nicotinic acetylcholine receptors during prenatal development and brain pathology in human aging, *Behavioural Brain Research*, 113, 159-168.

Olivier E., Baker S.N., Nakajima K., Brochier T., and Lemon R.N., 2001, Investigation into non-mono synaptic corticospinal excitation of macaque upper limb single motor units, *Journal of Neurophysiology*, 86, 1573-1586.

30

REFLECTIONS ON SPINAL REFLEXES

Douglas G. Stuart*

ABSTRACT

Over the past four decades, the understanding of proprioceptive spinal reflexes has advanced far more rapidly than generally considered. This problem could be largely obviated in undergraduate and graduate training programs if the topic of reflexes was introduced subsequent to the concept and mechanisms of pattern generation within the central nervous system. The key advantage would then be that the neuroscience community as a whole would gain appreciation of the fact that proprioceptive reflexes are not hard-wired but rather are context- and phase-dependent, with the central nervous system selecting input-output pathways appropriate for the task at hand.

INTRODUCTION

Nociceptive and other protective reflexes that operate at the spinal and brainstem level are relatively hard-wired when the central nervous system (CNS) is both passive (or anaesthetised) and active during the elaboration of posture and movement. Similarly robust, and of substantial clinical value, are the lower-threshold mechanosensory (proprioceptive) reflexes when the subject's CNS is relatively passive in both the anaesthetised and conscious state. When the subject is active, however, these latter reflexes become context and phase-dependent, with the CNS selecting motoneurone input-output pathways appropriate for the task at hand. Despite some exemplary models to the contrary (e.g., Shepherd, 1994; Windhorst, 1996; Pearson and Gordon, 2000), this flexibility is rarely treated as a major principle in introductory and even advanced textbooks of neuroscience. An unfortunate result is that early on in their training, many students of neuroscience gain the false impression that the study of proprioceptive reflexes is no longer an area of active neuroscientific enquiry. It behoves movement

* Department of Physiology, The University of Arizona College of Medicine, Tucson, AZ 85724-5051, USA. Email: dgstuart@u.arizona.edu

neuroscientists to correct this erroneous impression, because the present-day study of these reflexes (see Chapters 19, 36, 37, 40, and 43) is as intellectually and technically demanding as any in the field of systems neuroscience. To defend the above viewpoints, this chapter begins with some historical considerations (see also Prochazka et al., 2000), and then addresses why it is advantageous to begin consideration of CNS neurobiology with the nature and operation of central pattern generators (CPGs). Next, to emphasise the technical and theoretical demands of current research on spinal reflexes, some challenging issues for future research are considered.

From the outset, it deserves emphasis that the driving force in the developments to be reviewed has been the Swedish neuroscientist, Anders Lundberg, his long-time associate, Elzbieta Jankowska, and their Göteborg group. Their work since the late 1950s has continually emphasised the critical role of segmental ventral-horn interneurones, by virtue of their role in central pattern generation and by being the primary site of integration (convergence) of such activity, descending command signals, and sensory feedback. This work has stimulated several of the advances discussed in this volume (see Chapters 19, 33, 34, and 36).

HISTORICAL CONSIDERATIONS

In modern-day movement neuroscience, we rightly attribute to Lundberg the conceptual leap to the concept of alternative reflex pathways; the idea that led to the finding that a given sensory input can have either an excitatory or inhibitory effect on a given motoneurone as dependent on a CNS selection process related to the phase and intent of a movement. (Hultborn, 2001; McCrea, 2001). Lundberg would be the first to argue, however, that throughout three much-earlier decades, Sherrington ruminated (perhaps agonised) about this possibility. He did not actively promulgate it, however, just as he appears to have equivocated on spinal pattern generation (Stuart et al., 2001).

When discussing Sherrington's views on spinal reflexes, which still hold a surprisingly strong sway in their textbook description, it seems appropriate to hold him to his final "summing-up" position in Chapter VII of the influential Creed et al. (1932) volume. (Documentation that Sherrington alone wrote this chapter is provided in Eccles and Gibson [1979, p. 65]). In this chapter, Sherrington begins (p. 107) by seemingly stating that reflex reversals can occur under potentially "non-normal" (*author's interpretation*) situations which might be considered appropriate today, however, for demonstration of a true reflex reversal. The reports he cites include Sherrington (1900, 1905, 1908) and Sherrington and Sowton (1911a). Two pages later (pp.109-111), Sherrington emphasised the possibility of a "normal" reflex reversal, but he did not then promulgate the idea that the same sensory input could result in an "opposite" output, even though he discussed this possibility in earlier writings, like the above papers, and Brown and Sherrington (1912). To support normal reflex reversal, Sherrington strangely (at least to this author) cited two of his reports (1908, 1909) that addressed the "lengthening reaction" and the "shortening reaction," but not reflex reversal per se. Sherrington concludes his discussion of reflex reversal in his summing-up chapter with the statement: "...Each considerable afferent nerve presents a dominant reflex which may mask its competitors" (p. 111). For support of this statement, he cited Sherrington and Sowton (1911b) and Sherrington (1931), which again do not address reflex reversal, per se. Importantly, in some of the above papers, and others like his detailed, ever-impactful

1910 one, Sherrington cited evidence in favour of normal reflex reversal (see its p. 92 for discussion of "Umkehr" [turning round]) as having been provided in the reports of von Uexkull on the ray-arm of the schlangensterne (brittle-star, *Ophioglypha* (1897, 1904a-b), and Magnus on the tail of the spinal cat (1909, p. 251) and spinal hind-limb reflexes of the dog (1909, p. 219).

In summary, there does not appear to have been a Sherrington article that includes a section where he unequivocally espouses that a change in spinal "state" can elicit a reflex *reversal* of output to the *same* sensory input. Nonetheless, the concept of alternative spinal reflex pathways was indeed demonstrated and discussed at the turn of the late 19th and early 20th century.

For the current era, particular credit must be given to one of Lundberg's former Swedish trainees, Sten Grillner, for his collaborative work with a Canadian, Serge Rossignol. They extended Lundberg's 1960s' ideas and findings (particularly the DOPA-induced modulation of spinal circuitry; Jankowska et al., 1967) by demonstrating in the 1970s and early 1980s that phase-dependent reflex reversals not only occurred in moving animal preparations (decerebrate, spinal, etc.), but also when such preparations were paralysed and primed to generate a locomotor rhythm (fictive locomotion). They thereby showed that the spinal locomotor CPG has the capacity to enable and disable specific sensory reflex pathways on a time scale of hundreds of milliseconds, i.e., half-step-cycle by half-step-cycle. This is a seminal example of the reconfiguration of spinal neuronal circuitry on a very rapid time scale (for review, see Grillner, 1975; Rossignol, 1996).

The above Grillner-Rossignol observations, together with the preceding work of Mark Shik and Grigori Orlovsky and their Moscow colleagues (reviewed in Stuart and McDonagh, 1998), led to a flurry of work on fictive locomotion preparations that continues to advance the original Lundberg/Jankowska emphasis on alternative spinal circuitry (Hultborn, 2001; McCrea 2001).

SIGNIFICANCE OF SPINAL PATTERN GENERATION FOR REFLEXES

Particularly from the early 1960s onwards, work on a wide variety of animal species, including the human (Gurfinkel et al., 1998), has shown that interneuronal central pattern generators (CPGs) in the brainstem and spinal cord of vertebrates (and their analogues in invertebrates;Stein et al., 1997) have the intrinsic capacity (i.e., not dependent upon movement-related sensory feedback, and in some cases, like the respiratory CPG, not necessarily dependent upon descending command and/or sensory feedback signals) to generate rhythmical activation of motoneurones such as to produce the "bare bones" (see below) of repetitive movements like chewing, licking, scratching, and locomotion (Stein, 1995, 1999; Orlovsky et al., 1999). In a developmental sense, this is perhaps not surprising, particularly because spontaneous activity underlying the generation of co-ordinated movements appears in the embryonic spinal cord before the manifestation of reflexes (Coghill, 1929; Bekoff, 1981). Somewhat similar CNS pattern generation occurs in the hypothalamus for circadian rhythms (Herzog and Tosini, 2001). It can also be argued that since many telencephalic cells, including those in the neocortex, have voltage-gated channels in their dendritic membranes, the propensity for intrinsic activity is ever present in even the highest of brain structures. Note further the concept that rhythmic activity is at the core of higher functions, beginning with thalamocortical reverberations which lead to the three states of the human brain; waking, asleep, and asleep and dreaming (Llinás, 2001).

This past decade, it has been necessary to incorporate within central pattern generation the capability of neuromodulation to alter the connectivity patterns within a CPG's interneuronal circuitry (e.g., converting, an excitatory synapse into an inhibitory one) such that a CPG can produce an array of output patterns responsible for a set of motor behaviours (Harris-Warrick and Marder, 1991; also preface to Stein et al., 1997).

Once the above concepts are ingrained, the interplay between movement-induced sensory input and CPG activity is more easily grasped. This interplay must be presented as a two-edged sword! Sensory input has a substantial (but constrained) effect on the output of a CPG. Recall, for example, the now-classic comment that "...there is absence of a generator of locomotor rhythm independent of the movements actually performed" (Shik et al., 1966, p. 765). By this, the Moscow group meant that even though the CNS has the intrinsic ability to construct the sequence of muscle activation patterns required for stepping, it is nonetheless strongly influenced by movement-induced sensory input (McCrea, 2001; see also Chapter 40). In parallel, however, a CPG and/or descending command signal can reverse the output of a given sensory input! In regard to this latter principle, it is frustrating that many current textbooks do not emphasise alternative, spinal reflex pathways (excluding, however, Shepherd, 1994; Windhorst, 1996; Pearson and Gordon, 2000), particularly since the past 35 years' work on locomotor CPGs has repeatedly demonstrated phase-dependent reflex reversals during the elaboration of stepping (Rossignol, 1996; Hultborn 2001, McCrea, 2001). Another example of textbook refractoriness is the emphasis on the inhibitory action of sensory group Ib input from Golgi tendon organs onto homonymous motoneurones. This pathway was first demonstrated in the late 1950s, using barbiturate-anaesthetised cats. Using a variety of locomotion-primed, unaenesthetized preparations, however, it has been clearly demonstrated that this pathway can become excitatory to extensor motoneurons during the stance phase of the locomotor step cycle step (Duysens et al., 2000; Hultborn, 2001; McCrea, 2001).

Since Lundberg's 1969a article, his masterful companion essay on locomotion (1969b), the above-described Grillner/Rossignol contribution, and Grillner's broad conceptualizations on invertebrate-vertebrate movement neuroscience (1975, 1981), all of which work emphasised the advantage of interpreting reflexes on the basis of their interactions with ongoing spinal interneuronal activity, there has been a progressive elaboration of this approach to unravelling spinal cord circuitry in the adult cat (Hultborn, 2001; McCrea, 2001; present chapter 19) and several non-mammalian vertebrate species (in particular, the lamprey [Grillner et al., 2000], tadpole [Roberts et al., 1998], and turtle [Stein et al., 1998]). Virtuoso work has also appeared on the identification and functional morphology of segmental interneurones in the cat (Matsuyama and Mori, 1998; Huang et al., 2000; Jankowska, 2001), and the firing patterns of interneurones during voluntary movements of a non-human primate (see Chapter 32). Lundberg's approaches have also been extended to human spinal cord circuitry (see Chapter 33), while others have advanced human spinal neurobiology with other imaginative and demanding methods (see Chapters 28, 36, 37, 42 and 43).

SOME UNRESOLVED ISSUES

In parallel with the many advances on phase- and context-dependent reflexes reviewed recently by Hultborn (2001) and McCrea (2001), several challenging issues remain unresolved. For example, the association between interneurons' spinal connectivity and their actual repetitive firing patterns is largely unexplored (McDonagh et al., 1999). A more active area of enquiry is the current consideration at the cervical level of the spinal cord of the extent to which phylogeny has influenced interactions between descending command signals, propriospinal neurons, spinal interneurones, motoneurons, and sensory afferent input. As emphasised in Rothwell's present chapter 31, this is a noteworthy development in mammalian spinal neurobiology, with many parallels evident in invertebrate neuroscience (Stein et al., 1997) and other brain structures in vertebrates (e.g., Llinás et al., 1970; Ulinsky, 1997). It bears emphasis that when testing for a phylogenetically-driven emphasis or de-emphasis on a mammalian CNS pathway, comparative and evolutionary considerations are valuable (Strausfeld et al., 1998), including the the application of modern comparative statistics (Iwaniuk et al., 1999). Clearly, much valuable work remains to be done on the issues addressed in Chapters 33-35.

Another unresolved issue is that subtle, and sometimes-marked differences exist between the spinal circuitry controlling different muscle systems (e.g., limb, respiratory, head-neck; see, for example, Richmond et al., 1999). Differences between fore- and hindlimb reflex circuitry also remain relatively unexplored, not only in humans, with their phylogenetically advanced manipulative capacity of the hand (Wiesendanger, 1999), but also in lower mammals, who often use the forelimbs more for manoeuvrability, and the hindlimbs more for the generation of propulsive force. This latter difference is evident both behaviourally and neurally in the late-neonatal rat (Cazalets et al., 1990), and it has been demonstrated behaviourally in a marsupial (Hore et al., 1973), and non-mammalian vertebrates, as well (e.g., amphibians–Emerson and Koehl, 1990; reptiles–Carpenter and Ferguson, 1977; and birds–Gatesy and Dial, 1996). Clearly, much is yet to be learned about spinal reflexes, and among the many problems, two should prove to be particularly exciting to those of the next generation of movement neuroscientists who have a physical-science bent.

Hasan and I (1988) have argued that in both invertebrates and vertebrates, with an emphasis on the crab, cat and human, three roles for proprioceptive input arise from the mechanics of the musculoskeletal system and the need to smooth and stabilise internally generated motor programs: 1) correction for non-linear muscle properties; 2) compensation for lever-arm variations; and 3) correction of interjoint interaction effects. We postulated that three additional roles arise from interactions between the mechanics of the musculoskeletal system and the physical environment: 4) selection of appropriate responses to unexpected perturbations; 5) selection of appropriate synergies of response; and 6) assistance to external forces for movements requiring manoeuvrability rather than stability. To date, the near-exclusive focus has been on the spinal connectivity patterns associated with two (1, 4) of the above six functions. The key problems about the other four areas are both technical and due to the relatively weak interface between cellular neurophysiology and biomechanics. This situation is now changing rapidly, however (see Chapters 41, 53, and 54).

In a review now used widely in movement neuroscience training programs, Prochazka (1996) discussed application of modern control system theories (e.g., finite state [conditional]-, adaptive [self-organising]- and predictive networks; fuzzy logic) to accommodate the wealth of complex spinal circuitry that is know known to operate in spinal reflexes. He emphasised that each may play one or more roles, usually in combination with each other, in the control of multi-jointed movement during the elaboration of various motor tasks (see also Chapter 41). The marriage of these concepts to Lundberg/Jankowska-inspired alternative spinal cord circuitry is certain to engage major 21st century endeavour.

CONCLUSION

The summary message from the intense effort documented above should be featured in even the most introductory of textbooks: *during movement, the brain can selectively open appropriate afferent pathways to motoneurones such that their command signals to muscles are appropriate for the task at hand.* To emphasise this fundamental principle, textbooks should begin presentation of CNS control mechanisms with a chapter on central pattern generation, which is now known to occur from the spinal cord to the forebrain. Particularly in mammalian textbooks, this approach would have the added advantage of emphasising the need to keep abreast of findings on segmental control circuitry in both invertebrates and vertebrates (i.e., "interphyletic awareness"; e.g., Llinás et al., 1970; Stuart, 1985; Pearson, 1993 and his present chapter 40 [with Lam]; Hildebrand and Shepherd, 1997; Stein 1999). By this means, mammalian movement neuroscience trainees, including those who wish to focus on research on humans, will come to recognise the importance of evaluating when a finding illustrates a species specialisation or, alternatively, a conserved mechanism that can be considered to be a principle, like the phase/context-dependent reflex reversals emphasised in this chapter. Importantly, interphyletic awareness has merit in *all areas* of neuroscience, and from the cellular/molecular to the behavioural level of analysis.

ACKNOWLEDGEMENTS

I would like to thank Francois Clarac, Paul Stein and Gordon Shepherd for reviewing a draft of this chapter, and Edgar Garcia-Rill, George Goslow, Richard Levine, Robert Skinner, Nicholas Strausfeld and Mario Wiesendanger for help with some fine points. The chapter's preparation was supported by USPHS grants NS 20577, NS 07309 and NS 41876 (to D.G.S.), and GM O8400 (to Dr. W. H. Dantzler).

REFERENCES

Coghill, G. E ., 1929, *Anatomy And The Problem of Behavior*, Hafner, New York, pp. 86-87.

Bekoff, A., 1981, Embryonic development of the neural circuitry underlying motor coordination, in: *Studies in Developmental Neurobiology–Essays in Honor of Viktor Hamburger*, W. Maxwell Cowan, ed., Oxford University Press, New York, pp. 134- 170.

Brown, T. G., and Sherrington, C. S., 1912, The rule of reflex response in the limb reflexes of the mammal and its exceptions, *Journal of Physiology*, 44, 125-130.

Carpenter, C., and Ferguson, G. W., 1977, Variation and evolution of stereotyped behavior in reptiles, in: *Biology of the Reptilia, vol. 7, Ecology and Behavior,* C. Gans and D. W. Tinkle, eds., Academic Press, New York, pp. 335-403.

Cazalets, J. R., Menard, I., Crémieux, J., and Clarac, F., 1990, Variability as a characteristic of immature motor systems: an electromyographic study of swimming in the newborn rat, *Behavioural Brain Research*, 40, 215-225.

Duysens, J., Clarac, F., and Cruse, H., 2000, Load-regulating mechanisms in gait and posture: comparative aspects, *Physiological Reviews*, 80, 83-133.

Eccles, J. C., and Gibson, W. C., 1979, *Sherrington – His Life and Thought*, Springer International, Berlin.

Emerson, S. B, and Koehl, M. A. R., 1990, The interaction of behavioral and morphological change in the evolution of a novel locomotor type: "flying" frogs, *Evolution*, 44, 1931-1946.

Gatesy, S., and Dial, K .P., 1996, Locomotor modules and the evolution of avian flight, *Evolution*, 50, 331-340.

Grillner, S., 1975, Locomotion in vertebrates: Central mechanisms and reflex interaction. *Physiological Reviews*, 55, 247-304.

Grillner, S., 1981, Control of locomotion in bipeds, tetrapods and fish, in: *Handbook of Physiology, sec. 1, vol. II, pt. 2. The Nervous System: Motor Control*, (V. B. Brooks, vol. ed.), J. M. Brookhart and V. B. Mountcastle, eds., American Physiological Society, Bethesda, pp. 1179-1236.

Grillner, S., Cangiano, L., Hu, G., Thompson, R., Hill, R., and Wallén, P., 2000, The intrinsic function of a motor system--from ion channels to networks and behavior, *Brain Research*, 886, 224-236.

Gurfinkel, V. S., Levik, Y. S., Kazzenikov, O. V., and Selionov, V. A., 1998, Locomotor-like movements evoked by leg muscle vibration in humans, *European Journal of Neuroscience*, 10, 1608-1612.

Harris-Warrick, R. M., and Marder, E., 1991, Modulation of neural networks for behavior, *Annual Reviews of Neuroscience*, 14, 39-57.

Hasan, Z., and Stuart, D. G., 1988, Animal solutions to problems of movement control: The role of proprioceptors, *Annual Reviews of Neuroscience*, 11, 199-223.

Herzog, E. D., and Tosini, G., 2001, The mammalian circadian clock shop, *Seminars in Cell and Developmental Biology*, 12, 295-303.

Hore, J., Phillips, C. G., and Porter, R., 1973, The effects of pyramidotomy on motor performance in the brush-tailed possum (Trichosurus vulpecula), *Brain Research*, 49, 181-184.

Huang, A., Noga, B. R., Carr, P. A., Fedirchuk, B., and Jordan, L. M., 2000, Spinal cholinergic neurons activated during locomotion: localization and electrophysiological characterization, *Journal of Neurophysiology*, 83, 3537-3547.

Hildebrand, J. G., and Shepherd, G. M., 1997, Mechanisms of olfactory discrimination: Common principles across phyla, *Annual Reviews of Neuroscience*, 20, 595-631.

Hultborn, H.. 2001, State-dependent modulation of sensory feedback, *Journal of Physiology,* 533, 5-13.

Iwaniuk, A. N., Pellis, S. M., and Whishaw, I. Q., 1999, Is digital dexterity really related to corticospinal projections?: A re-analysis of the Heffner and Masterton data set using modern comparative statistics, *Behavioural Brain Research*, 101, 173-87.

Jankowska, E., 2001, Spinal interneuronal systems: identification, multifunctional character and reconfigurations in mammals, *Journal of Physiology*, 533, 31-40.

Jankowska, E., Jukes, M. G. M., Lund, S., and Lundberg, A., 1967, The effect of DOPA on the spinal cord. 5. Reciprocal organization of pathways transmitting excitatory action to alpha motoneurones of flexors and extensors, *Acta Physiologica Scandinavica*, 70, 369-388.

Llinás, R., 2001, *I of the Vortex, From Neurons to Self*, MIT Press, Cambridge.

Llinás, R., Hillman, D. E., and Precht, W., 1970, Functional aspects of cerebellar evolution, in: *The Cerebellum in Health and Disease*, W. S. Fields and W. D. Willis, eds., Green, St. Louis, pp.269-291.

Lundberg, A., 1969a, Convergence of excitatory and inhibitory action on interneurones in the spinal cord, in: *The Interneuron*, M. A. B. Brazier, ed., University of California Press, Los Angeles, pp. 231-236.

Lundberg, A., 1969b, *Reflex Control of Stepping. The Nansen Memorial Lecture V*, Universitetsforlaget, Oslo.

McCrea, D. A., 2001, Spinal circuitry of sensorimotor control of locomotion, *Journal of Physiology*, 533, 41-50.

McDonagh, J. C., Callister, R. J., Brichta, A. M., Reinking, R. M., and Stuart, D. G.,1999, A commentary on the properties of spinal interneurons vs. motoneurons in vertebrates, and their firing-rate behavior during movement, in: *Motor Control Today and Tomorrow,* G. Gantchev, ed., Academic Publishing House "Prof. M. Drinov," Sofia, pp. 3-29.

Magnus, R., 1909, On the control of movement by the central nervous system (in German), *Pflügers Archiv European Journal of Physiology*, 130, 219-269.

Matsuyama, K., and Mori, S., 1998, Lumbar interneurons involved in the generation of fictive locomotion in cats, *Annals of The New York Academy of Sciences*, 860, 441-3.

Orlovsky, G. N., Deliagina, T. G., and Grillner, S., 1999, *Neuronal Control of Locomotion – From Mollusc to Man*, Oxford University Press, New York.
Pearson, K. G., 1993, Common principles of motor control in vertebrates and invertebrates, *Annual Reviews of Neuroscience*, 16, 265-297.
Pearson, K., and Gordon, J., 2000, Spinal reflexes, in: *Principles of Neuroscience*, 4th ed., E. R. Kandel, J. H. Schwartz and T. M. Jessell, eds., McGraw-Hill, New York, pp. 713-36.
Prochazka, A., 1996, Proprioceptive feedback and movement regulation, in: *Handbook of Physiology, sec. 12. Exercise: Regulation and Integration of Multiple Systems,* L. B. Rowell and J. T. Shepherd, eds., Oxford University Press, New York, pp. 89-127.
Prochazka, A., Clarac, F., Loeb, G. E., Rothwell, J. C., and Wolpaw, J. R., 2000, What do reflex and voluntary mean? Modern views on an ancient debate, *Experimental Brain Research*, 130, 417-32.
Richmond, F. J. R., Cornell, B. D., and Singh, K., 1999, Animal models of motor systems: cautionary tales from studies of head movement, *Progress in Brain Research,* 123, 411-416.
Roberts, A., Soffe, S. R., Wolf, E. S., Yoshida, M., and Zhao, F. Y., 1998, Central circuits controlling locomotion in young tadpoles, *Annals of the New York Academy of Sciences,* 860, 19-34.
Rossignol, S., 1996, Neural control of stereotypic limb movements, in: *Handbook of Physiology, sec. 12. Exercise: Regulation and Integration of Multiple Systems,* L. B. Rowell and J. T. Shepherd, eds., Oxford University Press, New York, pp. 173-216.
Shepherd, G. M., 1994, *Neurobiology,* Oxford University Press, New York.
Sherrington, C. S., 1900, On the innervation of antagonistic muscles. Sixth note, *Proceedings of the Royal Society of London,* 67B, 66-67.
Sherrington, C. S., 1905, On reciprocal innervation of antagonistic muscles. Seventh note, *Proceedings of the Royal Society of London,* 76B, 160-163.
Sherrington, C. S., 1908a On the innervation of antagonistic muscles. Eleventh note: Further observations on successive induction, *Proceedings of the Royal Society of London,* 80B, 53-71.
Sherrington, C. S., 1908b On reciprocal innervation of antagonistic muscles. Twelth note: Proprioceptive reflexes, *Proceedings of the Royal Society of London,* 80B, 552-564.
Sherrington, C. S., 1909, On plastic tonus and proprioceptive reflexes, *Quarterly Journal of Experimental Physiology*, 2, 109-156.
Sherrington, C. S., 1910, Flexion-reflex of the limb, crossed extension-reflex, and reflex stepping and standing, *Journal of Physiology*, 40, 28-121.
Sherrington, C. S., 1931, Quantitative management of contraction in lowest level co-ordination. Hughlings Jackson Lecture, *Brain*, 54, 1-28.
Sherrington, C. S., 1932, VII. Lower reflex co-ordination, in: *Reflex Activity of the Spinal Cord*, R.S. Creed, D. Denny-Brown, J. C. Eccles, E. G. T. Liddell and C. S. Sherrington, eds., Clarendon Press, Oxford, pp 104-159.
Sherrington, C.S., and Sowton, S.C.M., 1911a, Chloroform and reversal of reflex effect, *Journal of Physiology,* 42, 383-388.
Sherrington, C.S., and Sowton, S.C.M., 1911b, Reversal of the reflex effect of an afferent nerve by altering the character of the electrical stimulus applied, *Proceedings of the Royal Society of London,* 83B, 435-446.
Shik, M. L., Severin, F. V., and Orlovsky, G. N., 1966, Control of walking and running by means of electrical stimulation of the mid-brain, *Biophysics,* 11, 756-765.
Stein, P. S. G., 1995, A multi-level approach to motor pattern generation, in: *Neural Control of Movement*, W. R. Ferrell and U. Proske, eds., Plenum Press, New York, pp. 159-165.
Stein, P. S. G., 1999, Central pattern generators and interphyletic awareness, *Progress in Brain Research*, 123, 259-271.
Stein, P. S. G., McCullough, M. L., and Currie, S. N., 1998, Spinal motor patterns in the turtle, *Annals of The New York Academy of Sciences,* 860, 142-154.
Stein, P. S. G., Grillner, S., Selveston, A. I., and Stuart, D. G., (eds.), 1997, *Neurons, Networks, and Motor Behavior,* MIT Press, Boston.
Strausfeld, N. J., Hansen, L., Li, Y., Gomez, R. S., and Ito. K., 1998, Evolution, discovery, and interpretations of arthropod mushroom bodies, *Learning and Memory*, 5, 11-37.
Stuart, D. G., 1985, Summary and challenges for future work, in: *Motor Control: From Movement Trajectories to Neural Mechanisms. Short Course Syllabus*, P.S.G. Stein, ed., Society for Neuroscience, Bethesda, pp. 95-105.
Stuart, D. G., and McDonagh, J. C., 1998, Reflections on a Bernsteinian approach to systems neuroscience: The controlled locomotion of high-decerebrate cats, in: *Progress in Motor Control: Bernstein's Traditions in Movement Studies,* M. L. Latash, ed., Human Kinetics, Champaign, pp. 21-49.

Stuart, D. G., Pierce, P. A., Callister, R. J., Brichta, A., and McDonagh, J. C., 2001, Sir Charles Sherrington: humanist, mentor, and movement neuroscientist, in: *Classical Papers in Movement Science*, M. L. Latash, and V. Zatsiorsky, eds., Human Kinetics, Champaign, pp. 317-374.
von Uexkull, R. R., 1897, On conditions underlying the initiation of secondary movement (in German), *Zeitschrift fur Biologie*, 35, 183-191.
von Uexkull, R. R., 1904a, Studies on posture II: movement of the brittle star (in German), *Zeitschrift fur Biologie*, 44, 1-37.
von Uexkull, R. R., 1904b, The origin of rhythmicity in the animal kingdom (in German), *Ergebnisse der Physiologie*, 3, 1-11.
Ulinsky, P. S., 1997, Vertebrate nervous system, in: *Comparative Physiology, sec. 13, vol. II,* W.H. Dantzler, ed., Oxford University Press, New York, pp. 17-53.
Wiesendanger, M., 1999, Manual dexterity and the making of tools - an introduction from an evolutionary perspective, *Experimental Brain Research*, 128, 1-5.
Windhorst, U, 1996, Spinal cord and brainstem: pattern generators and reflexes, in: *Comprehensive Human Physiology: From Cellular Mechanisms to Integration, vol. 1*, R. Greger and U. Windhorst, eds., Springer, Berlin, pp.1007-1032.

31

SPINAL INTERNEURONES

Re-evaluation and controversy

John C. Rothwell*

ABSTRACT

Signal interneurones have a crucial part to play in the control of motoneurones. Newer methods are being added to the ways in which they can be studied, both in animal and human experiments. This chapter provides some background to a re-evaluation of the role of interneurones in motor control.

INTRODUCTION

Millions of interneurones in the spinal cord have an influence on the activity of motoneurones, yet our knowledge of how they contribute to final movement is still rudimentary. The Sherringtonian approach to the study of interneuronal function was to characterise spinal reflex pathways. Briefly, stimulation of particular sets of sensory input produces particular reflex outputs, and in many cases, the neural circuits, and interneurones involved in each reflex are relatively separate. In some cases, such as the reciprocal inhibition of antagonist motoneurones evoked by large diameter muscle spindle input from the agonist muscle, the interneurones (Ia inhibitory interneurones) can even be identified anatomically within the grey matter. Thus identification of a reflex is one way of studying the excitability of subpopulations of spinal interneurones. As such it has proved remarkably successful in showing how spinal circuits are modified during natural movements and how they react after acute or chronic injury of the spinal cord.

However, this approach may tell us relatively little about how spinal interneurones modify the pattern of descending commands from supraspinal structures, in particular from the cerebral cortex. Apart from the monosynaptic corticomotoneuronal component of the corticospinal tract, all descending motor inputs to spinal cord terminate on

* Sobell Department of Neurophysiology, Institute of Neurology, Queen Square, London WC1N 3BG, UK. Email: j.rothwell@ion.ucl.ac.uk

interneurones. The existence of (a) synapse(s) usually indicates that some operation is performed on this input before it is transmitted through to the motoneurones. What is the nature of this transformation? Is it for example, to produce greater spatial selectivity of the output via a centre-surround contrast organisation, or is it a way of modulating the gain of the supraspinal input before it reaches motoneurones? Unfortunately knowing how spinal interneurones behave in reflex pathways does not necessarily tell us how they behave in descending supraspinal pathways.

Figure 1 gives a schematic summary of the problem. The horizontal arrows and grey boxes of spinal interneurones represent classic spinal reflex pathways. The vertical arrows above them indicate descending supraspinal inputs to the cord. If the latter were to use precisely the same subpopulations of interneurones as the reflexes (left panel), then study of reflexes alone would indeed give an insight into how descending motor commands would be modified by spinal interneurones. However, this would not be the case if some of the descending commands accessed interneurones not tested by known reflex pathways (right panel). At the extreme, one could imagine that in order to activate their target motoneurones, descending commands recruited subpopulations of interneurones separate from those participating in reflex pathways. The descending signals might also modulate excitability in spinal reflexes so that they would support the movement in an appropriate manner, but this would be secondary to the direct transmission of their own input to the muscle. If this happened, study of spinal reflexes would yield little information about the role of interneurones in processing supraspinal inputs.

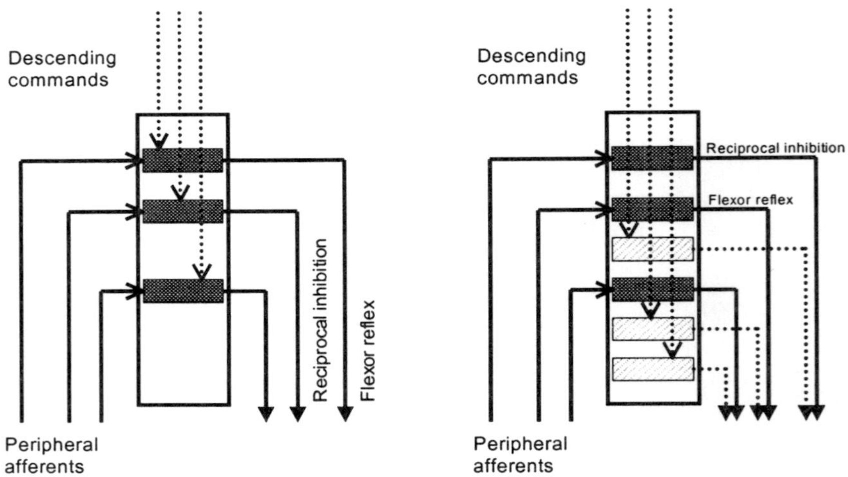

Figure 1. Schematic diagram to illustrate how peripheral reflex testing identifies subpopulations of spinal interneurones. These may be the same (left) or different (right) from those utilised by descending input to the cord. The large vertical rectangle represents the total population of spinal interneurones. Each smaller rectangle within it represents a subpopulation of interneurones interposed in a reflex or supraspinal pathway.

DIRECT RECORDINGS OF SPINAL INTERNEURONES IN BEHAVING ANIMALS

A different approach to studying the role of spinal interneurones has been pioneered by Fetz and colleagues (see Chapter 32 and Prut and Fetz, 1999; Prut et al., 2001;). They have succeeded in recording from the spinal cord of awake behaving monkeys and have shown for the first time how the activity of these cells changes during movement. Unfortunately at this stage it has not proved easy to identify individual neurones in terms of the older descriptions of reflex pathways (e.g. Renshaw cells, Ib interneurones etc), but experiments with implanted peripheral nerve stimulators that activate particular reflex pathways seem to be possible and if so this problem may be resolved in the future.

At the present time, the recorded neurones are identified in terms of whether or not they have demonstrable synaptic input to motoneurones (premotoneuronal interneurones), and the pattern of firing that they show during simple arm movements. This method is similar to that employed by others in the study of motor cortex or red nucleus neurones, and allows direct comparison between supraspinal and spinal processing.

One of the surprising findings is that most interneurones are active in most tasks, even if they involve antagonist muscles. Recordings made during movement cannot distinguish between activity due to afferent feedback as well as supraspinal drive, but the data suggest that the system operates through superimposed inhibition and excitation of motoneurones, rather than through the operation of small subpopulations devoted to particular tasks.

Perhaps the most surprising finding has been that many interneurones are also active in the warning period before a movement, and in many cases the activity parallels that seen in supraspinal centres.

THE C3-C4 PROPRIOSPINAL SYSTEM

Chapters 33 – 35 in this section concern the existence and function of a particular class of spinal interneurones that were identified by classical physiological techniques in the 1970s (see Alstermark and Isa, Chapter 34). These interneurones have a cell body in the C3-C4 cervical segments and project to motoneurones in lower cervical segments controlling arm and forearm muscles. In the cat they have been characterised extensively and receive a major input from corticospinal axons as well as extensive sensory feedback (facilitatory and inhibitory) from the arm. An unusual identifying feature is that a branch of the axon ascends to terminate in the lateral reticular nucleus.

In the cat, behavioural experiments showed that these neurones play a critical part in transmitting the commands for arm reaching to low cervical motoneurones. When the pathway was interrupted, the only function that did not recover well was accuracy of arm pointing movements (Alstermark and Lundberg, 1992).

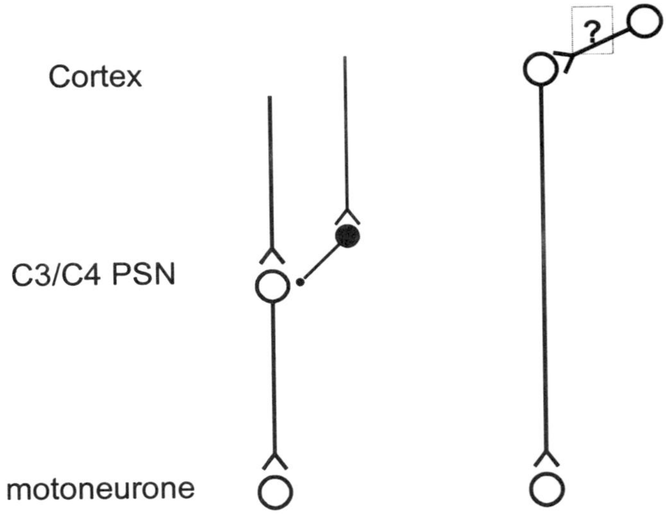

Figure 2. Schematic diagram of the proposed organisation of the C3-C4 propriospinal system. On the left is the known circuitry in the cat. The C3-C4 interneurone receives descending input from cortex (as well as other structures). In addition there is a feedforward inhibitory pathway that also receives strong peripheral afferent input (not shown). On the left is the proposed organisation without the C3-C4 interneurones. The cortical connection to lower cervical spinal segments is mainly monosynaptic. The neurone drawn at a cortical level (?) indicates that whatever processing was performed by the missing C3-C4 system must be performed by another mechanism at a supraspinal level.

The question that is addressed by the presentations is whether or not such a system exists in higher primates and in man (Fig. 2). At first sight, this may seem a little esoteric and relevant only to a small group of participants in the field. However, it is part of a much broader debate about what sort of neural processing is performed in spinal and supraspinal centres. In the cat, the C3-C4 interneurones presumably perform some operation on the input they receive from the corticospinal system. It has been suggested that this might involve focussing the command by using input from the periphery to facilitate transmission to muscles appropriate for the task, and to inhibit transmission to muscles not involved (Alstermark and Lundberg, 1992). The point is that if these neurones do not exist in higher primates, then this same function must be performed elsewhere (see Fig. 2). So the question whether these neurones exist or not becomes a question about why some forms of processing should be performed better in one part of the nervous system than another. It could equally well be asked, for example, why there is a subdivision of processing performed in different motor cortical areas.

As described in the following papers (Chapters 33-35), the debate turns around whether or not it is possible to observe disynaptic EPSPs in low cervical motoneurones after stimulation of the corticospinal tract above C3/C4. Such EPSPs are readily observed in the cat and are small in lower primates (Chapters 34 and 35). However, Kirkwood, Lemon and colleagues have rarely observed them in monkeys, where inputs are dominated by monosynaptic inputs (Maier et al., 1998). One possible explanation is that the system does not exist in the monkey, having been replaced by a direct monosynaptic connection. Another explanation revolves around the presence of inhibitory projections to the C3-C4 propriospinal interneurones. These receive inputs from corticospinal and peripheral afferent inputs and if they were particularly excitable in higher primates, then stimulation of the corticospinal tract above C3-C4 would activate them and might effectively "turn off" transmission through the pathway (see arguments in Pierrot-Deseilligny, 1996 and Chapters 33-35).

Indirect evidence from a large number of reflex studies in man has supported the suggestion that the C3-C4 propriospinal system is present but that it is under powerful inhibitory control (Pierrot-Deseilligny, 1996). However, such arguments are all necessarily indirect, and alternative explanations could always be proposed. Before this meeting the work of Alstermark and colleagues had suggested that inhibition might be prominent in the monkey. Administration of strychnine to reduce spinal inhibition revealed the presence of disynaptic EPSPs in motoneurones after corticospinal stimulation (Alstermark et al., 1999). However, this was not accepted as definitive proof because strychnine could have affected inhibition of many other pathways to motoneurones as well as the propriospinal pathway. Kirkwood, Lemon and colleagues called for much better evidence since in their experience, there was little evidence in either anaesthetised or awake animals for a disynaptic propriospinal pathway in monkey.

However, at the meeting, Alstermark presented a critical piece of new evidence and this seemed to swing an evenly balanced argument towards the existence of a C3-C4 pathway in all animals. Alstermark and colleagues had recorded directly from the cell bodies of C3-C4 propriospinal interneurones in monkeys and directly documented their projection to lower cervical motoneurones and the lateral reticular nucleus. Clearly the pathway existed in higher primates and the question seemed to move on to how important this pathway was in the control of skilled arm reaching in primates and man.

REFERENCES

Alstermark, B., Isa, T., Ohki, T., and Saito, T., 1999, Disynaptic pyramidal excitation in forelimb motoneurones mediated via C3-C4 propriospinal neurones in Macaca Fusata, *Journal of Neurophysiology,* 82, 3580-3585.

Alstermark, B., and Lundberg, A., 1992, The C3-C4 propriospinal system: target reaching and food taking, in: *Muscle Afferents and Spinal Control of Movement,* Jami, L., Pierrot-Deseillogny, E., and Zytnicki, D., eds., Pergamon Press, Oxford, pp. 327-354.

Pierrot_Deseilligny, E., 1996, Transmission of the cortical command for human voluntary movement through cervical premotoneurones, *Progress in Neurobiology,* 48, 489-517.

Prut, Y., and Fetz, E. E., 1999, Primate spinal interneurons show pre-movement instructed delay activity, *Nature,* 401, 590-594.

Prut, Y., Perlmutter, S. I., and Fetz, E. E., 2001, Distributed processing in the motor system: spinal cord perspective, *Progress in Brain Research,* 130, 276-278.

FUNCTIONAL PROPERTIES OF PRIMATE SPINAL INTERNEURONES DURING VOLUNTARY HAND MOVEMENTS

Eberhard E. Fetz, Steve I. Perlmutter, Yifat Prut, and Kazuhiko Seki*

ABSTRACT

The activity of cervical spinal interneurones (INs) was recorded in monkeys performing alternating hand movements. The contribution of INs to voluntary movement was determined by their response patterns during ramp-and-hold wrist movements and their postspike effects on forelimb muscle activity. Most INs were active during both flexion and extension, in contrast to the unidirectional activity of muscles and corticomotoneuronal cells. When recorded during performance of an instructed delay task, the activity of many INs was modulated during the delay period between the instruction cue and the subsequent go signal. Thus, spinal INs, like cortical neurones, participate in earliest stages of preparation for movement. The modulation of peripheral input to spinal INs was tested during an instructed delay task. The monosynaptic responses to electrical stimulation of a cutaneous nerve decreased during active movement, probably due to presynaptic inhibition. These results provide new insights into the role of spinal INs in preparation and execution of voluntary movement.

INTRODUCTION

The reflex responses of spinal cord interneurones (INs) have been extensively characterized in anesthetized animals (Baldissera et al., 1981; Jankowska, 1992), but little is known about the function of INs in normal voluntary movements. The responses evoked by stimulation of peripheral receptors and nerves as well as descending pathways have served to identify the convergent inputs to INs, but their output effects on different muscle groups remains less well understood. Much of our information derives from studies of lumbar spinal circuits related to the cat hindlimb, and much less is known about segmental circuits controlling forelimb movements in primates. We here review

* Department of Physiology & Biophysics and Regional Primate Research Center, University of Washington, Seattle WA 98195-7290, USA. Email: fetz@u.washington.edu

Sensorimotor Control of Movement and Posture
Edited by Gandevia *et al.*, Kluwer Academic/Plenum Publishers, 2002

briefly some recent work on the role of interneurones in cervical spinal cord of primates in preparation and execution of hand movements.

CONTRIBUTION TO VOLUNTARY WRIST MOVEMENT

To study the contribution of cervical INs to voluntary hand movement we documented their *activity* and their *correlational linkages* to muscles in monkeys performing ramp-and-hold flexion-extension torques about the wrist (Perlmutter et al., 1998; Maier et al., 1998). Single neurones were recorded with movable tungsten microelectrodes in C6-T1 spinal segments, along with EMG activity of multiple forearm muscles, while monkeys generated alternating flexion-extension torques in a step-tracking task. Task-related spinal INs increased their activity during flexion and/or extension, typically with a tonic or phasic-tonic response pattern in their preferred direction (Maier et al., 1998). The tonic components of these responses were usually an increasing function of static torque. Surprisingly, most INs (77%) were active during both flexion and extension, in unexpected contrast to the strictly unidirectional activity of corticomotoneuronal cells (Cheney & Fetz, 1980) and agonist motoneurones and the predominantly unidirectional activity of dorsal root afferents with post-spike effects (Flament et al., 1992). Interestingly, all rubromotoneuronal cells exhibit bidirectional activity (Cheney et al., 1991; Mewes and Cheney, 1994), like most INs.

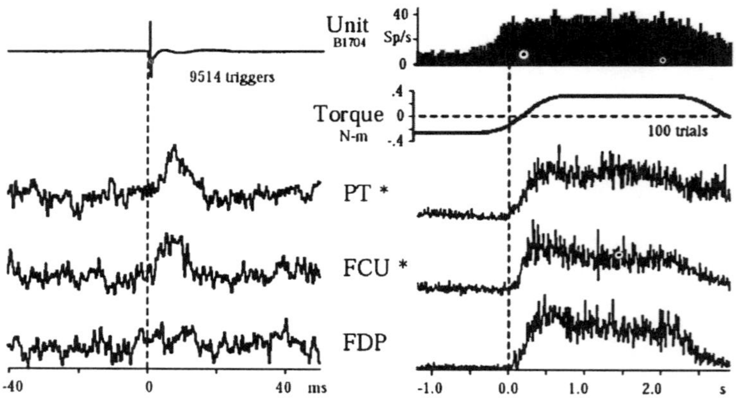

Figure 1. STAs and response averages of a facilitatory PreM-IN located in caudal C8. The STAs (left) show the triggering action potential [top] and average of rectified EMG in 3 independently recorded co-active muscles. Asterisks indicate muscles with significant post-spike effects. The average activity of the neurone during flexion torques is shown in the top right, above the average torque trajectory (positive torque in flexion, negative torque in extension) and average responses of the co-active muscles. This neurone exhibited tonic firing during static torques in both directions, with higher rates for flexion. (From Perlmutter et al., 1998.)

For about a fifth of the task-related INs spike-triggered averages (STA) of rectified EMG revealed significant spike-related features suggesting correlational linkages with forelimb muscles (Perlmutter at al., 1998). Some INs were identified as excitatory or inhibitory "premotor INs" (PreM-INs) on the basis of post-spike facilitation or suppression with appropriate onset latencies after the trigger spike (>3.5 ms, consistent with the latencies of post-stimulus facilitation evoked by single intraspinal microstimuli). Figure 1 illustrates an excitatory PreM IN that was more active during flexion and produced post-spike facilitation in two flexor muscles. The STAs of some INs showed broad peaks or troughs beginning earlier than 3.5 ms (usually before the trigger), attributed to synchronous firing of other PreM neurones. STA features were predominantly facilitatory (85%) and occurred twice as often in flexor muscles as in extensors. The muscle fields of PreM-INs (defined as the set of muscles showing postspike effects) were somewhat smaller than those of supraspinal PreM cells in cortex and red nucleus, and rarely involved reciprocal effects on antagonist muscles. In general the muscle fields of PreM-INs were functionally consistent with their firing properties: for example, excitatory PreM-INs were typically most active in movements that activated their target muscles (Fig. 1). However, many excitatory PreM-INs also fired when their facilitated muscles were silent. Some inhibitory PreM-INs were activated with antagonists of their suppressed target muscles, a pattern of reciprocal inhibition expected of Ia inhibitory INs. Other inhibitory PreM-INs were coactivated with the muscles they suppressed, a pattern expected of Renshaw cells.

These observations indicate that cortical and spinal PreM neurones represent movements and muscles in different ways. Corticomotoneuronal cells represent functional muscle synergies more distinctly than spinal INs: they produce postspike effects in larger muscle groups, including postspike suppression of antagonists of their facilitated target muscles (Kasser and Cheney, 1985), and they fire more specifically during particular movements (Cheney et al., 1991). In contrast, spinal PreM-INs affect one or a few particular muscles, and are broadly activated over a wider range of movements (Maier et al., 1998; Perlmutter et al., 1998). Thus, during normal voluntary movements motoneurones receive superimposed excitation and inhibition from populations of simultaneously active spinal PreM-INs. Excitation from these INs outweighs inhibition when the motoneurones are activated as agonists of a movement, and the balance drops below threshold when they become inactive as antagonists of the movement. In contrast, motoneurones receive more specific commands from corticomotoneuronal cells that are strictly consistent with their recruitment and inhibition.

INVOLVEMENT OF INs IN MOTOR PREPARATION

To document the role of spinal INs in *preparation* for voluntary movement we trained three monkeys to perform instructed delay tasks. Two monkeys performed isometric wrist flexion or extension after a visually cued delay period (Prut and Fetz, 1999; Perlmutter and Prut, 2001). During such instructed delays between a transient cue and a go signal many cortical neurones have been shown to change their activity as the monkey prepares for the subsequent movement. Figure 2 illustrates our instructed delay task and shows the activity of an IN that was inhibited during the delay period. About one third of the tested INs showed modulation of firing rate during the delay period following the visual cue, relative to the pre-cue rate. These *significant delay modulations* (SDM) were never accompanied by any EMG activity, and often differed from the INs' responses during the subsequent active torque period (Fig. 2). SDMs occurring during correct trials were usually absent when the monkey made directional errors. Many INs with SDM showed changes in firing during the delay period with the same polarity (increase or decrease) as the cell's activity during the active hold period, consistent with a subthreshold priming of INs in the direction they must fire during the subsequent movement. About two-thirds of the SDM were inhibitory, suggesting a global suppression of activity during the delay, preventing the overt expression of the movement.

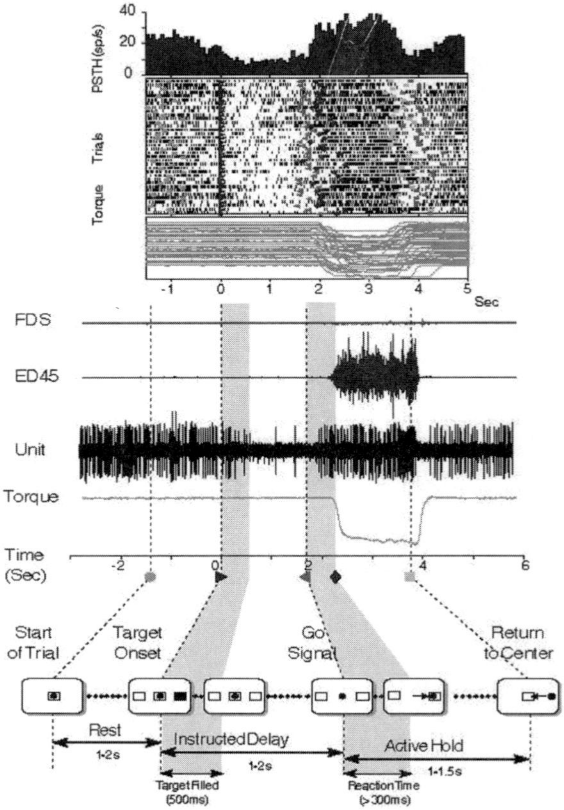

Figure 2. Activity of spinal IN during instructed delay task. Components of the behavioural task are schematically shown at bottom. Filled circle represents a cursor whose position is controlled by the monkey's wrist torque; squares represent targets. The instructed delay period begins with a transient visual cue (right target filled for 500 ms) and ends with a go signal (extinguishing of center hold target). Middle traces show activity of flexor digitorum sublimis (FDS) and extensor digitorum 4 & 5 (ED45) muscles and the IN, and isometric torque about the wrist. Top portion of figure shows responses during successive extension trials, aligned on cue onset (at time 0). From top down, peristimulus histogram (PSTH) of IN firing rate, rasters of IN spikes in successive trials and traces of torque trajectories. (Adapted from Prut and Fetz, 1999.)

A third monkey learned to interpret transient visual or proprioceptive cues to perform appropriate delayed wrist movements against an elastic load. The proprioceptive cue was a brief wrist perturbation whose direction and amplitude indicated the subsequent required movement. The perturbation evoked characteristic short-latency sensory responses in many INs that revealed two major groups of INs. Some INs gave bidirectional excitatory responses at the onset and offset of the perturbation pulses. These INs often fired transiently and bidirectionally with active movements, consistent with responses from cutaneous receptors. Other INs exhibited biphasic responses to perturbation pulses, being activated by the flexion component and inhibited by the extension component (or vice versa), consistent with proprioceptive input from muscle or joint receptors. These INs tended to exhibit sustained activity during the active hold in their preferred direction.

Excluding any brief response to the perturbation, the delay period activity for many INs was similar in visual and perturbation trials, consistent with motor preparation that is independent of the cue modality. However, other INs exhibited different SDM in the visual trials and the perturbation trials, similar to modality-specific SDM seen in cortical areas. These results suggest that segmental INs are involved, with cortex, in the earliest stages of movement preparation instructed by various cue modalities. The appearance of instructed delay period activity at spinal levels similar to that documented in motor cortical areas indicates a widely distributed network for movement preparation, mediated by the numerous descending and ascending pathways between cortex and spinal cord.

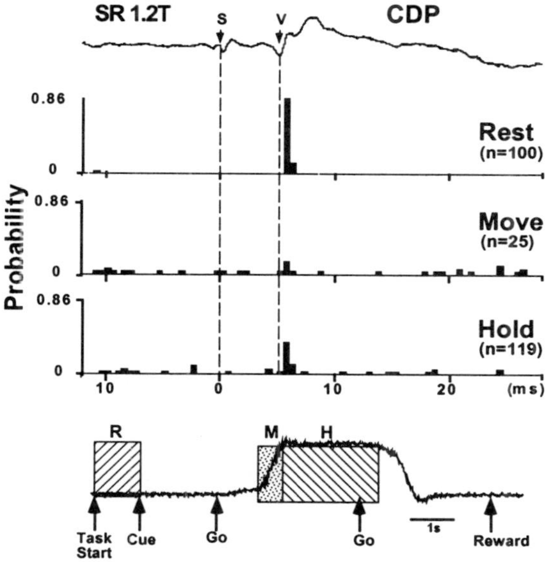

Figure 3. Modulation of monosynaptic response of cervical IN during task. Response to electrical stimulation of superficial radial nerve during performance of instructed delay task. Histograms show responses to electrical stimulation of nerve cuff compiled during task periods schematized at bottom: rest prior to cue (R), dynamic active movement (M) and static hold (H). Top trace shows cord dorsum potential (CDP) indicating time of stimulus (S) and afferent volley (V). The short latency of histogram peaks indicates a monosynaptic response, which is reduced during Move and Hold. Histograms plot probability of action potentials per stimulus per bin (0.5 ms).

MODULATION OF SENSORY INPUT DURING MOVEMENT

In other experiments we investigated the *task-dependent modulation* of neural responses to peripheral input. In a monkey performing wrist flexion/extension movements in a visually cued instructed delay task, cutaneous responses were evoked by electrical stimulation of the superficial radial (SR) nerve through a cuff electrode. SR stimulation of low-threshold afferents produced excitatory responses in many INs, sometimes at monosynaptic latencies from the afferent cord dorsum volley (<1.5 ms). Post-stimulus histograms compiled separately for different phases of the task (intertrial rest, instructed delay, dynamic movement and active hold period) revealed that the SR-evoked responses were typically suppressed or abolished during the dynamic movement phase (Fig. 3). In most cases the monosynaptic excitation from the SR was suppressed at the same time that task-related activity of the INs increased, indicating that the suppression was mediated by presynaptic rather than postsynaptic inhibition. The amplitudes of evoked monosynaptic field potentials in dorsal horn were also reduced during the dynamic movement period. Preliminary evidence from antidromic volleys in the nerve cuff evoked by intraspinal stimulation suggests that the excitability of SR afferents increased transiently during the dynamic component of active movement, consistent with primary afferent depolarization (Wall, 1958). These results indicate that presynaptic inhibition of cutaneous input increases significantly during the dynamic phase of active movement. This could function to gate out inappropriate reflex responses from peripheral receptors and is consistent with increased perceptual thresholds during active movement.

BEHAVIOURAL FUNCTIONS OF SPINAL INs

The observations summarized here suggest that under normal behavioural conditions many spinal INs have response properties similar to those previously documented for cortical neurones in behaving animals. The notion that cortical cells are involved in "higher order" behavioural functions while spinal INs are merely involved in transmitting sensory input or motor commands is based on a history of research involving fundamentally different experimental conditions. The reflex responses of spinal INs have been documented extensively in anaesthetized or decerebrate animals, while the activities of cortical cells have been documented during behavioural tasks that test their involvement in voluntary movement and cognitive functions. We suspect that examining spinal INs under appropriate behavioural conditions in awake animals will reveal that INs are also involved in many functions beyond simple sensory or motor relays (Fetz et al., 2000).

ACKNOWLEDGEMENTS

Supported by NIH grants NS12542, NS36781 & RR00166, and APA PBR2-9502 and Human Frontiers Science Program.

REFERENCES

Baldissera, F., Hultborn, H., and Illert, M., 1981, Integration in spinal neuronal systems, in: *Handbook of Physiology, Section 1: The Nervous System; Volume II: Motor Control, Part 2*, American Physiological Society, Bethesda, MD, pp. 509-595.

Cheney, P. D., Fetz, E. E., and Mewes, K., 1991, Neural mechanisms underlying corticospinal and rubrospinal control of limb movements, *Progress in Brain Research*, 87, 213-252.

Fetz, E. E., Perlmutter, S. I., and Prut, Y., 2000, Functions of mammalian spinal interneurons during movement, *Current Opinion in Neurobiology*, 10, 699-707.

Flament, D., Fortier, P. A., and Fetz, E. E., 1992. Response patterns and post-spike effects of peripheral afferents in dorsal root ganglia of behaving monkeys, *Journal of Neurophysiology*, 67, 875–889.

Jankowska, E., 1992, Interneuronal relay in spinal pathways from proprioceptors, *Progress in Neurobiology*, 38, 335-378.

Kasser, R. J., and Cheney, P. D., 1985, Characteristics of corticomotoneuronal postspike facilitation and reciprocal suppression of EMG activity in the monkey, *Journal of Neurophysiology*, 53, 959–978.

Maier, M. A., Perlmutter, S. I., and Fetz, E. E., 1998, Response patterns and force relations of monkey spinal interneurons during active wrist movement, *Journal of Neurophysiology*, 80, 2495-2513.

Mewes, K., and Cheney, P. D., 1994, Primate rubromotoneuronal cells: parametric relations and contribution to wrist movement, *Journal of Neurophysiology*, 72, 14–30.

Palmer, S. S., and Fetz, E. E., 1985, Discharge properties of primate forearm motor units during isometric muscle activity, *Journal of Neurophysiology*, 54, 1178–1193.

Perlmutter, S. I., Maier, M. A., and Fetz, E. E., 1998, Activity and output linkages of spinal premotor interneurons during voluntary wrist movements in the monkey, *Journal of Neurophysiology*, 80, 2475 2494.

Perlmutter, S. I., and Prut, Y., 2001, Transformation of descending commands into muscle activity by spinal interneurons in behaving primates, in: *Motor Neurobiology of the Spinal Cord*, T. C. Cope, ed., CRC Press pp. 193-213

Prut, Y., and Fetz, E. E., 1999, Primate spinal interneurons show pre-movement instructed delay activity, *Nature*, 401, 590-594.

Wall, P. D., 1958. Excitability changes in afferent fibre terminations and their relation to slow potentials, *Journal of Physiology*, 142, 1-21.

33

A CERVICAL PROPRIOSPINAL SYSTEM IN MAN

Emmanuel Pierrot-Deseilligny and Véronique Marchand-Pauvert*

ABSTRACT

Peripheral stimuli facilitate, at a pre-motoneuronal level, the responses elicited in human upper limb motoneurons (MNs) by transcranial magnetic stimulation over the motor cortex (TMS). Several features indicate that the relevant premotoneurones are distinct from segmental interneurones and located rostral to MNs. Thus, corticospinal volleys would have an indirect (propriospinal) pathway to upper limb MNs, in addition to the direct cortico-motoneuronal pathway. Slightly increasing the corticospinal input causes the facilitation to be reversed to inhibition. This is consistent with a well-developed system of inhibitory interneurones activated by corticospinal and afferent inputs inhibiting propriospinal neurones (PNs). Corticospinal activation of these inhibitory interneurones would explain why propriospinally-mediated corticospinal EPSPs are weak in primate MNs after artificial (electrical or magnetic) activation of the corticospinal system by itself. However, indirect evidence for descending facilitation of PNs can be provided during normal voluntary contractions, while using the modulation of the ongoing EMG or of the H reflex.

INTRODUCTION

In the cat, the descending command for target-reaching is transmitted disynaptically to forelimb MNs through a system of C3-C4 PNs, which are controlled by inhibitory interneurones (for review see Lundberg, 1999). It has been suggested that an analogous system might exist in man (see Pierrot-Deseilligny, 1996). However, this view has been challenged and an evolutionary decline in the importance of the propriospinal system in primates (Nakajima et al., 2000) has been proposed, based on the apparent weakness of disynaptic corticospinal excitation after stimulation of the pyramidal tract by itself in the macaque monkey (Maier et al., 1998).

* Neurophysiologie Clinique, Rééducation, La Salpêtrière, 756551, Paris, Cedex 13, France.
Email: emmanuel.pierrot-deseilligny@chups.jussieu.fr

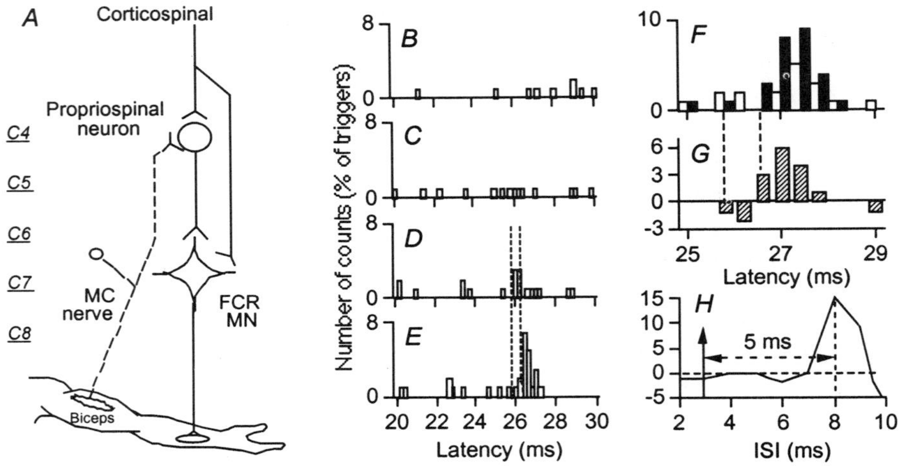

Figure 1. *A,* sketch of the presumed pathways. *B-G,* in PSTHs of single MUs, the number of counts (expressed as a % of the number of triggers) is plotted against the latency after TMS (even when peripheral stimulation is given alone). *B-E,* FDS MU (0.2 ms bins). *B,* background MU firing. *C-E,* changes in firing probability induced by isolated (*C-D*) or combined (*E*) stimulation. *B,* MC nerve stimulation (0.7 x motor threshold MT). C, TMS (26 % of the maximal stimulator output). *D,* combined MC and cortical stimulations at ISI = 8 ms. *F-G,* FCR MU (0.4 ms bins). *F,* the effect on combined stimulation (v) is compared to the algebraic sum of the effects of separate stimuli (□). *G,* subtraction histogram (hatched columns) showing the difference between the effect on combined stimulation and the algebraic sum of the effects of separate stimuli shown in *F.* Interrupted vertical lines in *D-G,* onset of the corticospinal peak and of the facilitation on combined stimulation. *H,* the difference between the effect on combined stimulation and the algebraic sum of the effects of separate stimuli is plotted against the ISI; vertical arrow at 3 ms indicates simultaneous arrival of MC and corticospinal volleys at MN level. The central delay of the facilitation is therefore 5 (8-3) ms. (Arranged from Pauvert et al., 1998.)

PERIPHERAL FACILITATION OF CORTICOSPINAL RESPONSES

Changes evoked in the post-stimulus time histogram (PSTH) of single motor units (MUs) were assessed (Pauvert et al., 1998; Nicolas et al., 2001). Figure 1 (panels B-E) shows the background firing of a flexor digitorum superficialis (FDS) MU (B), the changes evoked by separate stimulation of the musculo-cutaneous (MC) nerve (C) and of the motor cortex (D), and the large facilitation of the corticospinal peak on combined stimulation, when MC stimulation preceded TMS by 8 ms. This is also illustrated in a flexor carpi radialis (FCR, Fig. 1*F*) MU and in a flexor carpi ulnaris (FCU, Fig. 2B) MU, where the comparison is drawn between the algebraic sum of effects elicited by separate stimuli (MC nerve and TMS, □) and the effect on combined stimulation (v). In both cases extra facilitation on combined stimulation was clear. There were similar findings with 48/51 (94%) tested MUs from various upper limb muscles (biceps, FCR, extensor carpi radialis [ECR], FCU, FDS).

Peripheral facilitation of the corticospinal peak probably occurs primarily at a pre-motoneuronal level because: i) summation of two excitatory inputs at the MN produces little more than the sum of their effects in the PSTH (Pauvert et al., 1998); ii) in most MUs extra facilitation spared the initial part of the corticospinal peak (vertical interrupted lines in Fig. 1D-G), and the mean duration of this initial sparing was 0.8 ms, i.e. the delay required for transmission across an interneurone (Pierrot-Deseilligny et al., 1981).

Table 1. Mean ISI at which facilitation on combined stimulation first occurs.

MN pool	Number of MUs	Spinal rostro-caudal location	Mean ISI
Biceps	12	C5-C6	4.9 ± 0.2
ECR & FCR	17	C6-C8	6.1 ± 0.3
FCU & FDS	19	C7-T1	8.0 ± 0.2

This is what would be expected if cortical and peripheral volleys converged onto premotoneurones rather than directly onto the MN: because of the synaptic delay at the premotoneurones, this input would arrive at the MN after the direct fast conducting monosynaptic corticomotoneuronal (CM) input, and premotoneuronal facilitation would be unable to affect the onset of the monosynaptic CM response.

The mean (and SEM) inter-stimulus interval (ISI) at which the facilitatory interaction between corticospinal and peripheral volleys first appeared was calculated for MUs belonging to MN pools with the same rostro-caudal location in the spinal cord. The results shown in Table 1 indicate that the more caudal the MN pool the greater this ISI. This progressive increase in the ISI at which the facilitatory interaction first appeared must reflect a progressive increase in the central delay (see Nicolas et al., 2001). Thus, the more caudal the MN pool in the spinal cord, the greater the central delay of peripheral facilitation of the corticospinal peak. For these findings to be explicable by a segmental interneuronal pathway, one would have to postulate more interneurones in the pathway the more caudal the MN pool. A more parsimonious explanation would be that there was a longer intraspinal pathway for caudal MNs, and *this implicates premotoneurones located in segments above the afferent inflow, such as the C3-C4 PNs in the cat* (see Lundberg, 1999). Accordingly, these premotoneurones are referred to as presumed propriospinal neurones (PNs) in the following discussion.

FEEDBACK INHIBITION OF PRESUMED PROPRIOSPINAL NEURONES

In all MUs tested, increasing TMS intensity resulted in a *decrease in the peripheral facilitation* of the corticospinal peak (Nicolas et al., 2001). This is illustrated for a FCU MU in Figure 2 (B-D*)*: facilitation of the corticospinal peak with TMS at 26% (B) was reversed to inhibition with TMS at 28% (C) and 32% (D), both facilitation and inhibition sparing the first 0.5 ms bin. The graphs in Figure 2 (E-F*)* show the effects of increasing TMS intensity on the unconditioned cortical peak (E), and on the difference between the effect on combined stimulation and the sum of effects of separate stimuli (F). The mean increase in TMS intensity to cause facilitation to disappear was small (2.5 ± 0.3% of the maximal stimulator output) such that changes in complexity of the evoked corticospinal volley were probably negligible. In all MUs, the inhibition with stronger TMS occurred at the same ISI as the facilitation with weak TMS: e.g., in Figure 2*G-L*, significant facilitation of the corticospinal peak with TMS at 28% and significant inhibition with stronger TMS (31%,) both appeared at the 8 ms ISI (*H* and *K)* and disappeared at the 9 ms ISI *(J* and *L)*.

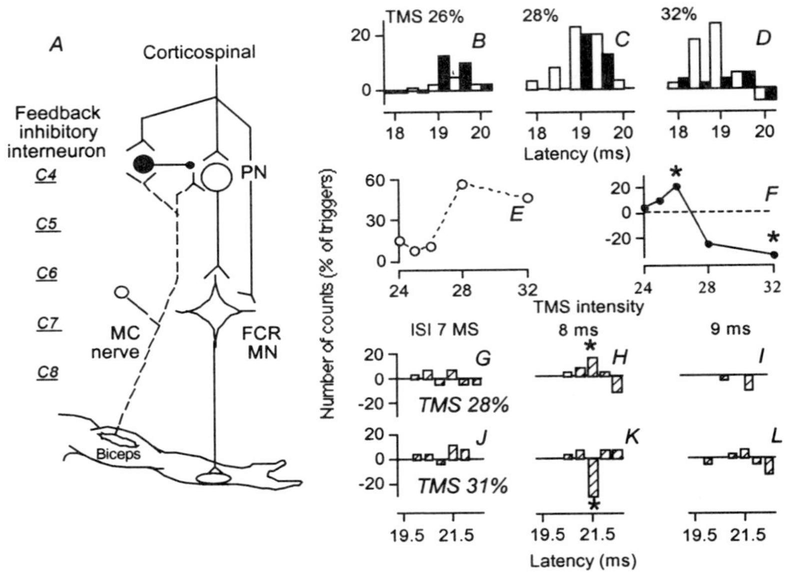

Figure 2. *A*, sketch of the presumed pathways. *B-D*, effects of varying TMS intensity in a FCU MU while the MC stimulus remained at 0.75 x MT and ISI 8 ms (□ and v as in Fig. 1*F*). Changes in the unconditioned corticospinal peak (*E*) and in the difference between conditioned and control peaks (*F*) are plotted against TMS intensity. *G-L*, subtraction histograms (difference between the effect on combined stimulation and the sum of effects of separate stimuli, as in Fig. 1*G*) showing in another FCU MU the effects of MC stimulation (0.75 x MT) on the corticospinal peak elicited by TMS at 28 (*G-I*) and 31 (*J-L*) % at variable ISIs (7 ms, *G,J;* 8 ms, *H,K;* 9 ms, *I,L*). The asterisks indicate $P < 0.05$. (Arranged from Nicolas et al., 2001.)

Occlusion could not explain why the corticospinal response decreased below its control level. It appears unlikely that suppression resulted from corticospinal facilitation of segmental inhibitory interneurones because: i) the central delay of the inhibition, which is the same as that of the facilitation (e.g., 5 ms in Figure 1*H*, see legend of Figure 1), is much longer than that observed in segmental pathways fed by low-threshold group I afferents (~0.8 ms, see Pierrot-Deseilligny et al., 1981); ii) inhibition due to corticospinal facilitation of segmental inhibitory interneurones projecting to MNs, should affect the entire corticospinal response, including the initial part due to the monosynaptic CM projection (see Pierrot-Deseilligny, 1996). That the peripheral suppression of the corticospinal peak spared the initial part of the corticospinal response (see Fig. 2*D* and *K*) in almost all MUs is *consistent with disfacilitation due to inhibition of premotoneurones transmitting corticospinal excitation.*

The mean duration of this initial sparing by inhibition was 0.73 ms. This suggests that inhibition is exerted at the premotoneuronal level of a disynaptic pathway mediating corticospinal excitation. The finding that suppression of the corticospinal response at higher TMS intensities had the same time course as the facilitation at low TMS intensity (which, as noted above, was greater the more caudal the MN pool) favours the view that the PreMNs in question are the putative PNs. This is consistent with a convergence of peripheral and corticospinal inputs onto both PNs and feedback inhibitory interneurones inhibiting them (see Fig. 2A), much as in the cat (see Lundberg, 1999): at very low TMS intensities, inhibitory interneurones are marginally activated and excitation of PNs can manifest itself, whereas at higher intensities, there is simultaneous activation of inhibitory projections and this activation prevents PNs from firing.

Corticospinal activation of inhibitory interneurones projecting to PNs can explain the negative results obtained after *artificial* (electrical or magnetic) stimulation of the pyramidal tract in macaque monkeys (Maier et al., 1998) or of the motor cortex in man (Martens de Noordhout et al., 1999). Indeed, such a stimulation will evoke unnaturally synchronised volleys with gross activation of inhibitory interneurones preventing a propriospinal discharge in response to corticospinal excitation. Accordingly, propriospinally-mediated disynaptic corticospinal EPSPs, that are rare and weak in MNs of the macaque monkey under control conditions, can be readily demonstrated when inhibition has been reduced by intravenous strychnine (Alstermark et al., 1999).

ACTIVATION OF PROPRIOSPINAL NEURONES DURING MOVEMENT

If corticospinal activation of PNs and of inhibitory interneurones cannot be dissociated readily when using artificial volleys, indirect evidence for corticospinal excitation of PNs during movement has been provided with indirect methods (see Fig. 3).

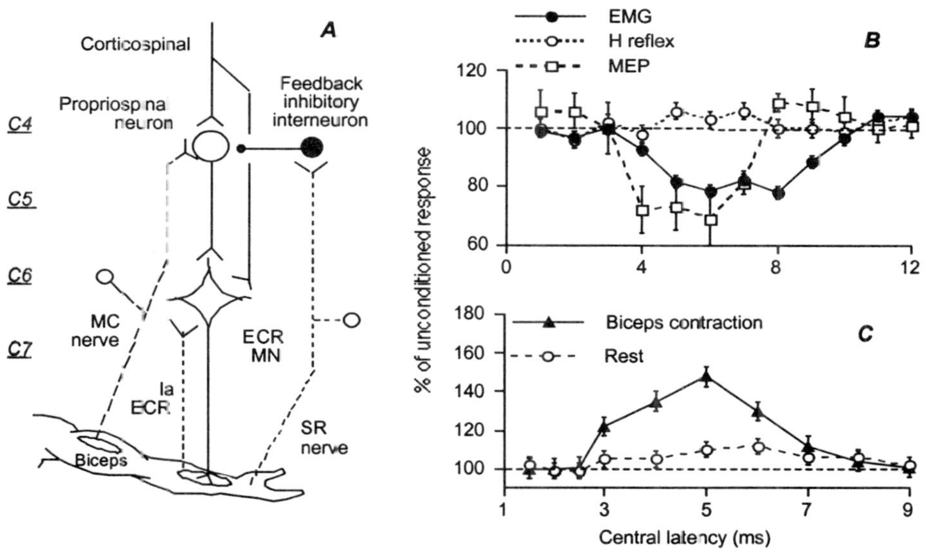

Figure 3. *A*, sketch of the presumed pathways. *B*, effects of a cutaneous volley to the superficial radial (SR) nerve on the ongoing EMG activity (●; average of 300 sweeps; % of unconditioned EMG), the MEP (□) and the H reflex (O) of the ECR. *C*, effects of a MC volley (0.8 x MT) on the ECR H reflex at rest (O) and at the onset of biceps contraction (σ). Conditioned responses for the H reflex and the MEP (mean values and SEM of 20 (*B*) or 120 (*C*) responses expressed as a % of unconditioned responses) are plotted against the central latency (i.e. the zero of abscissa corresponds to the arrival of the conditioning volley at the segmental level of MNs). (*B* arranged from Burke et al., 1992; 1994).

Figure 3*B* shows the effects of a cutaneous volley to the superficial radial nerve, which has been shown to inhibit presumed PNs (see Pierrot-Deseilligny, 1996). This resulted in a suppression of the ongoing EMG recorded in the ECR during voluntary wrist extension (●), which was accompanied by a depression of the MEP (□) but not of the H reflex (○) (Burke et al., 1994). Thus, the suppression of the EMG is not due to an inhibition exerted directly on MNs, but reflects an inhibition of premotoneurones mediating part of the natural descending excitation and acting in parallel with the monosynaptic CM pathway. Similar results were found in biceps and triceps, and here again, the more caudal the MN pool, the longer the central delay (see Pierrot-Deseilligny, 1996). This implies an inhibition of premotoneurones located rostral to MNs, i.e. of the putative PNs.

MC stimulation elicits a small facilitation of the FCR and ECR H reflex at rest (○ in Fig. 3*C*; ECR), which becomes much larger at the onset of biceps contraction (σ) (Burke et al., 1992; Mazevet and Pierrot-Deseilligny, 1994). This reflex facilitation has all the features of an effect mediated through presumed PNs (low threshold, long central delay, disappearance when the conditioning stimulus intensity is increased above 0.85 x MT). Thus, the large increase in the MC-induced facilitation of the H reflex of wrist muscles *at the onset* of biceps voluntary contraction reflects descending facilitation of the PNs receiving excitatory afferent feedback from the contracting muscle.

In contrast, feedback inhibitory interneurones receive more descending facilitation *at the offset than at the onset* of movement (see Pierrot-Deseilligny, 1996). This suggests that the corticospinal excitation of PNs and the cortical control of the inhibitory interneurones that adjusts the gain in the feedback loop can be controlled independently during "natural" movement.

REFERENCES

Alstermark, B., Isa, T., Ohki, T., and Saito, T., 1999, Disynaptic pyramidal excitation in forelimb motoneurons mediated via C3-C4 propriospinal neurons in the *Macaca Fuscata*, *Journal of Neurophysiology*, 82, 3580-3585.

Burke, D., Gracies, J. M., Mazevet, D., Meunier, S., and Pierrot-Deseilligny, E., 1992, Convergence of descending and various peripheral inputs onto common propriospinal-like neurones in man, *Journal of Physiology*, 449, 655-671.

Burke, D., Gracies, J. M., Mazevet, D., Meunier, S., and Pierrot-Deseilligny, E., 1994, Non monosynaptic transmission of the cortical command for voluntary movement in man, *Journal of Physiology*, 480, 191-202.

Lundberg, A., 1999, Descending control of forelimb movements in the cat, *Brain Research Bulletin*, 50, 323-324.

Maertens de Noordhout, A., Rapisarda, G., Bogacz, D., Gerard,. P., De Pasqua, V., Pennisi, G., and Delwaide, P. J., 1999, Corticomotoneuronal synaptic connections in normal man. An electrophysiological study, *Brain*, 122, 1327-1340.

Maier, M. A., Illert, M., Kirkwood, P. A., Nielsen, J., and Lemon R. N., 1998, Does a C3-C4 propriospinal system transmit corticospinal excitation in the primate? An investigation in the macaque monkey, *Journal of Physiology*, 511, 191-212.

Mazevet, D., and Pierrot-Deseilligny, E., 1994, Pattern of descending excitation of presumed propriospinal neurones at the onset of voluntary movement in man, *Acta Physiologica Scandinavica*, 150, 27-38.

Nakajima, K., Maier, M. A., Kirkwood, P. A., and Lemon R. N., 2000, Striking differences in transmission of corticospinal excitation to upper limb motoneurons in two primate species, *Journal of Neurophysiology*, 84, 698-709.

Nicolas, G., Marchand-Pauvert, V., Burke, D., and Pierrot-Deseilligny, E., 2001,Corticospinal excitation of presumed propriospinal neurones and its reversal to inhibition in humans, *Journal of Physiology*, 533, 903-919.

Pauvert, V., Pierrot-Deseilligny, E., and Rothwell, J. C., 1998, Role of spinal premotoneurones in mediating corticospinal input to forearm motoneurones in man, *Journal of Physiology,* 508, 301-312.
Pierrot-Deseilligny, E., 1996, Transmission of the cortical command for human voluntary movement through cervical premotoneurones, *Progress in Neurobiology,* 48, 489-517.
Pierrot-Deseilligny, E., Morin, C., Bergego, C., and Tankov, N., 1981, Pattern of group I fibre projections from ankle flexor and extensor muscles in man, *Experimental Brain Research,* 42, 337-350.

PREMOTONEURONAL AND DIRECT CORTICOMOTONEURONAL CONTROL IN THE CAT AND MACAQUE MONKEY

Bror Alstermark[1] and Tadashi Isa[2]

ABSTRACT

The literature on premotoneuronal and direct corticomotoneuronal (CM) control in the cat and macaque monkey is reviewed. The available experimental findings are not in accordance with a recently proposed hypothesis that direct CM connections have "replaced" the premotoneuronal pathways. Instead, we propose that premotoneuronal CM control plays an important role in motor control also in primates and that the direct CM connection has been added during phylogeny.

INTRODUCTION

This review covers the role of disynaptic and monosynaptic excitatory corticomotoneuronal (CM) pathways in the control of arm and digit movements in the cat and macaque monkey. This discussion starts out from a hypothesis put forward by Nakajima, Maier, Kirkwood and Lemon (2000) that spinal premotoneuronal systems involved in the control of voluntary movements like reaching and grasping have been gradually replaced by the direct CM connection through phylogeny from carnivores, lower and higher primates and completely so in man. These authors have focused their hypothesis on one group of interneurones, the C3-C4 propriospinal system, which is the only example so far in the spinal cord of a command mediating interneuronal system and which has been extensively investigated in the cat (Alstermark and Lundberg, 1992).

The cat lacks direct CM connection and Illert, Lundberg and Tanaka (1977) first described that disynaptic pyramidal excitation in forelimb motoneurones can be mediated via spinal interneurones (propriospinal neurones, PNs) with cell bodies located in the C3-C4 segments (see below and Figure 1A). They denoted this system as the C3-C4 propriospinal system. Behavioural studies later revealed that C3-C4 PNs could mediate

[1] Dept of Integrative Medical Biology, Section of Physiology, University of Umeå, S 901 87 Umeå, Sweden. Email: bror.alstermark@physiol.umu.se
[2] Dept of Integrative Physiol, National Institute for Physiological Sciences, Myodaiji, Okazaki 444-8585, Japan.

the descending command for forelimb target reaching to a tube with a morsel of food (Alstermark et al., 1981b; Alstermark and Kümmel, 1990). The cortical command for food taking (comprised of toe flexion, grasping, supination and bringing the food to the mouth) could be mediated via segmental interneurones in the forelimb segments (Alstermark et al., 1981b).

Attempts were made by Maier and colleagues (1998) to investigate if disynaptic pyramidal excitation could be mediated via C3-C4 PNs to forelimb motoneurones also in primates. In the anaesthetised macaque monkey, disynaptic pyramidal excitation was found in only 3% of the forelimb motoneurones, whereas monosynaptic excitation was found in 73% of the cells. They did not make a C2 corticospinal transection to test if the remaining disynaptic EPSPs were mediated via C3-C4 PNs or more rostrally located bulbospinal neurones. They concluded that, *"compared with the situation in the cat, transmission through a C3-C4 propriospinal system is far less effective or even absent in the macaque"*.

Figure 1A shows that the C3-C4 PNs receive disynaptic feed-forward and feed-back inhibition, which is mediated via separate groups of interneurones (see below). Common to these inhibitory interneurones is that they are strongly activated by corticospinal fibres (Alstermark et al., 1984b,c). Thus, in the experiments with electrical stimulation of corticospinal fibres in the pyramid, the C3-C4 PNs will be both excited and inhibited by the same descending pyramidal volleys. Selective cortical activation in the awake animal would not cause such an effect (Nicolas et al., 2001).

We repeated the study of Maier et al. (1998) and found (Alstermark et al., 1999) that after reduction of glycinergic inhibition with strychnine, disynaptic excitation was found in at least 95% of the forelimb motoneurones (Fig. 2A). These excitatory effects could also be observed after transection of the corticospinal tract in C5, indicating that they were mediated via interneurones located more rostrally. After a corresponding transection in C2, the frequency of disynaptic pyramidal EPSPs decreased to 8%. We therefore concluded that C3-C4 PNs can mediate disynaptic pyramidal excitation in high frequency of occurrence and that this transmission is under stronger inhibitory control than in the cat (Alstermark et al., 1999).

The C3-C4 propriospinal system is only one of several interneuronal systems, which can mediate disynaptic pyramidal excitation in motoneurones. The claim by Maier et al. (1998) and Nakajima et al. (2000) has far-reaching implications for the understanding of voluntary movements, as it touches upon important principles of control at premotoneuronal levels. These principles are not restricted to lower species of animals, but rather exist in primate and man as well. We therefore widen the discussion from the C3-C4 propriospinal system to describe other premotoneuronal systems, which can mediate disynaptic pyramidal excitation to motoneurones in the cat. A summary is then made of corticofugal pathways to motoneurones in the macaque monkey and of dexterity. Finally, we discuss disynaptic and monosynaptic CM connections and their functional roles in the control of arm and hand movements.

PREMOTONEURONAL SYSTEMS MEDIATING DISYNAPTIC PYRAMIDAL EXCITATION IN MOTONEURONES OF THE CAT

Forelimb Motoneurones

C3-C4 Propriospinal Neurones

Figure 1A shows a summary of the connections, which have been proven in electrophysiological and anatomical experiments. The C3-C4 PNs mediate disynaptic excitation to forelimb motoneurones (MNs) from the cortico-, rubro-, tecto- and reticulospinal tracts (Illert et al., 1977). The extensive convergence from subcortical descending tracts on the C3-C4 PNs was taken to suggest that the cortical command for a movement could be updated *en route* to the motoneurones, to changes occurring after the cortical command was initiated (Illert et al., 1978). In this way a new command does not have to be immediately generated in the motor cortex. In behavioural experiments the cortical command for ongoing visually guided reaching could be updated, at short latency, in response to a sudden change in location of the target and that the command for updating, could be mediated via the C3-C4 PNs (Pettersson et al., 1997). The results also suggested that the command for correction of the reaching movement could be given via the subcortical systems, which converged on the C3-C4 PNs. Subpopulations of C3-C4 PNs exist which have projections to different motoneuronal pools controlling the shoulder, elbow, wrist and digits of the forelimb (Tantisira et al., 1996) and to fast and slow motoneurones (Alstermark and Sasaki, 1986). A positive correlation was found between the strength of the excitatory projection from C3-C4 PNs to motoneurones and the monosynaptic excitation from group Ia afferents.

The C3-C4 PNs also receive feed-forward inhibition from descending tracts and feed-back inhibition from cutaneous and muscle afferents in the moving limb (Alstermark et al., 1984b,c). The function of the feed-forward inhibition is unknown, but one possibility could be that it is used in the selection of subpopulations of PNs, which are active during different phases of target reaching (see discussion). The feedback inhibition from forelimb afferents has been shown to be important for deceleration and termination of target reaching (Alstermark et al., 1986). The interneurones giving feedback inhibition receive monosynaptic corticospinal input (Alstermark et al., 1984c).

The origin of corticospinal neurones (CSNs) which evoke disynaptic excitation, mediated via C3-C4 PNs, in forelimb motoneurones is restricted to the lateral part of the anterior sigmoid gyrus and to the rostral part of the lateral sigmoid gyrus (Alstermark and Ohlson; 2000 a,b).

In addition to the descending projection to forelimb motoneurones, the C3-C4 PNs have ascending collaterals terminating in the lateral reticular nucleus (LRN), which is a major mossy fibre input to the cerebellum from the spinal cord (Illert and Lundberg, 1978; Alstermark et al., 1981a). In this way the cerebellum is provided with an efference copy of the descending command to the motoneurones, which could be used for fast cerebellar corrections of the converging descending systems to the C3-C4 PNs.

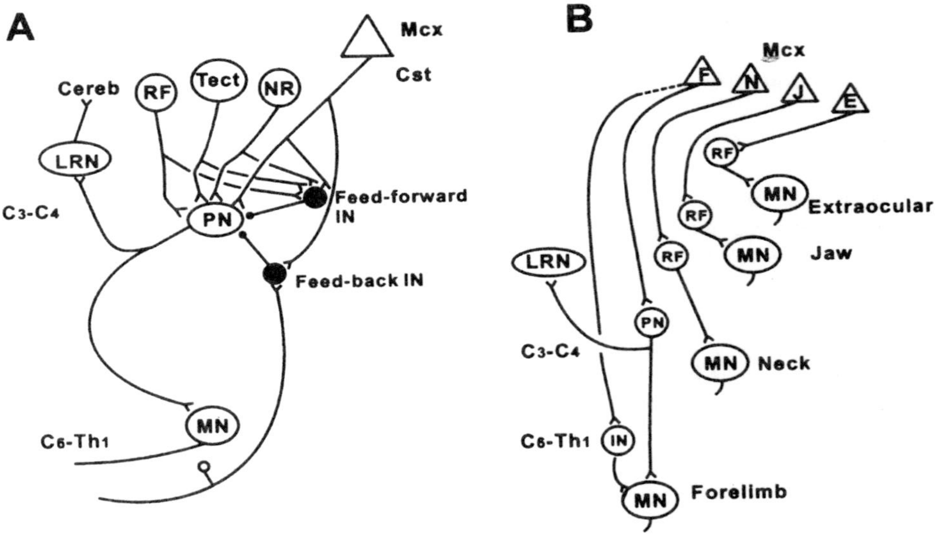

Figure 1. Schematic drawings of the disynaptic excitatory corticomotoneuronal pathways in the cat.
A: Summary of connections. Note that the neurones should be viewed as populations and not as single neurones. C3-C4 propriospinal neurones (PNs) mediating the corticospinal inputs to forelimb motoneurones (Illert et al., 1977). PNs in the C3-C4 segments receive convergent inputs from the motor cortex (MCx), the nucleus ruber (NR), the tectum (Tect) and the brain stem reticular formation (RF) and send excitatory commands to forelimb motoneurones (MN) in the C6-Th1 segments and also to the lateral reticular nucleus (LRN) in the medulla oblongata, which sends the mossy fiber inputs to the cerebellum. Two subtypes of inhibitory interneurones (feed-forward and feed-back inhibitory interneurones) located in the C3-C4 segments mediate inhibition of the C3-C4 PNs (Alstermark et al., 1984b,c). **B:** Disynaptic excitatory corticomotoneuronal pathways to the extraocular, jaw, neck and forelimb motoneurones. Interneurones mediating disynaptic excitation from the corticofugal neurones in extraocular (E), jaw (J), neck (N) and forelimb (F) areas of the motor cortex (MCx) tend to be located rostral to their target motoneurones. Note that the premotor interneurones of forelimb motoneurones are known to be located both in the mid-cervical (C3-C4, PN) and forelimb segments (C6-Th1, IN).

Interneurones in the Forelimb Segments C6-Th

Disynaptic pyramidal excitation can be evoked in forelimb motoneurones after transection in the ventral part of the lateral funiculus (corticospinal tract spared) of the propriospinal axons in C5/C6 (Illert and Wiedemann, 1984; Alstermark and Sasaki, 1985). This finding suggests that part of the disynaptic excitation can be mediated via interneurones in the forelimb segments as shown schematically in Figure 1B.

Recordings made directly from such segmental interneurones have revealed that they receive monosynaptic excitation from the corticospinal tract and that they project monosynaptically to forelimb motoneurones (Kitazawa et al., 1993). Some of these segmental interneurones can probably mediate the descending command for food-taking (Alstermark et al., 1981b). The cortical origin of the corticospinal neurones with projection to these interneurones is unknown.

Reticulospinal Neurones

Disynaptic EPSPs can be elicited by electrical stimulation in the contralateral tegmentum (Anderson et al., 1972) and stimulation in the medial part of the ipsilateral reticular formation evokes small monosynaptic EPSPs in forelimb motoneurones (Peterson et al., 1979; Alstermark and Sasaki 1986a). However, no disynaptic pyramidal EPSPs can be evoked in forelimb motoneurones after a corticospinal transection in C2, but pyramidal stimulation could facilitate disynaptic EPSPs evoked from the contralateral tectum after the C2 transection (Illert et al., 1977). Thus, the cortex can influence forelimb motoneurones via reticulospinal neurones (RSNs), but this pathway is not so strong as for neck motoneurones.

Neck Motoneurones

Disynaptic excitatory corticospinal pathway via RSNs to dorsal neck motoneurones has been demonstrated (Alstermark et al., 1985; 1992) as shown in Figure 1B. RSNs in the nucleus reticularis pontis caudalis (NRPC) and the nucleus reticularis gigantocellularis (NRG), which terminate in the cervical segments but do not descend to the lumbar segments, make monosynaptic connections with neck motoneurones (Iwamoto and Sasaki, 1990; Iwamoto et al., 1990). Single unit recordings and lesion studies showed that these RSNs are involved in the control of orienting head movements and possibly also in postural adjustments (Isa and Sasaki, 1988; Alstermark et al., 1991a; Isa and Naito, 1995; Sasaki, 1997).

NRPC-RSNs and NRG-RSNs seem to be differentially controlled from cortex (Isa and Sasaki, 1991). Following HRP injection into the NRPC, retrogradely labelled neurones were most densely located in the frontal oculomotor region (area 6aβ and medial half of area 4γ) bilaterally in the frontal cortex. In contrast, neurones in the lateral 2/3 of area 4γ were chiefly labelled following HRP injection into the NRG. Furthermore, it was found that cortical stimulation evoked monosynaptic EPSPs in the RSNs bilaterally (Isa and Sasaki, 1991).

It is noteworthy that neck motoneurones do not seem to receive any input from the C3-C4 PNs despite the fact that they are located in the same cervical segments (Alstermark et al., 1991b).

Extraocular Motoneurones

Disynaptic cortical excitation can be evoked in abducens motoneurones in the cat (Yamagata et al., 1988) as illustrated in Figure 1B. The intercalated last order interneurones are located in the NRPC and receive convergent tectal excitation. The corticofugal fibres were stimulated in the contralateral peduncle and the cortical origin is not known, but eye movements can be elicited from the frontal cortex and with shortest latencies from the coronary sulcus (Tamai et al., 1983).

Jaw Motoneurones

A disynaptic cortical pathway to trigeminal motoneurones exists from area 3a, ipsilateral side (Olsson and Landgren, 1980; Olsson et al., 1986) as shown in Figure 1B.

Last order interneurones were labelled bilaterally following HRP injection into the trigeminal motor nucleus (Landgren et al., 1986). These interneurones were located in the border zone of the trigeminal motor nucleus. The cortical projection to the trigeminal motor nucleus is bilateral and the origin appears to be restricted to the orbitofrontal cortical regions, which corresponds to the rostral part of area 3a, 3b and area 6 in the cat (Yasui et al., 1985). Thus, corticofugal neurones in the primary motor cortex of area 4γ do not seem to project directly to the trigeminal motor nucleus in the cat.

CORTICOFUGAL PATHWAYS TO MOTONEURONES IN THE MACAQUE MONKEY

Previous investigations in the baboon and monkey (Landgren et al., 1962; Clough et al., 1968; Shapovalov, 1975; Jankowska et al., 1976a,b; Fritz et al., 1985; Maier et al., 1998, Alstermark et al., 1999) have shown that monosynaptic pyramidal EPSPs can be evoked in motoneurones, which are followed by disynaptic IPSPs (Fig. 2A). In addition, as shown in Figure 2B, pyramidal disynaptic EPSPs in forelimb motoneurones can be mediated via C3-C4 PN (Alstermark et al., 1999).

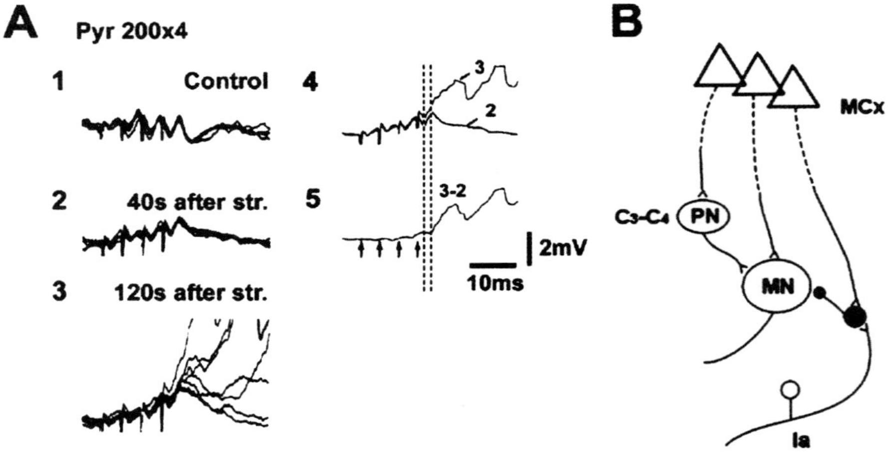

Figure 2. Non-monosynaptic corticomotoneuronal effects. *A*, Disynaptic excitatory corticomotoneuronal pathways in macaque monkeys (Alstermark et al., 1999). Intracellular recordings from forelimb motoneurones of the monkey. Stimulation of the pyramid (Pyr) with 4 pulses (x4) induced monosynaptic EPSPs and disynaptic IPSPs before injection of strychnine (Str, A1). Injection of strychnine (0.1 mg/kg) eliminated the IPSPs after 40 s (A2). 120 s after the strychnine injection, large di- and oligosynaptic EPSPs were induced following the pyramidal stimulation (A3). Averaged intracellular records from A2 and A3 (A4) and subtraction of the averaged records A2 from A3 (A5). Top arrows indicates the onset of the disynaptic EPSP and bottom arrows indicate the pyramidal stimuli. *B*, monosynaptic and disynaptic corticomotoneuronal pathways so far clarified. Note that the neurones should be viewed as populations and not as single neurones. Existing data suggest separate cortical control of the different neurones, but this has not been proven (dotted lines). In addition to the direct corticomotoneuronal pathways, disynaptic inhibition mediated by Ia inhibitory interneurones (Jankowska et al., 1976a) and propriospinal neurones in the C3-C4 segments (PN) have been shown to mediate corticofugal effects to motoneurones.

Strength of the Monosynaptic CM Projection

Motoneurone Type

Comparison of the monosynaptic pyramidal EPSP amplitude has revealed a correlation with motoneurone type. Motoneurones with small homonymous monosynaptic group Ia EPSPs have small pyramidal monosynaptic EPSPs, while motoneurones with large group Ia EPSPs have large pyramidal EPSPs (Clough et al. 1968). These authors also found that the monosynaptic EPSPs were large in motoneurones innervating intrinsic hand muscles and the long extensor of the fingers. In lumbar motoneurones, larger monosynaptic CM EPSPs in motoneurones were found for distal than more proximal muscles (Jankowska et al., 1976a). In a systematic comparison of monosynaptic pyramidal EPSP amplitudes of motoneurones innervating different muscles of the hand, wrist and elbow, it was found that more than 90% received monosynaptic pyramidal EPSPs (Fritz et al., 1985). However, whereas 100% of the intrinsic hand motoneurones showed monosynaptic pyramidal EPSPs, only 44% of the triceps motoneurones did so. It is interesting that a few of the intrinsic hand motoneurones received very small EPSPs (<0.5 mV) while the largest EPSPs were seen in this group (7.5 mV; Porter and Lemon, 1993). Thus, the amplitude may vary considerably even within a population of motoneurones to a single muscle because the motoneurones are not homogenous in their properties.

There are no electrophysiological data reported about corticofugal projections to motoneurones innervating neck and axial muscles, either for mono- or disynaptic inputs. Anatomically, there seems to be only sparse or no direct CM projection to neck and axial motoneurones (Bortoff and Strick, 1993). Anatomical studies have shown input to the reticular formation from area 6 (Catsman-Berrevoets and Kuypers, 1976). Furthermore, a direct reticulospinal projection to neck motor nuclei in C2 and extraocular motor nuclei in the brain stem have been revealed by retrograde labelling with WGA-HRP (Robinson et al., 1994). Thus, a disynaptic CM projection to neck motoneurones via RSNs is possible.

Type of Pyramidal Tract Neurones

A difference was found in the postspike facilitation from fast and slow corticospinal neurones to the same motor unit. Fast pyramidal tract neurones (PTNs) produced postspike facilitation (PSF) in 38-80% of the investigated motor units of 3 monkeys, whereas slow PTNs to the same motor units produced PSF in only 25-33% of the cases (Lemon et al., 1993).

Development

Anatomically, the corticospinal termination in the motor nuclei of the forelimb segments increases considerably during the first two postnatal years in the macaque monkey (Armand et al., 1997). In their Figure 8, it also is clear that the corticospinal termination in the intermediate zone increases in density during the first two postnatal years. Thus, the development of the corticospinal projection to the intermediate zone and the motor nuclei seem to occur in parallel.

Cortical Origin

Corticospinal fibres originate from a fairly wide region of the cerebral cortex through areas 1, 2, 3a, 3b, 4, 5 and 6 of Brodmann (Catsman-Berrevoets and Kuypers 1976; Coulter et al., 1976; Jones and Wise 1977; Biber et al., 1978; Kuypers 1981; Martino and Strick 1987; Dum and Strick 1991; Bortoff and Strick 1993; He et al., 1995; Dum and Strick 1996; Armand et al. 1997). With respect to the motor areas; area 4 gives rise to approximately 31%, area 6 to 29% and the parietal cortex to 40% of pyramidal tract fibres (Russel and DeMeyer, 1961). Interesting differences exist in the termination of the corticospinal fibres in the grey matter of the spinal cord. Corticospinal neurones (CSNs) in the primary motor cortex (M1) terminate both in the intermediate zone (70%) and in the motor nuclei (30%; Bortoff and Strick, 1993). Furthermore, these CSNs terminate both in the upper (C2-C4; 43%) and lower cervical and thoracic segments (C7-Th1; 50%; He et al., 1995). CSNs in the supplementary motor area (SMA) terminate in both upper (32%; includes cingulate motor area (CMA) projections) and lower cervical segments (66% includes CMA projections; He et al., 1995), but a higher proportion of their terminals were confined to the intermediate zone (87%; Dum and Strick, 1996) as compared to M1. SMA terminals in the motor nuclei was estimated to 11% and to 2% in the dorsal horn. Similarly, CSNs in the CMA (ventral and dorsal parts) had their terminals to a large extent in the intermediate zone (90%), while only 4% of the terminals were located in the motor nuclei and 6% in the dorsal horn. The projections from SMA and CMA (ventral and dorsal parts) were topographically organised, such that all three regions had largely separate regions with projections to the upper and lower cervical segments, respectively (He et al., 1995). CSNs in the lateral premotor area project in higher proportion to the upper (70%) cervical segments than to the lower cervical segments (26%; He et al., 1995).

Virtually all studies using electrical stimulation of the cortex have so far been devoted to M1. However, recently SMA was stimulated and at the same time the effect of M1 stimulation was compared while recording intracellularly from forelimb motoneurones (Maier et al., 1997; see Lemon et al., Chapter 48). While M1 stimulation evoked early (presumed monosynaptic) EPSPs in 90% of the forelimb motoneurones, stimulation of SMA elicited such EPSP in 57% of the cells. Furthermore, the EPSPs from SMA were much smaller in amplitude than those evoked from M1.

Dexterity

A close association has been suggested between the development of the corticospinal tract and precise control of the hand and digits during phylogeny (see Phillips, 1971; Phillips and Porter, 1977; Porter and Lemon, 1993). This proposal is based on the results of anatomical (see Armand, 1982; Bortoff and Strick, 1993) electrophysiological (see Porter and Lemon, 1993) and behavioural experiments (see Wiesendanger, 1981). Differences in the strength of the monosynaptic CM excitation in forelimb motoneurones belonging to various species of primates are taken to suggest differences in the dexterity. In this respect, Nakajima et al. (2000) compared the results from the squirrel monkey and the macaque monkey. On average the monosynaptic CM EPSPs were 0.6 mV in the squirrel monkey and 1.9 mV in the macaque monkey. Nakajima et al. (2000) mention that in the squirrel monkey, independent finger movements are limited and that there is no precision grip or capacity to manipulate objects within the hand (Costello and

Fragraszy, 1988). In contrast, the macaque monkey has the ability for making the tip-to-tip grip between the thumb and the index finger and uses the hand for tactile exploration and manipulation of objects (Lawrence and Hopkins, 1976; Heffner and Masterton, 1983).

In an extensive study, Lawrence and Kuypers (1968a) made pyramidotomies at the caudal border of the pons and the rostral tip of the inferior olivary nucleus. In this way the cortical input was eliminated both to the caudal brain stem and spinal cord. They found that the macaque monkeys lost permanently their ability to make individual finger movements, while they recovered the capacity to make independent limb movements.

DISCUSSION

Monosynaptic CM connections have been reported in the racoon, but the density of projection is weak in comparison to the projection in primates. In man the projection is even stronger than in the lower primates and most motor nuclei receive direct CM projection (Iwatsubo et al., 1990). Thus, there can be little doubt that this projection must be of importance. The association between dexterity and the corticospinal tract has been strongly emphasised with a special role given to the monosynaptic CM connections. The proposal by Nakajima et al. (2000) that the monosynaptic CM connection in primates and man has replaced the disynaptic pathway via interneurones is a far-reaching addition to this idea and gives a major role of the direct CM connections also for other movements such as reaching. Since we now know that disynaptic CM connections exist in the high primate (Alstermark et al., 1999), it seems necessary to compare the effects of these two pathways.

Strength of Connectivity

The strongest monosynaptic projection is to distal motoneurones innervating intrinsic digit muscles. Even so, intrinsic digit motoneurones show a range of fifteen times of EPSP amplitudes. By comparison between the squirrel monkey with a low dexterity and the macaque monkey with a high degree of dexterity, Nakajima et al. (2000) reports on average a three-fold increase in amplitude. The size of the monosynaptic CM EPSPs was correlated to motoneurone type; being small in fast motoneurones and large in slow motoneurones (Clough et al., 1968).

A positive correlation was also found for the monosynaptic LRN EPSPs amplitudes to motoneurone type, mediated by the C3-C4 PNs, in fast and slow cat forelimb motoneurones (Alstermark and Sasaki, 1986). Maier et al.(1998) and Nakajima et al. (2000) have stimulated in the LRN of the macaque monkey and the squirrel monkey, respectively. They report smaller LRN EPSPs in both species of monkey as compared to the cat. So far, the possibility of an ascending projection to the LRN of the C3-C4 PNs has not been investigated in the monkey. Also, it is not known to what extent effects evoked from the LRN in motoneurones are due to axon reflexes of other systems and how the compound EPSPs are influenced by concomitant IPSPs (Alstermark et al., 1984a).

Another factor, which must be taken into consideration, is the contribution of fast and slow PTNs and C3-C4 PNs to the various types of motoneurones. Differences in the CM projection from these systems may cause considerable variation in the efficacy of the synaptic strength. Therefore comparison of amplitudes must be done for similar types of motoneurones.

Also the age of the animals is important when comparison is made of the strength of CM innervation. The CM projection increases during the first two postnatal years in the monkey (Armand et al., 1997). Interestingly, the corticospinal termination in the intermediate zone increased in parallel with the termination within the motor nuclei. It may therefore be assumed that also the strength of disynaptic CM projection may increase postnatally. In humans, it has been shown that the ventral horn and intermediate zone receive corticospinal projection already by 24 weeks post-conception age and that transcranial magnetic stimulation could evoke presumed monosynaptic CM excitation and group Ia inhibition as early as 26 weeks post-conception age (Eyre et al., 2000). In the cat, the corticospinal tract develops considerably during the first three postnatal months (Li and Martin, 2000; Chakrabarty and Martin, 2000; Li and Martin, 2001).

Thus, at present it seems not meaningful to compare amplitudes between species. For the time being it may be more important to compare the monosynaptic and disynaptic CM EPSPs in the same motoneurone of the monkey. Such recordings have shown that the disynaptic CM EPSPs can be twice as large as the monosynaptic CM EPSPs as shown in Figure 2A.

Cortical Control

In the cat there is evidence for subregional specialisation of CSNs with projection to different premotoneuronal systems. In the monkey, the monosynaptic CM projection follows the somatotopic organisation although considerable overlap in M1 has been shown. It is claimed that SMA may also give monosynaptic CM excitation, but weaker and less frequently (Maier et al., 1997). Unfortunately, nothing is known electrophysiologically about the other cortical regions in the monkey as to the monosynaptic CM projections. When it comes to the disynaptic excitatory CM projections in the monkey, we have confirmed that disynaptic cortical EPSPs in cervical motoneurones can be evoked from the forelimb region of M1 (unpublished observations by Alstermark, Isa and Ohki). It is interesting that with the cortical expansion in the monkey, there seems to be a further differentiation as to the segmental level of termination compared with the cat. In the cat, it has been shown that the majority of corticospinal fibres which provide monosynaptic excitation to the C3-C4 PNs also project to the lower cervical segments (Illert et al., 1978). In the monkey, there is anatomical evidence, which suggest that each cortical region gives rise to separate corticospinal projections to the upper and lower cervical segments and that the SMA, cingulate and lateral premotor regions may give a preponderance to either the upper or lower cervical segments. M1 appears to give a balanced projection to both upper and lower cervical segments. Furthermore, when comparing the corticospinal termination in the intermediate zone and the motor nuclei, it is clear that M1 has its major termination in the intermediate zone and this preference for the intermediate zone is even more marked for both the SMA and CMA.

Strick and colleagues has emphasised that the M1, SMA, CMA and arcuate premotor area (APA) should be viewed as components of functionally distinct efferent systems, which are driven by mainly separate subcortical nuclei (Martino and Strick, 1987). According to this view, these systems operate largely in parallel, but interact at the cortical level where the motor areas are densely interconnected (Dum and Strick, 1991).

Furthermore, Martino and Strick (1987) proposed that the APA projection in the C2-C4 segments could be to a similar propriospinal system in primates as in cats. The findings by Dum and Strick (1996) were taken to suggest that the dorsal and ventral parts of the CMA might terminate on distinct sets of interneurones that project to motoneurones and thereby influence specific aspects of segmental motor control. Thus, from an anatomical point of view, the corticospinal projection to the intermediate zone seems not to have decreased during phylogeny. On the contrary, compared to the cat, the corticospinal projection in the monkey has apparently become more specialised both to the motor nuclei and to the intermediate zone.

Brain Stem and Spinal Control

In the cat, premotoneuronal control seems to be a common principle for many types of motor behaviour: forelimb, head, jaw and eye movements. So far, we know that premotoneuronal control also exists for the forelimb in the macaque monkey and it seems likely that at least neck and extraocular motoneurones also receive disynaptic CM excitation. Neck motoneurones receive direct input from reticulospinal neurones, which originate from the same region of the reticular formation that receives strong corticoreticular connections. Extraocular motoneurones seem to lack the direct monosynaptic CM projection even in man (Iwatsubo et al., 1990).

A unique feature of premotoneuronal as compared to direct CM control, is that fast updating of the cortical command can occur near the motoneurones. Such premotoneuronal integration takes advantage of the descending subcortical systems, which converge on common interneurones and of input from primary afferents (Illert et al., 1978). In the case of the C3-C4 propriospinal system a loop via the LRN and cerebellum also exists, which may give very fast corrections of the subcortical systems at the spinal level (Illert and Lundberg, 1978; Alstermark et al., 1981a). Consequently, if during phylogeny the disynaptic CM pathways have been replaced by the direct CM pathway, one must postulate that all of the premotoneuronal integration has moved to the cortical level. Indeed, it would be easier to reconcile with the view of Foster (1879): *"the cord contains a number of more or less complicated mechanisms capable of producing, as reflex result, co-ordinated movement altogether similar to those which are called forth by will. Now it must be an economy to the body that the will should make use of these mechanisms already present, by acting directly on their centres, rather than it should have recourse to a special apparatus of its own of a similar kind."*

It is often claimed that the monosynaptic CM projection is necessary for making fractionated movements with the fingers. This proposition is based on anatomical and electrophysiological findings in the monkey, which have shown a restricted projection pattern of single CSNs to a limited number of forelimb motoneurones (cf. Porter and Lemon, 1993). However, such restricted projection patterns are not unique for the CSNs. The same holds true for the C3-C4 PNs and for other premotoneuronal systems (Tantisira et al., 1996; Iwamoto and Sasaki, 1990).

There are other mechanisms that may contribute to fractionation of movements. One mechanism is feed-forward inhibition from the corticospinal tract, coupled to the excitatory command. The importance of inhibitory control was already noticed by Sherrington (1933) who stated: *"to refrain from an act is no less an act than to commit one, because inhibition is co-equally with excitation a nervous activity".* Feed-forward inhibition may act to select the appropriate groups of interneurones with the pattern of projections required for a given movement or phase of it. An example is the feed-forward inhibitory interneurones to C3-C4 with input from the same descending systems that project to the C3-C4 PN system (Alstermark et al., 1984b). In the monkey with a more elaborated motor control than in the cat, inhibition of the C3-C4 PNs is stronger. Several subpopulations of C3-C4 PNs have been shown to exist in the cat and with different projection patterns to forelimb motoneurones of shoulder, elbow, wrist and digit muscles. If this is also the case in the monkey, it is not difficult to understand that inhibitory control of the premotoneurones have developed during phylogeny. This idea is supported by recent findings of presumed PNs in man (Nicolas et al., 2001; see also Chapter 33).

It is also important to consider feed-back inhibition. In the cat, the behavioural work on the C3-C4 propriospinal system has shown that feed-back control by inhibition of the C3-C4 PNs is decisive for accurate slowing and termination of reaching. After interruption of the feed-back inhibition of the C3-C4 PNs, by transection of the dorsal column in C5, severe hypermetria occurs in reaching despite intact vision (Alstermark et al., 1986). It may take several weeks for the brain to compensate the loss of the feed-back inhibition by using vision. Thus, it seems as if the feed-back inhibition is part of the descending command for reaching.

Interestingly, a phylogenetic increase of descending inhibition has also been demonstrated of the reciprocal Ia-inhibition between antagonistic motoneurones in the baboon (Hongo et al., 1984). These authors proposed that such an increased inhibitory control could reflect an increasing need of primates to use the Ia-system in more flexible ways and that this phylogenetic adaptation is not in the connectivity of the interneurones, but in the descending control of an existing interneuronal pathway. We cannot exclude that species differences may also exist regarding tonic inhibition of the C3-C4 propriospinal system.

If the monkey would really lack disynaptic CM connections, then one would need to postulate that the feed-forward and feed-back inhibition has moved from the spinal cord and brain stem to the cortex.

Dexterity – Behavioural Experiments

There is no doubt that the monkey hand possesses a higher degree of dexterity than the forepaw of the cat. However, the idea of the capacity of the cat in terms of manipulative skill is generally underestimated. According to the scale of dexterity (Napier, 1960; Bishop, 1964; Napier and Napier, 1967) from 1 to 7 used by Heffner and Masterton (1975; 1983), the cat has an index of 2. According to this scale a rank of 2 refers to animals with *"separate digits that do not converge when flexed"* and such paw is specialised for locomotion. A rank of 3 refers to animals with *"convergent but not prehensile digits (not capable of holding an object in one hand)* and such paws possess the most primitive form of dexterity. Specialisation for manipulation starts at rank 4, which corresponds to animals with *"prehensile digits and non-opposable thumb"*, rank 5 is defined as *"prehensile digits, pseudo-opposable thumb"*. From the behaviour of the cat in a prehensile task (Górska and Sybirska, 1980) it is clear that the cat can grasp a morsel of food with one hand and bring it to the mouth (Alstermark et al., 1981b) and a more detailed analyses has shown that the morsel can be held by the claws or in between the footpads (Pettersson et al., 1998). Accordingly, following the scale of Heffner and Masterton (1975; 1983), the cat should have a ranking of 4. As to the degree of independent digit movements, the situation is not yet clarified in the cat although it seems to be able to move at least two digits together independently from the others (Boczek-Funcke et al., 1998; see Perfiliev et al., 1998). Thus, the issue of abilities and limitations of cats in performing dextrous digit movements should be approached with caution.

A reanalysis of the Heffner and Masterton data (1975; 1983), using modern comparative statistics (Iwaniuk et al., 1999), suggests that the conclusions about a phylogenetical development of dexterity, which is correlated with the monosynaptic CM connection are not safe because of a bias in the number of primate species. Iwaniuk et al. (1999) call for a new index of dexterity, since they noted that it is also giving too low a ranking of the rat's ability.

In the monkey, the best behavioural studies performed of the pyramidal system were made by Lawrence and Kuypers (1968a,b). Their conclusion was that *"the loss of individual finger movements was attributed to the interruption of direct cortico-motoneuronal connexions"*. However, it is important to note that they did not exclude the possibility that interruption of corticospinal input to spinal interneurones, may have contributed to their findings. Continuing the quotation, *"The initial impairment of independent movements of the distal parts of the limb might be related to the preferential distribution of the interrupted cortical fibres to* ***the interneurons and motoneurons*** *related to distal limb muscles. Recovery of these movements would then presumably be due to the readjustment of these interneurons so that they become increasingly responsive to the control excerted by brain-stem pathways"*. To our surprise, the suggestion that interruption of corticospinal projection to spinal interneurones could have contributed to the loss of individual finger movements in the monkey, has apparently been neglected over the decades.

Thus, in the monkey, it is not known to what extent interruption of cortical projections to the caudal brainstem and to spinal interneurones, may contribute to the deficits observed after a pyramidal lesion. To summarize, we can only agree with the statement made by Porter and Lemon (1993): *"It may thus be too sweeping a generalisation that CM projections are the sine qua non of independent digit movements"*.

CONCLUSIONS

We find that the replacement-hypothesis by Nakajima et al. (2000), which suggests a taking over of target reaching in addition to grasping by the monosynaptic CM, is not in accordance with the existing evidence.

Since premotoneuronal control of upper limb motoneurones is not exclusive to lower species of animals, but exists in high primates like the macaque monkey and most likely also in man, we propose that the monosynaptic CM connection has not replaced the disynaptic CM pathway, but has been "added" during evolution.

ACKNOWLEDGEMENTS

We would like to thank Professor Anders Lundberg, Drs. Y. Ohki and L-G. Pettersson for helpful comments on a previous version of this paper and Mrs. M. Seo for making the illustrations. Authors are supported by the Human Frontier Science Program.

REFERENCES

Alstermark, B., Górska, T., Johannisson, T., and Lundberg, A., 1986, Hypermetria in forelimb target-reaching after interruption of the inhibitory pathway from forelimb afferents to C3-C4 propriospinal neurones, *Neuroscience Research,* 3, 457-461.

Alstermark, B., Isa, T., Lundberg, A., Pettersson, L.G., and Tantisira, B., 1991, The effect of a low pyramidal transection following previous transection of the dorsal column in cats, *Neuroscience Research,* 11, 215-220.

Alstermark, B., Isa, T., Ohki, Y., and Saito, Y., 1999, Disynaptic pyramidal excitation in forelimb motoneurons mediated via C3-C4 propriospinal neurons in the Macaca fuscata, *Journal of Neurophysiology,* 82, 3580-3585.

Alstermark, B., Isa, T., and Tantisira, B., 1991a, Pyramidal excitation in long propriospinal neurones in the cervical segments of the cat, *Experimental Brain Research,* 84, 569-582.

Alstermark, B., Isa, T., and Tantisira, B., 1991b, Integration in descending motor pathways controlling the forelimb in the cat. 18. Morphology, axonal projection and termination of collaterals from C3-C4 propriospinal neurones in the segment of origin, *Experimental Brain Research,* 84, 561-568.

Alstermark, B., and Kümmel, H., 1990, Transneuronal transport of wheat germ agglutinin conjugated horseradish peroxidase into last order spinal interneurones projecting to acromio- and spinodeltoideus motoneurones in the cat. 2. Differential labelling of interneurones depending on movement type, *Experimental Brain Research,* 80, 96-103.

Alstermark, B., Lindström, S., Lundberg, A., and Sybirska, E., 1981a, Integration in descending motor pathways controlling the forelimb in the cat. 8. Ascending projection to the lateral reticular nucleus from C3-C4 propriospinal also projecting to forelimb motoneurones, *Experimental Brain Research,* 42, 282-298.

Alstermark, B., and Lundberg, A., 1992, Target-reaching and food-taking, in: *Muscle Afferents and Spinal Control of Movement,* L. Jami, E. Pierrot-Deseilligny and D. Zytnicki, eds., Pergamon, New York, pp. 327-354

Alstermark, B., Lundberg, A., Norrsell, U., and Sybirska, E., 1981b, Integration in descending motor pathways controlling the forelimb in the cat. 9. Differential behavioural defects after spinal cord lesions interrupting defined pathways from higher centres to motoneurones, *Experimental Brain Research,* 42, 299-318.

Alstermark, B., Lundberg, A., Pettersson, L.G., Tantisira, B., and Walkowska, M., 1987, Motor recovery after serial spinal cord lesions of defined descending pathways in cats, *Neuroscience Research,* 5, 68-73.

Alstermark, B., Lundberg, A., and Sasaki, S., 1984a, Integration in descending motor pathways controlling the forelimb in the cat. 10. Inhibitory pathways to forelimb motoneurones via C3-C4 propriospinal neurones, *Experimental Brain Research,* 56, 279-292.

Alstermark, B., Lundberg, A., and Sasaki, S., 1984b, Integration in descending motor pathways controlling the forelimb in the cat. 11. Inhibitory pathways from higher motor centres and forelimb afferents to C3-C4 propriospinal neurones, *Experimental Brain Research*, 56, 293-307.

Alstermark, B., Lundberg, A., and Sasaki, S., 1984c, Integration in descending motor pathways controlling the forelimb in the cat. 12. Interneurones which may mediate descending feed-forward inhibition and feed-back inhibition from the forelimb to C3-C4 propriospinal neurones, *Experimental Brain Research,* 56, 308-322.

Alstermark, B., and Ohlson, S., 2000a, Origin of corticospinal neurones evoking disynaptic excitation in forelimb motoneurones mediated via C3-C4 propriospinal neurones in the cat, *Neuroscience Research,* 37, 91-100.

Alstermark, B., and Ohlson, S., (2000b) Origin of corticospinal neurones evoking monosynaptic excitation in C3-C4 propriospinal neurones in the cat, *Neuroscience Research,* 38, 249-256.

Alstermark, B., Pinter, M.J., and Sasaki, S., 1985, Pyramidal effects in dorsal neck motoneurones of the cat, *Journal of Physiology,* 363, 287-302.

Alstermark, B., Pinter, M.J., and Sasaki, S., 1992, Tectal and tegmental excitation in dorsal neck motoneurones of the cat, *Journal of Physiology,* 454, 517-532.

Alstermark, B., and Sasaki, S., 1985, Integration in descending motor pathways controlling the forelimb in the cat. 13. Corticospinal effects in shoulder, elbow, wrist, and digit motoneurones, *Experimental Brain Research,* 59,353-364.

Alstermark, B., and Sasaki, S., 1986a, Integration in descending motor pathways controlling the forelimb in the cat. 14. Differential projection to fast and slow motoneurones from excitatory C3-C4 propriospinal neurones, *Experimental Brain Research*, 63, 530-542.

Anderson, M.E., Yoshida, M., and Wilson, V.J., 1972, Tectal and tegmental influences on cat forelimb and hindlimb motoneurons, *Journal of Neurophysiology,* 35, 462-470.

Armand, J., 1982, The origin, course and terminations of corticospinal fibers in various mammals, *Progress in Brain Research,* 57,329-360.

Armand, J., Olivier, E., Edgley, S.A., and Lemon, R.N., 1997, Postnatal development of corticospinal projections from motor cortex to the cervical enlargement in the macaque monkey. *Journal of Neuroscience,* 17, 251-266.

Biber, M.P., Kneisley, L.W., and LaVail, J.H., 1978, Cortical neurons projecting to the cervical and lumbar enlargements of the spinal cord in young and adult rhesus monkeys, *Experimental Neurology,* 59, 492-508.

Bishop, A., 1964, *Use of the hand in lower primates;* in Buettner-Janusch, *Evolutionary and genetic biology of primates,* Academic Press, New York, pp. 133-225.

Boczek-Funcke, A., Kuhtz-Buschbeck, J.P., Raethjen, J., Paschmeyer, B., and Illert, M., 1998, Shaping of the cat paw for food taking and object manipulation: an X-ray analysis, *European Journal of Neuroscience,* 10, 3885-3897.

Bortoff, G.A., and Strick, P.L., 1993, Corticospinal terminations in two new-world primates: further evidence that corticomotoneuronal connections provide part of the neural substrate for manual dexterity, *Journal of Neuroscience,* 13, 5105-5118.

Catsman-Berrevoets, C.E., and Kyupers, H.G.J.M., 1976, Cells of origin of cortical projections to dorsal column nuclei, spinal cord and bulbar medial reticular formation in the rhesus monkey, *Neuroscience Letters,* 3, 245-252.

Chakrabarty, S., and Martin, J.H., 2000, Postnatal development of the motor representation in primary motor cortex, *Journal of Neurophysiology,* 84,2582-2594.

Clough, J.F.M., Kernell, D., and Phillips, C.G., 1968, The distribution of monsynaptic excitation from the pyramidal tract and from primary spindle afferents to motoneurones of the baboon's hand and forearm, *Journal of Physiology,* 198, 145-166.

Costello, M.B., and Fragraszy, D.M., 1988, Prehension in cebus and saimiri: 1.grip type and hand preference. *American Journal of Primatology,* 15, 235-245.

Coulter, J.D., Ewing, L., and Carter, C., 1976, Origin of primary sensorimotor cortical projections to lumbar spinal cord of cat and monkey, *Brain Research,* 103,366-372.

Dum, R.P., and Strick, P.L., 1991, The origin of corticospinal projections from the premotor areas in the frontal lobe, *Journal of Neuroscience,* 11(3), 667-689.

Dum, R.P., and Strick, P.L., 1996, Spinal cord terminations of the medial wall motor areas in macaque monkeys, *Journal of Neuroscience,* 16(20), 6513-6525.

Eyre, J.A., Miller, S., Clowry, G.J., Conway, E.A., and Watts, C., 2000, Functional corticospinal projections are established prenatally in the human foetus permitting involvement in the development of spinal motor centres, *Brain,* 123, 51-64.

Foster, M.A., 1879, *A textbook of physiology,* London. Cited by E.G. T. Liddell: *The discovery of reflexes,* Clarendon, Oxford, 1960, pp. 101.

Fritz, N., Illert, M., Kolb, F.P., Lemon, R.N., Muir, R.B., van der Burg, J., Wiedemann, E., and Yamaguchi, T., 1985, The cortico-motoneuronal input to hand and forearm motoneurones in the anaesthetized monkey, *Journal of Physiology,* 366, P20.

Górska, T., and Sybirska, E., 1980, Effects of pyramidal lesions on forelimb movements in the cat, *Acta Neurobiologiae Experimentalis,* 40, 843-859.

He, S.Q., Dum, R.P., Strick, P.L., 1995, Topographic organization of corticospinal projections from the frontal lobe: motor areas on the medial surface of the hemisphere, *Journal of Neuroscience,* 15, 3284-3306.

Heffner, R., and Masterton, B., 1975, Variation in form of the pyramidal tract and its relationship to digital dexterity, *Brain, Behavior and Evolution,* 12, 161-200

Heffner RS, Masterton RB (1983) The role of the corticospinal tract in the evolution of human digital dexterity, *Brain, Behavior and Evolution,* 23, 165-183.

Hongo, T., Lundberg, A., Phillips, C.G., and Thompson, R.F., 1984, The pattern of monosynaptic Ia-connections to hindlimb motor nuclei in the baboon: a comparison with the cat, *Proceedings of the Royal Society of London, Series B: Biological Sciences,* 21(1224), 261-289.

Illert, M., Lundberg, A., 1978, Collateral connections to the lateral reticular nucleus from cervical propriospinal neurones projecting to forelimb motoneurones in the cat, *Neuroscience Letters,* 7, 167-172.

Illert, M., Lundberg, A., and Tanaka, R., 1977, Integration in descending motor pathways controlling the forelimb in the cat. 3. Convergence on propriospinal neurones transmitting disynaptic excitation from the corticospinal tract and other descending tracts, *Experimental Brain Research,* 29, 323-346.

Illert, M., Lundberg, A., Padel, Y., and Tanaka, R., 1978, Integration in descending motor pathways controlling the forelimb in the cat. 5. Properties of and monosynaptic excitatory convergence on C3-C4 propriospinal neurones, *Experimental Brain Research,* 33,101-130.

Illert, M., Wiedemann, E., 1984, Pyramidal actions in identified radial motornuclei of the cat, *Pflügers Archives,* 401, 132-142.

Isa, T., Naito, K., 1995, Activity of neurons in the medial pontomedullary reticular formation during orienting movements in alert head-free cats, *Journal of Neurophysiology*, 74, 73-95.

Isa, T., and Sasaki, S., 1988, Effects of lesion of paramedian pontomedullary reticular formation by kainic acid injection on the visually triggered horizontal orienting movements in the cat, *Neuroscience Letters*, 87, 233-239.

Isa, T., and Sasaki, S., 1991, Differential cortical control of pontine and medullary reticulospinal neurons in the cat, *3rd IBRO Congress (Montréal)*, P63.18.

Iwamoto, Y., and Sasaki, S., 1990, Monosynaptic excitatory connexions of reticulospinal neurones in the nucleus reticularis pontis caudalis with dorsal neck motoneurones in the cat, *Experimental Brain Research*, 80, 277-289.

Iwamoto, Y., Sasaki, S., and Suzuki, I., 1990, Input-output organization of reticulospinal neurones, with special reference to connexions with dorsal neck motoneurones in the cat, *Experimental Brain Research*, 80, 260-276.

Iwaniuk, A.N., Pellis, S.M., and Whishaw, I.Q., 1999, Is digital dexterity really related to corticospinal projections?: a re-analysis of the Heffner and Masterton data set using modern comparative statistics, *Behavioural Brain Research*, 101, 173-187.

Iwatsubo, T., Kuzuhara, S., Kanemitsu, A., Shimada, H., and Toyokura, Y., 1990, Corticofugal projections to the motor nuclei of the brainstem and spinal cord in humans, *Neurology*, 40, 309-312.

Jankowska, E., Padel, Y., and Tanaka, R., 1976a, Projections of pyramidal tract cells to alpha-motoneurones innervating hind-limb muscles in the monkey, *Journal of Physiology*, 249, 637-667.

Jankowska, E., Padel, Y., and Tanaka, R., 1976b, Disynaptic inhibition of spinal motoneurones from the motor cortex in the monkey, *Journal of Physiology*, 258, 467-487.

Jones, E.G., and Wise, S.P., 1977, Size, laminar and columnar distribution of efferent cells in the sensory-motor cortex of monkeys, *Journal of Comparative Neurology*, 175, 391-438.

Kitazawa, S., Ohki, Y., Sasaki, M., Xi, M., and Hongo, T., 1993, Candidate premotor neurones of skin reflex pathways to T1 forelimb motoneurones of the cat, *Experimental Brain Research*, 95, 291-307.

Kuypers, H.G.J.M., 1981, Anatomy of the descending pathways, in: *Handbook of Physiology, Section 1, Vol. II*, V. B. Brooks, ed., American Physiological Society, Chicago, pp. 597-666.

Landgren,S., Olsson, K.A., and Westberg, K.G., 1986, Bulbar neurones with axonal projections to the trigeminal motor nucleus in the cat, *Experimental Brain Research*, 65, 98-111.

Landgren, S., Phillips, C.G., and Porter, R., 1962, Minimal synaptic actions of pyramidal impulses on some alpha motoneurons of the baboon's hand and forearm, *Journal of Physiology*, 161, 91-111.

Lawrence, D.G., and Hopkins, D.A., 1976, The development of motor control in the rhesus monkey: evidence concerning the role of corticomotoneuronal connections, *Brain* 99, 235-254.

Lawrence, D.G., and Kuypers, H.G., 1968a, The functional organization of the motor system in the monkey. I. The effects of bilateral pyramidal lesions, *Brain*, 91, 1-14.

Lawrence, D.G., and Kuypers, H.G., 1968b, The functional organization of the motor system in the monkey. II. The effects of lesions of the descending brain-stem pathways, *Brain*, 91, 15-36.

Lemon, R.N., Werner, W., Bennett, K.M.B., and Flament, 1993, The proportion of slow and fast pyramidal tract neurones producing post-spike facilitation of hand muscles in the conscious monkey, *Journal of Physiology*, 459, 166P.

Li ,Q., and Martin, J.H., 2000, Postnatal development of differential projections from the caudal and rostral motor cortex subregions, *Experimental Brain Research*, 134, 187-198.

Li, Q., and Martin, J.H., 2001, Postnatal development of corticospinal axon terminal morphology in the cat, *Journal of Comparative Neurology*, 435, 127-141.

Maier, M.A., Davis, J.N., Armand, J., Kirkwood, P.A., and Lemon, R.N., 1997, Anatomical and electrophysiological comparison of cortico-motoneuronal projections from primary motor cortex and supplementary motor area in the macaque monkey, *Journal of Physiology*, 505(P), 84-85.

Maier, M.A., Illert, M., Kirkwood, P.A., Nielsen, J., and Lemon, R.N., 1998, Does a C3-C4 propriospinal system transmit corticospinal excitation in the primate? An investigation in the macaque monkey, *Journal of Physiology*, 511, 191-212.

Martino, A.M., and Strick, P.L., 1987, Corticospinal projections originate from the arcuate premotor area, *Brain Research*, 404, 307-312.

Nakajima, K., Maier, M.A., Kirkwood, P.A., and Lemon, R.N., 2000, Striking differences in transmission of corticospinal excitation to upper limb motoneurons in two primate species, *Journal of Neurophysiology*, 84, 698-709.

Napier, J.R., 1960, Studies of the hand of living primates, *Proceedings of the Zoological Society of London*, 134, 647-657.

Napier, J.R., and Napier, P.H., 1967, *A Handbook of Living Primates*, Academic Press, New York.

Nicolas, G., Marchand-Pauvert, V., Burke, D., and Pierrot-Deseilligny, E., 2001, Corticospinal excitation of presumed cervical propriospinal neurones and its reversal to inhibition in humans, *Journal of Physiology*, 533, 903-919.

Olsson, K.A., and Landgren, S., 1980, Facilitation and inhibition of jaw reflexes evoked by electrical stimulation of the cat's cerebral cortex, *Experimental Brain Research*, 39, 149-164.

Olsson, K.A., Landgren, S., and Westberg, K.G., 1986, Location of, and peripheral convergence on, the interneuron in the disynaptic path from the coronal gyrus of the cerebral cortex to the trigeminal motoneurons in the cat, *Experimental Brain Research*, 65, 83-97.

Perfiliev, S., Pettersson, L.G., Lundberg, A., 1998, Control of claw movements in cats, *Neuroscience Research*, 31, 337-342.

Peterson, B.W., Pitts, N.G., and Fukushima, K., 1979, Reticulospinal connections with limb and axial motoneurons, *Experimental Brain Research*, 36, 1-20.

Pettersson, L.G., Blagovechtchenski, E., Perfiliev, S., Krasnochokova, E., and Lundberg, A., 2000, Recovery of food-taking in cats after lesions of the corticospinal (complete) and rubrospinal (complete and incomplete) tracts, *Neuroscience Research*, 38, 109-112.

Pettersson, L.-G., Lundberg, A., Alstermark, B., Isa, T., and Tantisira, B., 1997, Effect of spinal cord lesions on forelimb target-reaching and on visually guided switching of target-reaching in the cat, *Neuroscience Research*, 29, 241-256.

Pettersson, L.G., Perfiliev, S., Zotova, E., and Lundberg, A., 1998, Role of claws and pads in taking and holding food in cats. *Neuroscience Research*, 31,343-346.

Phillips, C.G., 1971, Central control of movement. How can one discover the relative functional roles of pyramidal inputs to alpha as compared to gamma motoneurons? *Neurosciences Research Program Bulletin*.

Phillips, C.G., and Porter, R., 1977, Corticospinal Neurones. Their Role in Movement, *Monographs of the Physiological Society*.

Porter, R., and Lemon, R., 1993, Corticospinal Function and Voluntary Movement, *Monographs of the Physiological Society*.

Robinson, F.R., Phillips, J.O., and Fuchs, A., 1994, Coordination of gaze shifts in primates: Brain stem inputs to neck and extraocular motoneuron pools, *Journal of Comparative Neurology*, 346, 43-62.

Russel, J.R., Demeyer, W., 1961, The quantitative cortical origin of pyramidal axons of Macaca rhesus, with some remarks on slow rate of axolysis, *Neurology*, 11, 96-108.

Sasaki, S., 1997, Axonal branching and termination of cervical reticulospinal neurons in the cat brachial segments, *Neuroscience Letters*, 228, 83-86.

Shapovalov, A.I., 1975, Neuronal organization and synaptic mechanisms of supraspinal motor control in vertebrates, *Reviews of Physiology, Biochemistry and Pharmacology*, 72, 1-54.

Sherrington, C.S., 1933, The *Brain and its Mechanism*, Cambridge University Press.

Tamai, Y., Fujii, T., Nakai, M., Komai, N., and Tsujimoto, T., 1983, Eye movements evoked by electrical stimulation of the frontal cortex in the coronary sulcus of the cat, *Japanese Journal of Physiology*, 33, 305-308.

Tantisira, B., Alstermark, B., Isa, T., Kümmel, H., and Pinter, M., 1996, Motoneuronal projection pattern of single C3-C4 propriospinal neurones, *Canadian Journal of Physiology and Pharmacology*, 74, 518-530.

Tomasch, J., 1969, The numerical capacity of the pontine cell and fibre systems, *Journal of Anatomy*, 104, 187.

Wiesendanger, M., 1981, The pyramidal tract. Its structure and function, in: *Handbook of Physiology - The Nervous System II*, A. L. Tomer and E. S., eds., Luschei, Plenum Press, New York, pp. 401-491.

Yamagata, Y., Matsuno, K., Taoka, N., and Shimo-oku, M., 1988, Common interneurones mediating cortical and tectal excitation of abducens motoneurones in the cat, *Experimental Brain Research*, 71, 279-290.

Yasui, Y., Itoh, K., Mitani, A., Takada, M., Mizuno, N., 1985, Cerebral cortical projections to the reticular regions around the trigeminal motor nucleus in the cat, *Journal of Comparative Neurology*, 241, 348-356.

INTERSPECIES COMPARISONS FOR THE C3-C4 PROPRIOSPINAL SYSTEM: UNRESOLVED ISSUES

Peter A. Kirkwood[1], M A. Maier[2] and Roger N. Lemon[1]

ABSTRACT

There are conflicting views on the functional importance of the system of C3-C4 propriospinal neurones in the macaque, although the two sets of observations from the opposing laboratories are actually quite similar, both making the system appear much weaker than its well-known equivalent in the cat One group asserts, mainly via evidence derived from experiments with strychnine, that this is a consequence simply of inhibition of the propriospinal neurone. However, we ague here that this judgement is premature and that much more needs to be known about the neurones involved and their connectivity before the analogy with the system in the cat is safe. This is particularly important because of a similar analogy which has been made with respect to measurements in human subjects.

INTRODUCTION

There are thousands of interneurones in the mammalian spinal cord, and we have very little idea about what most of them do. Moreover, the majority of those for which a function has been identified have been classified primarily by their peripheral afferent inputs and/or their roles in spinal reflexes. An outstanding example for which rather more is known is the system of C3-C4 propriospinal neurones (PNs) in the cat. For these interneurones, central inputs form part of their definition (convergence from corticospinal and other descending tracts) and an important functional role for them in normal movements has been identified (Alstermark and Lundberg, 1992). They additionally receive convergent excitation from low threshold muscle and cutaneous afferents of the forelimb.

[1] Sobell Department for Motor Neuroscience and Movement Disorders, Institute of Neurology, University College London, Queen Square, London WC1N 3BG, UK. Email: pkirkwoo@ion.ucl.ac.uk
[2] INSERM U483, Université Pierre et Marie Curie, Paris, France.

This system was therefore the obvious model for Pierrot-Desilligny and colleagues to choose when they needed to interpret their measurements of non-monosynaptic excitation in cortical pathways to upper limb motoneurones in man, pathways where convergence from low threshold peripheral afferents could also be demonstrated (Pierrot-Desilligny, 1996). However, such an interpretation, as an analogy, is fraught with danger. We know that even in the cat large numbers of other interneurones are present, within the same segment as the motoneurones, as are PNs within segments other than C3 or C4 (Alstermark and Kümmel, 1990a). We have very little idea which of these neurones are active in the awake animal or human subject, except that many undoubtedly are, and many of these show firing related to motor tasks (Perlmutter et al., 1998, Prut and Fetz, 1999). Thus, the assumption that in man, the one clear system that has been demonstrated in an anaesthetised cat is the one responsible for a given effect needs some supporting evidence. One such piece of evidence specifically in support of a human C3-C4 PN system has been cited, that the central delay of peripheral facilitation of the corticospinal excitation is greater the more caudal the location of the motor nucleus concerned (Pauvert et al., 1998; Nicolas et al., 2001). However that evidence is indirect and the explanation that the authors chose, that of a rostral location for the interneurones involved was only distinguished from the alternative (more synapses on the pathway to the caudal motoneurones) by it being "more parsimonious" (Nicolas et al., 2001).

It was with the expectation that we would be able to provide more supporting evidence for these ideas that, a few years ago, we and our colleagues set out to repeat in the macaque the original experiments which demonstrated the C3-C4 PN system in the cat. These experiments consisted of intracellular recordings from upper limb motoneurones with stimulation of the contralateral pyramidal tract (PT) in the medullary pyramid. To avoid complications from segmental pathways (direct excitation or disynaptic inhibition), the important measurements were done with an ipsilateral dorsolateral funicular lesion at C5, which interrupts the lateral corticospinal tract and prevents a large proportion of corticospinal excitation below this level (Fig. 1). We found, contrary to our expectations, that the corticospinal excitation transmitted by such a system was very much weaker than in the cat (Maier et al., 1998). This finding, together with a subsequent study (Nakajima et al., 2000) that showed an intermediate situation in a more primitive primate, the squirrel monkey, became controversial when its conclusions were challenged by Alstermark et al. (1999). These authors made the same basic observation of weak transmission through the C3-C4 system in the macaque, but then repeated the measurements following systemic administration of strychnine. The results led them to conclude that the weakness of the excitation in the macaque was due only to stronger inhibition of the PNs than in the cat. At the meeting in Cairns, Australia (2001) that led to this book, this topic provided a lively debate. In the course of discussion, it became clear that some of our conclusions, in particular their limited nature, were not fully understood by a number of those present. This chapter is intended to restate these, in an appropriate functional context (also see Lemon et al., 2002), and to attempt to provide pointers to future experiments that might resolve the differences between our two laboratories. In fact, at this meeting Alstermark presented preliminary data from some such (current) experiments. Perhaps the results from these may be published by the time this book appears. If so, then this chapter could also help in the interpretation of such data.

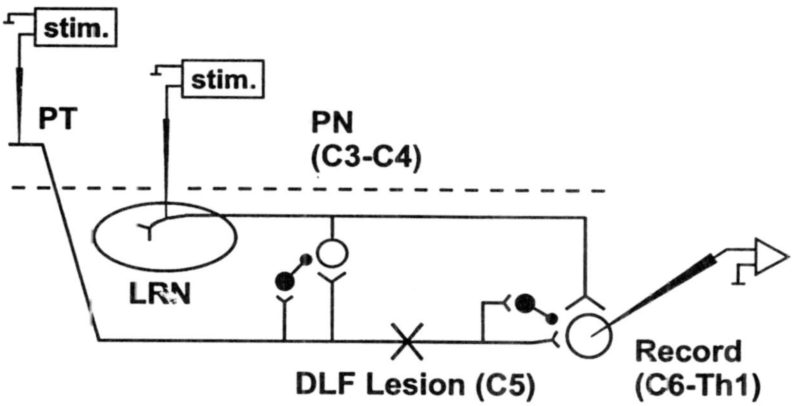

Figure 1. Experimental arrangement and basic circuit discussed in text. Large open symbol, an upper limb motoneurone, in which intracellular recording is indicated; small open symbol, presumed C3-C4 PN. Filled symbols represent inhibitory neurones. PT, pyramidal tract; DLF, dorsolateral funiculus; LRN, lateral reticular nucleus. The inhibitory neurone shown in C3-C4 represents either a feedback interneurone, which receives peripheral afferent inputs, or a feedforward interneurone, which does not. The DLF lesion removes most of the segmental corticospinal inputs to the motoneurones (disynaptic inhibition and, in the primate, monosynaptic excitation).

INTERPRETATION OF QUANTITATIVE INTERSPECIES COMPARISONS

Disynaptic EPSPs from the PT

One important question is which features of the systems being studied can or should be compared between species? Features such as the amplitude (perhaps also the occurrence) of di- or oligo-synaptic EPSPs were felt by some to be unwise or inappropriate. The dangers are obvious and admitted (Maier et al., 1998): in addition to the basic difficulty of interspecies comparison, the macaque data, being necessarily derived from far fewer experiments, do not allow for motoneurone size or type to be taken into account, as has been possible in the cat (Alstermark and Sasaki, 1986a,b). However, the difference in amplitude between the species is considerable, around a factor of 4 for the disynaptic EPSPs from PT stimulation. We maintain that this is in itself worthy of note, but it was also coupled with a difference in occurrence of at least a factor of 5 (depending on which EPSPs were accepted as disynaptic), making an overall difference of a factor of around 20. The difference is also made more significant if the suggestion of Nakajima et al. (2000) can be accepted, namely that because the squirrel monkey (a more primitive primate) gave responses intermediate between the macaque and the cat, an extrapolation to the human would give an even weaker linkage in that species. The question of motoneurone size is of course important, but, as already pointed out (Maier et al., 1998), even if all the macaque recordings were from large motoneurones, the EPSP amplitudes would still be considerably lower than for the equivalent group in the cat.

More generally, however, it is worth restating the logic underlying our most basic conclusion. The establishment of the C3-C4 PN system in the cat has depended on a very long series of experiments of several kinds, each with a number of important controls (18 papers in the main series, "Integration in descending motor pathways controlling the forelimb in the cat", in Experimental Brain Research from 1976 to 1991, and several more since). It is highly unlikely that this number of experiments will be possible in the macaque. Thus, the argument that the same system is present in the macaque as in the cat is itself one of analogy and depends on showing measurements that are really as much like the data from the cat as possible in a few key experiments. If there is any doubt about the similarity that is demonstrated, then new controls are needed for the macaque experiments, and we are back in the situation of large numbers of experiments being needed in the macaque to provide independent proof. If cat data is to be of real use in interpreting the human experiments, then the analogy between macaque and the cat must be a sound one, with most of the features of the PN system being the same in these two lower species. It was because of doubts that this was so that the most basic conclusion of Maier et al. (1998) (repeated in Nakajima et al., 2000) was stated simply as "*deductions based on data from the cat must be interpreted with great caution*".

Monosynaptic EPSPs from the LRN

It is also worth emphasising that previous publications from this laboratory have not claimed that C3-C4 PNs are absent in the macaque, nor necessarily that there are fewer of them than in the cat. Indeed, Maier et al. (1998) went to some trouble to provide evidence that such a population *was* present. This evidence came from experiments using stimulation in the lateral reticular nucleus (LRN) (Fig. 1). In the cat 84% of C3-C4 PNs projecting to lower cervical levels (but not to below T9) also have an ascending collateral to the LRN (Alstermark et al., 1981), used in some of the cat studies as one of their identifying features (e.g. Alstermark et al., 1990). Stimulation in this region, particularly at the dorsomedial border where the collaterals concerned enter the nucleus (Alstermark et al., 1981), activates antidromically most of the C3-C4 PN axons and therefore gives a monosynaptic EPSP in motoneurones with an amplitude indicative of the size of the overall PN-motoneurone connection. Generally, in the cat this EPSP is larger than the disynaptic EPSP from PT stimulation (measured from the response to the third of three stimuli at 3ms intervals), but not much larger, the difference being ascribed by Alstermark and Sasaki (1986*a*) to the degree of inhibition (feedforward or feedback) pertaining at the PN. In the macaque, stimulation of a directly homologous region gave rise to an extracellular field potential within upper limb motor nuclei and to monosynaptic EPSPs in upper limb motoneurones, both with properties similar to those in the cat (Maier et al., 1998; Nakajima et al., 2000). Also, as in the cat, the amplitudes of the EPSPs (mean 1.2 mV) were generally slightly larger than the corresponding disynaptic EPSPs from PT stimulation (when these latter were present). These EPSPs, like the disynaptic EPSPs from PT stimulation, were thus also about 4 times smaller than the typical EPSPs from the LRN in the cat. On the assumption that, as in the cat, the EPSP from the LRN can reasonably represent the strength of the PN connections to motoneurones (see below), this evidence allowed us to ascribe all the difference in amplitude of the disynaptic PT EPSP between cat and macaque, to a weaker connection at the PN-motoneurone synapse. There need be no difference in the excitation of the C3-C4 PNs from the PT.

Inhibition at the C3-C4 PNs

In fact the conclusions of Maier et al. (1998) and Nakajima et al. (2000) were based on both amplitude and occurrence of the disynaptic EPSPs. The issue of occurrence, but not amplitude was taken up by Alstermark et al. (1999), who showed that following systemic strychnine administration, the frequency of occurrence in the macaque rose to at least 95%, implying that this factor was strongly controlled by inhibition, presumably at the C3-C4 PN. The authors go on to assert that this inhibitory control was stronger than in the cat, i.e. that there is a difference in the excitation of the PNs. However, they provided no evidence for this. If we assume the main difference between cat and macaque is that individual PNs contact fewer motoneurones in the macaque, then both the smaller amplitudes and the lower occurrence would be expected, again without needing fewer PNs to be excited by the PT stimulus. We cannot deny that fewer PNs are excited in the macaque than in comparable experiments in the cat, but we believe the observations to date do not provide evidence either way. This issue is of great interest, with respect to how the cortex controls spinal cord circuits in general, so it must be addressed in the future. It cannot really be resolved from observations simply of the disynaptic EPSPs but needs a much fuller approach, including detailed studies both of the inhibition of the interneurones and characterisation of those interneurones, in terms of their targets, projection frequencies etc.

Thus the observations of Alstermark et al. (1999) are not in conflict with those of Maier et al. (1998) and Nakajima et al. (2000). Indeed in one aspect they further support the conclusions from the Lemon laboratory, in that the *disynaptic* components of the published examples of the EPSPs evoked from the PT after strychnine were all of the order of 1mV in amplitude (Alstermark et al., 1999). This low amplitude is just what would be predicted from the sizes of the LRN evoked EPSPs in Maier et al. (1998) and is entirely consistent with the above suggestion of low connectivity between PNs and motoneurones. However, we are aware that this argument cannot be taken too far. It is uncertain how complete the blocking of inhibition by strychnine in these experiments was, both because the effect was transient (no steady state was demonstrated) and because a non-glycinergic component of inhibition might be present. Nevertheless, there is as yet no positive evidence for *large* disynaptic EPSPs from the PT in the macaque.

Recent measurements in man (Nicolas et al., 2001) were also interpreted as supporting the case made by Alstermark et al. (1999) for strong inhibition of the C3-C4 PNs. The critical feature of their data was a facilitatory effect from peripheral afferents on the non-monosynaptic excitation of upper limb motoneurones by transcranial magnetic stimulation (TMS), which was only evident at the lowest intensities of stimulation, but which was replaced by inhibition at higher intensities of TMS. The authors suggested that this represented convergence of cortical and peripheral excitation at the PNs, which was replaced by inhibition when the strength of the cortical excitation was sufficiently increased to recruit feedback inhibitory interneurones and thus to suppress the PN discharge. This interpretation fits well with the known circuits in the cat (Alstermark et al., 1984a,b), but there is no direct evidence for them to actually operate in this way. Indeed, in a small sample of disynaptic EPSPs evoked by repetitive PT stimulation in the macaque after a C5 lesion for which an increasing series of intensities

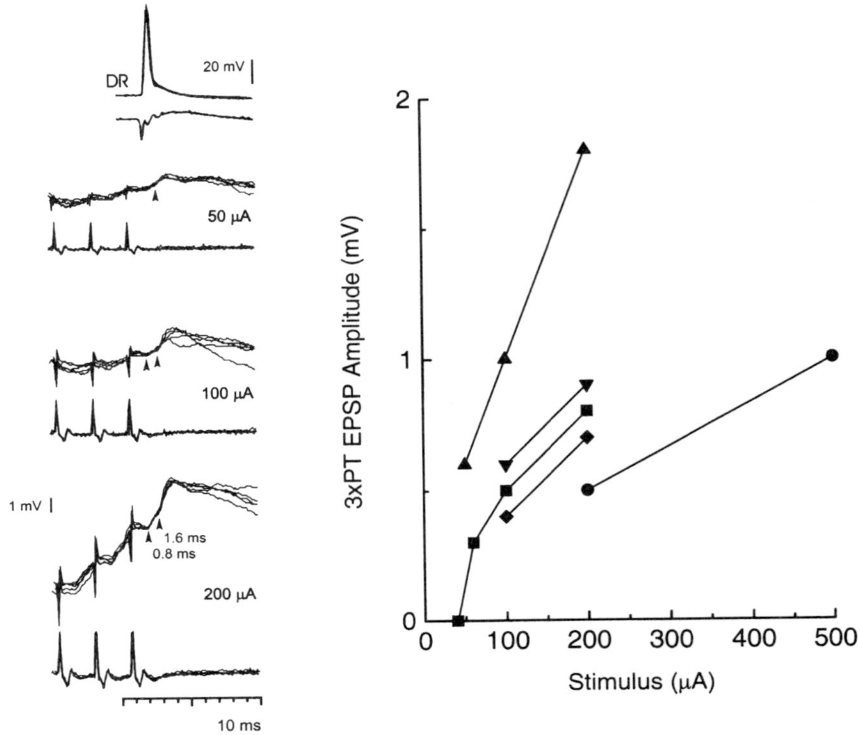

Figure 2. Variation in disynaptic EPSP with stimulus strength following a C5 lesion in the macaque. Left, from top: antidromic identification of motoneurone (deep radial); EPSPs in this motoneurone evoked by three stimuli to the contralateral pyramid at three different stimulus strengths, as indicated. There is a small monosynaptic EPSP remaining after the lesion (arrowhead at segmental latency of 0.8 ms) and a disynaptic one at 1.6 ms (second arrowhead). Lower trace for each panel is a cord dorsum recording; note the very small PT volley (negative upwards) following the stimulus artifact in each of the lower panels. Right, amplitude vs. stimulus strength for the disynaptic components in five different macaque upper limb motoneurones (different symbols for each). Note that in all examples monotonic increases were observed. (Unpublished data from Maier et al., 1998.)

was used, no indications were found of inhibition appearing at some critical intensity. All EPSPs showed a monotonic increase in amplitude with stimulus strength (Fig. 2).

The Way Ahead

The deciding factor that might allow a choice between the different interpretations made by the two groups is the effect of stimulation in the LRN. If the EPSPs presently reported are reliable as an indication of the size of the overall PN-motoneurone connection, then the interpretation of the Lemon group would be strongly supported over that of Alstermark and colleagues, so this is worth considering in some detail.

Firstly, it is an assumption that the C3-C4 PNs in the macaque should have the collateral to the LRN. If it turns out that in the macaque relatively few of the C3-C4 PNs that project to the upper limb motor nuclei have this collateral, then the measurements of the EPSP from stimulating the region of the LRN could not be used to indicate the strength of the C3-C4 PN projection to motoneurones. However, absence of this collateral should not be regarded as a trivial difference between macaque and cat. We presume that this collateral gives these PNs an important role in signalling efference copy to the cerebellum (Alstermark and Lundberg, 1992). A set of C3-C4 PNs without this should not therefore necessarily be regarded as sufficiently equivalent to those in the cat for it to be safe to base interpretations of the oligosynaptic pathway in man on particular properties of this disynaptic pathway in the cat.

Secondly, it is possible that other pathways in addition to this collateral may be stimulated at the LRN and give rise to the observed monosynaptic EPSPs in motoneurones. These pathways could be either ascending or descending, but also having a collateral to the LRN, or could be other neurones in the nearby brainstem, as discussed in Maier et al. (1998). However, our argument that the EPSPs from the LRN are rather small is mostly unaffected by this argument; it is a more of a danger for those who wish to say that the EPSPs we demonstrated should really be larger. We were aware of some of these dangers when we chose a stimulation strength of 200 μA as our standard, as opposed to the 500 μA used by Alstermark and Sasaki (1986a,b). We found that 500 μA quite often evoked a monosynaptic IPSP in addition to the EPSP, this IPSP apparently being evoked by elements located rather more dorsomedially. A conservative approach of using only 200 μA was thus preferable, despite the modest underestimate this was likely to give of the LRN-evoked EPSP amplitude (see Nakajima et al., 2000).

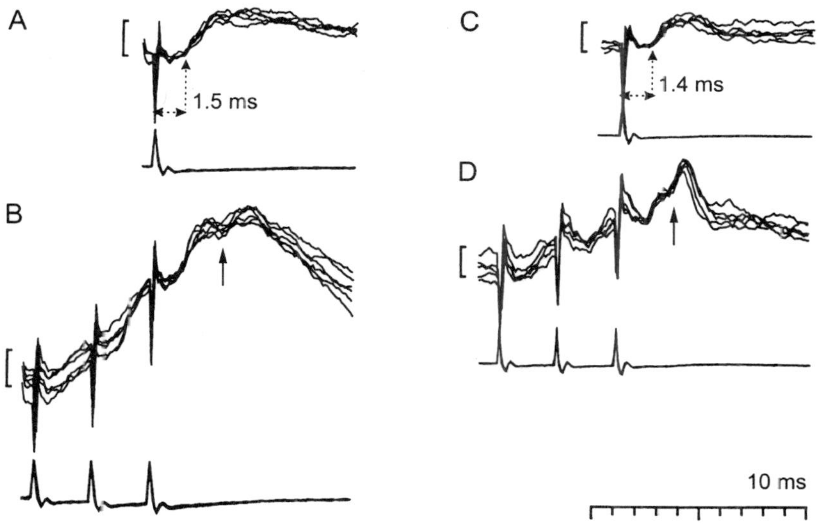

Figure 3. Responses recorded in two different macaque motoneurones (A,B, biceps; C,D, median) to single (A,C) and triple (B,D) stimuli (200 μA) to the LRN. In each motoneurone the single stimulus evoked mainly a monosynaptic EPSP, as evidenced by the latency (indicated), whereas a triple stimulus revealed in addition a later excitatory component (arrowed) but little sign of disynaptic inhibition. Voltage calibrations, 1 mV. (Data replotted from Maier et al., 1998.)

Finally, Alstermark and Sasaki (1986a) pointed out that in order for the measurement of an LRN EPSP amplitude to be valid, examples involving a disynaptic IPSP should be avoided, such an IPSP arising at lower cervical levels by excitation from PN axons of both motoneurones and inhibitory interneurones (most notably Ia inhibitory neurones) (Illert and Tanaka, 1978; Alstermark et al., 1981). The possible occurrence of an IPSP can be investigated by comparing the EPSP at depolarised and hyperpolarised membrane potential. Although this was checked for some of the EPSPs in Maier et al. (1998), it was not done for all. However, an alternative measurement was made for most motoneurones, which was to use both single and repetitive stimulation of the LRN. Just as three stimuli to the PT bring out disynaptic responses not evoked by a single stimulus, so this should have exaggerated any disynaptic inhibition from the LRN. In some motoneurones this was indeed observed, but the effects were modest. For tests made after a C5 lesion, a likely di- or oligosynaptic IPSP was detected after the 3rd stimulus of a train of 3 stimuli in only 19/77 motoneurones. Moreover, in some motoneurones, not only was there little sign of a disynaptic IPSP, but in its place there was a di-or trisynaptic EPSP (Fig.3). In the same tests, such potentials were seen in 11/77 motoneurones.

This observation is particularly interesting because in Alstermark et al (1999) the illustrated responses from stimulation of the PT after strychnine also contained late components, but in these circumstances the later components could be considerably larger than the relatively small disynaptic EPSP. These later components could have come via one or several unknown pathways, including, we suggest, oligosynaptic excitation from the C3-C4 PNs. Any of the interneurones on these pathways could be controlled by inhibition. This explanation fits just as well into the teleological argumant in Alstermark et al (1999) for "focused excitatory control" but it also becomes another complication if measurements of EPSPs from the LRN are made after strychnine. Alstermark showed preliminary data at the recent meeting in Cairns, in which LRN-evoked EPSPs in the macaque were considerably larger under strychnine than they had been without, and suggested that the reason was the unmasking of a large monosynaptic EPSP by the blockade of a latent disynaptic inhibitory component. The alternative explanation for these large EPSPs, that we now suggest needs investigation, is the presence of a di- or trisynaptic *excitatory* component released from tonic, feedforward or feedback inhibition.

In summary, as regards EPSPs from the LRN, we admit that there are a number of complications involved in interpreting such responses. At present the evidence is still in favour of our original interpretation, but that could be changed by further investigations. We suggest that these investigations may need to be quite extensive to establish the properties of the PN system(s) of the macaque in their own right. These systems may or may not turn out to be closely similar to that in the cat

In the end, a simple way of looking at the situation is to remember that the connections seen in an acute *in vivo* experiment are always dependent upon the particular preparation used. The classic example of this is the reversal of inhibition to excitation seen in the responses to the stimulation of flexor reflex afferents when a decerebrate animal with a mid-pontine lesion is made spinal (Holmqvist and Lundberg, 1961). In the present context we suggest that the chloralose-anaesthetised macaque with strychnine is just as much a different preparation from the chloralose-alone macaque and certainly different from the chloralose cat. This could be argued *a priori*, but the evidence for it is already present in the responses to PT stimulation illustrated in Alstermark et al (1999).

As mentioned above, either with or without a C5 DLF lesion, these responses were dominated by tri- or oligosynatic excitation over and above the relatively small disynaptic components. We suggest that, although the chloralose macaque with strychnine can be used to demonstrate the existence of C3-C4 PNs in the primate, if this preparation is to be taken as a model for parallel pathways controlling the human upper limb, then these oligosynaptic components must figure just as largely in our thinking as the disynaptic ones, including their inhibitory control. Thus not only would the original conclusion of Maier et al. (1998) "*deductions based on data from the cat must be interpreted with great caution*" be further strengthened, but also the argument used by Pauvert et al. (1998) and Nicolas et al. (2001) in order to base their interpretation on a disynaptic pathway with a rostrally located relay in preference to a multi-synaptic pathway would fail. If both pathways are present, then choosing one is not more parsimonious than choosing the other.

Our view therefore is that the human experiments should stand in their own right until much more evidence is obtained about the descending pathways in the macaque. Some further progress will be made if transmission can be tested in the macaque without use of either chloralose or strychnine (cf. Olivier et al., 2001), an approach which could be particularly useful for testing the specific motor acts in which C3-C4 PN transmission might be significant. Of course it is still possible that the disynaptic C3-C4 PN pathway in the macaque may yet be proved stronger than we presently are suggesting, and undoubtedly it is essential to understand the inhibitory controls on this and other descending pathways, but at present the jury is still out and, we suggest, probably will be for some time to come.

ACKNOWLEDGEMENTS

This work was carried out with support from the Wellcome Trust and the International Spinal Cord Trust

REFERENCES

Alstermark,B., Isa, T., Ohki, Y., and Saito., 1999, Disynaptic pyramidal excitation in forelimb motoneurons mediated via C_3-C_4 propriospinal neurons in the *Macaca fuscata, Journal of Neurophysiology,* 82, 3580-3585.

Alstermark, B., and Kümmel, H., 1990, Transneuronal transport of wheatgerm agglutinin conjugated horseradish peroxidase into last order spinal interneurones projecting to acromio- and spinodeltoideus motoneurones in the cat. 1. Location of labelled interneurones and influence of synaptic activity on the transneuronal transport, *Experimental Brain Research,* 80, 83-95.

Alstermark, B., Kümmel, H., Pinter, M.J., and Tantisira, B., 1990, Integration in descending motor pathways controlling the forelimb in the cat. 17. Axonal projection and termination of C3-C4 propriospinal neurones in the C6-Th1 segments, *Experimental Brain Research,* 81, 447-461.

Alstermark, B., Lindström, S., Lundberg, A., and Sybirska, E., 1981, Integration in descending motor pathways controlling the forelimb in the cat. 8. Ascending projection to the lateral reticular nucleus from C3-C4 propriospinal neurones also projecting to forelimb motoneurones, *Experimental Brain Research,* 42, 282-298.

Alstermark, B., and Lundberg, A., The C3-C4 propriospinal system: target-reaching and food-taking, in: *Muscle Afferents and Spinal Control of Movement,* L. Jami, E. Pierrrot-Deseilligny and D. Zytnicki, eds., Pergamon, Oxford, pp. 327-354.

Alstermark, B., Lundberg, A., and Sasaki, S., 1984a, Integration in descending motor pathways controlling the forelimb in the cat. 11. Inhibitory pathways from higher motor centres and forelimb afferents to C3-C4 propriospinal neurones, *Experimental Brain Research,* 56, 293-307.

Alstermark, B., Lundberg, A., and Sasaki, S., 1984b, Integration in descending motor pathways controlling the forelimb in the cat. 12. Interneurones which may mediate descending feed-forward inhibition and feed-back inhibition from the forelimb to C3-C4 propriospinal neurones, *Experimental Brain Research,* 56, 308-322

Alstermark, B., and Sasaki, S., 1986a, Integration in descending motor pathways controlling the forelimb in the cat. 14. Differential projection to fast and slow motoneurones from excitatory C3-C4 propriospinal neurones, *Experimental Brain Research,* 63, 530-542.

Alstermark, B., and Sasaki, S., 1986b, Integration in descending motor pathways controlling the forelimb in the cat. 15. Comparison of the projection from excitatory C3-C4 propriospinal neurones to different species of forelimb motoneurones, *Experimental Brain Research,* 63, 543-556.

Holmqvist, B. and Lundberg, A., 1961, Differential supraspinal control of synaptic actions evoked by volleys in the flexion reflex afferents in alpha motoneurones, *Acta Physiologica Scandinavica,* 54, suppl. 186.

Illert, M., and Tanaka, R., 1978, Integration in descending motor pathways controlling the forelimb in the cat. 4. Corticospinal inhibition of forelimb motoneurones mediated by short propriospinal neurones, *Experimental Brain Research,* 31, 131-141.

Lemon, R.N., Kirkwood, P.A., Maier, M.A., Nakajima, K., and Nathan, P., 2002, Direct and indirect pathways for corticospinal control of upper limb motoneurones in the primate, in, *Brain Mechanisms for the Integration of Posture and Movement,* S. Mori, D.G. Stuart, M. Wiesendanger, eds., *Progress in Brain Research,* Elsevier, Amsterdam (Submitted).

Maier, M.A., Illert. M., Kirkwood., P,A., Nielsen, J., and Lemon, R.N., 1998, Does a C3-C4 propriospinal system transmit corticospinal excitation in the primate? An investigation in the macaque monkey, *Journal of Physiology,* 511, 191-212.

Nakajima, K, Maier, M.A., Kirkwood, P.A., and Lemon, R.N., 2000, Striking differences in transmission of corticospinal excitation to upper limb motoneurons in two primate species, *Journal of Neurophysiology,* 84, 698-709.

Olivier, E., Baker, S.N., Nakajima, K., Brochier, T., and Lemon, R.N., 2001, Investigation into non-monosynaptic corticospinal excitation of macaque upper limb single motor units, *Journal of Neurophysiology,* 86, 1573-1586.

Pauvert, V., Pierrot-Desilligny, E., and Rothwell, J.C., 1998, Role of spinal premotoneurones in mediating corticospinal input to forearm motoneurones in man, *Journal of Physiology,* 508, 310-312.

Perlmutter, S.I., Maier, M.A., and Fetz, E.E., 1998, Activity of spinal interneurons and their effects on forearm muscles during voluntary wrist movements in the monkey, *Journal of Neurophysiology,* 80, 2475-2494.

Pierrot-Desilligny, E., 1996, Transmission of the cortical command for human voluntary movement through cervical premotoneurones, *Progress in Neurobiology,* 48, 489-517.

Prut, Y., and Fetz, E.E., 1999, Primate interneurons show pre-movement instructed delay activity, *Nature,* 401, 590-594.

36

CENTRAL NERVOUS SYSTEM LESIONS AND SEGMENTAL ACTIVITY

Sabine Meunier[1], Rose Katz[1] and Marion Simonetta-Moreau[2]

ABSTRACT

Study of activity in segmental pathways can help in understanding the pathophysiology of clinical disorders due to basal ganglia damage. Disynaptic Ia reciprocal inhibition (DI) acts by actively inhibiting antagonistic motoneurones and reducing the inhibition of agonist ones. During movement, activity of interneurones mediating DI is significantly modulated by descending inputs. In Parkinsonian patients, this descending modulation almost completely vanished. Lack of modulation was not dependent on L-DOPA as it occurred in treated patients and was not modified when patients were off medication. A potent heteronymous group II excitation of quadriceps MNs has been recently demonstrated in normal subjects after stimulation of the common peroneal nerve. This group II excitation was significantly enhanced in the rigid lower limb of Parkinsonian patients. We propose that enhanced group II excitation could contribute to rigidity in PD and result from a change in a tonic noradrenergic descending inhibitory control from locus coeruleus.

INTRODUCTION

Studying consequences onto segmental pathway activity of damages at supraspinal level may be of particular interest. Indeed, abnormal activity of motoneurones (MN) (which drives abnormal muscular tone or movements) is not only due to dysactivation in direct cortico-motoneuronal pathways but also (and mainly in some diseases) to dysactivation in descending pathways projecting to spinal interneurones upstream of the MNs. We take as an example the dysactivation of one inhibitory and one excitatory spinal pathway accompanying damage to the basal ganglia in Parkinson's disease (PD).

---

[1] Laboratoire de Neurophysiologie Clinique, Hôpital de la Salpêtrière, Paris, France.
Email: meunier@chups.jussieu.fr

[2] INSERM U 455, CHU Purpan, Toulouse, France.

PARKINSON'S DISEASE

Reciprocal Inhibition

Disynaptic Ia reciprocal inhibition (DI) acts, at spinal level, by actively inhibiting antagonist motor neurones and reducing the inhibition of agonist ones. A schematic diagram showing the organisation of this pathway at the level of wrist muscles in man is presented in Figure 1 A. To explore DI in humans, stimulation of the radial nerve in the spiral groove is used to activate extensor-coupled Ia interneurones and evoke a disynaptic inhibition of FCR MNs (Day et al., 1984). Amount of the DI was assessed by comparing the size of an H reflex, evoked in wrist flexor muscles by stimulating the median nerve, before and after the radial nerve stimulation. To explore the only disynaptic part of the inhibition, the delay between radial and median nerve stimulation was kept between ± 1 ms.

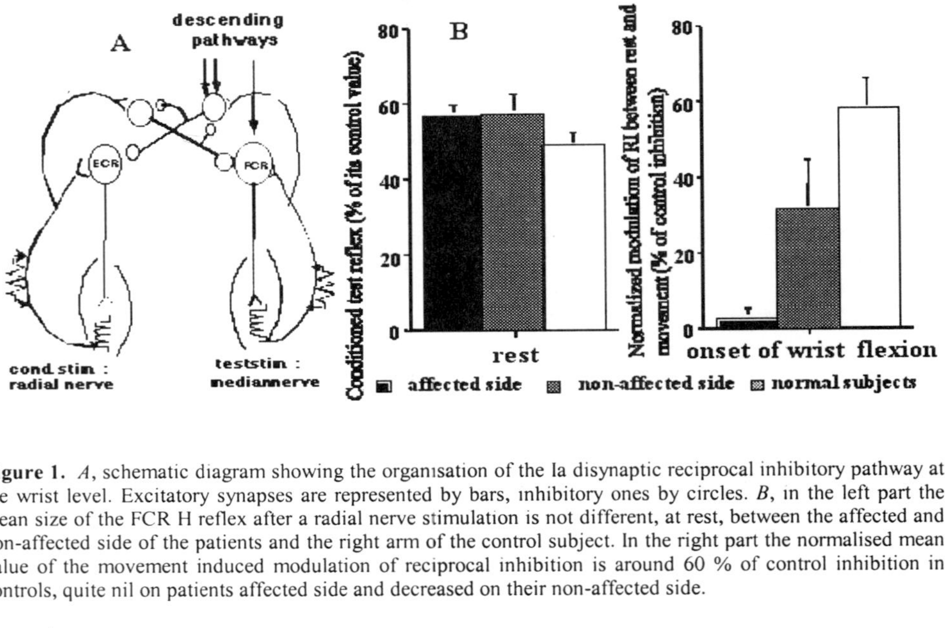

Figure 1. *A*, schematic diagram showing the organisation of the Ia disynaptic reciprocal inhibitory pathway at the wrist level. Excitatory synapses are represented by bars, inhibitory ones by circles. *B*, in the left part the mean size of the FCR H reflex after a radial nerve stimulation is not different, at rest, between the affected and non-affected side of the patients and the right arm of the control subject. In the right part the normalised mean value of the movement induced modulation of reciprocal inhibition is around 60 % of control inhibition in controls, quite nil on patients affected side and decreased on their non-affected side.

Sixteen patients with moderately severe Parkinson's disease (Hoehn and Yahr stages I-II) were compared with 15 age-matched healthy subjects. All patients were treated (L-dopa). They were clinically evaluated (UPDRS) at the time when the effect of the treatment was maximal. For each patient's arm, we calculated (from the items of the UPDRS score -part III) a subscore of akinesia, and rigidity. Parkinsonian signs were unilateral in 11 patients, asymmetrical in 4 and bilateral in one.

When measured at rest, the amount of DI from wrist extensors to flexors was not different between patients and controls (see Fig. 1B, left part).

At the very onset of a phasic wrist flexion amount of DI was largely decreased in control subjects (see Fig. 1B, right part, white column). As such a decrease was observed in the 60 ms preceding a willed wrist movement (Day et al., 1983) it has been argued that it is supraspinal in origin. It is thought that the descending command simultaneously activates the agonist α motor neurones and their "corresponding" Ia interneurones. If the descending drive is strong enough to make the agonist-coupled interneurones fire, the antagonist-coupled interneurones will be inhibited because of mutual inhibition between antagonist Ia interneurones (Fig. 1A). So, at the same time, agonist MNs are excited and antagonist ones inhibited; in addition inhibition of the antagonist coupled-interneurones prevents any later inhibition of the agonist MNs when the antagonist muscle is stretched during the movement.

On the more affected side in patients, DI did not decrease at the onset of a phasic wrist flexion (Fig. 1B, right part, black column). On patients' less or non-affected sides the decrease of DI was still present but less than in control subjects (Fig. 1B, right part, grey column).

Control experiments ruled out the possibility that such a lack of decrease of DI at the onset of movement was due to the fact that patients performed slower or weaker movements or were not fully relaxed between each movement.

Loss of an adapted supraspinal control of interneurones mediating DI in PD patients during movement leads to insufficient inhibition of antagonist MNs and to excessive inhibition of agonist ones when the antagonist muscle is stretched during movement.

No correlation was found between the subscores of akinesia or rigidity and the degree of abnormalities of DI. There was a contrast between the good clinical response to L-dopa of the patients and the severe impairment (most often a loss) of supraspinal modulation of DI on their more affected sides. Two patients were reassessed 18 hours after withdrawal of all antiparkinsonian therapy, at the period of maximal disability. Results were exactly the same as during treatment: loss of the central modulation on the most affected side. It indicates that loss of dopamine in the substantia nigra is not, per se, the relevant factor for abnormalities of DI. Abnormal modulation of DI during movement seems not to be the neurophysiological correlate of rigidity or akinesia but could participate in difficulty of patients to perform fine precise hand movements.

Group II Excitation

While it has been shown that long-latency reflexes for intrinsic hand muscles are transmitted through a transcortical long loop fed by group I afferents (Douglas et al., 1990), convincing evidence has been recently accumulated that the long-latency response evoked by stretch in ankle and foot muscles of normal subjects is a segmental reflex mediated by group II muscle afferents (Schieppati et al., 1995; Corna et al., 1995). A potent heteronymous group II excitation of quadriceps MNs has also been demonstrated in normal subjects after stimulation of the common peroneal (CP) nerve (Simonetta et al., 1999). Excitatory group II interneurones (transmitting disynaptic group II excitation to MNs) are thought to transmit some of the motor command. The brain would select subsets of interneurones with convergence of group II afferents from muscles participating in the ongoing movement.

In PD, the monosynaptic stretch reflex is not modified, but long-latency stretch reflexes are often enhanced (Berardelli et al., 1983). We investigated whether an enhanced group II excitation contributes to rigidity in PD.

Twenty "de novo" patients with PD were compared with 20 age-matched control subjects. Seventeen of the 20 patients had unilateral Parkinsonian symptoms, and 3 were affected bilaterally . For each explored lower limb subscore of akinesia and rigidity was calculated from the items of UPDRS scale (part III).

A quadriceps (Q) H reflex was elicited by stimulating the femoral nerve in the femoral triangle. A conditioning stimulation, applied to the CP nerve at an intensity of twice that for the motor threshold in pretibial flexors, induced a twofold facilitation of the Q H reflex (Fig. 2, triangles). The early and late facilitations have been respectively attributed to non-monosynaptic group I and group II excitations (Simonetta et al., 1999).

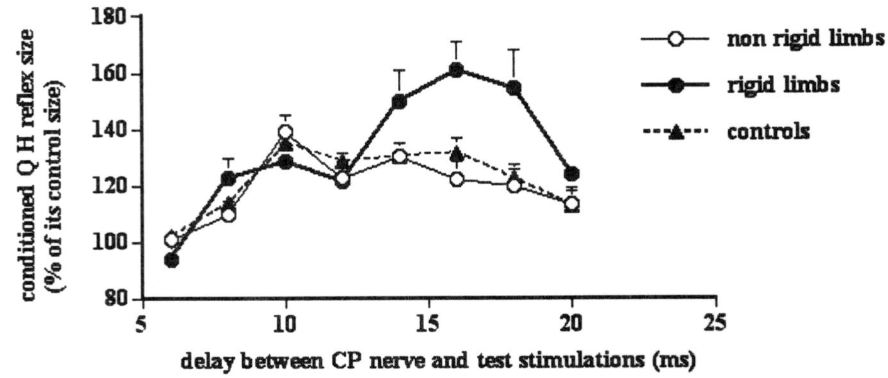

Figure 2. Mean time course of the modifications of the Q H reflex size after a CP nerve stimulation (2 MT). Two peaks are clearly seen: the first one due to group I afferent stimulation, the second later one to group II afferent stimulation. There was no difference in the size of the first peak between patients rigid or non-rigid lower limbs and controls limbs. The group II peak is enhanced exclusively for patients rigid limbs.

There was no significant effect (multivariate analysis of variance-MANOVA) either of group (patient versus control), or side (affected or non-affected side of patients, right or left side of controls), or duration of the illness, or akinesia on group I and group II excitation. There was a highly significant effect of rigidity and this effect was exclusively observed for group II induced facilitation. This result is illustrated in Figure 2. Q H reflex group I facilitation was the same in controls and patients whatever the clinical status of the patient's limbs: rigid or not rigid. By contrast, in patients rigid lower limbs group II facilitation of the Q H reflex was clearly increased (153.2 ± 7 % of control H reflex size) versus patients non-rigid limbs (124 ± 4) and controls limbs (126.1 ± 4).

Enhanced late facilitation of Q H reflex observed in rigid lower limbs of the patients is probably due to a descending facilitation (or disinhibition) of transmisssion in the group II pathways activated by CP afferents.

In the cat, interneurones fed by group II afferents and located in the lumbar spinal cord receive a descending inhibitory monoaminergic control from locus coeruleus and medullary raphe nuclei (Noga et al., 1992). We propose that disinhibition in such a noradrenergic descending pathway in humans participates in pathophysiology of

Parkinsonian rigidity indeed: (1) the long-latency stretch reflex evoked in ankle and foot muscles is selectively depressed by tizanidine an α_2 agonist; (2) it is known that locus coeruleus plays a role in the control of tone and posture (Pompeiano, 2001); (3) neuropathological studies in PD have disclosed a significant cell loss in LC and especially in its caudal part which projects to cerebellum and spinal cord (Chan-Palay and Asan, 1989); (4) a dramatic depletion of noradrenergic concentrations contrasting with normal dopamine levels has been observed in lumbar spinal cord of Parkinsonian patients (Scatton et al., 1986).

CONCLUSION

Study of activity in spinal pathways in Parkinsonian patients has put forward the role of non-dopaminergic descending pathways in this disease. In Parkinson's disease, significant abnormalities (cell loss and also abnormal neuronal activity) are described in pedunculopontine nucleus and locus coeruleus, two structures which have descending projections down to the spinal cord. Glutamatergic or cholinergic neurons of the PPN and noradrenergic neurones of the LC would be the relevant cells.

REFERENCES

Berardelli, A., Sabre, A. F., and Hallett, M., 1983, Physiological mechanisms of rigidity in Parkinson's disease, *Journal of Neurology, Neurosurgery and Psychiatry*, 46, 45-53.

Chan-Palay, V., and Asan, E., 1989, Alterations in catecholamine neurons of the locus coeruleus in senile dementia of the Alzheimer type and in Parkinson's disease with and without dementia and depression, *Journal of Comparitive Neurology*, 287, 373-392.

Corna, S., Grasso, M., Nardone, A., and Schieppati, M, 1995, Selective depression of medium-latency leg and foot muscle responses to stretch by an α2-agonist in humans, *Journal of Physiology*, 484, 803-809.

Day, B. L., Marsden, C. D., Obeso, J.A., and Rothwell, J. C., 1984, Reciprocal inhibition between the muscles of the human forearm, *Journal of Physiology*, 349, 519-534.

Day, B. L., Rothwell, J. C., and Marsden, C. D., 1983, Transmission in the spinal reciprocal Ia inhibition pathway preceding willed movements of the human wrist, *Neuroscience Letters*, 37, 245-250.

Douglas, S., Goodin, D. S., Aminoff, M. J., and Shih, P., 1990, Evidence that the long latency stretch responses of the human wrist extensor muscle involve a transcerebral pathway, *Brain*, 113, 1075-1091.

Noga, B. R. R., Bras, J., and Jankowska, E., 1992, Transmission from group II muscle afferents is depressed by stimulation of locus coeruleus/subcoerulus, Kölliker-Fuse and raphe nuclei in the cat, *Experimental Brain Research*, 88, 502-516.

Pompeiano, O., 2001, Role of the locus coeruleus in the static and dynamic control of posture, *Archives Italiennes de Biologie*, 139, 109-124.

Scatton, B., Dennis, T., L'Heureux, R., Montfort, J. C., Duyckaerts, V., and Javoy-Agid, F., 1986, Degeneration of noradrenergic and serotonergic but not dopaminergic neurones in the lumbar spinal cord of parkinsonian patients, *Brain Research*, 380, 181-185.

Schieppati, M., Nardone, A., Siliotto, R., and Grasso, M., 1995, Early and late stretch responses of human foot muscles induced by perturbation of stance, *Experimental Brain Research*, 105, 411-422.

Simonetta-Moreau, M., Marque, Ph., Marchand-Pauvert, V., and Pierrot-Deseilligny, E.,1999, The pattern of excitation of human lower limb motoneurones by probable group II muscle afferents, *Journal of Physiology*, 517, 287-300.

REFLEX MECHANISMS FOR MOTOR IMPAIRMENT IN SPINAL CORD INJURY

Brian D. Schmit[1], Ela N. Benz[2] and William Z. Rymer[2]

ABSTRACT

Spasticity is common feature of human spinal cord injury. It contributes to motor impairment and it also promotes joint deformity in patients who have sustained such injury. The classical definition of spasticity highlights the increased resistance of a joint to externally imposed motion. This resistance is attributable largely to changes in stretch reflex excitability, and it is manifested primarily in those muscles being stretched by the motion. Under this definition, there would be little activity in muscles crossing other joints. In spinal cord injury, however, muscles innervated from distal spinal segments often exhibit little hypertonia, yet patients report the occurrence of disabling spasms. These spasms appear as coordinated patterns of muscle activation throughout the limb, involving either limb flexors or extensors. These patterns are therefore quite different from those of classical spasticity. The receptor origins and neural pathways responsible for the spasms in spinal cord injury will be addressed.

INTRODUCTION

Spasticity is defined as an increase in muscle tone. It is usually detected clinically as an increased resistance to externally imposed joint motion (Lance, 1980). This physical sign is a prominent clinical feature following brain injury, including stroke, and it is also believed to be a major factor associated with motor impairment in spinal cord injury. However, while spasticity may be present in some spinal cord injured persons, in many others, muscular hypertonia is often quite modest and tendon tap reflexes may be reduced or even absent. Despite the absence of hyperexcitable stretch

[1] Rehabilitation Institute of Chicago, and Department of Physical Medicine and Rehabilitation at Northwestern University Medical School, and the Department of Biomedical Engineering, Marquette University, Milwaukee Wisconsin, USA.

[2] Rehabilitation Institute of Chicago, and Department of Physical Medicine and Rehabilitation at Northwestern University Medical School, USA. Email: zevric@casbah.acns.nwu.edu

reflexes, many spinal cord injured patients still complain of disturbances in muscle tone, which they describe as widespread involuntary muscle spasms (Schmit et al., 2000). These spasms arise either as flexion or extension of the legs, and sometimes both patterns may be present within the same patient. It is our assertion that these spasms are the primary source of altered tone in many spinal cord patients, rather than classical spasticity, and that the mechanisms of these spasms are different from those arising in supraspinal injuries (such as stroke).

The origins of the flexion spasms in spinal cord injured patients are relatively well understood. These spasms consist of broadly distributed activation of flexor muscles, resulting in coordinated flexion of all joints of the leg, and sometimes including activation of both legs. These spasms are likely to be analogues of the flexion withdrawal response that has been studied extensively in both intact and spinal cord injured animals (Eccles and Lundberg, 1959; Baldissera et al., 1981).

As has been reported in studies of the flexion withdrawal response in animal models, flexion responses in man can be activated by several classes of receptors, including mechanoreceptors and nociciceptors in skin, muscle, and joint (Shahani and Young, 1971). Indeed, flexion reflexes may even be triggered by visceral afferent activation, such as may arise during bowel or bladder distension. While the involvement of these multimodal sensory systems is not surprising, given the known physiology of the flexion reflex pathways, the particular pathways responsible for flexion spasms in man have not been well described, and it is our intent to briefly consider the their origins.

In contrast to the relatively mature state of knowledge about flexion spasms, there is very little known about the origins of extension spasms, which also occur frequently in spinal cord injured persons. Accordingly, it is also our intent to briefly consider the factors responsible for triggering extension spasms in spinal cord injured persons and to speculate on the receptor mechanisms that might be responsible based upon measurements of extension reflexes made in a controlled laboratory setting. Finally, we will also briefly consider the mechanisms that could be responsible for releasing these multi-joint reflexes in spinal cord injury and will examine the important regulatory role of monoamines released by descending brain stem pathways.

METHODS

Subjects

Subjects were recruited from the clinics of the Rehabilitation Institute of Chicago. Inclusion criteria included a history of chronic spinal cord injury (SCI) of greater than four months duration, and reports of spasms by the subject or physical therapist. Participants included clinically complete and incomplete individuals with cervical or thoracic SCI. Exclusionary criteria included: multiple central nervous system lesion sites, the presence of significant complications or other concurrent illness limiting the capacity to conform with study requirements, the inability to give informed consent, significant osteoporosis or a cardiac pacemaker. Subject's consent was obtained and the Institutional Review Board of Northwestern University approved all procedures.

Figure 1. Experimental apparatus for testing the extensor reflexes. A custom-designed, instrumented brace was constructed to hold the leg and to provide measurements of the ankle and knee torques. A torque cell was located within the brace, aligned with the axes of rotation of the joints. The entire brace fit onto the motor of a Biodex System, which provided the controlled hip extension.

Extensor Spasm Measurements

Reflex responses that are analogous to extensor spasms were recorded in the laboratory as the muscle activity and joint torques following imposed movement of the hip. In order to record the multijoint torque associated with the reflex response, a custom-designed apparatus was constructed. This apparatus, shown in Figure 1, consisted of a leg brace with torque transducers aligned with the axes of rotation of the knee and ankle. The foot was placed in a footplate and secured using a clamp placed on the dorsum of the foot and a strap around the heel as described previously (Schmit et al., 2000). The hip-knee and knee-ankle links were adjustable to fit a wide range of leg sizes. The entire leg brace was affixed to the motor of a Biodex Rehabilitation Testing System 2 (Biodex Medical Systems, Inc., Shirley, NY), with the axis of the motor aligned with the axis of rotation of the hip.

Hip torque, position, and velocity signals were recorded from the transducers on the Biodex. In addition, ankle and knee torques were recorded from the torque cells of the leg brace. All signals were low-pass filtered (450 Hz), and sampled at 1000 Hz using a data acquisition card (National Instruments, Austin TX) on a personal computer. Custom LabVIEW software (National Instruments) was used for controlling data acquisition and timing of imposed movement perturbations. Surface EMGs were recorded from the tibialis anterior, soleus, medial gastrocnemius, vastus medialis, rectus femoris, biceps femoris and adductors. Active Delsys electrodes (model DE2.1, Delsys Inc., Boston MA) were applied to lightly abraded, skin over the respective muscle bellies. The signal was amplified (10,000X) and filtered (20-450 Hz; Bagnoli 4, Delsys Inc.). The signals were sampled at 1000 Hz using the same computer system used for acquiring the position/torque data.

The hip was aligned with the axis of the Biodex motor and the brace was adjusted to align the ankle and knee with the appropriate torque transducer. Alignment of the hip was confirmed by a lack of leg translation during imposed flexion and extension of the hip. The motor rotated the hip at a constant velocity of 30 degrees/s from the start to end position. The movement was held at end position for ten seconds, during which the isometric reflex torques and EMG activity were measured. After this period, the motor rotated the hip back to the start position, again at 30 degrees/s. A delay of 2-5 min was allowed between trials, and ten trials were repeated for each subject.

Flexor Spasm Measurements

In a different set of experiments, flexion reflex analogues of flexion spasms were measured using a different experimental apparatus. For these experiments, the foot was clamped to a footplate, which, in turn was coupled to the motor of the Biodex System. The axis of rotation of the ankle was aligned with the motor axis and the flexion/extension torque responses of the ankle, knee and hip were measured using multiaxis load cell placed beneath the footplate. EMG recordings were made of the same muscles that were measured during extensor spasms.

Flexion spasms were triggered by externally imposed angular perturbations of the ankle or by electrocutaneous stimuli applied to the foot. Angular perturbations were again imposed as constant velocity stretches by the Biodex motor into both plantarflexion and dorsiflexion. Electrocutaneous stimuli were comprised of a pulse train of 10 pulses administered at 500 Hz to a pair of surface gel electrodes placed on the dorsal surface of the foot near the metatarsals. The stimulation intensity was set to 50 mA with a 1ms pulse width and delivered using a Digitimer DS7AH (Digitimer Ltd., Hertfordshire, England), triggered from a personal computer.

Analysis

Isometric torque data was obtained for the hip, knee, and ankle following the movement. For extensor spasms the responses were recorded with the hip in the maximum extended position. For the ankle movements, responses were measured with the ankle near the end of range of motion in either dorsiflexion or plantarfelxion. The gravitational-passive torque offset was removed by subtracting a torque measurement in the end position, when there was no muscle activity. The signals were then low-pass filtered (25 Hz) using a 4th order Butterworth filter (MatLab; The Math Works, Inc., Natick, MA) and plotted against time. The pattern of muscle activity was identified for each subject and related to the measured reflex torque. Peak ankle, knee, and hip torque responses were identified for each subject.

Electromyograms were band-stop filtered at 57.5-62.5 Hz (8th-order Butterworth filter) and rectified. Rectified EMGs were evaluated to detect the timing of muscle activity during and following imposed hip extension.

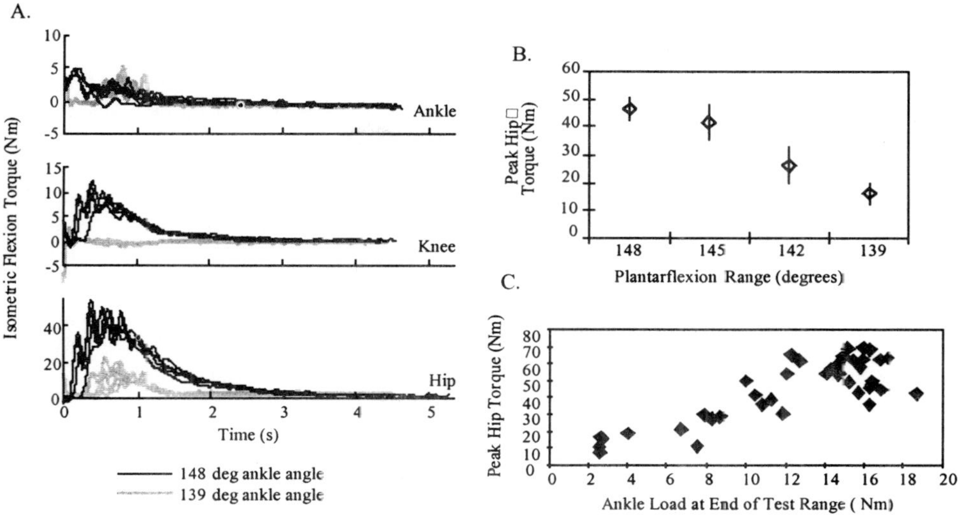

Figure 2. Data from a patient with incomplete spinal cord injury. *A*, Plantar flexion movements of the ankle induce flexion torques at each of the three major joints. *B*, The larger torques arise as the ankle angle approaches the maximum, as reflected by the increase in passive torque. *C*, The peak hip torque from the flexion reflex increased with the associated increase in the passive ankle torque that was measured at the end of the imposed ankle movement.

RESULTS

Flexion Spasms

Flexion spasms of the leg could be elicited readily by electrical stimulation of the skin of the foot or other cutaneous regions of the leg. More interestingly, in 19 subjects, we were able to elicit spasms consistently by imposing large amplitude ramp and hold angular deflections of the ankle joint.

We found that when ankle joint plantar flexion angle was of sufficient magnitude to increase passive force in the anterior compartment muscles (such as the tibialis anterior), there was an increased likelihood of triggering a flexion response. The resulting response was characterized by a multijoint response including ipsilateral flexion torques at hip, knee and ankle in a pattern similar to that observed in response to an electrocutaneous stimulus.

As shown in Figure 2, the onset of hip flexion torque in response to ankle plantar flexion motion appears to be closely linked to ankle flexion torque. In addition to increasing the likelihood of eliciting a movement-triggered response, there is a substantial increase in the magnitude of the flexion torque at the hip response as the ankle is driven to near its maximum (plantar flexed) position. Indeed, when imposed peak hip flexion torque (associated with the flexion reflex) is plotted against the passive ankle torque, measured at the end of the movement perturbation, the relationship that emerges is almost linear (Fig. 2C). Because small changes in joint angle in this range induce substantial changes in hip torque response, it seems less likely that flexion withdrawal is simply dependent on ankle position. Rather, it appears to be a reflection of the magnitude of the ankle load, even when that load is passively mediated.

In addition, end-range dorsiflexion movements also triggered flexion reflexes in 3 subjects, although greater loads were required to trigger such reflexes using dorsiflexion movements (compared to plantarflexion movements). These responses also appeared to be load dependent, evidenced by one subject that demonstrated flexion reflexes in response to force in the plantar flexors elicited by a stretch reflex. Furthermore, the flexion reflex response to imposed ankle movements was not dependent on the velocity of the imposed movements. In 11 subjects, we found comparable peak hip torque responses to 30, 60, 90 and 120 deg/s movements.

Based upon these combined observations, we propose that excitation of muscle and joint-related afferents, such as those activated by end-range movements, may be responsible for exaggerated flexion reflex responses in spinal cord injury.

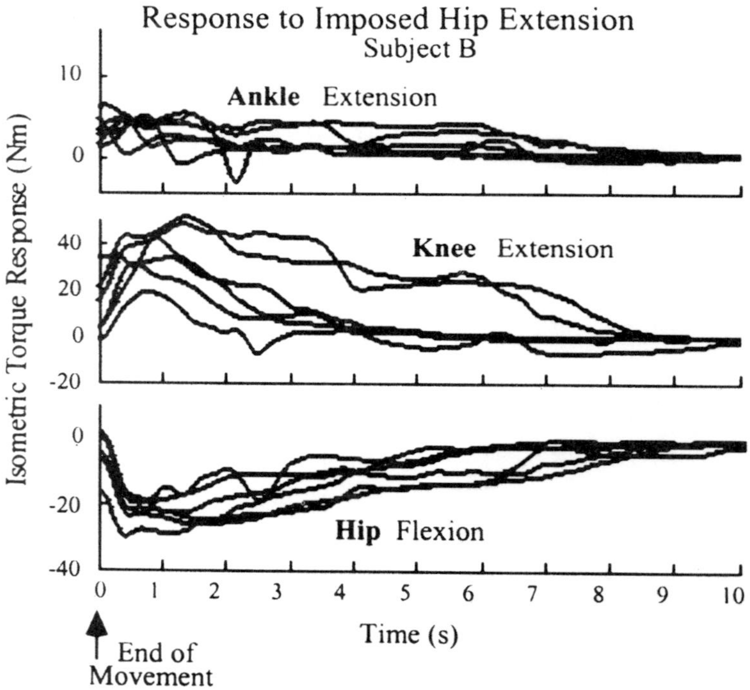

Figure 3. Variation in the pattern of torque response to hip extension. Ankle, knee and hip torques are shown from top to bottom, beginning at the end of the hip extension movement. Note that there is extension torque at the ankle and knee, but flexion torque elicited at the hip.

Extension Spasms

The factors triggering extension spasms appear to be quite different from those involved in inducing flexion withdrawal. In particular, there seems to be a key role for receptors responding to changes in hip angle, which has been suggested by clinical observations that sit to supine movements (bilateral hip extension) are a strong trigger for extensor spasms.

We began investigating the possible role that hip proprioceptors may play in extensor spasms using an imposed unilateral extension movement of the hip, with the knee and ankle held isometric. Figure 3 shows the response of the ipsilateral leg beginning at the end of the movement perturbation (end of range for hip extension). The resulting response is characterized by ankle and knee extension torque and a hip flexion torque, each lasting for up to 10 seconds following the end of the movement.

Interestingly, the responses that followed imposed hip extension were not consistent with uniform spread of activation to extensor muscles, with a flexion torque elicited at the hip. The net torque pattern and posture at the end of the movement are consistent with the late stance phase of gait. However, there was also significant EMG activity in the ankle and knee flexors that slightly preceded and overlapped with the activity seen in the extensors. As a result, the responses may have included a swing to stance transition, rather than solely a stance reaction.

DISCUSSION

We have found that both flexion and extension reflex activation patterns can be elicited readily in many spinal cord injured persons, and that these patterns are derived from large scale and coordinated activity in many muscle groups of the lower extremity. In this respect, the responses appear to be quite different from those recorded in spastic hypertonia, which is usually concentrated in those muscles that are subjected to length changes around one joint.

Flexion Withdrawal

It is well known that many classes of receptors can elicit flexion withdrawal (Eccles and Lundberg, 1959; Baldissera et al., 1981; Cleland et al., 1990). Our findings indicate that there may be an especially important role for free nerve ending mechanoreceptors located in ankle dorsiflexors, such as the tibialis anterior (Cleland et al., 1990; Schmit et al., 2000). This assertion is based on the fact that angle (and therefore length) changes alone were not strongly correlated with the onset of flexion withdrawal response; rather there appeared to be a need to induce loading of passive tissues. While loading could also induce activity in Golgi tendon organs (Houk et al., 1980), these are not believed to be responsive in passive muscles, although the evidence in human subjects is currently very limited.

Furthermore, it may be premature to dismiss significant contributions from muscle spindle afferents, especially from secondary spindle receptors and their central (group II) connections. While these afferents would normally be strongly activated well before the extremes of ankle dorsiflexion, it is likely that fusimotor innervation in the spinal cord injured patient may be lower than normal, so the threshold for onset of discharge in these muscle afferents may be abnormally high.

Finally, it is difficult to exclude some role of skin afferents, although in our experience, non-noxious manipulation of skin does not replicate flexion withdrawal to the extent of visible during ankle dorsiflexion.

Extension Spasms

Our findings highlight the pivotal role of hip proprioceptors in triggering extension activity in a pattern comparable to that elicited by hip extension in animal models (Grillner and Rossignol, 1978; Andersson and Grillner, 1981). The receptors responsible presumably lie in the large hip flexors, such as the psoas and iliacus, although joint mechanoreceptors could also contribute.

Interestingly, the reflex responses elicited by hip motion do not involve uniform activation of limb extensor muscles as a whole. Instead, there seem to be muscle combinations that regularly include hip flexion coupled with knee and ankle extension. This combination of joint torques is consistent with a pattern of muscle activation that is usually evident in the late stance phase of gait, although it will also be important to assess the sequence when muscle activation is triggered by other means.

In this context, it has been shown in various reduced cat preparations that force feedback pathways from Golgi tendon organs (carried via Ib afferents), which are normally inhibitory to extension motor neurons in the resting situation, become suppressed during locomotion, and are even switched to different neuronal motor pathways to augment extensor excitation (Hiebert et al., 1996). It is likely that similar sequences are responsible for eliciting extensor spasms in man.

Role of Monoamines

While many different descending pathways are likely to participate in control of segmental circuits, and may therefore contribute to the resulting alterations that follow their interruption in spinal cord injury, there appears to be a special and unique role for monoaminergic systems. Indeed, under normal (intact) conditions, many spinal interneuronal pathways are held in a state of reduced excitability because of the actions of monamines on spinal noradrenergic receptors, and there also appear to be important roles for serotonergic pathways as well. Interruption of such monoaminergic inputs may lead to the release of interneuronal control. This may be reflected in the activation of flexion withdrawal interneuronal sequences, and may also underlie the release of interneurons associated with extension spasms as well.

ACKNOWLEDGEMENTS

This work was carried out with support from the NIH NS40901-01S1 (BDS), NIH 28076 (WZR) and from the Christopher Reeve Paralysis Foundation (WZR).

REFERENCES

Andersson, O., and Grillner, S., 1981, Peripheral Control of the Cat's Step Cycle I. Phase dependent effects of ramp-movements of the hip during "fictive locomotion", *Acta Physiologica Scandinavica*, 113, 89-101.

Baldissera, F., Hultborn, H., and Illert, M., 1981, Integration in spinal neuronal systems, in: *Handbook of Physiology - The Nervous System*, Brookhart, J.M., and Mountcastle, V.B., eds., American Physiological Society, Baltimore, pp. 509-595.

Cleland, C. L., Hayward, L., and Rymer, W. Z., 1990, Neural mechanisms underlying the clasp-knife reflex in the cat. II. Stretch-sensitive muscular-free nerve endings, *Journal of Neurophysiology*, 64, 1319-1330.

Eccles, R.M., and Lundberg, A., 1959, Synaptic actions in motoneurones by afferents which may evoke the flexion reflex, *Archives Italiennes de Biologie*, 9, 199-221.

Grillner, S., and Rossignol, S., 1978, On the initiation of the swing phase of locomotion in chronic spinal cats, *Brain Research*, 146, 269-277.

Hiebert, G. W., Whelan, P. J., Prochazka, A., and Pearson, K. G., 1996, Contribution of hind limb flexor muscle afferents to the timing of phase transitions in the cat step cycle, *Journal of Neurophysiology*, 75, 1126-1137.

Houk, J. C., Crago, P. E., and Rymer, W. Z., 1980, Functional properties of the golgi tendon organs, in: *Spinal and Supraspinal Mechanisms of Voluntary Motor Control and Locomotion*, Desmedt, J. E, ed., Basel, Karger, pp. 33-43.

Kumazawa, T., and Mizumura, K., 1977, Thin-fibre receptors responding to mechanical, chemical, and thermal stimulation in the skeletal muscle of the dog, *Journal of Physiology*, 273, 179-194.

Lance, J. W., Pathophysiology of spasticity and clinical experience with baclofen, in: *Spasticity: Disordered Motor Control*, Feldman, R. G., Young, R. R., and Koella, W. P., eds., Year Book, Chicago, pp. 185-203.

Mense, S., and Meyer, H., 1985, Different types of slowly conducting afferent units in cat skeletal muscle and tendon, *Journal of Physiology*, 363, 403-417.

Schmit, B. D., McKenna-Cole, A., and Rymer, W. Z., 2000, Flexor reflexes in chronic spinal cord injury triggered by imposed ankle rotation, *Muscle and Nerve*, 23, 793-803.

Shahani, B. T., and Young, R. R., 1971, Human flexor reflexes, *Journal of Neurology, Neurosurgery and Psychiatry*, 34, 616-627.

SECTION VI

Locomotion

In recent years important progress has been made on the subject of afferent feedback involved in the segmental control of locomotion. This subject was well represented at the conference and included consideration of locomotor mechanisms in humans and animals. This Section covers these presentations.

A. Taylor (Chapter 38) makes a plea for muscle receptors as proprioceptors, in the sense that they signal the body's own actions. We currently consider two roles for muscle spindles. The first is their contribution to kinaesthesia, which includes the sense of position and movement of our limbs. However the role alluded to by Taylor concerns their second major function, that of providing feedback regulation for the unconscious, automatic control of movements.

Taylor describes experiments on the decerebrate cat in which the aim was to derive some clues about the role of the fusimotor system in locomotion. The general idea was that the γ system was activated according to a temporal template of the expected movement. The γ_s patterns were arranged to give an afferent response that closely followed the unloaded shortening of the muscle. γ_D activity during shortening was thought to contribute to the γ_S evoked response by spread of generator current from afferent terminals excited subliminally by the stimulation, and during lengthening resembled the differentiated length signal, that is, it appeared to be a template of the velocity of lengthening.

In a closely related talk, P. Ellaway (Chapter 39) provides some direct supporting evidence for the inferences drawn from afferent recordings by recording directly from identified fusimotor neurones. This brings new complexities to the overall picture. There were two distinct types of γ_S activation pattern. The more common one conformed with the movement template concept. The other remains somewhat mysterious, resembling more closely the antagonist EMG pattern than that of the test muscle. Firing of γ_D, unlike γ_S, was discontinuous during the step cycle. The γ_D axon fell silent early during muscle lengthenings to re-start just before shortening. It is suggested that γ_D activity sensitises the spindle to stretch (by its after-affects?) but allows it to fall again afterwards. It also sensitised the spindle to unexpected lengthening during the shortening phase of locomotion, presumably, as well as providing subliminal electrotonic support to any simultaneous γ_S activity.

The way in which the central nervous system uses afferent feedback during locomotion is addressed in studies in the cat by K. Pearson (Chapter 40). A variety of approaches based on unexpected placement of the hindfoot in a "hole" and use of deafferentation have been used to reveal the potency of afferent feedback in generation of extensor activity during walking in the cat. Such feedback, it has been estimated, may be sufficient to generate about half of the electromyographic activity in muscles which plantarflex the ankle, a finding which appears to hold for human walking. Roles for monosynaptic paths, paths via "phase-dependent" excitatory interneurones, and via the central pattern generator continue to be identified (e.g. McCrea, 2001). Recent studies are now providing support for the view derived initially from the classical studies of Shik and Orlovsky that the swing phase of walking is also highly dependent upon afferent feedback, particularly from flexor muscle afferents (see also Shik and Orlovsky, 1976). A hierarchical arrangement of afferent "pathways" from specific muscles to motoneurones may be accessed during particular circumstances encountered during locomotion.

In his presentation, A. Prochazka (Chapter 41) takes the provocative view that in some circumstances sensory feedback during locomotion may not be important. In the absence of sensory feedback, the central nervous system can generate a coordinated pattern on muscle activation. Making use of the spring-like properties of muscle, if synergist muscles are activated appropriately, locomotion is possible. A model is used to show that stretch reflexes are important when activation levels are low, but become less so at higher levels. Perhaps an important additional role of sensory feedback is for higher-level planning of anticipated movements and the associated activation patterns.

The subject of task-dependent activation patterns in human locomotion is considered by the contribution from J. Nielsen (Chapter 42). He describes recent studies on the strength of spindle-mediated facilitation of ankle dorsiflexor and plantarflexor muscles during human walking. Surprisingly, the strength of the stretch reflex of tibialis anterior was largest when the muscle was quiescent during the swing phase of walking. Inputs travelling over transcortical paths and also via Group II muscle afferent paths have been identified

P. Dyhre-Poulsen (Chapter 43) reports some unexpected results on changes in the H reflex during human walking under various loading conditions. A controversial finding is that in the stance phase the reflex appears larger for greater speeds of locomotion, a conclusion which is opposite to that described by others (e.g. Capaday and Stein, 1987). While the reason for this discrepancy is uncertain, it appears that methodological factors may be crucial. The conditions for recording and eliciting the complex afferent volley that underlies the H reflex are not stable during locomotion. Furthermore, the amplitude of the H reflex is not determined purely by the strength of group Ia projections (e.g. Burke et al., 1984).

REFERENCES

Burke D., Gandevia S.C., and McKeon B., 1984, Monosynaptic and oligosynaptic contributions to human ankle jerk and H-reflex, *Journal of Neurophysiology*, 52, 435-448.

Capaday C., and Stein R.B., 1987, Difference in the amplitude of the human soleus H reflex during walking and running, *Journal of Physiology*, 392, 513-522.

McCrea D.A., 2001, Spinal circuitry of sensorimotor control of locomotion, *Journal of Physiology*, 533, 41-50.

Shik M.L., and Orlovsky G.N., 1976, Neurophysiology of locomotor automatism, *Physiological Reviews*, 56, 465-501.

GIVE PROPRIOCEPTORS A CHANCE

Anthony Taylor*

ABSTRACT

The theme of this review is that it is inappropriate to regard proprioceptors as general purpose transducers of system variables associated with movements. We should not try to describe their properties by general expressions derived by testing with a wide range of externally applied disturbances, in the way that is customary in engineering practice. Instead, if study is concentrated on their behaviour during natural active movements such as locomotion, then the significance of the signals which they feed back to the CNS is much easier to understand. This idea is developed briefly for tendon organs, and then in more detail for muscle spindles in locomotion.

INTRODUCTION

It would not be disputed that proprioceptors are essential to the expression of normal movement patterns, yet it has proved very difficult to establish precisely how they are used. The purpose of this review is to suggest that some of the difficulty arises from the fact that their properties have usually been investigated using controlled external disturbances over a wide range of frequencies and amplitudes rather than to see how they behave in the restricted conditions of natural movements. They have, after all, evolved to be efficient in these conditions, rather than as wide range linear transducers of the sort that an engineer might create by prospective design. The most widely studied proprioceptors are the Golgi tendon organs and the muscle spindles. Attention will be restricted to these two, but the same approach should be appropriate for any of the other receptor types which qualify as proprioceptors.

TENDON ORGANS

When the responses of muscle receptors were first studied, the effects were recorded of applying known fixed tensions via the muscle tendon. Later, when it was appreciated

---

* Division of Neuroscience, Imperial College School of Medicine, London, W6 8RF, UK.
Email: ataylor@ic.ac.uk

that spindles were best suited to monitor length changes and tendon organs to monitor force, controlled patterns of displacement or force were used to assess their properties. This seems reasonable until we reflect that the term 'proprioceptor' as introduced by Sherrington (1906), emphasises the role of these receptors in reporting on the animal's own active movements, rather than those applied passively from the outside. New insight into the relevance of this concept was provided by studies on tendon organs (reviewed by Matthews, 1972). Early work showed that substantial tensions had to be applied passively to make them discharge continuously (B.H.C. Matthews, 1933). Later observations confirmed this (Hunt and Kuffler, 1951) and led to the view that the function of tendon organs was to act as overload detectors. The importance of considering actively generated forces as the adequate stimuli, rather than passive, was suggested by Jansen and Rudjord (1964) and tested by Houk and Henneman (1967b). These authors showed that even the small forces generated by the active contraction of single motor units highly effective in exciting tendon organs and that this was due to their anatomical arrangement in relation to individual muscle fibres. At once it was evident that these receptors had to be regarded as potential controllers of natural muscle contractions under most conditions. Some doubts have been expressed as to how well they can signal active force linearly and monotonically, but on balance it does seem that an ensemble can do so very effectively (Hulliger et al., 1995). Telling evidence to this effect was also provided by Prochazka and Gorassini (1998) who constructed ensemble averages of tendon organ firing as recorded from hindlimb muscles of the cat during natural locomotion. There was a clear parallel between the ensemble patterns and the whole muscle EMG signal. The potential importance to normal control of tendon organs working to signal active force has been further emphasised by recent experiments on locomotion in cats (Prochazka et al., 1997; Hiebert and Pearson, 1999). These insights would not have been possible without the recognition that tendon organs operate as proprioceptors, as originally defined by Sherrington.

MUSCLE SPINDLES

The identification of muscle spindles as muscle length detectors at first sight seems to justify attempts to characterise their behaviour by imposing length changes passively and recording the resultant afferent discharge. Spindle afferents, both primary and secondary, do indeed respond well to passive stretches and their behaviour is modified by contractions of the intrafusal muscle fibres mediated by the gamma static (γ_S) and gamma dynamic (γ_D) motoneurones. The most obvious function of spindles was seen from the beginning to be to provide the signal for negative feedback control of movement via the segmental stretch reflex (e.g. Merton, 1953; Houk and Henneman, 1967a). It followed that they should be regarded as displacement/velocity transducers, with properties regulated by the CNS through the fusimotor system. In this way the sensitivity and the dynamics of the feedback loop would be adjusted to suit different tasks and loads. Accordingly, much research was directed to finding general ways of describing the effects of γ_S and γ_D activity on the parameters of spindle response to stretch, either by generalising from simple descriptions or by using techniques developed for engineering systems. This subject has been extensively reviewed (e.g. Matthews, 1972; Hulliger, 1984; Boyd and Gladden, 1985; Prochazka, 1996). The facts that emerge are that the presence of non-linearities and the variation of properties between afferents make it difficult to construct general mathematical models. The actions of the γ_S and γ_D systems are not easily defined, for example in determining the length and velocity sensitivities. Rather they seem to interact in complex ways which depend on the amplitude of stretch. Maintaining various constant levels of γ_S and γ_D stimulation and applying standard procedures such as sinusoidal testing over a range of frequencies has not led to the hoped for generalised expressions which could be used in dynamic systems modelling.

Spindle Behaviour During Locomotion

With the above considerations in mind, we have re-examined the behaviour of spindle afferents in cat hindlimb muscles in the restricted and specific situation of relatively natural locomotor movements in the high decerebrate preparation. The methods have been fully described elsewhere (Taylor et al., 2000a), but are briefly as follows. Animals were prepared under halothane anaesthesia. They were supported so as to walk on a treadmill with three legs. One hindlimb was denervated save for the medial gastrocnemius (MG) and tibialis anterior (TA) muscles. It was clamped so that the only movement allowed was ankle rotation, and this was recorded without significant loading. Muscle spindle afferents from MG and TA were recorded from dorsal. Each afferent was characterised by its conduction velocity as primary or secondary and, at the end of the experiment, by the use of succinylcholine to assess the existence of sensory contacts on bag$_1$ (b_1) and bag$_2$ (b_2) fibres (Taylor et al., 2000a,b). After pre-mammillary decerebration and cessation of halothane anaesthesia, regular locomotion would usually occur on running the treadmill and sometimes with the addition of stimulation in the midbrain locomotor region. Spindle activity from six identified afferents was recorded first during active movements. Fusimotor activity was then suppressed with barbiturate and the recorded ankle movement patterns replayed through a servo mechanism to exactly reproduce the normal muscle length changes. From regular stepping sequences cycle averages could be computed of movement and of spindle firing frequency. The difference frequency patterns computed by subtracting the firing frequencies recorded during passive movements from those recorded during active movements were assumed to be due to the fusimotor activity present during the latter.

Secondary Afferents

The results from secondary afferents are most easily interpreted, because they are influenced (with few exceptions) purely by the static fusimotor system. Figure 1 shows recordings from two secondary afferents from TA. In both cases (A & B) the frequency during active movements rises during the shortening period, contrasting with the fall during the same period in the passive state. The difference records (C & D) thus show a progressive rise throughout the shortening, which could only be accounted for by a rise in static fusimotor activity during that period. Other experiments have shown that sinusoidal modulation of γ_S firing frequency produces a close to linear modulation of secondary afferent firing frequency (Andersson et al., 1968; Durbaba et al., 2001). It follows, therefore that the difference signals in Figure 1 are likely to be closely related to the form of the fluctuating γ_S firing pattern occurring during locomotion. The difference records in Figure 1 C and D are quite similar to the smoothed EMG records, but actually resemble even more closely the record of muscle *shortening*. This is shown in Figure 2 A in which the ensemble difference record from the two units in Figure 1 is superimposed on a scaled record of muscle length inverted to show shortening upward. This direct relationship is emphasised by the plot of length against difference frequency in B. The less direct connection with EMG is shown in Figure 2 C and D. The most straightforward interpretation of these observations is that in these particular movements there is a modulation of the firing frequency of γ_S motoneurones with a time course closely resembling that of the active unloaded muscle shortening. It also seems likely that, within the restricted ranges and speeds of natural movements, the secondary spindle afferents signal the simple algebraic summation of muscle length and γ_S firing frequency. Similar observations have also been made for MG spindle secondary afferents, but in this case the difference signal is phase advanced relative to the shortening record. Confirmation of this interpretation has been provided by direct recordings from γ_S neurons, as described in the Chapter by Ellaway and colleagues (Chapter 39).

Primary Afferents

Because of the additional contribution of the γ_D firing which acts *via* the bag_1 intrafusal fibres to determine the activity of primary afferents, the interpretation of their firing patterns is expected to be much more difficult than that of the secondaries.

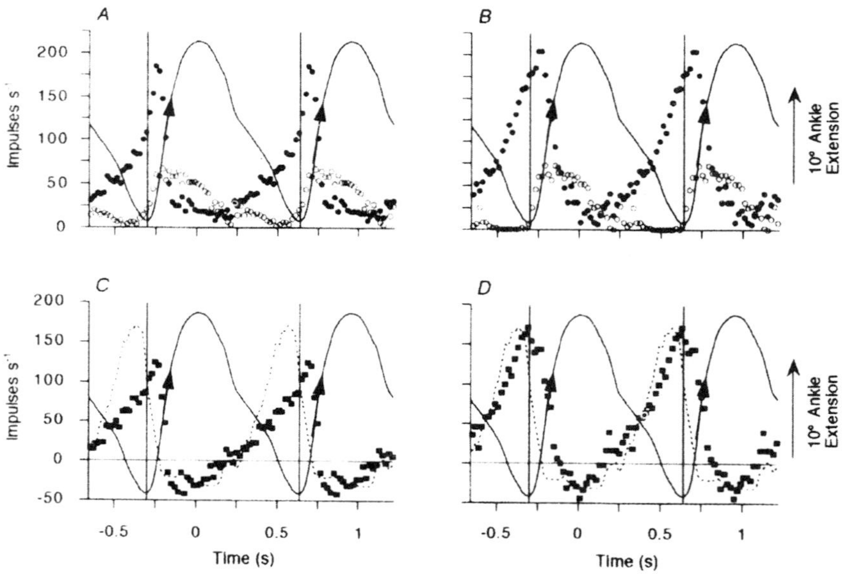

Figure 1. Behaviour of two TA secondary afferents in active and passive movements. In all four panels the ankle rotation is shown as a continuous line with an arrowhead indicating direction of muscle lengthening. *A* and *B* show for each of the two units, cycle histograms of firing frequency in active (•) and in passive (o) movement. In *C* and *D* the mean difference records are shown, active minus passive (■) and the cycle means of rectified EMG in TA (dashed line). Vertical lines indicate transition from muscle shortening to lengthening. Records are means of 25 cycles. (Modified from Taylor et al., 2000a, with permission.)

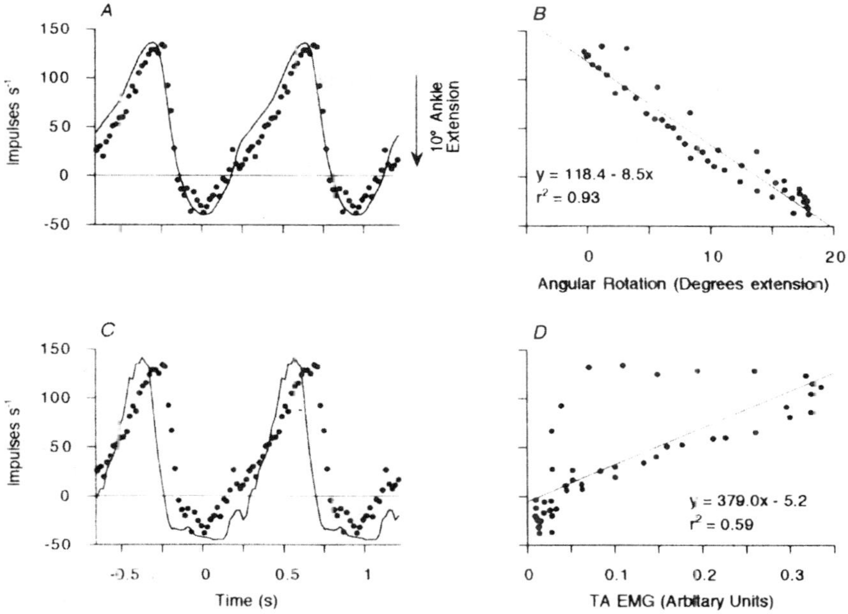

Figure 2. Ensemble means of TA secondary difference records related to movement and EMG. In *A* and *C* the ensemble mean of the difference records from Fig. 1 *C* and *D* is plotted (●) superimposed on the ankle rotation record (*A*) and the mean rectified EMG record (*C*). Muscle lengthening downward. The difference signal is also plotted at against ankle angular rotation in *B* and mean EMG in *D*. In *B* and *D* the straight lines are the linear regressions, the parameters of which are inset. (Modified from Taylor et al., 2000a, with permission.)

However, it is known that there is a minority set of primary afferents which have no terminations upon bag_1 intrafusal fibres and they can be detected by virtue of their response to succinylcholine. They are designated as b_2c primaries to distinguish them from the usual b_1b_2c type. Figure 3 illustrates data obtained from an experiment in which two b_2c and three b_1b_2c type primaries from TA were recorded simultaneously. In A the ensemble difference records of the two types are shown superimposed. It is evident that the difference records are very similar throughout the muscle shortening phase, but immediately lengthening begins they diverge. The difference between the two (Fig. 3 B) then must represent the contribution of the bag_1 fibres, which are evidently contracted during active movements. The contraction of the bag_1 fibres under the influence of γ_d firing confers strong velocity sensitivity on the primaries during muscle lengthening as shown in Figure 3 B, in which the velocity record, derived from the length record, matches the profile of the bag_1 fibre component quite well during the muscle lengthening phase. The fact that the difference signals for the b_2c and b_1b_2c type endings are so similar during the shortening phase indicates that the underlying modulated pattern of γ_S firing influences the primaries in much the same way as the secondaries, i.e. it summates

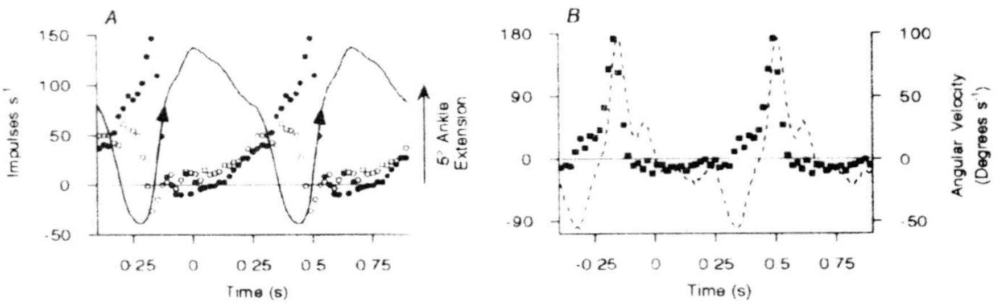

Figure 3. Prediction of the component of primary afferent firing due to the bag$_1$ fibre. In *A* the ensemble mean active minus passive difference signals are shown for three TA b_1b_2c primary afferents (●) and from two b_2c primary afferents (o) recorded at the same time (31 cycles). Note the close similarity of the two difference records during the muscle shortening phase and their divergence at the onset of lengthening. In *B* the symbols (■) plot the difference of the b_1b_2c mean minus the b_2c mean. The ankle angular velocity (dashed line) has been obtained by differentiating the ankle angle record (continuous line in *A*). (Modified from Taylor et al., 2000a, with permission.)

algebraically with the muscle length changes. It also follows that the bag$_1$ fibre contributes little if at all to the primary discharge during the shortening phase. It is known that the primary afferent divides to innervate the bag$_1$ fibre on the one hand and the bag$_2$ and chain fibres on the other. Each branch has its own impulse initiation site, and they interact so that the primary afferent firing is determined by whichever site is firing at the higher frequency at the time (Banks et al., 1997). One can thus visualise switching of output between the bag$_2$ and chain fibres during active muscle shortening and the bag$_1$ fibre during lengthening in the presence of γ_D activity. In addition, there will probably be a small contribution due to the electrotonic spread of depolarisation to the dominant impulse initiation site from the other. In Figure 4 this idea is illustrated using the ensemble mean of five b_1b_2c type primary afferents recorded at the same time from TA. Figure 4 A shows the active minus passive difference record (•) superimposed on the ankle angle record (muscle lengthening indicated by the arrowhead). The velocity record has been derived from the latter and when suitably scaled (o) matches the difference signal during muscle lengthening, showing that the primary afferent signal is dominated by velocity sensitivity during the lengthening phase of active movements (see also Prochazka and Gorassini, 1998). When this velocity component is subtracted from the difference signal (Fig. 4B) it leaves a component (▪) which matches the shortening record quite faithfully. This is precisely what would have been predicted from the secondary afferents and the b_2c primary afferents described above. (Note that in Figure 4 a small allowance has been made for electrotonic spread from the bag$_1$ ending to the bag$_2$ and chain ending). Thus it appears that during locomotor movements primary afferent firing can be predicted during lengthening as a measure of stretch velocity and during shortening as the algebraic summation of γ_S firing and muscle length. Similar recordings from MG primary and secondary afferents have shown that the same interpretations can be applied (Taylor et al., 2000a). The one difference consistently found between spindles in the flexor and extensor muscles is that in MG the γ_S firing profile appears to be phase advanced with respect to the muscle shortening record.

DISCUSSION

Muscle spindles, when tested with passive stretches and without fusimotor activity, are non-linear except for very small movements. In the presence of constant levels of static or dynamic fusimotor activity their behaviour may be more linear, but is still too complex to allow any simple general description. To add to the apparent difficulties, it has become clear that the firing of γ_S and γ_D motoneurones fluctuates with distinct patterns during natural movements and there may even be two classes of static gamma motoneurone (Murphy et al., 1984; Taylor et al., 2000b). Prediction of the expected afferent signals under these conditions, based on what we have learned of the effects of constant fusimotor drive is altogether too difficult, but as shown above, direct recordings from muscle spindles during natural locomotor movements show that their firing can be explained on quite simple principles. In essence, secondary afferents show linear summation between static fusimotor drive and muscle length. Primary afferents show similar behaviour with respect to static fusimotor drive, but in addition show a strong positive linear velocity component during stretch. These are not properties which would be obvious if the spindles were characterised by standardised passive movements with no fusimotor activity or with constant fusimotor activity. The rather simple properties described here emerge as a result of distinct static and dynamic fusimotor firing patterns appropriately timed in relation to and not directly as a result of the spindle properties as passive stretch detectors.

The lesson is that when tendon organs and muscle spindles are given the chance to behave truly as proprioceptors in natural movements, their behaviour is predictable experimentally and explicable in terms of their controlling function. They have evolved selectively through competitive pressures to provide ever better control of movement in specific restricted conditions. It makes sense to study them in these conditions, rather than as general purpose transducers. Analytical methods which are appropriate to devices designed prospectively by an engineer can be useful in studying receptor properties and biophysics, but have to be used cautiously to avoid unjustified assumptions regarding the evolved control strategy.

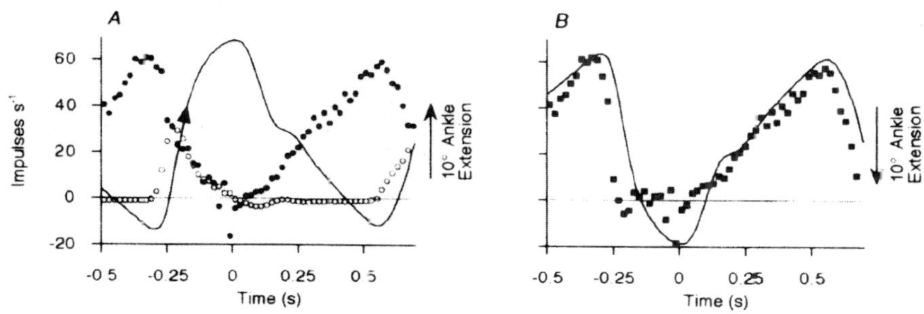

Figure 4 The component of firing in b_1b_2c TA primary afferents due to static fusimotor firing. In *A* the mean active minus passive difference record (●) is shown for an ensemble of five b_1b_2c afferents recorded simultaneously. Muscle lengthening shown by arrowhead on ankle angle record. Superimposed is the differentiated angular displacement trace (○) scaled to match the difference trace during extension and further scaled by a factor of 0.2 during flexion. In *B* the result (■) of subtracting the scaled velocity. The ankle angle record (continuous line) has been inverted. (Modified from Taylor et al., 2000a, with permission.)

ACKNOWLEDGEMENTS

Supported by the UK Medical Research Council. The research was in collaboration with P.H. Ellaway, R. Durbaba and S. Rawlinson, with technical support by O.D. Taylor.

REFERENCES

Andersson, B. F., Lennerstrand, G., and Thoden, U., 1968, Response characteristics of muscle spindle endings at constant length to variations in fusimotor activation, *Acta Physiologica Scandinavica,* 74, 301-318.

Banks, R. W., Hulliger, M., Scheepstra, K. A., and Otten, E., 1997, Pacemaker activity in a sensory ending with multiple encoding sites: the cat muscle spindle primary ending, *Journal of Physiology,* 498, 177-199.

Boyd, I. A., and Gladden, M. H., 1985, *The Muscle Spindle*, Macmillan, London.

Durbaba, R., Taylor, A., Ellaway, P. H., and Rawlinson, S., 2001, Effectiveness of static γ-axons in modulating the firing of secondary spindle afferents, *Journal of Physiology,* 531, 143-144P.

Hiebert, G. W., and Pearson, K. G., 1999, Contribution of sensory feedback to the generation of extensor activity during walking in the decerebrate cat, *Journal of Neurophysiology,* 81, 758-770.

Houk, J., and Henneman, E., 1967a, Feedback control of skeletal muscles, *Brain Research,* 5, 433-451.

Houk, J., and Henneman, E., 1967b, Responses of Golgi tendon organs to active contractions of the soleus muscle of the cat, *Journal of Neurophysiology,* 30, 466-481.

Hulliger, M., 1984, The mammalian muscle spindle and its central control, *Reviews in Physiology, Biochemistry and Pharmacology,* 101, 3-93.

Hulliger, M., Sjolander, P., Windhorst, U. R., and Otten, E., 1995, Force coding by populations of cat Golgi tendon organ afferents: the role of muscle length and motor unit pool activation strategies, in: Alpha and Gamma Motor Systems, A. Taylor, M. H. Gladden and R. Durbaba, ed., Plenum Press, New York, pp. 302-308.

Hunt, C. C., and Kuffler, S. W., 1951, Stretch receptor discharges during muscle contraction, *Journal of Physiology,* 113, 298-315.

Jansen, J. K. S., and Rudjord, T., 1964, On the silent period and Golgi tendon organs of the soleus muscle of the cat, *Acta Physiologica Scandinavica,* 62, 364-379.

Matthews, B. H. C., 1933, Nerve endings in mammalian muscle, *Journal of Physiology,* 78, 1-53.

Matthews, P. B. C., 1972, *Mammalian Muscle Receptors and their Central Actions*, Edward Arnold, London.

Merton, P. A., 1953, Speculations on servo control of movement, in: *The spinal cord*, G. E. W. Wolstenholme, ed., Churchill, London, pp. 247-255.

Prochazka, A., 1996, Proprioceptive feedback and movement regulation, in: *Handbook of Physiology: Section 12. Exercise: Regulation and Integration of Multiple Systems*, L. B. Rowell and J. T. Sheperd, ed., American Physiological Society, New York, pp 89-127.

Prochazka, A., Gillard, D., and Bennett, D. J., 1997, Implications of positive feedback in the control of movement, *Journal of Neurophysiology,* 77, 3237-3251.

Prochazka, A., and Gorassini, M., 1998, Ensemble firing of muscle afferents recorded during normal locomotion in cats, *Journal of Physiology,* 507, 293-304.

Sheerington, C. S., 1906, *The Integrative Action of the Nervous System*, Constable, London.

Taylor, A., Durbaba, R., Ellaway, P. H., and Rawlinson, S., 2000a, Patterns of fusimotor activity during locomotion in the decerebrate cat deduced from recordings from hindlimb muscle spindles, *Journal of Physiology,* 522, 515-532.

Taylor, A., Ellaway, P. H., Durbaba, R., and Rawlinson, S., 2000b, Distinctive patterns of static and dynamic gamma motor activity during locomotion in the decerebrate cat, *Journal of Physiology,* 529, 825-836.

ROLE OF THE FUSIMOTOR SYSTEM IN LOCOMOTION

Peter Ellaway, Anthony Taylor, Rade Durbaba and Stephen Rawlinson*

ABSTRACT

The contribution of muscle spindles to the control of locomotion depends on the patterns of discharge that occur in static and dynamic gamma motoneurones (γ_S and γ_D). Discharges of γ-axons to the MG muscle were studied during treadmill locomotion in pre-mammillary, decerebrated cats. All γ_S-efferents increased their rate of discharge at onset of locomotion and were modulated with movement. Type-1 increased their rate with muscle shortening whereas type-2 γ_S efferents increased their discharge rate during muscle lengthening. The type-1 γ_S pattern appears to be a temporal template of the intended movement. The type-2 γ_S pattern may be appropriate for afferent biasing through bag_2 intrafusal fibres. The γ_D axons showed an interrupted discharge pattern with sudden onset of firing at the start of muscle shortening and falling quiet shortly after the beginning of lengthening. The γ_D discharge would prepare primary endings to respond with high sensitivity at the start of muscle lengthening.

INTRODUCTION

Our recent work on locomotion in the cat (Taylor et al., 2000a,b) was designed to reveal some of the unresolved issues regarding the roles of primary and secondary spindle afferents and their complex system of efferent control through γ motoneurones. In another chapter, Taylor describes an analysis of primary and secondary spindle afferent recordings during locomotion strongly indicating that static fusimotor drive is modulated and that the pattern may represent a "temporal template" of the expected movement. An increase in dynamic fusimotor drive is also evident during active locomotion but it was not possible to say with afferent recording whether this fluctuated during the step cycle. There have been few previous attempts to record directly from γ motoneurones in behaving preparations (Lund et al., 1979; Appenteng et al., 1980; Murphy et al., 1984)

* Division of Neuroscience, Imperial College School of Medicine, London W6 8RF, UK.
Email: p.ellaway@ic.ac.uk

Sensorimotor Control of Movement and Posture
Edited by Gandevia *et al.*, Kluwer Academic/Plenum Publishers, 2002

and there have been difficulties in identifying individual axons as static or dynamic. This, coupled with our own uncertainties regarding the precise interactions between γ_S and γ_D activity and muscle length changes, prompted us to make simultaneous recordings from several single γ-axons and spindle afferents during locomotion.

PREPARATION

This study of the discharge characteristics of γ motoneurones has been carried out in cats decerebrated at a pre-mammillary level. Full descriptions of the preparation, the electrophysiological procedures and the induction of locomotion are given in another chapter (Taylor) and have been published recently (Taylor et al., 2000a,b).

In addition, the activity of several single γ-motor axons was recorded from small fascicles separated from the MG nerve, leaving much of the nerve intact so ensuring a path for other active α and γ efferents to the muscle. Efferent axons were confirmed as γ-motor if their conduction velocity lay between 12 and 45 m s^{-1}.

RESULTS

One of the principal advances made in this study has been to combine recording the discharge patterns of γ motoneurones with those of spindle afferents, and to record simultaneously from a number of both efferents and afferents. This has allowed us to gauge more surely what the populations of these neurones are doing and not to be misled by sampling possibly unrepresentative behaviour of single units.

Identification of γ_S and γ_D Axons

The identification of γ motor recordings was achieved using electrical stimulation of the midbrain. An area of the midbrain was sought that would produce a marked static or dynamic fusimotor effect. Effects were monitored by simultaneous recordings from γ efferents and spindle afferents with known bag_1 (b_1), bag_2 (b_2) and chain fibre contacts. The MG muscle was subjected to repeated ramp and hold stretch during which central stimulation was applied for several cycles. Figure 1 illustrates the identification of three γ axons. In response to stimulation in the midbrain at a point dorsal to the midbrain locomotor region (MLR) two γ efferents showed an immediate and marked increase in impulse frequency. The third γ efferent showed a decrease in discharge rate after a short delay. Also displayed are the discharges of a primary and a secondary spindle afferent. It is evident in both cases that their baseline discharge rates (bias) are increased during midbrain stimulation and that there was no increase in the dynamic sensitivity of the primary ending to muscle lengthening. The action must have been due to an increase in γ_S output since both afferents, and four others recorded at the same time (not shown), all showed a marked increase in bias during stimulation. Our interpretation is that the two γ axons showing an increase in discharge rate are γ_S efferents and that the third is a γ_D efferent.

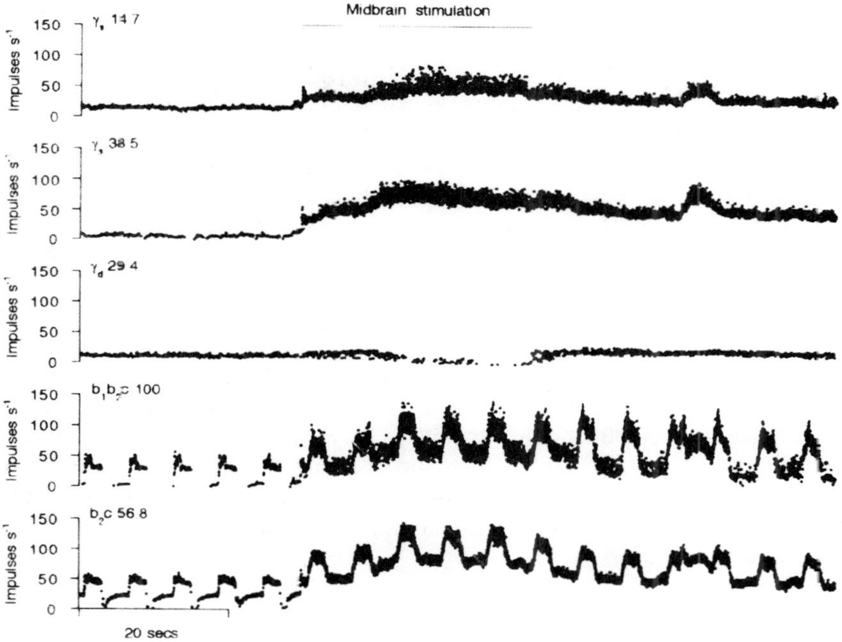

Figure 1. Identification of γ_S and γ_D efferents. Upper three traces show discharges of three single γ-axons recorded simultaneously from a small filament of the MG nerve, which is otherwise intact. Type and conduction velocity in m s^{-1} are shown inset. The lower two traces show discharges of a primary and a secondary MG spindle afferent recorded simultaneously with the γ-axon activity. Intrafusal fibre contacts and conduction velocity are shown inset. All records are instantaneous frequency plots. Ramp and hold stretches of MG (not shown) were applied continuously every 6 s. The spindle records show that midbrain stimulation (20 Hz, 0.2 ms, 1.3 V, signaled by the horizontal bar) increased static and decreased dynamic fusimotor activity. (From Taylor et al., 2000b, with permission.)

γ_S Firing Patterns

The behaviour of identified γ_S efferents was monitored during locomotion elicited by turning on the treadmill. The induced stepping rate was approximately 1.5 steps s^{-1}. The left ankle was free to move without loading. The relationship of γ_S firing to the movement is displayed in cycle averages in Figure 2. The point of maximum ankle flexion has been taken as time zero and the time axis has been normalised with respect to cycle duration. For one type of γ_S unit (filled symbols) there was a noticeable, abrupt increase in rate of discharge approximately 15% of the step cycle before the onset of ankle extension (MG muscle shortening). The discharge then remained elevated throughout that shortening phase falling rapidly at the beginning of the lengthening phase. Another type of unit (open symbols) was less strongly modulated and simply showed a slow, steady rise in discharge rate, starting at the peak of ankle extension and continuing throughout the flexion phase before falling again during ankle extension (muscle shortening). The discharge patterns of γ_S efferents recorded in other animals during locomotion fell into one or other of these two patterns. Thirteen γ_S neurons consistently increased their firing during extension (type-1) and seven others increased their firing during flexion (type-2).

Cursory inspection of the recordings might suggest that the type-1 γ_S efferent discharge is an example of alpha-gamma co-activation. Certainly the type-2 discharge cannot be so described. Indeed, the discharge of the type-2 MG γ_S efferents more closely parallels that of the antagonist TA alpha motoneurone activity. Even the type-1 γ_S efferent discharge does not closely match that of the MG alpha activity. The type-1 γ_S activity is phase advanced with respect to the EMG and is a far better match of the secondary difference record (difference between the active and passive discharge of secondary endings, see Taylor et al., 2000b). The profiles of both the type-1 γ_S and the secondary difference record match the movement of the ankle (albeit with phase advance) better than they do the EMG of the MG muscle. The present work reveals only the discharge patterns of γ fusimotor neurones and provides no insight into the involvement of mixed skeletofusimotor (beta) neurones that provide obligatory anatomical alpha-gamma linkage and functional co-activation of intrafusal and extrafusal muscle fibres. Prochazka and Gorassini (1998) have found that predicted models of primary afferent discharge during locomotion are fitted better if some phasic component of γ_S is added linked to the EMG. However, the present results appear to relate better to a γ_S profile that is closely related to length changes.

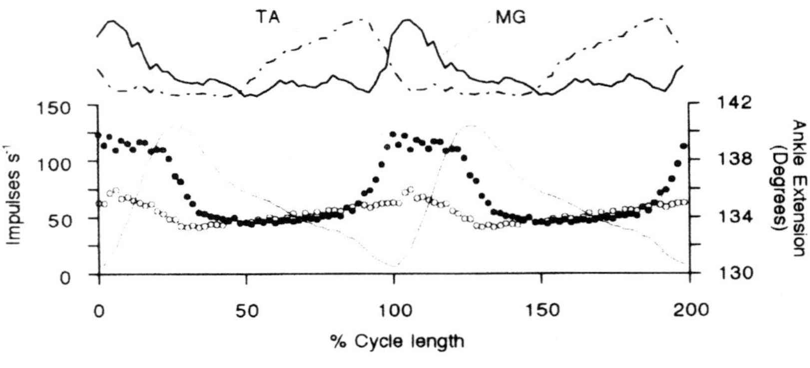

Figure 2. Behaviour of γ_S efferents during locomotion. Averages (n = 14) computed for locomotor cycles of regular amplitude of ankle movement for the γ_S efferents identified in Figure 1. The uppermost two traces are the smoothed, rectified EMG signals from MG and TA. The lower three records are the averaged frequency records for a type-1 γ_S efferent (filled circles), a type-2 γ_S efferent (open circles) and the averaged ankle angle (continuous line). In the horizontal scale, time is normalised with respect to the individual stepping cycle duration (mean cycle period 0.65 s). (Modified from Taylor et al., 2000b, with permission.)

γ_D Firing Patterns

Our initial studies (Taylor et al., 2000a) revealed an increased dynamic fusimotor drive during active locomotion but could shed no light on the timing of that discharge. Previously, the idea of fusimotor set (Prochazka et al., 1985) had led to the expectation of a maintained rise in γ_D activity during locomotion but this was contradicted by direct recordings of a fluctuating drive (Murphy et al., 1984).

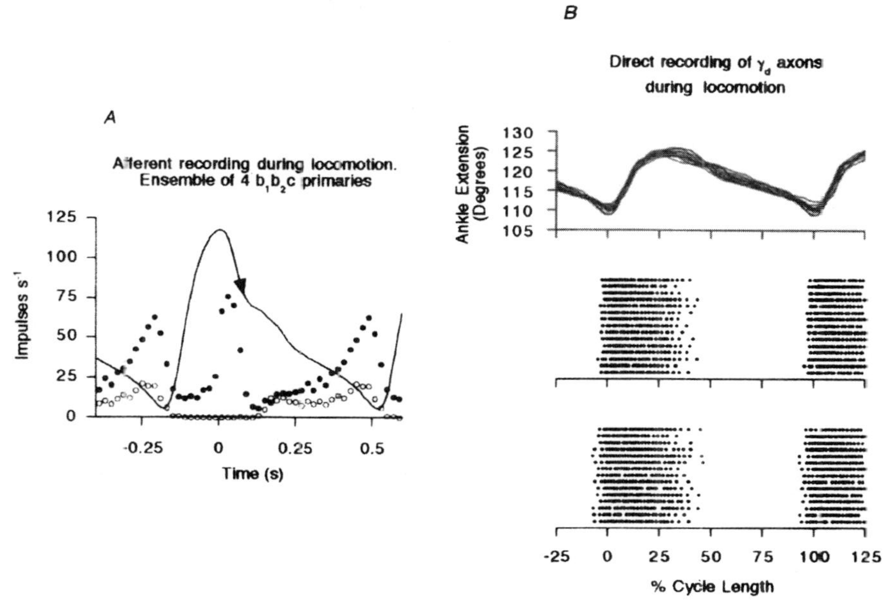

Figure 3. Primary spindle activity and the timing of γ_D firing during locomotion. *A*, Cycle histograms of the discharges of an ensemble of four b_1b_2c primaries during rhythmic locomotion of regular amplitude. The active (filled circles) and the passive (open circles) discharges and ankle angle (continuous line, arrow indicates muscle lengthening) are shown. *B*, separate experiment in which 2 γ_D efferents were recorded simultaneously. Above: 15 superimposed ankle movement traces, normalised with respect to cycle length. The mean cycle period was 0.96 s. Below: dot raster displays of the discharges of the γ_D efferents. Note the abrupt onset of the γ_D discharge at the beginning of ankle extension and the cessation of discharge shortly after the beginning of ankle flexion (muscle lengthening). Interrupted vertical lines mark the transition from ankle flexion to extension (left) and *vice versa* (right) in both panels. (Modified from Taylor et al, 2000a,b with permission.)

The discharge patterns of two MG γ_D axons recorded simultaneously during well-sustained locomotion are shown as raster displays in Figure 3B. The regular ankle movements are displayed above the rasters. The firing of both γ_D efferents fluctuates during each locomotor cycle in the same specific way. The γ_D units are silent during most of the ankle flexion phase, start firing suddenly just before the start of muscle shortening (ankle extension) and then stop early during ankle flexion. This burst-like pattern with peak frequencies ranging from 18 to 90 impulses s^{-1} was typical of the 13 dynamic units recorded. The burst-like nature of the γ_D discharge suggests that its timing may be critically important for the dynamic sensitivity of spindle primary endings. Knowing the pattern of γ_D discharge now makes it easier to interpret primary discharges recorded during locomotion. Figure 3A shows a cycle histogram of the ensemble discharge of four b_1b_2c primary endings both during active locomotion and passively when all γ activity was blocked with anaesthetic. Superimposed is the ankle angle with ankle flexion indicated by the downward arrow. The increase in active primary firing that reaches a peak at the end of ankle flexion, i.e. MG muscle extension, is much greater than the passive response and can be explained by the modulation of γ_S activity described above. Another, high frequency, burst of primary firing occurs just at the beginning of ankle flexion but does not continue despite the continuing MG muscle extension. Evidently, the persistence of the elevated γ_D discharge just into the phase of ankle flexion accounts for this burst of primary activity. This conclusion was reinforced by experiments in which direct stimulation of γ_D axons was employed in different phases of sinusoidal stretch of the muscle (Taylor et al., 2000b).

DISCUSSION

We have recently reviewed previous evidence that locomotion is supported by specific patterns of γ_S and γ_D activity and discussed how the present results have advanced our understanding of the role of the fusimotor system in locomotion (Taylor et al., 2000a,b).

Behaviour of γ_S Motor Activity

In summary, the present results extend the findings of Murphy et al., (1984) but emphasise that γ_S efferents may be deeply modulated in time with the locomotor cycle. Additionally, our approach has revealed two distinct patterns of γ_S activity. Approximately 65% of units (type-1) have a smooth and deeply modulated pattern of activity that closely resembles the time course of the difference of active minus passive firing of the spindle secondary afferents. Both the γ_S efferent and the secondary afferent patterns of activity match the profile of active muscle shortening, but with some phase advance. The type-1 MG γ_S modulation is superimposed on a tonically raised level of static activity. The other 35% of γ_S efferents (type-2) have a distinctly different pattern of modulation. They show an increase in discharge rate aligned more closely with TA contraction. Since the two types of discharge pattern are qualitatively different we have speculated that they may reflect preferential innervation of bag_2 fibres by type-2 and mixed innervation of chain and bag_2 fibres by type-1 γ_S axons (Taylor et al., 2000b). Although the question of the degree of specificity of innervation of bag_2 and chain fibres by distinct axons is not settled (Celichowski et al., 1994; Taylor et al., 1998), the different γ_S firing patterns suggests that the distinct properties of the bag_2 and chain fibres may indeed be exploited by the CNS (Gladden, 1981). The identification of two types of γ_S activity would also argue against any strict alpha-gamma linkage in the static system even without the evidence that the patterns of neither type match well that of the EMG recorded during locomotion.

Behaviour of γ_D Motor Activity

The present findings have emphasised the very sudden onset and rapid fall to zero of γ_D firing in each locomotor cycle. These abrupt changes are qualitatively different from the continuously modulated firing of both types of γ_S motoneurones neither of which stop discharging at any point in the locomotor cycle. The timing of the γ_D discharge has two functional implications. First, it would sensitise spindle primary afferents to the initial rapid component of muscle stretch during the ankle flexion phase (Durbaba et al., 2000), but allow the sensitivity to stretch to fall immediately afterwards. The γ_D firing thus appears to be responsible for the vigorous primary afferent burst at the onset of muscle lengthening in the decerebrated preparation and also in normal locomotion (Prochazka, 1996). That afferent discharge might constitute an appropriate timing signal fed back to the central pattern generator to entrain the natural motion of the limb (Whelan and Pearson, 1997). Second, the presence of γ_D firing during muscle shortening would make primary endings sensitive to any unexpected lengthening that might result from obstruction of the limb's trajectory.

The burst of γ_D firing is closely associated in time with the EMG observed in the MG muscle and raises again the question whether such apparent co-activation is due to alpha-gamma linkage as envisaged by Bessou and colleagues (1990). The recordings made by Murphy et al., (1984) also showed co-activation of γ_D and EMG activity during locomotion. However, later studies of cutaneous reflexes revealed that sural nerve input excited γ_D axons but simultaneously inhibited EMG activity (Murphy and Hammond, 1991) suggesting that tight alpha-gamma linkage is unlikely to be operating for the dynamic system during locomotion.

Comparison with Natural Locomotor Patterns

The relevance of our data to the natural behaviour in an intact animal has been discussed (Taylor et al., 2000a,b) but was raised again at this symposium by Pearson (see Lam and Pearson, Chapter 40). He also described recordings made from muscle spindle afferents in a pre-mammillary, decerebrated cat with ipsilateral denervation except for one head of the gastrocnemius-soleus (GS). However, the muscle was held isometric at a length where passive tension would begin to arise if the muscle were to be stretched. During locomotion, Pearson recorded vigorous GS spindle discharge that shut down during muscle contraction. The reduction during contraction was greater than observed in our preparation where the muscle was allowed to shorten during locomotion. Since the reverse would be expected from passive spindle properties, the difference in results suggests that γ-discharge is inhibited when the muscle contracts isometrically, possibly due to afferent force feedback (Stein et al., 2000). The wider question is whether circuits generating locomotion in such decerebrated and partially denervated animals resemble normal processes (Prochazka et al., 1976; Prochazka and Gorassini, 1998). Although input from the other limbs would be expected to compensate for the denervation to some extent, reflex effects on γ efferents will have been reduced. One obvious behavioural difference with normal cats is the absence in our preparation of external loading and foot contact with the ground. Afferent feedback from tendon organs (Ellaway and Murphy, 1980) and cutaneous receptors (Johansson and Sojka, 1985; Ellaway et al., 1997) changes fusimotor output and this may be task dependent (Murphy and Hammond, 1991).

What has clearly emerged is the recognition of different and modulated activity patterns in the dynamic and in two types of static fusimotor neurones. These varied patterns do not support the classical concepts of α-γ linkage. It appears rather that full use may be made by the CNS of the three different kinds of intrafusal muscle fibres to adapt the properties of the muscle spindle to suit the various demands made on skeletal muscle during locomotion.

ACKNOWLEDGEMENTS

This research was supported by a grant from the UK Medical Research Council. We thank Mrs. O.D. Taylor for technical assistance.

REFERENCES

Appenteng, K., Morimoto, T., and Taylor, A., 1980, Fusimotor activity in masseter nerve of the cat during reflex movements, *Journal of Physiology,* 305, 415-431.

Bessou, P., Dupui, J. M., Cabelguen, M., Joffroy, M., Montoya, R., and Pagès, B., 1990, Discharge patterns of γ-motoneurones of extensor and flexor hindlimb muscles during walking in the thalamic cat, *Progress in Brain Research,* 80, 37-45.

Celichowski, J., Emonet-Dénand, F., Laporte, Y., and Petit, J., 1994, Distribution of static γ-axons in cat peronius tertius spindles determined by exclusively physiological criteria, *Journal of Neurophysiology,* 71, 722-732.

Durbaba, R., Taylor, A., Ellaway, P. H., and Rawlinson S., 2000, The importance of the timing of dynamic gamma discharge for its effects on the primary muscle spindle afferent response to stretch, *Journal of Physiology,* 527, 136-137P.

Ellaway, P. H., Davey N. J., and Ljubisavljevic, M., 1997, Organization of the sural cutaneous input regulating the discharge of triceps surae γ-motoneurones in the cat, *Experimental Physiology,* 82, 121-138.

Ellaway, P. H., and Murphy, P. R., 1980, Autogenetic effects of muscle contraction on extensor gamma motoneurones in the cat, *Experimental Brain Research,* 38, 305-312.

Gladden, M. H., 1981, The activity of intrafusal muscle fibres during central stimulation in the cat, in: *Muscle Receptors and Movement*, A. Taylor, and A. Prochazka, eds., Macmillan, London, pp. 109-122.

Johansson, H., and Sojka, P., 1985, Actions on gamma motoneurones elicited by electrical stimulation of cutaneous afferent fibres in the hindlimb of the cat, *Journal of Physiology,* 366, 343-364.

Lund, J. P., Smith, A. M., Sessle, B. J., and Murakami, T., 1979, Activity of trigeminal alpha and gamma motoneurones and muscle afferents during performance of a biting task, *Journal of Neurophysiology,* 42, 710-725.

Murphy, P. R., and Hammond, G. R., 1991, The role of cutaneous afferents in the control of γ-motoneurones during locomotion in the decerebrate cat, *Journal of Physiology,* 434, 529-547.

Murphy, P. R., Stein, R.B., and Taylor, J., 1984, Phasic and tonic modulation of impulse rates in γ-motoneurones during locomotion in premammillary cats, *Journal of Neurophysiology,* 52, 228-243.

Prochazka, A., 1996, Proprioceptive feedback and movement regulation, in: *Handbook of Physiology. Section 12. Exercise: Regulation and integration of multiple systems*, L B. Rowell, and J. T. Sheperd, eds., American Physiological Society, New York, pp. 89-127.

Prochazka, A., and Gorassini, M., 1998, Ensemble firing of muscle afferents recorded during normal locomotion in cats, *Journal of Physiology,* 507, 293-304.

Prochazka, A., Hulliger, M., Zangger, P., and Appenteng, K., 1985, "Fusimotor set": new evidence for γ-independent control of γ-motoneurones during movement in the awake cat, *Brain Research,* 339, 136-140.

Prochazka, A., Westerman, R., and Ziccone, S. P., 1976, Discharges of single hindlimb afferents in the freely moving cat, *Journal of Neurophysiology,* 39, 1090-1104.

Stein, R. B., Misiaszek, J. E., and Pearson, K. G., 2000, Functional role of muscle reflexes for force generation in the decerebrate walking cat, *Journal of Physiology,* 525, 781-791.

Taylor, A., Durbaba, R., Ellaway, P. H., and Rawlinson, S., 2000a, Patterns of fusimotor activity during locomotion in the decerebrate cat deduced from recordings from hindlimb muscle spindles, *Journal of Physiology,* 522, 515-532.

Taylor, A., Ellaway, P. H., and Durbaba, R., 1998, Physiological signs of the activation of bag_2 and chain intrafusal muscle fibers of gastrocnemius muscle spindles in the cat, *Journal of Neurophysiology,* 80, 130-142.

Taylor, A., Ellaway, P. H., Durbaba, R., and Rawlinson, S., 2000b, Distinctive patterns of static and dynamic gamma motor activity during locomotion in the decerebrated cat, *Journal of Physiology,* 529, 825-836.

Whelan, P. J., and Pearson, K. G., 1997, Comparison of the effects of stimulating group I afferents on cycle period during walking in conscious and decerebrate cats, *Experimental Brain Research,* 117, 444-452.

THE ROLE OF PROPRIOCEPTIVE FEEDBACK IN THE REGULATION AND ADAPTATION OF LOCOMOTOR ACTIVITY

Tania Lam[1] and Keir G. Pearson [1,2]

ABSTRACT

Feedback from muscle afferents is essential for locomotion to be functional under changing external conditions. In this article, we review the role of afferent feedback in adapting locomotor activity to transient and more sustained changes in sensory input in reduced and walking cat preparations. Much of the work on muscle afferent regulation of locomotion has focused on the regulation of stance phase activity. Proprioceptive feedback from extensor muscles during the stance phase ensures that the leg does not go into swing when loaded and that the magnitude of extensor activity is adequate for support. Proprioceptive feedback from flexor muscles towards the end of the stance phase facilitates the initiation of the swing phase of walking. Evidence that muscle afferent feedback also contributes to the magnitude and duration of flexor activity during the swing phase has been demonstrated recently. The regulation of the magnitude and duration of extensor and flexor activity during locomotion is mediated by monosynaptic, disynaptic, and polysynaptic muscle afferent pathways in the spinal cord. In addition to allowing for rapid adaptation in motor output during walking, afferent feedback from muscle proprioceptors is also involved in longer-term adaptations in response to changes in the biomechanical or neuromuscular properties of the walking system.

INTRODUCTION

It is well established that central pattern generators (CPG) located in the spinal cord produce the basic pattern of alternating flexor and extensor activity during mammalian locomotion (reviewed in Rossignol, 1996). This basic pattern of locomotor activity is strongly regulated by input from sensory afferents. Shaping of the basic locomotor pattern by sensory input is essential for meaningful locomotion to take place in changing

[1] Centre for Neuroscience and [2] Department of Physiology, University of Alberta, Edmonton, AB T6G 2S2, Canada. Email: kpearson@ualberta.ca

conditions. Afferent feedback is also important for the adaptive modification of the locomotor system after changes or injury to the musculoskeletal or nervous system. The first two sections of this review specifically focus on the regulation of extensor and flexor burst activity during locomotion by muscle proprioceptors. The final section focuses on the role of proprioceptive input in mediating adaptive modifications after biomechanical changes or nervous system injury. Some recent reviews covering these and related topics are also available (Pearson, 1995; Rossignol, 1996, McCrea, 1998; Pearson et al., 1998, Burke, 1999, Zehr and Stein 1999, Duysens et al., 2000, McCrea, 2001).

AFFERENT REGULATION OF THE TIMING AND MAGNITUDE OF EXTENSOR BURST ACTIVITY

Timing

Much of the recent work on the proprioceptive regulation of stepping has been directed at extending ideas drawn from initial studies by Duysens and Pearson (1980) and Grillner and Rossignol (1978) on afferent mechanisms regulating stance phase duration. Duysens and Pearson (1980) proposed that unloading of the support limb promotes the transition from stance to swing whereas Grillner and Rossignol (1978) proposed that extension of the hip joint is an important factor in promoting this transition. It is now quite apparent that both load- and length-related sensory inputs are involved.

Numerous observations have demonstrated that load-sensitive afferents influence the timing of the locomotor rhythm in cats (Duysens and Pearson, 1980; Conway et al., 1987; Pearson et al., 1992; Pearson and Collins 1993; Whelan et al., 1995a). The most direct evidence is that rhythmic force pulses delivered to ankle extensor muscles via electrical stimulation of the L7 ventral roots entrains the locomotor rhythm in acute spinal cats (Pearson et al., 1992). Entrainment of the locomotor rhythm in spinal cats can also be achieved by applying large amplitude stretches to the ankle extensor muscles but not by the application of low amplitude muscle stretches or vibrations, which would preferentially only activate muscle spindles (Conway et al., 1987; Pearson et al., 1992; Pearson and Collins, 1993). In addition, electrical stimulation of group I afferents from the ankle extensor muscles, particularly from the lateral gastrocnemius/soleus (LGS), increases stance phase duration and cycle period in decerebrate walking cats (Whelan et al., 1995a). All these observations indicate that force-sensitive Golgi tendon organs strongly influence the locomotor rhythm generating network.

An important role for length-sensitive muscle spindle afferents in the timing of locomotor activity was first suggested by the observation that sinusoidal hip movements could entrain the locomotor rhythm in the hindlimb of spinal (Andersson and Grillner, 1983) and decerebrate cats (Kriellaars et al., 1994). Denervation of afferents from the hip joint capsule did not affect the entrainment while progressive elimination of afferent input from various hip muscles led to a weakening of the entrainment (Kriellaars et al., 1994). These findings, along with the fact that Golgi tendon organ responses are weakened in paralyzed preparations (Prochazka and Wand, 1980), led to the conclusion that the entrainment by sinusoidal hip movements was largely dependent on spindle afferents arising from different muscles around the hip.

Subsequent experiments by Hiebert and colleagues (1996) provided direct evidence that spindle afferents from hip muscles influence the timing of the locomotor rhythm.

They found that activation of Ia afferents from the hip flexor muscle iliopsoas (IP) by ramp stretches or vibration promoted an earlier onset of flexor activity in walking decerebrate cats and could also entrain the locomotor rhythm. Spindle afferents from ankle flexor and extensor muscles also influence the timing of the locomotor rhythm (Guertin et al., 1995; Hiebert et al., 1996). The fact that ankle extensor group Ia afferents can influence the locomotor rhythm in decerebrate cats (Guertin et al., 1995) but not in spinal cats (see above, Conway et al., 1987; Pearson et al., 1992) may reflect a difference between the decerebrate and spinal cat preparations. Secondary muscle spindles from flexor muscles are also effective in influencing the locomotor rhythm since activation of group II muscle afferents from the tibialis anterior (TA) during extension promoted flexor activity in decerebrate walking cats (Hiebert et al., 1996). On the other hand, in fictive decerebrate cats, stimulation of group II afferents from the sartorius muscle during extension prolonged the extensor phase (Perreault et al., 1995). Notwithstanding the latter observation, the main functional implication of recent findings on the role of muscle spindles in flexor muscles is that lengthening of the flexor muscles at the end of stance facilitates the termination of extensor activity thus contributing to the triggering of the stance-to-swing transition.

The question of whether afferent feedback is important for regulating the timing of extensor activity during human walking has also been examined (Stephens and Yang 1999; Stein et al., 2000; reviewed in Duysens et al., 2000). Stephens and Yang (1999) reported a small effect on stance and cycle duration with the application of an additional 30% of body weight load to the legs. Similarly, Misiaszek et al. (2000) observed slight changes in the timing of the stance to swing transition when load was applied along the long axis of the leg during supported walking (by holding a rail) in adult humans. The absence of marked effects on timing may be due to the fact that subjects were forced to walk at a constant rate on a treadmill. When the subjects did not support themselves, a prolongation of stance duration occurred, but this was accompanied by complex responses in other muscles. Thus, the effect on timing could not be completely accounted for by loading of the limb itself and may involve other factors related to the control of equilibrium and volitional responses (Misiaszek et al., 2000). One way these factors can be eliminated is by using human infants. Infants provide a means to study human stepping with minimal interference from descending input and postural responses (Forssberg, 1985). A recent study has shown that phasic sensory cues such as unloading of the support limb or extension of the hip joint are strong stimuli for the initiation of the swing phase, similar to that reported in reduced preparations of the cat (Pang and Yang, 2000). Developmental changes, the nature of bipedal walking, and volitional responses may be factors that account for the differences in the results obtained from cats and adult humans.

Magnitude

Proprioceptive input plays an important role in regulating the magnitude of extensor activity during locomotion. One demonstration of this comes from a series of "foot-in-hole" experiments in cats (Gorassini et al., 1994; Hiebert et al., 1994; Hiebert et al., 1995). With the unexpected loss of ground support, there is a shortening of extension and a marked reduction in the magnitude of extensor activity before the limb is lifted out of the hole (Gorassini et al., 1994). Hiebert and colleagues (1994) proposed that these responses were due to the lack of Ia and Ib afferent activity from extensor muscles that should normally reinforce extensor muscle activity during stance. Evidence for this proposal was later obtained by Hiebert et al. (1995) who showed that the corrective response was suppressed and extensor activity was maintained with electrical stimulation of extensor group I muscle afferents delivered when the foot entered the hole, thus indicating that feedback from extensor group I afferents reinforces extensor activity during locomotion. This conclusion supported previous findings that stimulation of extensor group I afferents enhanced the amplitude of extensor bursts during stance in spinal (Conway et al., 1987; Gossard et al., 1994) and decerebrate cats (Guertin et al., 1995; McCrea et al., 1995).

More recently, the extent to which afferent feedback contributes to extensor activity during walking has been examined quantitatively. Using a variety of approaches, Hiebert and Pearson (1999) estimated that between 50 and 80% of knee and ankle extensor activity is produced by afferent feedback during walking in decerebrate cats. First, the overall contribution of afferent feedback to extensor burst activity during locomotion was estimated by measuring the reduction in ankle and knee extensor activity during the "foot-in-hole" trials. Electromyographic (EMG) activity in these muscles was reduced by about 70%. If the ankle extensors were loaded as the leg entered the hole, the reduction in extensor activity was prevented. Second, unloading of the hindlimbs (by lifting the hindquarters) during treadmill locomotion resulted in a large decrease in the magnitude of knee extensor EMG. Finally, transection of the dorsal roots that carry afferent input from the knee extensor muscles resulted in a reduction of knee extensor muscle activity by about 50%. Transection of more caudal dorsal roots that carry afferent input from ankle extensor muscles resulted in a reduction of ankle and knee extensor activity by between 50 and 80%.

In a subsequent study, Stein and colleagues (2000) quantified the contribution of reflexes, evoked by phasic length changes of the ankle extensors, to the force generated by the ankle extensor muscles during the stance phase. In walking decerebrate cats, one hindlimb was immobilized and the triceps surae muscles were attached to a mechanical puller. With this puller, the triceps surae could either be held at a constant length or made to go through the changes in length that occur normally during the E2 and E3 phases of locomotion during walking. By using this approach and comparing muscle forces with the forces generated in deafferented triceps surae muscles, it was estimated that 35% of the force generated in ankle extensors during stance comes from feedback related to length changes that occur during walking.

The importance of afferent feedback from extensor muscles in regulating extensor burst magnitude during stance has also been shown in humans. In both healthy and spinal cord injured human adults, afferent feedback during the stance phase of locomotion contributes to the magnitude of extensor activity (Yang et al., 1991; Harkema et al., 1997; Stephens and Yang 1999; Sinkjaer et al., 2000). Sinkjaer et al. (2000) reported that up to 50% of soleus muscle activity during the stance phase of human walking is due to afferent feedback. When the ankles were unloaded (using a motorized mechanical joint strapped around the ankle) during treadmill walking, a significant reduction (50%) in soleus EMG activity was observed (Sinkjaer et al., 2000). Group II afferents may contribute to the magnitude of soleus EMG during the stance phase in human walking (Grey et al., 2001). Their conclusions were based on the fact that the medium-latency component of the stretch reflex response evoked during the stance phase was not influenced by ischemic block of soleus Ia afferents. In addition, after subjects ingested tizanidine (a drug known to depress transmission in group II afferent pathways), the medium-latency component of the corrective response to ankle extensor stretch during the stance phase was depressed (Grey et al., 2001). It is not known whether tizanidine has an effect on Ib afferent pathways, so the possibility that load-sensitive afferents also contribute to the medium-latency stretch reflex response has not been excluded.

Segmental Pathways

With the demonstration that proprioceptive feedback regulates the timing and magnitude of extensor activity during locomotion, an important issue is the identity of the segmental reflex pathways by which these afferents exert their action on the locomotor system. It is now apparent that multiple pathways are involved in shaping the output of the locomotor CPG and that input from different afferents converge onto interneurons thus yielding common effects on locomotor activity. In addition, transmission in many of these reflex pathways is modulated in a task- and phase-dependent manner (reviewed in McCrea, 2001). For example, the inhibitory influence from group Ib extensor afferents to extensor motoneurons becomes excitatory during locomotion (Pearson and Collins, 1993, McCrea et al., 1995).

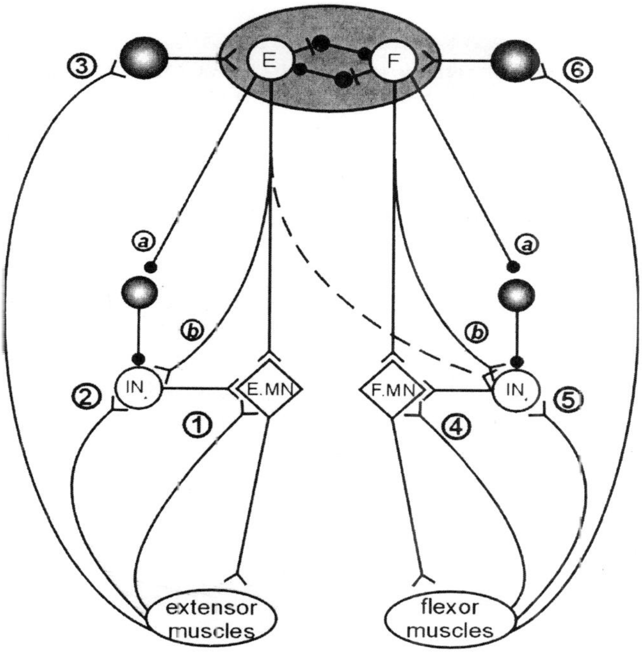

Figure 1. Muscle afferent pathways regulating extensor and flexor activity during locomotion. Excitatory connections are represented by inverted arrows and inhibitory connections by black dots. The locomotor central pattern generator is represented by the network contained in the grey shaded ellipse. Shaded circles represent putative interneuron(s) in the connections made to and from the locomotor CPG. Interneurons INe (left) and INf (right) are the interposed interneurons in the identified disynaptic reflex pathway from extensor and flexor group I afferents to extensor and flexor motoneurons, respectively. Extensor and flexor motoneurons are represented by diamonds. Pathways 1 and 4 are the monosynaptic pathway from group Ia afferents to their respective motoneurons. Pathways 2 and 5 represent the disynaptic pathway from group I afferents to the motoneurons. Pathways 3 and 6 represent muscle afferent pathways which directly affect the locomotor central pattern generator. (Modified from Pearson, 1995; Quevedo et al., 2000; McCrea, 2000.)

Figure 1 illustrates the various pathways by which muscle afferents affect the locomotor output. As shown in the left side of the figure, extensor group I afferents can exert their effects on extensor motoneurons by at least three pathways. First, extensor muscle afferents can exert a direct effect on extensor motoneurons via the well-known group Ia monosynaptic pathway (Eccles et al., 1957) (*pathway 1*, Fig. 1). Second, extensor group I muscle afferents can excite extensor motoneurons via disynaptic excitatory pathways (McCrea et al., 1995, Angel et al., 1996) (*pathway 2*, Fig. 1). A feature of this disynaptic excitatory pathway is that it is open only during locomotion (Angel et al., 1996, McCrea et al., 1995). The locomotor-dependent disynaptic excitatory connection may be opened by either a disinhibition (*pathway a*, Fig. 1) or excitation from the extensor half-centre (*pathway b*, Fig. 1) of the intervening interneuron (IN_e in Fig. 1). Recently, an interneuron population that may mediate the disynaptic excitation of extensor motoneurons during locomotion was identified (McCrea et al., 2001). These interneurons are located in laminae 4-6 and close to the motoneurons to which they project (see McCrea 2001). A third pathway by which extensor group I afferents can shape locomotor output is by direct action onto the locomotor CPG. The ability for a given stimulus to reset or entrain the locomotor rhythm is evidence for its effect on the locomotor CPG (Hultborn et al., 1998). As discussed in the preceding sections, activation of extensor group I afferents by various stimuli (e.g. large amplitude ramp stretches, force pulses, electrical stimulation) entrains the locomotor rhythm (Conway et al., 1987; Pearson et al., 1992; Gossard et al., 1994) and also prolongs the duration of extensor activity (Guertin et al., 1995; Whelan et al., 1995a) (*pathway 3*, Fig. 1). The fact that activation of primary and secondary muscle spindles from flexor muscles also resets the locomotor rhythm indicates a pathway from these afferents to locomotor CPG (not shown in Fig. 1) (Perreault et al., 1995, Hiebert et al., 1996).

AFFERENT REGULATION OF THE TIMING AND MAGNITUDE OF FLEXOR BURST ACTIVITY

An important issue in the regulation of swing phase activity during walking is how the nervous system ensures that the trajectory of the limb and subsequent foot placement in preparation for support is accurately achieved. The significant role of cutaneous afferents (reviewed in Rossignol, 1996) and supraspinal control (reviewed in Drew et al., 1996) during the swing phase has been established. The first evidence for the influence of proprioceptive feedback during the swing phase of locomotion came from the experiments of Orlovskii and Shik (1965). During treadmill locomotion in dogs they found that when a braking force was applied to the forelimb at the elbow, the forelimb was able to quickly compensate (within 30 ms) and resume forward movement with the same speed and inter-joint coordination as during control steps. The short time delay of the compensatory response led Orlovskii and Shik (1965) to conclude that segmental mechanisms exist which monitor and appropriately respond to changes in proprioceptive input during the swing phase.

Timing and Magnitude

Recently, more evidence for the functional importance of proprioceptive input during the swing phase of walking has emerged. It is now apparent that, as with the extensor system, both the timing and magnitude of flexor activity are regulated by proprioceptive feedback from receptors in flexor muscles. This was clearly demonstrated in a recent study in decerebrate walking cats (Lam and Pearson, 2001a). Proprioceptive disturbances applied during the swing phase strongly influenced the duration and magnitude of activity in the hip flexors IP and sartorius (Sart). Assisting hip flexion during swing shortened and diminished IP and Sart burst activity while resisting hip flexion strongly increased IP and Sart burst activity during swing. Afferent feedback from the sartorius muscles was found to be particularly important for this response while input from IP had only a weak influence (Lam and Pearson, 2001a).

Although the specific receptors controlling flexor burst activity in these experiments could not be identified, the results suggest that Golgi tendon organs and muscle spindles could be important receptors for monitoring the load and length changes imposed on the muscles during the various perturbations. Indeed, a significant advance over the past few years has been the accumulation of direct information on the role of feedback from these large muscle afferents in regulating the duration and magnitude of flexor activity during locomotion. During fictive locomotion in decerebrate cats, stimulation of group I afferents from TA, EDL, IP, Sart, and posterior biceps and semitendinosus (PbST) during the flexion phase generally resulted in a prolongation of flexor bursts, although the effects from TA stimulation were variable (Perreault et al., 1995, McCrea et al., 2000). In the case of Sart group I afferent stimulation, a prolongation of the step cycle duration was also produced (Perreault et al., 1995). Additionally, stimulation of group I muscle afferents from the IP and sartorius muscles enhances the magnitude of flexor bursts (McCrea et al., 2000).

Because stimulation of muscle afferents at group I strengths recruits afferents from both muscle spindles and Golgi tendon organs, the relative contribution of Ia and Ib afferents to flexor motoneuron activation during locomotion is presently unclear. A role for Ia afferents is likely since small stretches to the EDL muscle produced disynaptic excitation of EDL and TA motoneurons, similar to that obtained with electrical stimulation (Quevedo et al., 2000). On the other hand, stretch of the IP tendon during flexion produced little effect on cycle period or flexor burst duration (Hiebert et al., 1996; Lam and Pearson, 2001a). Thus, the extent to which group I flexor muscle afferents have an effect on the locomotor pattern appears to depend on the muscle from which these afferents arise.

The influence of feedback from flexor group II muscle afferents on the flexor phase of locomotion has also been investigated (Perreault et al., 1995, McCrea et al., 2000). Thus far, the findings have been inconsistent. Initially, Perreault et al. (1995) reported that stimulation of TA, PbST, and sartorius nerves at group II strength during flexion reset the locomotor rhythm to extension during fictive locomotion in decerebrate cats. More recently however, McCrea et al. (2000) found that during flexion, stimulation of group II muscle afferents from EDL, IP, and Sart prolonged the duration and enhanced the magnitude of flexor activity while stimulation of group II afferents from TA reset the locomotor rhythm to extension. Activation of group II afferents from the EDL and TA muscles also prolongs flexor burst duration in walking decerebrate cats (Hiebert et al., 1996). The inconsistency between reports of the action of group II muscle afferents may be partly explained by recent findings from our laboratory (Lam and Pearson, 2001b). During treadmill walking of decerebrate cats, group II stimulation of sartorius muscle afferents during flexion initially resulted in inhibition of IP bursts (decreased IP magnitude) followed by a resetting of the locomotor rhythm. However, with prolonged exposure to this stimulus (>5 trials), this inhibitory effect was suppressed and an excitatory response on IP burst duration and amplitude emerged (Lam and Pearson 2001b). Thus, it appears that the time-course and history of exposures to group II stimulation should be taken into account when considering the influence of group II flexor muscle afferents on locomotor activity and rhythm.

The effects of group I and II flexor muscle afferent stimulation on flexor activity during the swing phase of decerebrate cat locomotion could account for the effects observed in conscious cats undertaking different walking tasks. For example, in cats walking uphill, hip flexor muscles produce greater activity during the swing phase (Carlson-Kuhta et al., 1998). This increased activity could be mediated by increased Golgi tendon organ activity due to loading of the flexor muscles as the leg has to move more against gravity. Another example comes from our unpublished observations that the addition of an extra load strapped around a cat's hindlimb results in greater flexor burst magnitude and duration during swing (Lam and Pearson, unpublished data). Thus, feedback from Golgi tendon organs could be functionally important in situations where greater force generation is required in flexor muscles.

Segmental Pathways

The identity of the segmental pathways by which muscle afferents from flexor muscles exert an affect on flexor motoneurons during locomotion is beginning to be addressed. Data available so far indicate that the regulation of flexor burst activity by flexor group I afferent pathways bears a close resemblance to those from extensor afferents regulating extensor burst activity during locomotion (Fig. 1). Afferent feedback from flexors influences both timing and magnitude of flexor burst activity by direct action on flexor motoneurons (Lundberg 1981; Quevedo et al., 2000) (*pathway 4*, Fig. 1), by disynaptic excitatory pathways to flexor motoneurons (Degtyarenko et al., 1998; Quevedo et al., 2000) (*pathway 5*, Fig. 1), and by direct action on the locomotor pattern generator (Perreault et al., 1995; McCrea et al., 2000) (*pathway 6*, Fig. 1). Other similarities are convergence of different afferent inputs to produce common effects and reorganization of reflex pathways in a state-dependent manner. For example, the switch from an inhibitory to an excitatory connection in the disynaptic pathway from flexor group I afferents to flexor motoneurons is an example of reflex reorganization that occurs in the presence of locomotion (Quevedo et al., 2000). Disynaptic excitatory pathways from flexor group I afferents to flexor motoneurons are opened during locomotion (via *pathways a and b* onto IN_f in Fig. 1) but not at rest, with transmission strongest during the flexion phase and only weakly present during the extension phase (indicated by dotted line in Fig. 1). In addition, the pathways appear to provide input mainly to homonymous motoneurons. Input to synergists is present, but to a lesser extent (Quevedo et al., 2000).

ADAPTIVE PLASTICITY IN THE WALKING SYSTEM

The previous two sections have highlighted the importance of afferent pathways in regulating the motor output during locomotion and the significant advances that have been made in the identification of the pathways by which afferent feedback shapes locomotor output. The important concept is that proprioceptive inputs provide cues about the mechanical state of the musculoskeletal apparatus and this information is used to control the timing and magnitude of motor activity. One implication of this concept is that proprioceptive signals can also provide information about persistent changes in the mechanical properties of the system and therefore be used to adapt the motor output to meet the requirements of the altered limb mechanics. Since the musculoskeletal properties of the leg change throughout life, mechanisms must be in place for matching the motor output with current leg mechanics. There are a number of recent reviews on plasticity in the system generating the locomotor pattern (Rossignol, 2000; Pearson, 2000; Edgerton et al., 2001). This final section will focus on the role of proprioceptive feedback in mediating adaptive modifications to the locomotor network after peripheral nerve injury.

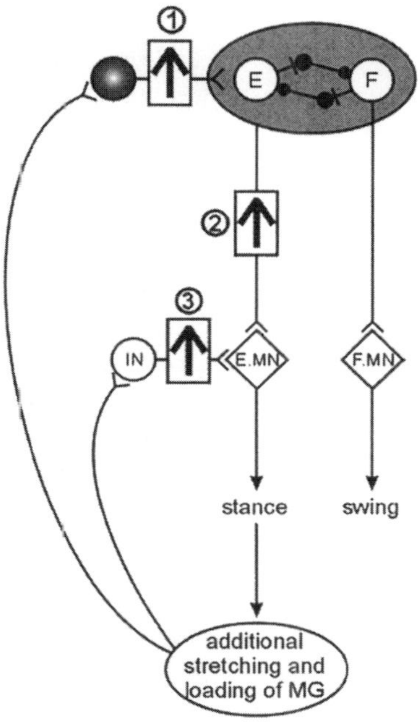

Figure 2. Adaptive modifications of locomotor pattern following partial denervation of ankle extensors. Schematic diagram of central and reflex pathways from MG group I afferents that regulate extensor activity during locomotion. After LGS and plantaris denervation, adaptive increases in MG activity occur via an increase in the influence in the oligosynaptic pathway from group I afferents to the locomotor central pattern generator (1), an increase in the central drive from the locomotor CPG to extensor motoneurons (2), and an increase in reflex activation of extensor motoneurons via the disynaptic excitatory pathway (3).

The first indication that adaptive plastic changes occur in the reflex pathways that regulate locomotor output came from studies on the influence of group I afferents on extensor duration (Whelan et al., 1995b, 1997). An interesting feature of the regulation of extensor burst duration by group I afferents from the ankle extensor muscles is that there is an imbalance in the relative contribution from afferents from different muscles. Stimulation of group I afferents from the LGS nerve is normally more effective than stimulation of group I afferents from the medial gastrocnemius (MG) nerve for regulating extensor burst duration during stance (Whelan et al., 1995a). After sectioning all of the ankle extensor nerves, except those from MG, the effectiveness of MG nerve stimulation became stronger in the days following the neurectomy and was able to exert a more powerful effect on extensor burst duration (Whelan et al., 1995b, Whelan and Pearson, 1997). The increase in efficacy of MG nerve stimulation on modifying extensor burst duration is most likely mediated by an increased gain in the pathway from MG group I afferents to the locomotor CPG (Whelan and Pearson, 1997) (*pathway 1*, Fig. 2).

Denervation of the synergists to MG also caused a progressive adaptive increase in the magnitude of MG activity during walking (Pearson et al., 1999, Pearson and Misiaszek, 2000). The increase in the late component of MG EMG (which is centrally and peripherally driven) occurred rapidly, within the first few days after denervation (Pearson et al., 1999). On the other hand, the magnitude of the early component of MG EMG (which is centrally driven) increased gradually over a 1-week period (Pearson et al., 1999). Both of these adaptive increases were correlated with functional improvement, as indicated by a decrease in the amount of ankle flexion during stance. The differential time course of recovery between the early and late components of MG EMG indicates separate mechanisms that mediate the adaptive increases in MG activity. Since the early component is generated by central mechanisms (Gorassini et al., 1994), the gradual increase in this component of MG EMG is likely mediated by an increase in central drive (*pathway 2*, Fig. 2) (Pearson et al., 1999, Gritsenko et al., 2001).

What are the mechanisms that underlie the rapid adaptive increase in the late component of MG EMG after denervation of MG synergists? Pearson and Misiaszek (2000) found a use-dependent increase in the slope of the relationship between MG magnitude and the amount of yield (flexion) at the ankle during stance. These observations are consistent with the idea that an increase in gain of the reflex pathways from MG afferents to MG motoneurons occurs after denervation of MG synergists (Pearson et al., 1999; Pearson and Misiaszek, 2000). This increase in gain appears to be driven by the greater afferent input from the isolated MG muscle since they do not occur if the ankle is immobilized after the neurectomy (Pearson et al., 1999). The adaptive increase in the magnitude of MG EMG could be due to facilitation of disynaptic reflex pathways from extensor group I afferents to MG motoneurons (*pathway 3*, Fig. 2). On the other hand, Gritsenko et al. (2001) contend that the adaptive increase in the late component of MG EMG after neurectomy could be accounted for by an increase in central drive. This conclusion was based on the fact that the increase in the initial component of MG EMG occurred in proportion to an increase in the MG stretch reflex response, thereby indicating that the adaptive increase in the stretch reflex was due to higher levels of central drive to MG motoneurons (Gritsenko et al., 2001). Methodological differences could account for the different conclusions made by Gritsenko et al. (2001) and Pearson and Misiaszek (2000). Gritsenko et al. (2001) commenced post-denervation recordings after 12 hours, during which time the cats were free to move about in their cages while Pearson et al. (1999) commenced post-denervation recordings after only 5 hours, during which time the cats were confined to a small kennel, allowing little movement. The largest increases in the magnitude of the late component were observed by Pearson et al. (1999) to occur early in the recovery phase. Thereafter, the magnitudes of the early and late components increased relatively in proportion, similar to the findings of Gritsenko et al. (2001). The combination of allowing the animals to exercise after the denervation coupled with a slightly longer time delay before the first recordings were made could account for Gritsenko et al. (2001) finding no evidence for an increase in gain in stretch reflex pathways.

After denervations of the lateral gastrocnemius and soleus muscles, chronic spinal cats also show similar functional improvement (decreased ankle flexion during stance) as observed in intact cats (Bouyer et al., 2001). This further indicates that neural mechanisms at the spinal level can detect and make compensations for deficits induced by ankle extensor neurectomy. However, chronic spinal cats do not consistently show the differential time course of adaptive increase in the early- and late-components of MG EMG. Furthermore, the increased effectiveness of electrical stimulation of MG group I muscle afferents on extensor burst duration seen in intact cats (Whelan and Pearson, 1997) does not necessarily occur in chronic spinal cats after neurectomy (Bouyer et al., 2001). Thus, it remains unclear whether the mechanisms used by chronic spinal cats to recover function after neurectomy are the same as those used by intact cats.

CONCLUSIONS

This review focused on the role of feedback from muscle afferents in regulating locomotor activity and the involvement of these afferents in the adaptive modifications to the locomotor pattern that take place after injury and persistent change to limb mechanics. Substantial progress has been made in the identification of the afferents and their segmental pathways in the regulation of locomotor activity, particularly those regulating stance phase activity. The identification of the interneurons interposed in these afferent pathways is beginning to occur. Further work is needed to elucidate the specific afferents and their pathways in the regulation of swing phase activity. Nevertheless, data available so far indicate that the pathways are analogous to those regulating the stance phase. Together, data on the afferent regulation of extensor and flexor activity provide an important framework for examining the mechanisms underlying adaptive modifications of locomotor activity in response to changes in the nervous or musculoskeletal systems. Indeed, it has been found that some of the same pathways that regulate extensor activity during locomotion can also undergo changes in gain to compensate for changes in the neuromuscular apparatus of the lower leg. Knowledge of the reflex pathways that regulate locomotor output and their role in adaptive modifications after injury have important implications for the development and improvement of artificial control systems as well as rehabilitative techniques in individuals after central and peripheral nervous system injury.

ACKNOWLEDGEMENTS

We thank Dr. John Misiaszek for his valuable comments on this manuscript.

REFERENCES

Andersson, O., and Grillner, S., 1983, Peripheral control of the cat's step cycle. II. Entrainment of the central pattern generators for locomotion by sinusoidal hip movements during "fictive locomotion", *Acta Physiologica Scandinavica,* 118, 229-239.

Angel, M. J., Guertin, P., Jimenez. T., and McCrea, D. A., 1996, Group I extensor afferents evoke disynaptic EPSPs in cat hindlimb extensor motoneurones during fictive locomotion, *Journal of Physiology,* 494, 851-861.

Bouyer, L. J. G., Whelan, P. J., Pearson, K. G., and Rossignol, S., 2001, Adaptive locomotor plasticity in chronic spinal cats after ankle extensors neurectomy, *Journal of Neuroscience,* 21, 3531-3541.

Burke, R. E., 1999, The use of state-dependent modulation of spinal reflexes as a tool to investigate the organization of spinal interneurons, *Experimental Brain Research,* 128, 263-277.

Carlson-Kuhta, P., Trank, T. V., and Smith, J. L., 1998, Forms of forward quadrupedal locomotion. II. A comparison of posture, hindlimb kinematics, and motor patterns for upslope and level walking, *Journal of Neurophysiology,* 79, 1687-1701.

Conway, B. A., Hultborn, H., and Kiehn, O., 1987, Proprioceptive input resets central locomotor rhythm in the spinal cat, *Experimental Brain Research,* 68, 643-656.

Degtyarenko, A. M., Simon, E. S., Norden-Krichmar, T., and Burke, R. E., 1998, Modulation of oligosynaptic cutaneous and muscle afferent reflex pathways during fictive locomotion and scratching in the cat, *Journal of Neurophysiology,* 79, 447-463.

Drew, T., Jiang, W., Kably, B., and Lavoie S., 1996, Role of the motor cortex in the control of visually triggered gait modifications, *Canadian Journal of Physiology and Pharmacology,* 74, 426-442.

Duysens, J., Clarac, F., and Cruse, H., 2000, Load-regulating mechanisms in gait and posture: comparative aspects, *Physiological Reviews,* 80, 83-133.

Duysens, J., and Pearson, K. G., 1980, Inhibition of flexor burst generation by loading ankle extensor muscles in walking cats, *Brain Research,* 187, 321-332.

Eccles, J. C., Eccles, R. M., and Lundberg, A., 1957, The convergence of monosynaptic excitatory afferents on to many different species of alpha motoneurones, *Journal of Physiology,* 137, 22-50.

Edgerton, V. R., Leon, R. D., Harkema, S. J., Hodgson, J. A., London, N., Reinkensmeyer, D. J., Roy, R. R., Talmadge, R. J., Tillakaratne, N. J., Timoszyk, W., and Tobin, A., 2001, Retraining the injured spinal cord, *Journal of Physiology,* 533, 15-22.

Forssberg, H., 1985, Ontogeny of human locomotor control. I., Infant stepping, supported locomotion and transition to independent locomotion, *Experimental Brain Research,* 57, 480-493.

Gorassini, M. A., Prochazka, A., Hiebert, G. W., and Gauthier, M. J., 1994, Corrective responses to loss of ground support during walking. I. Intact cats, *Journal of Neurophysiology,* 71, 603-610.

Gossard, J-P., Brownstone, R.M., Barajon, I., and Hultborn, H., 1994, Transmission in a locomotor-related group Ib pathway from hindlimb extensor muscles in the cat, *Experimental Brain Research,* 98, 213-228.

Grey, M. J., Ladouceur, M., Andersen, J. B., Nielsen, J. B., and Sinkjaer, T., 2001, Group II muscle afferents probably contribute to the medium latency soleus stretch reflex during walking in humans, *Journal of Physiology,* 534, 925-933.

Grillner, S., and Rossignol, S., 1978, On the initiation of the swing phase of locomotion in chronic spinal cats, *Brain Research,* 146, 269-277.

Gritsenko, V., Mushahwar, V., and Prochazka, A., 2000, Adaptive changes in locomotor control after partial denervation of triceps surae muscles in the cat, *Journal of Physiology,* 533, 299-311.

Guertin, P., Angel, M. J., Perreault, M. -C., and McCrea, D. A., 1995, Ankle extensor group I afferents excite extensors throughout the hindlimb during fictive locomotion in the cat, *Journal of Physiology,* 487, 197-209.

Harkema, S. J., Hurley, S. L., Patel, U. K., Requejo, P. S., Dobkin, B. H., and Edgerton, V.R., 1997, Human lumbosacral spinal cord interprets loading during stepping, *Journal of Neurophysiology,* 77, 797-811.

Hiebert, G. W., Gorassini, M. A., Jiang, W., Prochazka, A., and Pearson, K. G., 1994, Corrective responses to loss of ground support during walking. II. Comparison of intact and chronic spinal cats, *Journal of Neurophysiology,* 71, 611-622.

Hiebert, G. W., and Pearson, K. G., 1999, Contribution of sensory feedback to the generation of extensor activity during walking in the decerebrate cat, *Journal of Neurophysiology,* 81, 758-770.

Hiebert, G. W., Whelan, P. J., Prochazka, A., and Pearson, K. G., 1995, Suppression of the corrective response to loss of ground support by stimulation of extensor group I afferents, *Journal of Neurophysiology,* 73, 416-420.

Hiebert, G. W., Whelan, P. J., Prochazka, A., and Pearson, K. G., 1996, Contribution of hind limb flexor muscle afferents to the timing of phase transitions in the cat step cycle, *Journal of Neurophysiology,* 75, 1126-1137.

Hultborn. H., Conway, B. A, and Gossard, J. -P., 1998, How do we approach the locomotor network in the mammalian spinal cord? *Annals of the New York Academy of Sciences,* 860, 70-82.

Kriellaars, D. J., Brownstone, R. M., Noga, B. R., and Jordan, L. M., 1994, Mechanical entrainment of fictive locomotion in the decerebrate cat, *Journal of Neurophysiology,* 71, 2074-2086.

Lam, T., and Pearson, K. G., 2001a, Proprioceptive modulation of hip flexor activity during the swing phase of locomotion in decerebrate cats, *Journal of Neurophysiology,* 86, 1321-1332.

Lam, T., and Pearson, K. G., 2001b, Stimulation of group I and II sartorius muscle afferents during the swing phase of walking in decerebrate cats, *Society for Neuroscience Abstracts*, 402.9.

Lundberg, A., 1981, Half-centres revisited. In: *Regulatory Functions of the CNS. Motion and Organization Principles, vol.1, Advances in Physiological Sciences*, J. Szentagothai, M. Palkovits, J. Hamori, eds., Pergamon Press, Budapest, pp. 155-167.

McCrea, D. A., 1998, Neuronal basis of afferent-evoked enhancement of locomotor activity, *Annals of the New York Academy of Sciences,* 860, 216-225.

McCrea, D. A., Stecina, K., Quevedo, J., and Gosgnach, S., 2000, Flexor group II muscle afferents can enhance flexor activity during fictive locomotion, *Society for Neuroscience Abstracts,* 26, 1233.

McCrea, D. A., Shefchyk, S. J., Stephens, M. J., and Pearson, K. G., 1995, Disynaptic group I excitation of synergist ankle extensor motoneurones during fictive locomotion in the cat, *Journal of Physiology,* 487, 527-539.

McCrea, D. A., 2001, Spinal circuitry of sensorimotor control of locomotion, *Journal of Physiology,* 533, 41-50.

Orlovskii, G. N., and Shik, M. L., 1965, Standard elements of cyclic movement, *Biophysics,* 10, 935-944.

Pang, M. Y. C., Yang, J. F., 2000, The initiation of the swing phase in human infant stepping: importance of hip position and leg loading. *Journal of Physiology* 528, 389-404.

Pearson, K. G., 1995, Proprioceptive regulation of locomotion, *Current Opinion in Neurobiology,* 5, 786-791.

Pearson, K. G., 2000, Neural adaptation in the generation of rhythmic behavior, *Annual Review in Physiology,* 62, 723-753.

Pearson, K. G., and Collins, D. F., 1993, Reversal of the influence of group Ib afferents from plantaris on activity in medial gastrocnemius muscle during locomotor activity, *Journal of Neurophysiology,* 70, 1009-1017.

Pearson, K. G., and Misiaszek, J. E., 2000, Use-dependent gain change in the reflex contribution to extensor activity in walking cats, *Brain Research,* 883, 131-134.

Pearson, K. G., Fouad, K., and Misiaszek, J. E., 1999, Adaptive changes in motor activity associated with functional recovery following muscle denervation in walking cats, *Journal of Neurophysiology,* 82, 370-381.

Pearson, K. G., Misiaszek, J. E., and Fouad, K., 1998, Enhancement and resetting of locomotor activity by muscle afferents, *Annals of the New York Academy of Sciences,* 860, 203-215.

Pearson, K. G., Ramirez, J. M., and Jiang, W., 1992, Entrainment of the locomotor rhythm by group Ib afferents from ankle extensor muscles in spinal cats, *Experimental Brain Research,* 90, 557-566.

Perreault, M. -C., Angel, M. J., Guertin, P., and McCrea, D. A., 1995, Effects of stimulation of hindlimb flexor group II afferents during fictive locomotion in the cat, *Journal of Physiology,* 487, 211-220.

Prochazka, A., and Wand, P., 1980, Tendon organ discharge during voluntary movements in cats, *Journal of Physiology,* 303, 385-390.

Quevedo, J., Fedirchuk, B., Gosgnach, S., and McCrea, D. A., 2000, Group I disynaptic excitation of cat hindlimb flexor and bifunctional motoneurones during fictive locomotion, *Journal of Physiology,* 525, 549-564.

Rossignol, S., 1996, Neural control of stereotypic limb movements, in: *Handbook of Physiology*, L.B. Rowell and J.T. Shepherd, eds., American Physiological Society, New York, p. 173-216.

Rossignol, S., 2000, Locomotion and its recovery after spinal injury, *Current Opinion in Neurobiology,* 10, 708-716.

Sinkjaer, T., Andersen, J.B., Ladouceur, M., Christensen, L. O. D., and Nielsen, J. B., 2000, Major role for sensory feedback in soleus EMG activity in the stance phase of walking in man, *Journal of Physiology,* 523, 817-827.

Stein, R. B., Misiaszek, J. E., and Pearson, K. G., 2000, Functional role of muscle reflexes for force generation in the decerebrate walking cat, *Journal of Physiology,* 525, 781-791.

Stephens, M. J., and Yang, J. F., 1999, Loading during the stance phase of walking in humans increases the extensor EMG amplitude but does not change the duration of the step cycle, *Experimental Brain Research,* 124, 363-370.

Whelan, P. J., Hiebert, G. W., and Pearson, K. G., 1995a. Stimulation of the group I extensor afferents prolongs the stance phase in walking cats, *Experimental Brain Research,* 103, 20-30.

Whelan, P. J., Hiebert, G. W., and Pearson, K. G., 1995b, Plasticity of the extensor group I pathway controlling the stance to swing transition in the cat, *Journal of Neurophysiology,* 74, 2782-2787.

Whelan, P. J., and Pearson, K. G., 1997, Plasticity in reflex pathways controlling stepping in the cat, *Journal of Neurophysiology,* 78, 1643-1650.

Yang, J. F., Fung, J., Edamura, M., Blunt, R., Stein, R. B., and Barbeau, H., 1991, H-reflex modulation during walking in spastic paretic subjects, *Canadian Journal of Neurological Sciences,* 18, 443-452.

Zehr, E. P., and Stein, R. B., 1999, What functions do reflexes serve during human locomotion? *Progress in Neurobiology,* 58, 185-205.

41

SENSORY CONTROL OF LOCOMOTION: REFLEXES VERSUS HIGHER-LEVEL CONTROL

Arthur Prochazka, Valeriya Gritsenko, and Sergiy Yakovenko*

ABSTRACT

In the absence of sensory input, the central nervous system can generate a rhythmical pattern of coordinated activation of limb muscles. Contracting muscles have spring-like properties. If synergistic muscles are co-activated in the right way, sustained locomotion can occur. What is the role of sensory input in this scheme? In this chapter we first discuss the implications of positive force feedback control in hindlimb extensor reflexes in the cat. We then raise the question of whether the sensory-evoked responses, which are modest in size and quite delayed in the stance phase, contribute to any significant extent. A locomotor model is used to show that when centrally generated activation levels are low, stretch reflexes can be crucial. However, when these levels are higher, stretch reflexes have a less dramatic role. The more important role for sensory input is probably in mediating higher level control decisions.

INTRODUCTION

Muscles differ from most man-made robotic actuators in that they are essentially springs whose stiffness and viscosity varies with activation level (Hogan, 1985). Furthermore, the stretch reflex pathways providing feedback control of individual muscles differ from those in most man-made robotic control systems in that they incorporate positive feedback loops interlaced with negative feedback loops (Prochazka and Yakovenko, 2001). Finally, the overall control of rhythmical movements such as locomotion appears to combine prediction, central rhythm generation, proportional feedback control and finite state control. It is only recently that some of these unusual features of sensorimotor control have been recognized, partly because they have only recently been found to be effective in the control of "biomimetic" robots. In this Chapter we will discuss some of the implications of these new ways of looking at sensorimotor control. First, we will identify the positive feedback loops in stretch reflex pathways and discuss how they remain stable by interacting with the negative

* Centre for Neuroscience, University of Alberta, Edmonton, Alberta, T6G 2S2, Canada.
Email: arthur.prochazka@ualberta.ca

feedback loops. Second, we will ask the more important question, do these reflexes contribute significantly to locomotor load compensation? Third, we will use biomechanical models to test some of the control schemes that have been suggested for animal locomotion. The models reveal a surprising ability of the intrinsic properties of the skeletomuscular machinery, driven by an invariant centrally generated pattern of muscle activation profiles, to adapt to speed, slope and small irregularities in terrain without sensory feedback. However they also show that although the stretch reflex contribution to load compensation in the stance phase can play an important role when the amplitudes of the centrally generated activation profiles are close to threshold for generating stable locomotion, their effects are more modest at higher central activation amplitudes. Finally, they reveal the overriding importance of prediction and finite-state control (IF-THEN rules for phase-switching) when the terrain and cadence are variable.

POSITIVE FEEDBACK LOOPS

Figure 1 shows a simple model of the neural feedback loops controlling muscles at the spinal segmental level. The first thing to notice is that the intrinsic properties of the muscle actuator are represented by a negative feedback loop involving length and velocity (loop A in the figure). This merely describes the fact that as a muscle is stretched, the force it develops is basically a product of muscle length, velocity and activation level. The fact that this can be represented in terms of a feedback loop was recognized many years ago (Partridge, 1966).

The second thing to notice is that the excitatory action of the tendon organ pathway on the motoneuron element represents positive force feedback (loop B in Fig. 1). Until fairly recently, tendon organ feedback to homonymous motoneurons was assumed to be inhibitory (negative feedback), but in 1987 it was shown that in the cat locomotor system, when locomotion starts, there is a switch from inhibition of extensors by their tendon organ afferents to excitation and this has been confirmed by other groups (Conway et al., 1987;

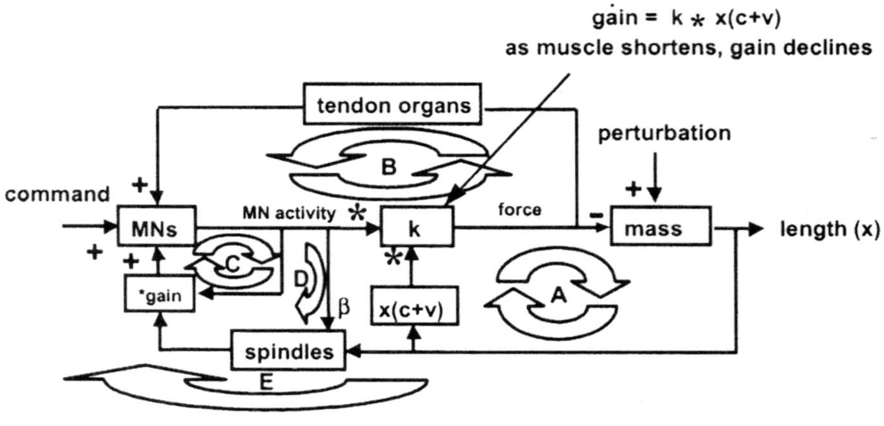

Figure 1. Reflex model of load-moving muscle. Loop A represents intrinsic muscle stiffness. Pathways B and E represent tendon organ and spindle feedback. Loop C represents automatic gain control due to motoneuron recruitment and pathway D represents β-skeletofusimotor action. Loops B, C and D are positive feedback loops.

Pearson and Collins 1993; Guertin et al., 1995; Prochazka et al., 1997b). Normally one would expect that a positive feedback loop would become unstable when the open loop gain in the loop exceeded unity, however in digital simulations of the operation of the system in Figure 1, stability was maintained even though the open loop gain of loop B was set to start at values greater than unity. The reason turned out to be that even though the loop gain exceeded unity at a given initial muscle length, provided the muscle was free to shorten, the gain in this loop rapidly declined to unity as the muscle shortened. This is because muscles produce less force for a given neural input the shorter they become. In the model in Figure 1, shortening is represented by a decline in the length variable in loop A. Because forward gain in loop B depends on the product of motoneuron activation level and muscle length, this gain therefore declines and when it reaches unity, stability is restored. Negative feedback loop A thus stabilizes the interlaced positive feedback loop B. Spindle afferent feedback (pathway E) excites homonymous motoneurons, which causes the receptor-bearing muscle to resist lengthening, i.e. negative feedback. Positive feedback loops are never included intentionally in linear control systems by control engineers, so the above explanation of how stability is maintained in the presence of positive feedback, though simple, was not obvious from the perspective of linear control theory.

The final thing to notice is that as more motoneurons are recruited, the response to a given synaptic input increases. This is represented by positive feedback loop C. Yet another positive feedback loop can also be identified (D), representing β-skeletofusimotor drive to muscle spindles. Evidently because the gains in all the positive feedback loops involved (B, C and D) are held in check by the operation of the interlaced negative feedback loops A and E (the spindle stretch reflex loop), the system as a whole is surprisingly stable.

ARE STRETCH REFLEXES IMPORTANT IN LOCOMOTOR CONTROL?

Having said this, we will now argue that in locomotion in the cat at least, the gains of the stretch reflex loops appear to be rather low and reflex action is surprisingly delayed duringload compensation in the stance phase. Some years ago, we designed the so-called "foot-in-hole" experiment to separate the reflex and centrally-generated components of ankle

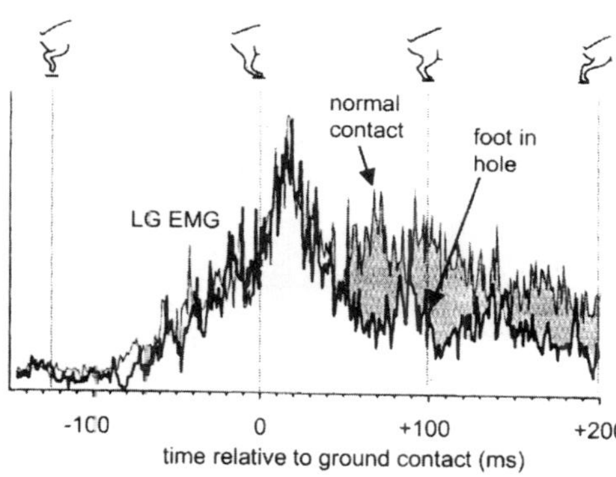

Figure 2. Mean EMG of lateral gastrocnemius (LG) muscle. Data from 9 cats for 29 steps with normal ground contact (thin traces) and 29 steps in the absence of ground contact (thick trace). Grey area represents the component of EMG attributable to sensory input to CNS signaling ground contact and stretch of ankle extensor muscles.

extensor muscle activation in cat locomotion (Gorassini et al., 1994). Many skin and muscle afferents of the foot and lower leg generate high frequency bursts when the paw touches the ground. We reasoned that if ground contact were absent, the sensory bursts and the responses to them would be absent, leaving just the centrally-generated components of activity. A walkway was built with a hidden spring-loaded trapdoor that could be triggered to descend a few milliseconds before footpad contact, i.e. at precisely the time the sensory guard hairs between the toes of the hind paw would have signaled first contact to the spinal cord. The foot then continued on into the hole, usually for at least 40 to 50 ms before an adaptive

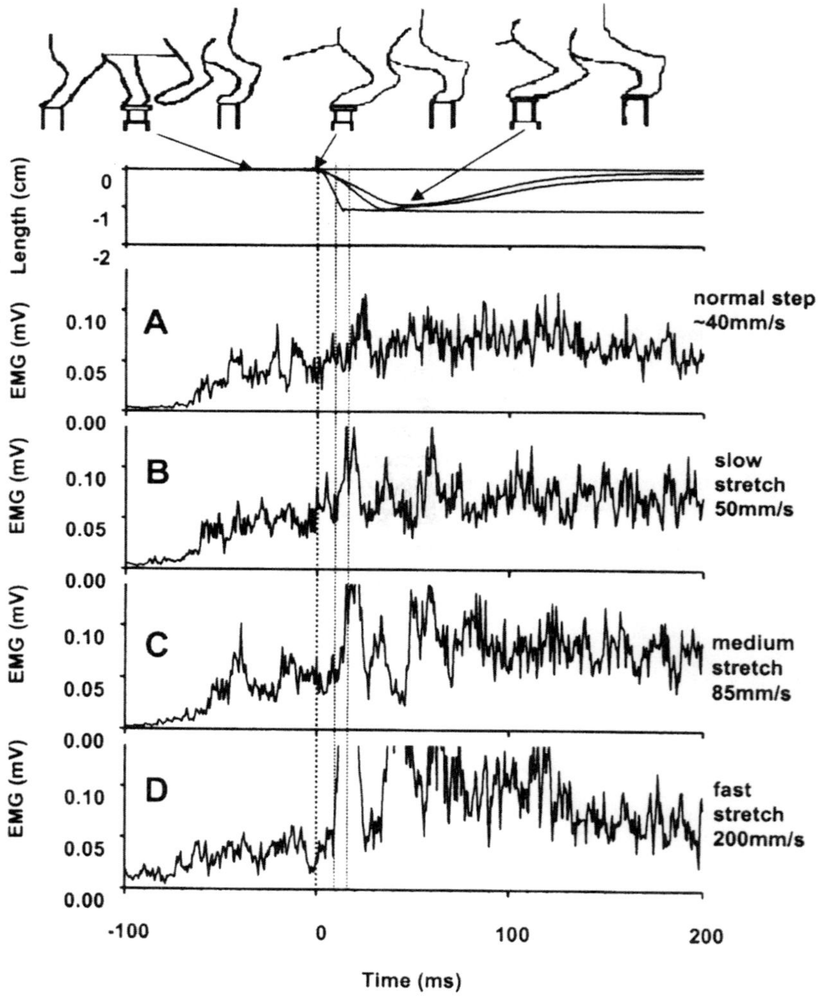

Figure 3. Rapid upward displacement of ground support pegs triggered at moment of contact of left hindlimb. LG EMG responses for four peg velocities. Estimated stretch velocities of triceps surae are shown on right. *A*, normal ground contact (no peg displacement). *B-D*, increasing rates of stretch (corresponding displacements of pegs shown in top panel).

flexion response occurred. We compared averaged electromyograms (EMGs) of ankle extensor muscles in trials in which the trapdoor remained locked in place, providing normal ground contact and support, with trials in which the trapdoor was triggered (foot-in-hole trials). The result was surprising. The averaged EMG signals were virtually identical for the first 40 ms or so after the trigger signal (Fig. 2). We had expected to see a clear difference commencing at about 9-10 ms, the latency of the monosynaptic reflex arc in cat extensor muscles and we had posited that the peak of stance-phase EMG at around 20 ms after ground contact was reflexive in origin (Prochazka et al., 1976; Trend 1987). In retrospect, we should probably have anticipated the long latency, because in a previous study of EMG responses in ankle extensors to landing from falls, even though the ankle extensors are stretched at velocities up to 500 mm/s, there is a delay in this occurring, that we attributed to an initial dorsiflexion of the toes (Prochazka et al., 1977).

To shed light on the elusive reflex component, we did the opposite experiment. A walkway was built which consisted of a row of pegs, some of which were spring-loaded. These could be triggered to pop up, dorsiflexing the ankle (stretching the ankle extensor muscles) at the moment the cat's hind paw made contact. Figure 3 shows the averaged ankle extensor EMGs for trials with normal ground support (A) and with stretch at three rates (B, C and D). As the stretch rate increased, the reflex response emerged clearly. In D it had a latency of about 10 ms, as expected of the monosynaptic response. In the slower stretch responses of B and C the latency was 15 to 20 ms and in the normal contact trials, it is arguable whether there was a clear stretch reflex at all, given the existence of an EMG peak at about 20 ms latency in the foot-in-hole trials of Figure 2 (where no reflex could have been elicited). To make the comparison easier we have re-scaled and combined the traces of Figures 2 and 3 in Figure 4. Not only was the activity attributable to sensory responses of unexpectedly long latency, it was also a rather modest component of the overall time course of extensor EMG in the step cycle. From time to time, researchers have tried to estimate the percentage contribution of stretch reflexes to overall muscle EMG and force in locomotion and other motor tasks (Allum et al., 1982; Bennett et al., 1994; Bennett et al., 1996; Stein et al., 2000). The estimates were in the range 25% (Bennett et al., 1996) to 35% (Stein et al., 2000). However, as we have seen above, a clear reflexive contribution only emerges 30 to 40 ms after ground contact, about 110 ms after EMG commences, and after peak EMG has been reached. Most if not all of the load compensation in the first third of the stance phase is therefore attributable to EMG of non-stretch-reflex origin that commences about 70 ms before foot contact and the onset of load -bearing.

The activation of EMG prior to expected load-bearing is a well-known feature in the control of postural muscles of the trunk (Massion, 1994), leg muscles in locomotion (Engberg and Lundberg, 1968), arm and hand muscles in tasks such as catching a ball (Lacquaniti et al., 1991). Our laboratory has previously compared the yield at the human elbow caused by impact of a heavy ball with three types of compensation: intrinsic stiffness of the steadily activated flexors), intrinsic stiffness modulated by stretch reflexes, intrinsic stiffness plus reflexes plus predictive activation (Bennett et al., 1994). Net yield was less when reflexes modulated the intrinsic stiffness and it became near-zero when prediction was allowed (Lang and Bastian, 1999). The crucial factor for such an effective compensation was that biceps EMG started 100 ms prior to first loading.

If extensor stretch reflexes including the positive force feedback responses mediated by tendon organs were absent, would this make a big difference to the kinematics of the limb during cat locomotion? De-afferentation experiments have been equivocal on this issue. In

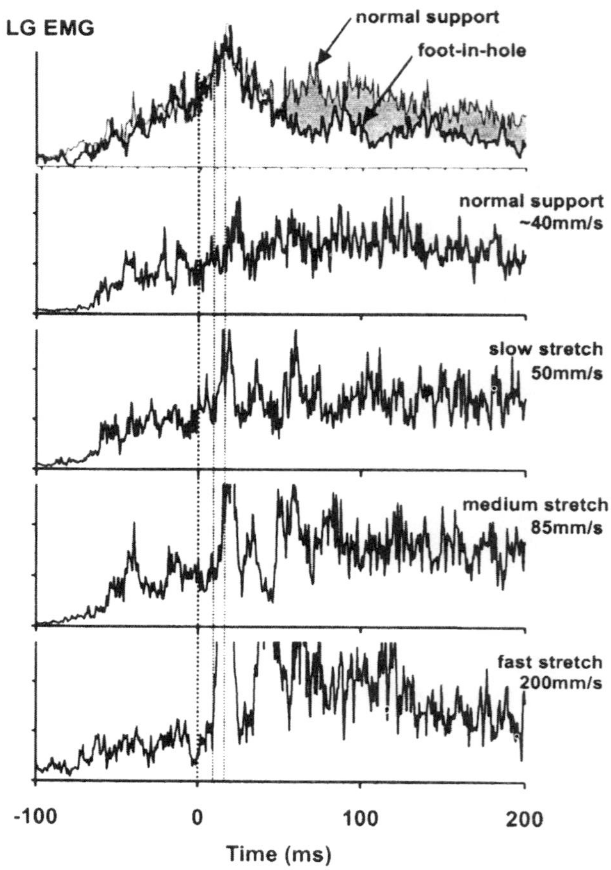

Figure 4. Data of Figs. 2 and 3 superimposed. This highlights timing of presumed stretch reflex part of EMG.

the first days and weeks after de-afferentation there is usually a greatly increased yield of the limb in the stance phase, which manifests as a pronounced limp. But this may be because extensor EMG activity, including pre-ground-contact EMG, is generally reduced. After some weeks, little difference is noticed between de-afferented and normal limbs (Wetzel et al., 1976; Goldberger, 1977; Rasmussen et al., 1986; Giuliani and Smith, 1987) unless specialized tasks are performed (Abelew et al., 2000). The ideal experiment would be somehow to abolish sensory input suddenly in single step cycles with normal ground support, but it is hard to see how this could be done. However, it can certainly be simulated in models.

Figures 5 and 6 show the results of this experiment performed with a biomechanical locomotor model (Prochazka and Yakovenko, 2001). A full description of the model and analytical methods will be published elsewhere. Briefly, the model comprises a simplified skeletal structure with a representative set of leg muscles (Fig. 5C) characterized by Hill-type length-force-velocity relationships. The model is based on the anatomy of the cat but as it is intended as a test bed for general hypotheses of locomotion across species we did not strive for an exact parametric representation. A point near the front of the body is supported on a frictionless rail. The model was constructed and simulated using Matlab version 5.3 software coupled to 2D Model version 5 software.

Locomotion was driven by a set of "EMG" activation patterns of the muscles of the model. These were based on known EMG profiles (Yakovenko et al., 2000). After some trial and error adjustments of these profiles we were able to optimize them to produce stable locomotion on a flat surface in the absence of sensory input. The EMG patterns may therefore be viewed as the centrally generated, or "default" output of the central pattern generator (CPG) in the spinal cord in the absence of sensory input. Each step was kinematically unique, indicating that the intrinsic muscle properties compensated for small variations in the kinematic and kinetic variables involved. Similar results have been obtained before using inverse dynamics or neural networks to optimize activation patterns (Taga et al., 1991; Taga, 1995b; Taga, 1995a; Yamazaki et al., 1996; Neptune et al., 2001; Ogihara and Yamazaki, 2001).

Spindle Ia and tendon organ Ib response properties are represented by the following equations derived from the literature (Prochazka, 1999).

Ia model $\quad Ia(t) = K_{Ia} * (65 * velocity^{0.5} + 200 * length + 50)$

Ib model: $\quad Ib(s) = K_{Ib} * force * (s+0.15)(s+1.5)(s+16) / (s+0.2)(s+2)(s+37)$

where Ia(t) is the time function of the Ia signal and Ib(s) is the tendon organ response in the frequency domain, s = frequency domain operator. K_{Ia} and K_{Ib} represent gain coefficients.

The Ia and Ib reflex signals were set to have a latency (delay) of 35 ms, in accordance with the latency of the EMG components attributable to reflexes in Figure 3 (top panel). They were active only when the receptor-bearing muscles were active, i.e. only when the CPG EMG profile of the corresponding muscle was non-zero. The gain coefficients K_{Ia} and K_{Ib} were adjusted so that the Ia and Ib signals each added a mean of 15% to the CPG EMG profile. The value of 15% was chosen because the sum, 30%, corresponded to the proportion of net EMG attributable to reflexes in Figure 3.

At the meeting in Cairns, Prochazka predicted from Figures 2 and 3 above that the size and timing of the reflexes were such that they could have little kinematic effect on the step cycle. This prediction has since been tested as illustrated in Figures 5 and 6. In Figure 5, the amplitudes of the CPG patterns were scaled down to about 90% of the level required to just produce stable locomotion. The net reflexive components of EMG are shown as black caps on top of each EMG profile in the first two step cycles of Figure 5A. Locomotion was stable in the presence of the reflex contributions. Reflex transmission was suddenly reduced to zero after the second cycle. The resulting reduction in weight support and forward thrust were such that the hindquarters collapsed over the next two cycles (Fig. 5). This was of course expected, because the CPG pattern had been deliberately set at 90% of the level needed to sustain locomotion. The surprise was that in the first two steps, the stretch reflexes clearly provided enough extra activation to make gait possible.

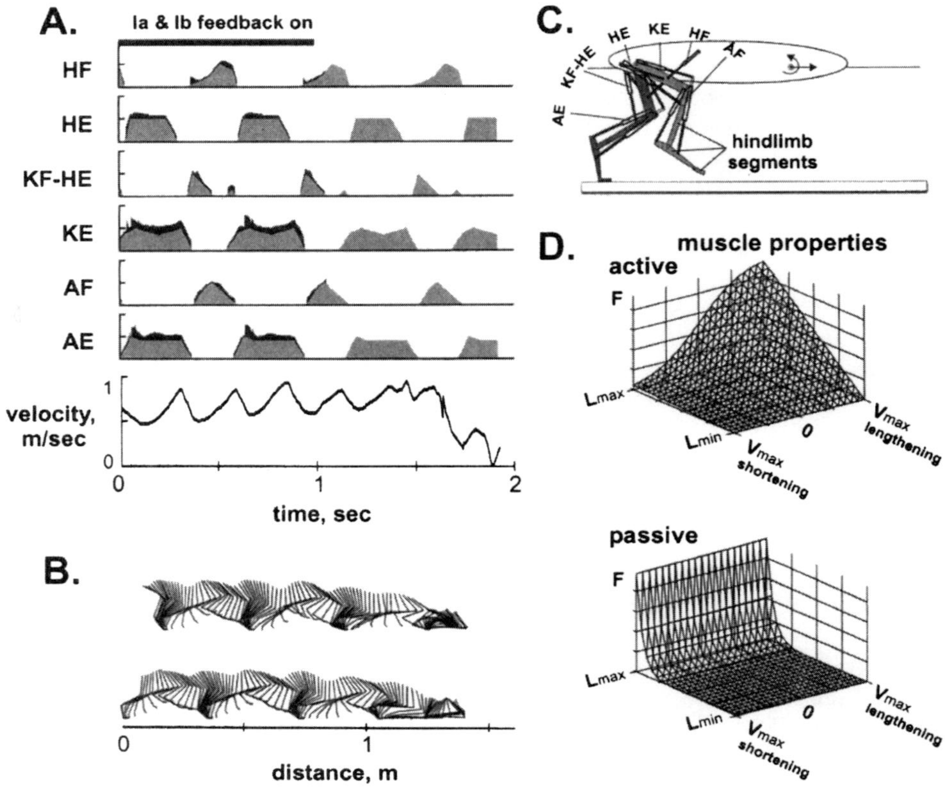

Figure 5. Model of control of quadruped hindlimbs during locomotion with and then without stretch reflexes mediated by muscle spindle Ia and tendon organ Ib afferents. ***A,*** black bar indicates reflexes present. Basic EMG profiles due to central patter generator (CPG) shown in grey, additional reflex components in black: hip flexors (HF), hip extensors (HE), knee flexors (KF), knee extensors (KE), ankle flexors (AF), ankle extensors (AE). Bottom traces shows velocity. ***B,*** stick figures of left and right legs. ***C,*** Physical structure of model. ***D,*** muscle properties: length-force and velocity-force.

The next question was, if the basic CPG profiles were adequate to sustain gait, would the addition of the reflexes make any significant difference? At the Cairns meeting (2001), Prochazka had suggested they would not. Figure 6 shows that adding the stretch reflexes after the first few cycles, again set to add about 30% to the underlying CPG activation profiles, caused a modest but significant increase in the velocity of gait.

We conclude that even though the reflex contributions are delayed in the cycle and add only about 30% to the centrally generated extensor EMGs, they can play a role in sustaining and controlling the speed of gait. This outcome was not obvious from qualitative judgements, though the modulation of locomotor speed by gain control of positive force feedback had been proposed from a simpler single-muscle analysis (Prochazka et al., 1997a).

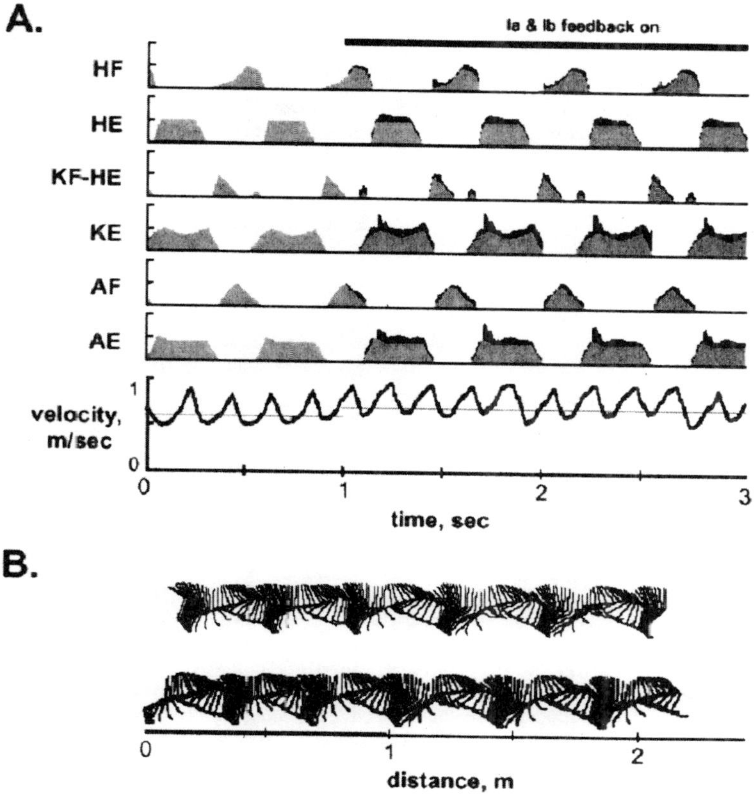

Figure 6. Adding stretch reflexes to a stable locomotor pattern. Similar simulation to that in Figure 5, except that amplitude of CFG EMG patterns was sufficient to sustain locomotion without reflexes. *A*, reflexes added as shown by black bar. The result was a small increase in velocity. *B*, stick figures of left and right legs.

HIGHER LEVEL CONTROL

The modulation of load compensation and speed described above, though significant, still seems a rather modest role for sensory input to the CNS given that muscle afferents are the fastest-conducting axons in the body and that axons involved in proprioception and sensation far outnumber motor axons innervating extrafusal muscle (Matthews, 1972). Another crucial role for sensory input is to allow for higher-level decisions, for example those based on conditional logic in which IF-THEN rules determine state transitions such as phase-switching in the step cycle and the prediction of global EMG levels required for future movements "one-step-ahead" control (Granat et al., 1993; Prochazka 1993).

The biomechanical modeling described above, and also the accelerating effort being put into the design of control systems for biomimetic robots (Quinn and Ritzmann 1998; Nelson and Quinn 1999) has led to a number of general conclusions about the overall roles of sensory feedback that are in line with the concepts presented in this article and will serve as a fitting conclusion:

1. The intrinsic stiffnesses of limb muscles, when activated with optimized cyclical patterns can generate stable locomotion in the face of small variations in speed and terrain. Stretch reflexes contribute to load compensation within a given phase of the step cycle, and provide a limited means of changing gait speed and posture.
2. Larger adjustments in speed and terrains require higher-level control strategies such as finite-state logic.
3. Global rules that use multisensory input are required for movement selection, predictions about upcoming movements and overall balance.

ACKNOWLEDGEMENTS

This work was supported by the Canadian Institutes of Health Research and the Alberta Heritage Foundation for Medical Research.

REFERENCES

Abelew, T. A., Miller, M. D., Cope, T. C., and Nichols, T. R., 2000, Local loss of proprioception results in disruption of interjoint coordination during locomotion in the cat, *Journal of Neurophysiology,* 84, 2709-2714.

Allum, J. H., Mauritz, K. H., and Vogele, H., 1982, The mechanical effectiveness of short latency reflexes in human triceps surae muscles revealed by ischaemia and vibration, *Experimental Brain Research,* 48, 153-156.

Bennett, D. J., De Serres, S. J., and Stein, R. B., 1996, Gain of the triceps surae stretch reflex in decerebrate and spinal cats during postural and locomotor activities, *Journal of Physiology,* 496, 837-850.

Bennett, D. J., Gorassini, M., and Prochazka, A., 1994, Catching a ball: contributions of intrinsic muscle stiffness, reflexes, and higher order responses, *Canadian Journal of Physiology and Pharmacology,* 72, 525-534.

Conway, B. A., Hultborn, H., and Kiehn, O., 1987, Proprioceptive input resets central locomotor rhythm in the spinal cat, *Experimental Brain Research,* 68, 643-656.

Engberg, I., and Lundberg, A., 1968, An electromyographic analysis of muscular activity in the hindlimb of the cat during unrestrained locomotion, *Acta Physiologica Scandinavica,* 75, 614-630.

Giuliani, C. A., and Smith, J. L., 1987, Stepping behaviors in chronic spinal cats with one hindlimb deafferented, *Journal of Neuroscience,* 7, 2537-2546.

Goldberger, M. E. 1977Locomotor recovery after unilateral hindlimb deafferentation in cats, *Brain Research,* 123, 59-74.

Gorassini, M. A., Prochazka, A., Hiebert, G. W., and Gauthier, M. J., 1994, Corrective responses to loss of ground support during walking. I. Intact cats, *Journal of Neurophysiology,* 71, 603-610.

Granat, M. H., Heller, B. W., Nicol, D. J., Baxendale, R. H., and Andrews, B. J., 1993, Improving limb flexion in FES gait using the flexion withdrawal response for the spinal cord injured person, *Journal of Biomedical Engineering,* 15, 51-56.

Guertin, P., Angel, M. J., Perreault, M. C., and McCrea, D. A., 1995, Ankle extensor group I afferents excite extensors throughout the hindlimb during fictive locomotion in the cat, *Journal of Physiology,* 487, 197-209.

Hogan, N., 1985, The mechanics of multi-joint posture and movement control, *Biological Cybernetics,* 52, 315-331.

Lacquaniti, F., Borghese, N. A., and Carrozzo, M., 1991, Transient reversal of the stretch reflex in human arm muscles, *Journal of Neurophysiology,* 66, 939-954.

Lang, C. E., and Bastian, A. J., 1999, Cerebellar subjects show impaired adaptation of anticipatory EMG during catching, *Journal of Neurophysiology,* 82, 2108-2119.

Massion, J., 1994, Postural control system, *Current Opinion in Neurobiology,* 4, 877-887.

Matthews, P.B.C., 1972, *Mammalian Muscle Receptors and Their Central Actions,* Arnold, London.

Nelson, G. M., and Quinn, R. D., 1999, Posture control of a cockroach-like robot, *IEEE Transactions on Control Systems,* 19, 9-14.

Neptune, R. R., Kautz, S. A., and Zajac, F. E., 2001, Contributions of the individual ankle plantar flexors to support, forward progression and swing initiation during walking, *Journal of Biomechanics*, in press.

Ogihara, N., and Yamazaki, N., 2001, Generation of human bipedal locomotion by a bio-mimetic neuro-musculo-skeletal model, *Biological Cybernetics*, 84, 1-11.

Partridge, L. D., 1966, Signal-handling characteristics of load-moving skeletal muscle, *American Journal of Physiology* 210, 1178-1191.

Pearson, K. G., and Collins, D. F., 1993, Reversal of the influence of group Ib afferents from plantaris on activity in medial gastrocnemius muscle during locomotor activity, *Journal of Neurophysiology*, 70, 1009-1017.

Prochazka, A., 1993, Comparison of natural and artificial control of movement. *IEEE Transactions on Rehabilitation Engineering*, 1, 7-16.

Prochazka, A., 1999, Quantifying proprioception, in: *Peripheral and spinal mechanisms in the neural control of movement*, M. D. Binder ed., Elsevier, Amsterdam.

Prochazka, A., Gillard, D., and Bennett, D.J., 1997a, Implications of positive feedback in the control of movement, *Journal of Neurophysiology*, 77, 3237-3251.

Prochazka, A., Gillard, D., and Bennett, D. J., 1997b, Positive force feedback control of muscles, *Journal of Neurophysiology*, 77, 3226-3236.

Prochazka, A., Schofield, P., Westerman, R. A., and Ziccone, S. P., 1977, Reflexes in cat ankle muscles after landing form falls, *Journal of Physiology*, 272, 705-719.

Prochazka, A., Westerman, R. A., and Ziccone, S. P., 1976, Discharges of single hindlimb afferents in the freely moving cat, *Journal of Neurophysiology*, 39, 1090-1104.

Prochazka, A., and Yakovenko, S., 2001, Locomotor control: from spring-like reactions of muscles to neural prediction, in: *The Somatosensory System: Deciphering The Brain's Own Body Image*, R. Nelson, ed., CRC Press, Boca Raton.

Quinn, R. D., and Ritzmann, R. E., 1998, Biologically based distributed control and local reflexes improve rough terrain locomotion in a hexapod robot, *Connection Science*, 10, 239-254.

Rasmussen, S. A., Goslow, G. E., and Hannon, P., 1986, Kinematics of locomotion in cats with partially deafferented spinal cords: the spared-root preparation, *Neuroscience Letters*, 65, 183-188.

Stein, R. B., Misiaszek, J. E., and Pearson, K. G. 2000, Functional role of muscle reflexes for force generation in the decerebrate walking cat, *Journal of Physiology*, 525, 781-791.

Taga, G., 1995a, A model of the neuro-musculo-skeletal system for human locomotion. I. Emergence of basic gait, *Biological Cybernetics*, 73, 97-111.

Taga, G., 1995b, A model of the neuro-musculo-skeletal system for human locomotion. II Real-time adaptability under various constraints, *Biological Cybernetics*, 73, 113-121.

Taga, G., Yamaguchi, Y., and Shimizu, H., 1991, Self-organized control of bipedal locomotion by neural oscillators in unpredictable environment, *Biological Cybernetics*, 65, 147-159.

Trend, P., 1987, *Gain control in proprioceptive pathways*, Ph.D., London, UK, University of London, 280.

Wetzel, M. C., Atwater, A. E., Wait, J. V., and Stuart, D. G., 1976, Kinematics of locomotion by cats with a single hindlimb deafferented, *Journal of Neurophysiology*, 39, 667-678.

Yakovenko, S., Mushahwar, V., Vanderhorst, V., Holstege, G., and Prochazka, A., 2002, Spatiotemporal activation of lumbosacral motoneurons in the cat locomotor step cycle, *Journal of Neurophysiology*, 87, 1542-1553.

Yamazaki, N., Hase, K., Ogihara, N., and Hayamizu, N., 1996, Biomechanical analysis of the development of human bipedal walking by a neuro-musculo-skeletal model, *Folia Primatologica*, 66, 253-271.

REFLEX EXCITATION OF MUSCLES DURING HUMAN WALKING

Jens Bo Nielsen[1] and Thomas Sinkjær[2]

ABSTRACT

Sensory activity may contribute to the control of human walking in two different ways. It may contribute to the pre-programmed drive to the motoneurones and to the reactions to unexpected external perturbations. Some recent findings in relation to these two different roles of sensory activity will be reviewed. When unloading the ankle plantarflexors in the stance phase of walking a drop in the soleus EMG activity is seen at a latency of around 60 ms. This drop is likely caused by the removal of the contribution of Gp II afferents from the ankle plantarflexors to the motoneuronal drive. When stretching plantarflexor muscles in the stance phase three reflex responses are generally observed. These responses may be caused by the spinal monosynaptic Ia reflex pathway, a spinal Gp II pathway and a transcortical reflex pathway, respectively. The reflex responses are modulated with the background EMG activity and may not be evoked in the swing phase when the plantarflexors are not active. In contrast, stretch of the ankle dorsiflexor muscles evoke relatively small responses in the swing phase when these muscles are active, but very large responses in the stance phase when the muscles are silent. Part of these responses may have a transcortical nature. These findings illustrate the complexity with which sensory input may contribute to the ongoing muscle activity during walking and may also mediate adequate responses to sudden external perturbations.

INTRODUCTION

The role of muscle afferent activity in the control of human walking is not yet fully clarified, but several new insights have been gained in recent years. In this chapter we will focus on two different roles of muscle afferent activity in the control of walking: i) muscle afferent feedback may contribute to the activation of the muscles during normal unperturbed walking and ii) muscle afferent activity may mediate some of the reactions

[1] Department of Medical Physiology, University of Copenhagen, 2200 Copenhagen N, Denmark. Email: j.b.nielsen@mfi.ku.dk
[2] Center for Sensory-Motor Interaction, University of Aalborg, 9220 Aalborg, Denmark.

to sudden external perturbations during gait. We will argue that different afferents and different central pathways are involved in these two different control paradigms. We will also argue that there are important differences in the control of different muscles depending on these functions.

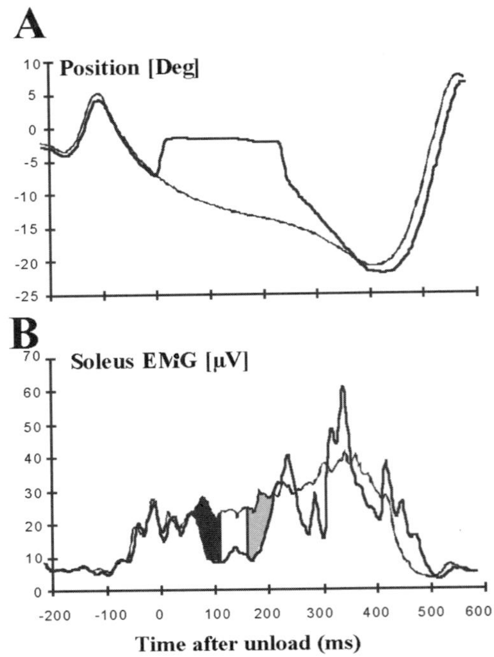

Figure 1. Unloading of ankle plantar flexors during walking. Example of averaged recorded data during control steps (heavy lines) and steps with unloading (thin lines) of the ankle extensors in the stance phase. *A*, Ankle angle positions. *B*, Rectified and filtered soleus muscle EMG. The black, white, and grey areas represent the different time windows used to characterise the response. 0° equals standing position. Positive degrees are plantar flexion direction. 10 trials per average. (Modified from Sinkjær et al., 2000.)

MEDIUM DIAMETER MUSCLE AFFERENT FEEDBACK CONTRIBUTES TO MUSCLE ACTIVATION IN THE STANCE PHASE OF WALKING

When we walk our muscles change length and develop tension, our joints change position, skin receptors are cyclically activated and the sensitivity of the muscle spindles is modulated by changes in gamma-drive. Through direct or - more often - indirect positive feedback pathways, this afferent activity may contribute to the muscle activity, thereby reducing the amount of supraspinal drive necessary to activate the muscles. By temporarily removing the afferent activity, it would therefore be expected that the EMG activity from the active muscles would drop. As shown in Figure 1 this is also what happens. In the experiment illustrated in the figure the ankle plantarflexors were suddenly unloaded in the stance phase of walking (Sinkjaer et al., 2000). When the muscle is unloaded the muscle spindles go slack and the activity of Gp Ia and Gp II afferents decreases abruptly. At the same time the force in the muscle decreases, leading also to a decreased discharge of Gp Ib afferents. This removal of the afferent discharge in Gp I and Gp II afferents is the likely explanation of the drop in EMG activity recorded from the soleus muscle, which is observed in Figure 1.

It could be argued that when the ankle plantarflexors are shortened, the ankle dorsiflexors are simultaneously stretched and that the observed drop in EMG activity could be explained by central inhibition of soleus motoneurones evoked by this stretch. However, the same drop in EMG activity could be observed when transmission in the nerve from the ankle dorsiflexors (common peroneal nerve) was blocked by local lidocaine injection. The observed drop in EMG activity thus must originate from removal of the afferent activity from the ankle plantarflexors.

It is not possible from these experiments to decide conclusively which afferents were responsible for the observed drop in the EMG activity, but the rather long latency and the fact that the drop in EMG activity was also observed during ischaemia of the leg, during which transmission in large diameter afferents is effectively blocked, suggests that Gp I afferents were not involved. If so, this would suggest that transmission in Gp II afferents makes a major contribution to the ongoing extensor EMG activity in the stance phase of human walking, whereas Gp I afferents play no or only a minor role. This would be consistent with the observation that presynaptic inhibition of Gp Ia afferents seems to be increased during human walking (Capaday and Stein, 1986; Faist et al., 1996). This would effectively hinder the central effects of transmission in these afferents as has also been suggested previously (Dietz et al., 1985).

Ia MONOSYNAPTIC STRETCH REFLEXES ARE PROMINENT IN THE SOLEUS MUSCLE IN THE STANCE PHASE OF WALKING

Given that the central effects of Gp Ia afferents seem to be depressed by presynaptic inhibition during walking in human subjects, it may seem surprising that signficant stretch reflexes at monosynaptic Gp. I latency may nevertheless be evoked in the soleus muscle in the stance phase of walking (Sinkjær et al., 1996; Fig. 2). Indeed, these reflexes are not depressed during walking as compared to tonic contraction in standing subjects with an equal amount of background EMG activity in the soleus muscle. One possible explanation of this is that there is an increased gamma-drive during walking, which antagonizes the depressive effects of presynaptic inhibition. Another possible explanation is that presynaptic inhibition may not have the same effect on a stretch reflex response in which the Ia afferents discharge at high frequency as it has on the Ia afferent feedback evoked by the ongoing movement in which discharge rates may be somewhat lower. At least there is data from the cat, which suggest that the efficacy of presynaptic inhibition in depressing synaptic potentials may depend on the impulse traffic in the pathway, which receives the inhibition (Nielsen et al., 1999). Stretch reflexes evoked by external perturbations may thus be rather significant, although the on-going contribution to motoneuronal drive of sensory activity in the same pathway is limited.

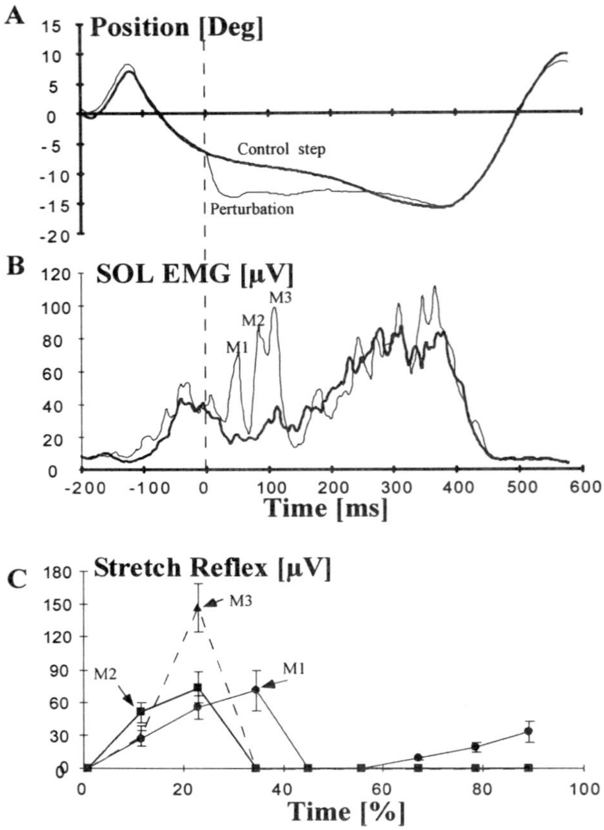

Figure 2. Soleus stretch reflexes during walking. *A*, Position of the ankle joint during the stance phase. The thin line shows an average of ten individual steps with an 8° stretch elicited at time zero. The onset of the stretch corresponds to 250 ms after heel contact. The thick line is an average of eight control steps. *B*, Averaged soleus EMG reflexes to the imposed dorsiflexion movement shown in A. The thin line shows the rectified and filtered soleus EMG with a short latency reflex (labelled M1) followed by two long-latency reflexes (M2, M3). The soleus EMG activity is superimposed on the EMG activity from a control step (thick line). C: Modulation of M1, M2, and M3 soleus stretch reflexes during the entire step. The stretch is kept constant at a displacement of 8° and a displacement velocity of 280°/s. (Modified from Sinkjær et al., 1999.)

DURING WALKING STRETCH REFLEXES IN THE TIBIALIS ANTERIOR ARE LARGEST WHEN THE MUSCLE IS SILENT

It is also seen from Figure 2 that the stretch responses in the soleus muscle are greatly modulated with the gait cycle. Significant responses are thus only seen in the stance phase when the muscle is active, whereas no responses are observed in the swing phase when the muscle is silent. The stretch reflexes thus seem to be modulated with the excitability of the spinal motoneurones. One would therefore also expect stretch reflexes in other muscles to behave in a similar way, but as seen from Figure 3 this is not the case. Stretch reflexes in the tibialis anterior muscle would be expected to be largest in the swing phase when the muscle is active, but actually by far the largest responses were observed in the stance phase when the muscle was silent.

Such findings suggest that stretch reflexes may mainly play a role in the stabilization of the supporting limb during walking, whereas they seem to play less of a role in the correction of external perturbations in the swing phase. This was also recently suggested by Zehr and Stein (1999).

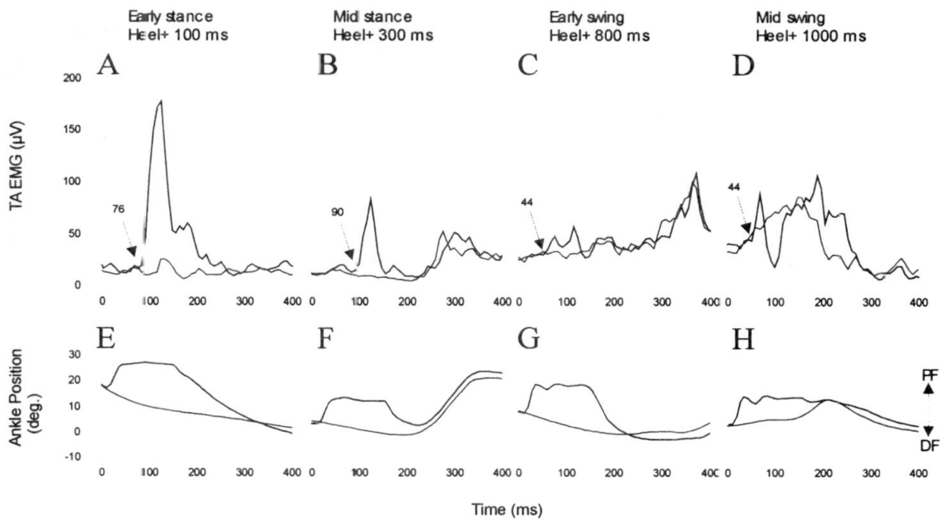

Figure 3. Modulation of TA stretch responses during the walking cycle. Stretches (amplitude 8°, velocity 250 °/s and a hold phase of ~ 120 ms) were applied to the ankle dorsiflexors by a portable stretching device at different times during the walking cycle. *A-D*, The EMG responses in TA to the stretches in early and mid stance (*A* and *B*) and in early and mid swing (*C* and *D*) The onset of the responses are marked by arrows. The latency of the responses is given in ms. The changes in the position of the ankle joint are shown below (*E-H*). The thin lines show the EMG activity and the ankle joint position in control steps without stretch. All the traces consist of an average of 10 sweeps. PF and DF (for plantar and dorsiflexion, respectively) signify the movement direction. (Modified from Christensen et al., 2001.)

SEVERAL DIFFERENT PATHWAYS CONTRIBUTE TO STRETCH REFLEX RESPONSES

Closer inspection of Figures 2 and 3 reveals that several different reflex responses may be observed at different latencies. In addition to the short-latency monosynaptic Ia mediated reflex in the soleus muscle (Fig. 2), two further responses are observed at longer latencies throughout the stance phase. These have been denoted M2 and M3, whereas the earlier monosynaptic reflex is denoted M1. The origin of these later responses is not yet fully clarified, but recent data suggest that M2 may be mediated by a Gp II pathway (Grey et al., 2001). Whereas the M1 response is linearly related to the velocity of the stretch this is not the case of M2. During ischaemia, which blocks transmission in large diameter afferents, only M1 and M3 are abolished, whereas M2 is preserved. During cooling of the nerve M2 is more delayed than M1, which is also consistent with transmission in a pathway with lower conduction velocity. Finally, administration of tizanidine selectively abolishes the M2 reflex. Tizanidine has been shown in animal experiments to block transmission in Gp II pathways.

The later M3 response in soleus in contrast seems to be mediated by Gp I afferents, since this response, like the M1 response, is depressed during ischaemia. The response is very small or absent in patients with lesions of the corticospinal tract (Sinkjaer et al 1999) and it seems as a reasonable possibility that the response is at least partly mediated by a transcortical reflex pathway.

The evidence that the response at a similar long latency in the tibialis anterior muscle (see Fig. 3) is mediated by a transcortical mechanism is more convincing (Christensen et al., 2001). Like the M3 response in soleus, the late response in TA is absent or very small in patients with corticospinal lesions (personal observations). Furthermore, the latency of the reflex response fits rather well to the latency of a transcortical reflex pathway calculated from the sum of the afferent conduction time from the muscle receptors to the cortex (measured by somatosensory evoked potentials) and the efferent conduction time from the cortex to the muscle (measured from MEPs evoked by TMS). Corresponding to the late reflex response there is also evidence of increased excitability in the motor cortex. MEPs evoked by TMS are thus strongly facilitated corresponding to the M3 reflex response, whereas this is not the case of MEPs evoked by electrical cortical stimulation (TES). TES differ from TMS in activating the corticospinal tract at a more distal site, bypassing the influence of cortical excitability changes on the evoked responses. Finally, recent experiments have demonstrated that it is possible to selectively depress the late M3 response in TA by TMS at an intensity at which only cortical inhibitory networks are activated (Johan Van Doornik, Yoshihisa Masakado, Thomas Sinkjær and Jens Bo Nielsen, unpublished observations). Such weak stimuli may thus depress the transmission of the transcortical reflex through the motor cortex.

These findings thus demonstrate that several different reflex pathways may contribute to the reaction to external perturbations during walking. Spinal as well as supraspinal pathways contribute as do interneurones activated from both Gp I and Gp II afferents. Most importantly the modulation of the reflexes is not simply determined by the excitability of the spinal motoneurones, but rather by gating of transmission through the various spinal and supraspinal pathways.

A full understanding of the contribution of these reflex responses to the adequate correction of an external perturbation requires that the responses in several muscles on both sides of the body are taken into account. In addition, the sensory activity evoked by the perturbations also mediate less reflex-like reactions, which work on a time scale only slightly longer than the reflexes. It is the combined pattern of all these reflex-, pre-programmed and voluntary reactions, which determine the actual functional correction of the perturbation. We do not yet know all the pieces of this puzzle.

ACKNOWLEDGEMENTS

This work was carried out with support from the Danish Health Research Council.

REFERENCES

Capaday, C., and Stein, R. B., 1986, Amplitude modulation of the soleus H-reflex in the human during walking and standing, *Journal of Neuroscience*, 6, 1308-1313.

Christensen, L. O. D., Andersen, J. B., Sinkjær, T., and Nielsen, J. B., 2001, Transcranial magnetic stimulation and stretch reflexes in the tibialis anterior muscle during human walking, *Journal of Physiology*, 531, 545-557.

Dietz, V., Quintern, J., and Berger, W., 1985, Afferent control of human stance and gait: evidence for blocking of group I afferents during gait, *Experimental Brain Research*, 61, 153-163.

Nielsen, J. B., Enriquez-Denton, M., Sinkjaer, T., Morita, H., Christensen, L. O. D., and Petersen, N., 1999, Presynaptic inhibition of excitatory postsynaptic potentials evoked by muscle stretch and electrical nerve stimulation in cat lumbar motoneurones, *Neuroscience Abstracts*, pp. 123.

Faist, M., Dietz, V., and Pierrot-Deseilligny, E., 1996, Modulation, probably presynaptic in origin, of monosynaptic Ia excitation during human gait, *Experimental Brain Research*, 109, 441-449.

Grey, M., Ladouceur, M., Andersen, J. B., Nielsen, J. B., and Sinkjær, T., 2001, Group II muscle afferents probably contribute to the medium latency soleus stretch reflex during walking in humans, *Journal of Physiology*, 534, 925-933.

Sinkjær, T., Andersen, J. B., and Larsen, B., 1996, Soleus stretch reflex modulation during gait in humans, *Journal of Neurophysiology*, 76, 1112-1120.

Sinkjær, T., Andersen, J. B., Nielsen, J. F., and Hansen, H. J., 1999, Soleus long latency stretch reflexes during walking in healthy and spastic humans, *Clinical Neurophysiology*, 110, 951-959.

Sinkjær, T., Andersen, J. B., Ladouceur, M., Christensen, L. O. D., and Nielsen, J. B., 2000, Major role for sensory feedback in soleus EMG activity in the stance phase of walking in man, *Journal of Physiology*, 523, 817-827.

Yang, J. F., Stein, R. B., and James, K. B., 1991, Contribution of peripheral afferents to the activation of the soleus muscle during walking in humans, *Experimental Brain Research*, 87, 679-687.

Zehr, E. P., and Stein, R. B., 1999, What functions do reflexes serve during human locomotion? *Progress in Neurobiology*, 58, 185-205.

H REFLEXES RECORDED DURING LOCOMOTION

Poul Dyhre-Poulsen and Erik B Simonsen*

ABSTRACT

We recorded H reflexes and the biomechanics of movement during locomotion. The soleus H reflex was strongly modulated during normal walking, depressed during the swing phase and modulated with the EMG in the stance phase. The amplitude of the H reflex increased with the EMG activity and was larger during running than walking. There were individual differences in the modulation pattern covariant with the biomechanics of walking. Interpretation of the results requires knowledge of the method used and assessment of the stimulus and recording conditions.

INTRODUCTION

The importance of sensations arising from receptors around joints for the control of movement is not well understood. Kinaesthesia, or the sense of movement involves sense of position of body parts, that is joint angles and position of body segments in space. It also involves the sensations arising from both active and passive movements, and finally the sense of force. The sensation of position is clearly distinct from the sensation of movement and the sensations arise from more than one receptor type in contrast to other senses such as touch, hearing etc. For further discussion see chapters by Gandevia et al. (Chapter 8) and Kerr and Worringham (Chapter 10).

Application of a vibratory stimulus to a tendon or muscle belly introduces an illusion of movement and pulling the antagonist tendon introduces motor errors. Patients with knee joint replacements have kinesthetic deficiencies and lesions to ligaments normally cause proprioceptive deficits (see also Refshauge, Chapter 12). Therefore the 'muscular sense' as Sherrington called it arises both in muscle and joint and possibly involves spinal mechanisms and corollary discharge. The current consensus attributes a predominant role to the muscle spindle receptors providing the proprioceptive signals for kinesthetic sensation and lesser role to the Golgi tendon organ and receptors in joint-capsules and skin. The numerical distribution of the receptors, with plenty of receptors in the muscles

* Institute of Medical Physiology and Institute of Anatomy, Panum Institute, University of Copenhagen, Copenhagen, Denmark. Email: dyhre@mfi.ku.dk

and in the skin in contrast to the sparse amount of receptors in the joint structures also locates the dominant responsibility for proprioception to the muscle receptors.

Studies of the muscle spindle function during locomotion in human beings have repeatedly utilized the Hoffmann (H) reflex evoked by electrically stimulating the afferent nerve instead of mechanical stretch of the muscle spindle. The modulation of the H reflex has been extensively studied in humans for different tasks (Zehr and Stein, 1999) The study of H reflexes instead of the stretch reflex has both limitations and advantages because it bypasses the fusimotor activation and the muscle spindle and focuses on the handling of the input from the Ia afferents. Thus H reflex modulation may not elucidate the spindle function, but it allows examination of the spinally mediated reflex modulation. The amplitude of the H reflex is normally diminished when shifting from sedentary behaviour to any type of movement reflecting the idea that the brain increases the control during activity. During muscular activity usually a higher EMG level will coincide with a higher H reflex amplitude due to increased excitability of the motor neurone pool at the same reflex gain. However, previous studies on H reflex modulation during human locomotion have concluded that soleus H reflex gain was greater for walking than running (Capaday and Stein, 1986) despite the lower amplitude of the EMG during walking.

The purpose of our investigations has been to assess soleus H reflex gain during walking and running over a wide range of muscle activity levels. We used walking and running at different speeds, inclinations and gravity levels to modify soleus activity levels during human locomotion while recording the H reflex gain. We hoped to reveal the mechanisms behind the utilisation of afferent feedback, spinal circuits and supraspinal control during locomotion. The amplitude of the H reflex is very dependant of stimulus strength and placement of the recording electrodes relative to the underlying muscle. Thus the stimulus and recording methods may affect the recordings substantially especially when the H reflex is recorded during locomotion.

METHODS

The experiments were performed on a treadmill. EMG was recorded from leg muscles while the soleus H-reflex was elicited by stimulating the tibial nerve in the fossa poplitea. An automatic tracking mechanism controlled the stimulus strength so that the stimulus elicited an M wave, which measured 25% of the amplitude of the maximal M wave. The soleus H reflex was normalized to the maximal M wave recorded 60 ms after the H reflex. The method compensates for errors elicited both from the stimulus site and from the recording site (Simonsen and Dyhre-Poulsen, 1999). Simulated low gravity influences the EMG amplitude but not the overall EMG pattern during locomotion (Ferris et al., 2001). The H reflex gain was calculated as the amplitude of the normalized H reflex divided by the EMG amplitude both during walking and running on a treadmill at different simulated gravity levels. Footswitches recorded heel strike and toe off. Heel strike triggered a timer that was used to control the position of the stimulus in the gait cycle. The kinematics were recorded while walking over a force plate and the kinetics were calculated from the force plate recordings and the kinematics using standard inverse dynamics methods.

RESULTS

Level Walking

The soleus H reflex is profoundly modulated during normal walking. This was evident whether we normalized the reflex amplitude to a reference excitability curve, normalized to the maximal M-wave measured during the gait cycle or measured by averaging of multiple sweeps. In general the reflex excitability was suppressed during the swing phase and facilitated during the stance phase, in which the foot is in contact with the ground (Fig. 1) Some subjects show a very fast rise in reflex excitability just after heel strike and before onset of EMG activity in the soleus muscle. Other subjects show a more gradual rise in H reflex amplitude during the stance phase (Fig. 2). During the swing phase some subjects show an almost completely suppressed H reflex during the swing phase while others show a suppression immediately after heel strike followed by a gradual increase towards heel strike as also seen during running. Grouping subjects into one group with a suppressed H reflex (S) during the swing phase and another group with a less suppressed reflex (LS) showed that the subjects with the suppressed reflex (S) had higher EMG activity in the quadriceps muscle than group (LS). They exhibited also higher muscle forces about the knee joint and lower muscle forces about the ankle joint (Simonsen et al., 2002). It is therefore suggested that group LS relies on the stretch reflex of the plantar flexors to recover from unexpected perturbations around heel strike while group S relies on the knee extensors.

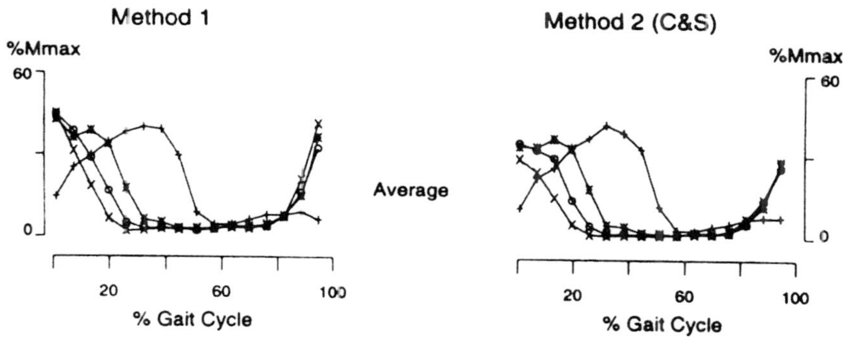

Figure 1. H reflex during walking and running. The modulation of the soleus H reflex during walking at 4.5 kmh^{-1} (+), running at 8 kmh^{-1}(*), running at 12 kmh^{-1}(o) and running at 15 kmh^{-1}(×). Average of 7 subjects. In the left panel the data were normalized to the M_{max} recorded in every sweep (Method 1). In the right panel the same data were obtained by the method described by Capaday and Stein (1986, 1987). The gait cycle was divided into 16 time slices and normalized.

Uphill and Downhill Walking

During uphill walking the modulation pattern changes so that the reflex amplitude increases gradually almost following the pattern of EMG activity in the soleus muscle. During downhill walking the reflex excitability rises immediately at heel strike to an almost constant level during the rest of the stance phase. The peak H reflex amplitude was significantly lower during downhill than uphill walking while the ratio between H reflex and EMG was lower during uphill walking mostly due to higher EMG activity during uphill walking (Simonsen et al., 1995). It is suggested that downhill walking represents a more 'dangerous' task around hill strike as compared to uphill walking and therefore the stretch reflex is switched on at heel strike in the downhill situation.

Running

The excitability and the modulation pattern of the soleus H reflex were first investigated by Capaday and Stein (1986, 1987) and later by Edamura et al. (1991). In the first case running at 8 km/h was compared to walking at 4.5 km/h and it was reported that the peak soleus H reflex amplitude was significantly lower during running. In the study of Edamura et al. walking and running at identical speeds were investigated and it was found that the soleus H reflex was always lower during running. Using our own measuring technique as well as the technique of Capaday and Stein (1987) we were unable to reproduce the results mentioned for walking and running at 4.5 and 8 km/h, respectively. We found that the peak H reflex amplitude was unchanged between walking at 4.5 km/h and running at 8, 12 and 15 km/h when using the technique of sweep averaging as described by Capaday and Stein (1987). However, using our own method correcting for fluctuations in Mmax we found significantly higher peak amplitude during running at 12 and 15 km/h than during walking. Despite the methods used we were never able to find a reduced reflex excitability during running (Simonsen and Dyhre-Poulsen, 1998). It is suggested that the H reflex can only attain moderately higher levels than during walking and that it makes sense to facilitate the stretch reflex during running as the duration of the stance phase fits nicely to the latency of the stretch reflex and the electro-mechanical delay of skeletal muscles. In all cases the soleus H reflex was suppressed during the swing and flight phases but started to rise before heel strike.

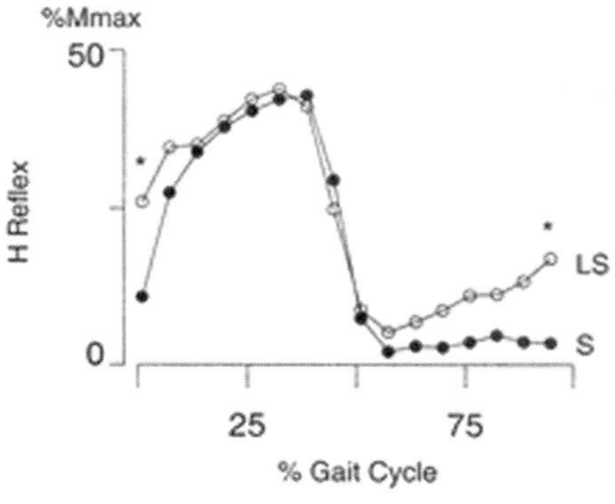

Figure 2. Different H reflex modulation patterns. Two types of soleus H reflex modulation during human walking. Group S (n=6; *filled circles*) showed on average an almost completely suppressed reflex during the entire swing phase, while group LS (n=9; *open circles*) showed a gradually increasing H reflex during the swing phase. The reflex amplitude was normalized to the maximal M wave (M_{max}) measured during walking. The gait cycle was divided into 16 time slices and normalized with heel strike at 0% and 100%. The swing phase starts at approximately 50%. *Asterisks* The first slice (P=0.022) and the last slice (P=0.001) differed statistically between the two groups; the remaining slices were not tested. The mean reflex amplitude during the swing phase also showed a significant difference (P=0.008).

EMG / H-reflex Gain

Capaday and Stein (1986) related the H reflex amplitude to the EMG amplitude during standing and walking and later during walking and running (Ferris et al. 2001). This ratio, called reflex gain, was highest during standing, lower during walking and even lower during running. We examined the same ratio during walking and running and during levels of simulated reduced gravity of 1.0, 0.75, 0.5 and 0.25 relative to normal gravity. Linear regression fits were applied on relative EMG and H reflex amplitudes showing that the so-called reflex gain, i.e. the slope of the regression line, was similar despite reduced gravity and similar between walking and running. However, the y-intercept of the regression line was significantly higher during walking than during running. This indicates that the nervous system adjusts H reflex threshold but not H reflex gain between walking and running (Ferris et al., 2001).

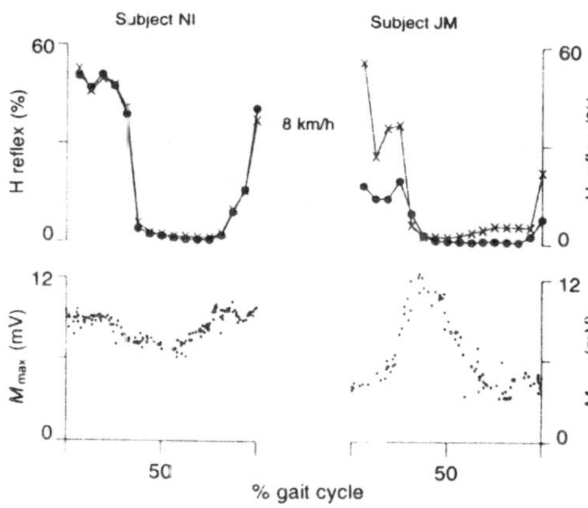

Figure 3. H reflex and M_{max} during running. Two subjects with large differences in the variation of M_{max} during running at 8 km h^{-1}. The subject on the left panels showed a rather small variation in the M_{max} during the gait cycle, while the subject depicted at the right showed large variations. The H reflex is expressed as the percentage of the M_{max} measured in every sweep (×) and during standing (•). For the subject shown on the right this variation of the M_{max} during gait resulted in lower H reflex amplitude during the stance phase.

The reflex amplitude rises with EMG activity regardless of reason for the increased EMG amplitude. Walking uphill, running and walking during high gravity all increase EMG levels and H reflex activity (Simonsen et al., 1995, Simonsen and Dyhre-Poulsen, 1999; Ferris et al., 2001). In contrast to previous findings we found that the H reflex gain is vaguely influenced by gait pattern but that the threshold of the reflex is greater during running than during walking (Fig. 4). Furthermore, the reflex activity is principally not controlled by the kinematics of locomotion (Simonsen et al., 2002), as individual modulation patterns (Fig. 2) exist even though the kinematics looks the same.

DISCUSSION

The soleus H reflex is remarkably modulated during locomotion. The reflex is almost fully depressed during the swing phase but present during the stance phase where it may supplement the push off activity. The reflex amplitude rises with EMG activity regardless of reason for the increased EMG amplitude. Walking uphill, running and walking during high gravity all increase EMG levels and H reflex activity. (Simonsen et al., 1995, Ferris et al., 2001) In contrast to previous findings (Capaday and Stein, 1986, 1987) we found that the H reflex gain is vaguely influenced by gait pattern but that the threshold of the reflex is greater during running than during walking. Furthermore, the reflex activity is principally not controlled by the kinematics of locomotion, as individual modulation patterns exist even though the kinematics looks the same. Most studies present averages of EMG, biomechanics and reflex modulations, and therefore obscure the underlying individual mechanisms.

During locomotion both the stimulus and recording electrodes unavoidably move in relation to the underlying tissue. We therefore elicited maximal M-waves just after the H reflex recordings and normalized all recordings to the maximal M wave. The stimulus efficacy was monitored by the small M wave preceding the H reflex. An automatic tracking mechanism then controlled the stimulus intensities so that the M wave preceding the H reflex was held constant at about 25% of the amplitude of the maximal M wave recorded during the same phase of movement. We found that the amplitude of the maximal M wave varied considerably during locomotion in many of the subjects and that normalization to the maximal M wave influenced the interpretation of the results (Fig. 3).

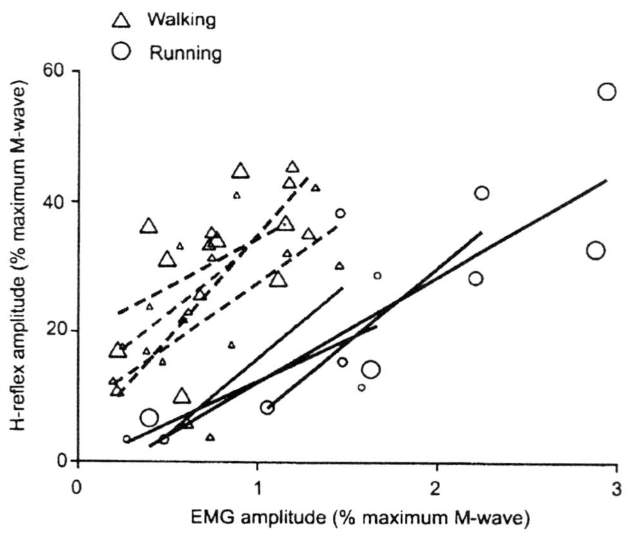

Figure 4. H reflex gain. Triangles denote walking and circles denote running. The size of the symbol reflects the gravity level (Smallest symbol 0.25 g, largest 1.0 g). Each data point reflects data from a single time period taken from the stance phase. Linear least-squares regression lines were fitted for each gravity level and gait. The calculated slope is a measure of H reflex gain. The y (ordinate) intercept denotes the threshold of the H reflex. Walking has a lower H reflex threshold than running but the gain is the same regardless of gait.

Functional Role of the Stretch Reflex

The functional significance of the stretch reflex during voluntary dynamic movement is not easy to establish and much controversy still exists in the literature. We have reported that inhibition of the stretch reflex is necessary when a muscle is required to act as a damper absorbing kinetic energy and dissipating it as heat (Dyhre-Poulsen and Laursen, 1984). In support for this the soleus H reflex is inhibited during landings when jumping downward (Dyhre-Poulsen et al., 1991). During jumping in place, which requires elastic muscle contractions, the H reflex and the stretch reflex facilitated (Dyhre-Poulsen et al., 1991). All studies hitherto have shown that the soleus H reflex is facilitated during the stance phase of walking, which implies that the central component of the stretch reflex loop is open. However, it has been questioned to what extent afferent input through the Ia fibres contributes to the total soleus output (EMG) activity. In a recent study Sinkjær et al. (2000) found by use of ischaemia that the Ia afferents did not influence the soleus activity during walking while group II and/or Ib afferents may play an important part in regulating the soleus activity. It is possible that the relatively slow stretching velocity of the soleus muscle during most of the stance phase of walking only gives rise to Ia volleys too weak to contribute to the soleus motor output, whereas a strong volley from a severe perturbation would still elicit a stretch reflex as suggested by Sinkjær and colleagues (2000). It is therefore likely that movement like jumping in place and running produce afferent volleys of sufficient strength to elicit a stretch reflex. Moreover, the cyclic duration of these tasks fits the use of the stretch reflex with a latency of approximately 35 ms and an electromechanical delay of approximately 50 ms. In this way the stretch reflex may be elicited during the initial lengthening (eccentric) contraction just after touch down and then contribute to activate the muscles during the following shortening (concentric) contraction during push off from the ground.

REFERENCES

Capaday, C., and Stein, R. B., 1986, Amplitude modulation of the soleus H-reflex in the human during walking and standing, *Journal of Neuroscience*, 6, 1308-1313.

Capaday, C., and Stein, R. B., 1987, Difference in the amplitude of the human soleus H reflex during walking and running, *Journal of Physiology*, 392, 513-522.

Dyhre-Poulsen, P., and Laursen, A. M., 1984, Programmed electromyographic activity and negative incremental muscle stiffness in monkeys jumping downward, *Journal of Physiology*, 350, 121-136.

Dyhre-Poulsen, P., Simonsen, E. B., and Voigt, M., 1991, Dynamic control of muscle stiffness and H reflex modulation during hopping and jumping in man, *Journal of Physiology*, 437, 287-304.

Edamura, M., Yang, F., Stein, R. B., 1991, Factors that determine the magnitude and time course of human H-reflexes in locomotion, *The Journal of Neuroscience*, 11, 420-427.

Ferris, D. P., Aagaard, P., Simonsen, E. B., Farley, C. T., and Dyhre-Poulsen, P., 2001, Soleus H-reflex gain in humans walking and running under simulated reduced gravity, *Journal of Physiology*, 530, 167-180.

Simonsen, E. B., Dyhre-Poulsen, P., and Voigt, M., 1995, Excitability of the soleus H reflex during graded walking in humans, *Acta Physiologica Scandinavica*, 153, 21-32.

Simonsen, E. B., and Dyhre-Poulsen, P., 1999, Amplitude of the soleus H reflex during walking and running, *Journal of Physiology*, 515, 929-939.

Simonsen, E. B., Dyhre-Poulsen, P., Alkjaer. T., Aagaard, P., and Magnusson, S. P., 2002, Interindividual differences in H reflex modulation during normal walking, *Experimental Brain Research*, 142, 108-115

Sinkjær, T., Andersen, J. B., Ladouceur, M., Christensen, L. O. D., Nielsen, J. B., 2000, Major role for sensory feedback in soleus EMG activity in the stance phase of walking in man, *Journal of Physiology*, 523, 817-827.

Zehr, E. P., and Stein. R. B., 1999, What functions do reflexes serve during human locomotion? *Progress in Neurobiology*, 58, 185-205.

SECTION VII

Supraspinal Control of Movement

The central control of movement is based on the appropriate combination of sensory inputs and descending inputs to motoneurones, with the central controller having access to models of how the whole system such as the upright body or the hand behaves. This simply restates the familiar notions of corollary discharge and efference copies in which feedback and feedforward mechanisms are combined in task-dependent ways. This Section covers aspects of how this combination may be achieved at supraspinal levels.

H-J Freund provides a summary of recent studies of functional imaging of the brains of human subjects during grasping, scanning movements by the fingers, imagination of the performance of movements, and even the observation of movements (Chapter 44). While the latter and movement imagery use neural circuits that overlap those used during real performance, there is new evidence that the parietal cortex as well as the dorsolateral prefrontal cortex is powerfully activated. The parietal activation is proposed to specify "how" the hand movement is to be performed. R. Johansson (Chapter 45) dissects the actual dynamics of how an object is grasped according to its size, shape, surface texture and apparent weight. The actual grasping itself employs both feedback and feedforward mechanisms. The various receptors in the digits provide important information about surface texture and slips between the digit and object, while vision allows the prediction of the shape and weight of the object. The overall result is that the grasp is achieved efficiently with an appropriate safety margin between forces generated at the tip of the finger and thumb, and the tangential forces required for the task. The afferent input from the pulp of the finger provides a complex "array", spatially and temporally changing, which encodes the direction of forces acting at different locations. Based on a model which can be rapidly updated (even during a lift), any mismatch between expected and observed behaviour of the object being lifted leads to a quick compensatory response, which is likely to use both spinal and supraspinal paths (see also Chapters 6 and 9).

The properties of neurones in the primary motor cortex that are active during cued grasping movements in trained monkeys are described by M. Schieber (Chapter 46). The discharge of cells in the primary motor cortex can be associated with the voluntary (instructed) movement of multiple fingers, so that the cells are unlikely to "address" only muscles acting on single digits. Furthermore, the location of these cells is distributed across the cortical surface in the mediolateral direction. A diffuse network of neuronal activity appears to underlie movements of the digits in the primate. A functional consequence is that there are some central (as well as peripheral) limits to the production of independent movement of single fingers in the flexion-extension plane (e.g., Kilbreath and Gandevia, 1994; Li et al., 1998). Further evidence on this point will come from studies in which spike-triggered averaging is used to identify the cells making direct corticomotoneuronal connections. The presence of afferent inputs that drive the motor cortical cells may complicate the interpretation, however.

The cerebral cortex has several output areas with direct projections to the spinal cord in addition to the well-known output from the primary motor cortex. J. Tanji and colleagues (Chapter 47) reveal one role for the "non-primary" motor cortical areas in the higher order planning of movement. They analyse the behaviour of cells in the cingulate motor areas in trained monkeys in terms of an association between the reward provided to the monkey during the task and its successful accomplishment. These areas are critically placed to link the internal "drives" for actions and the selection of motor commands to achieve them. When the GABA agonist, muscimol, is topically applied to the rostral (but not caudal) cingulate motor area, the local inactivation impairs the monkey's ability to select the movement which provides the greater reward (drops of juice). Some cells in this area change their discharge in the critical period after a reduced reward had been given and before the final selection of the next movement. R. Lemon and colleagues examine the detail of the properties of the corticospinal output from the primary motor cortex and the supplementary motor area in anaesthetized monkeys (Chapter 48). Their electrophysiological studies show that motoneurones involved in hand movements receive larger and more frequent EPSPs from the primary motor cortex than from the supplementary motor area. These EPSPs often include a monosynaptic input from direct (rather than transsynaptic) activation of corticospinal neurones. Anatomical studies of the density of spinal projections also exposes differences between outputs from the primary motor cortex and the supplementary motor area and thereby supports the electrophysiological results. These authors' experimental paradigms may provide an important means of assessing the potential functional role of the projections from different cortical areas to spinal motoneurones.

J. Taylor and colleagues (Chapter 49) present novel evidence that the properties of the connections between the human motor cortex and spinal motoneurones can vary with usage. They use single transmastoid stimuli to activate descending corticospinal axons (Ugawa et al., 1991; Gandevia et al., 1999) before and after strong voluntary contractions of the human elbow flexor muscles. Based on the response to single corticospinal stimuli, there appears to be strong post-contraction depression of the effectiveness of the descending volley. Evidence is presented that this may involve corticospinal synapses and that the change occurs at presynaptic sites. Their results suggest the value of combining this form of stimulation with the more commonly used transcranial magnetic stimulation. A dual approach should help resolve the relative extent of change at cortical versus spinal levels during cortical stimulation. Collision studies have shown that responses to transcranial magnetic stimulation and transmastoid stimulation traverse many common corticospinal axons. It follows that the latter stimulus must be a better index of changes in spinal excitability than responses evoked in the H-reflex – the latter procedure involves different spinal circuitry and synapses.

Movements of the distal extremities are usually combined with movements of the axial parts of the body. T. Miles reviews new features of the cortical and reflex control of the human masticatory system (Chapter 50). The motor nuclei for the different muscles involved in chewing receive strong corticobulbar projections, with the strength of these projections being far from uniform. For example, the contralateral motor cortex sends fast-conducting, excitatory corticobulbar projections to motoneurones supplying the masseter muscle. In contrast such projections are slower conducting and inhibitory from the ipsilateral cortex, with the inhibition presumably accomplished by spinal interneurons.

Much is written about the plasticity of the cerebral cortex controlling movement. In many such studies, the properties of cortical neurones are assessed by recording their postsynaptic responses to different afferent inputs. A review by M. Calford (Chapter 51) covers the substantial evidence on adaptive changes in the behaviour of cells in primary sensory cortical areas (somatosensory, visual, auditory) when the input they receive is markedly reduced (e.g., by deafferentation) or increased by usage. Some of these adaptations occur immediately the input changes. Calford argues that conventional measurements of a test cell's post-synaptic response to stimulation of the sensory surface underestimates the peripheral area which can influence its discharge. Furthermore, there appears to be a central inhibition (perhaps driven by tonic afferent activity) which limits the extent of the central receptive fields.

To conclude this Section, J. Pettigrew provides a reminder about the higher-order control of movement, based on the visual system (Chapter 52). They use Bonneh's illusion, in which motion of a series of dots induces transient blindness for static coloured shapes, to reveal two salient points for movement neuroscience. First, processing of afferent information about the external world may be influenced by events at high as well as low levels within the afferent pathways. Second, even though the Bonneh illusion concerns the apparent appearance and disappearance of a bright target shape, the rhythmicity of this event, in effect an oscillation, is linked to the current mood of the subject. These authors argues that this oscillation reflects inter-hemispheric switching and that such switching is likely to have been highly conserved in evolution.

REFERENCES

Gandevia S.C., Petersen N., Butler J.E., and Taylor J.L., 1999, Impaired response of human motoneurones to corticospinal stimulation after voluntary exercise, *Journal of Physiology*, 521, 749-759.

Kilbreath S.L., and Gandevia S.C., 1994, Limited independent flexion of the thumb and fingers in human subjects, *Journal of Physiology*, 479, 487-497.

Li Z.M., Latash M.L., and Zatsiorsky V.M., 1998, Force sharing among fingers as a model of the redundancy problem, *Experimental Brain Research*, 119, 276-286.

Ugawa Y., Rothwell J.C., Day B.L., Thompson P.D., and Marsden C.D., 1991, Percutaneous electrical stimulation of corticospinal pathways at the level of the pyramidal decussation in humans, *Annals of Neurology*, 29, 418-427.

fMRI STUDIES OF THE SENSORY AND MOTOR AREAS INVOLVED IN MOVEMENT

Hans-Joachim Freund*

ABSTRACT

A wide range of natural hand movements such as grasping, exploring or manipulating objects activates a parietal-premotor network upstream of motor cortex. The specific representations of each motor act are embedded in this circuitry and reflect the demands imposed by the sensory and motor processes involved in these motor behaviours including oculomotor and attentional control and memory processes. Further, the same network is activated during the observation or imagination of these movements. These complex intertwined and partially overlapping functional maps can be segregated in the time domain by means of real time techniques such as MEG that allow to disentangle the sequential processing stages.

INTRODUCTION

Functional activation studies can provide two types of information. When fMRI for a given experimental condition is compared with rest the whole network of cortical areas involved is displayed. The subcortical contributions are more difficult to recognise, in particular the brainstem. When only those activations shall be identified that are specific for a particular task the substraction of two active conditions that are different only for one aspect is the method of choice. Figure 1a shows a typical example for the first approach. A reaching for grasping gesture minus rest activates the sensorimotor areas and a parietal-premotor network including midline areas bilaterally. Figure 1b illustrates the second approach. Subtracting pointing from the same reaching for grasping task shown in Figure 1a reveals the grasp specific activation of the anterior part of the intraparietal sulcus (IPS), whereas the reach component that is common to both tasks is subtracted.

The advantage of imaging studies for the human as compared with experiments in monkeys is the specification of sensorimotor associations areas upstream primary sensorimotor cortex that are involved in different sensorimotor behaviours. In the

* Department of Neurology, Heinrich Heine University Duesseldorf, Duesseldorf, Germany.
Email: freund@rz.uni-duesseldorf.de

following a few prototypical examples for activities of the hand and eye will be given. They will be briefly discussed in the context of experimental results but also of data from lesion analysis in patients. The comparison between functional activations and small focal lesions is similar to the experimental approach where the correlation of neural discharges with behaviour provide a positive image of the deficits in behaviour caused by ablation or deactivation.

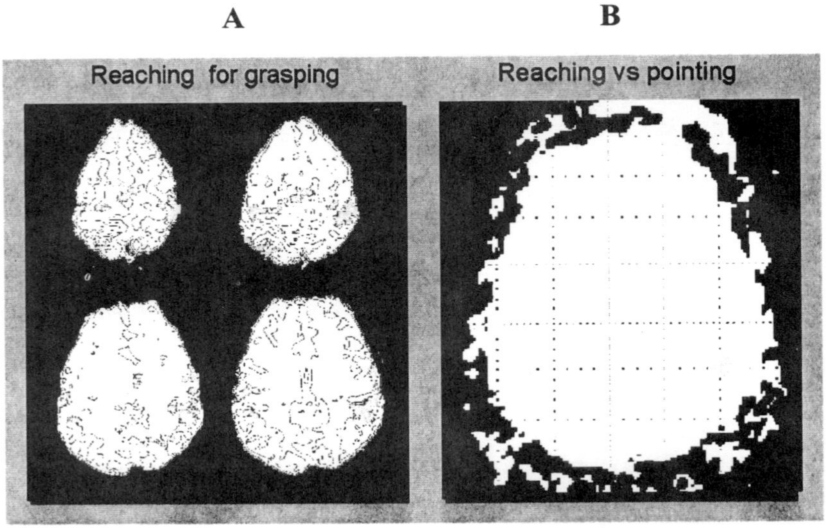

Figure 1. A: Reaching for grasping versus rest in a single subject. B: Reaching for grasping versus pointing in the same subject.

Grasping

Visuomotor processing has been extensively investigated for the act of reaching for grasping. Here a wealth of information is available for the neuronal mechanisms underlying these behaviours in the monkey that can be compared with lesion and activation studies in the human. Experimentally it has been shown that sensory signals from many modalities – visual, vestibular, auditory and somatosensory – converge onto parietal areas 7a and lateral intraparietal sulcus and the medial superior temporal area to code for the spatial location of the goals for movement. Hereby the multimodal signals can build a common reference frame, but can also concurrently code target locations in multiple coordinate frames required for different behaviours such as the control of gaze, arm or of navigation.

In human activation studies the parietal-premotor network is typically activated bilaterally also when the task is performed unimanually. This bilateral activation stands in contrast to lesion data associated with purely contralesional deficits. More specifically small lesions scattering around the intraparietal sulcus (IPS) interfere with hand shaping and the formation of hand aperture during grasping on the contralesional side (Binkofski et al., 1998). This location corresponds to the area specifically activated during grasping as shown in Figure 1b.

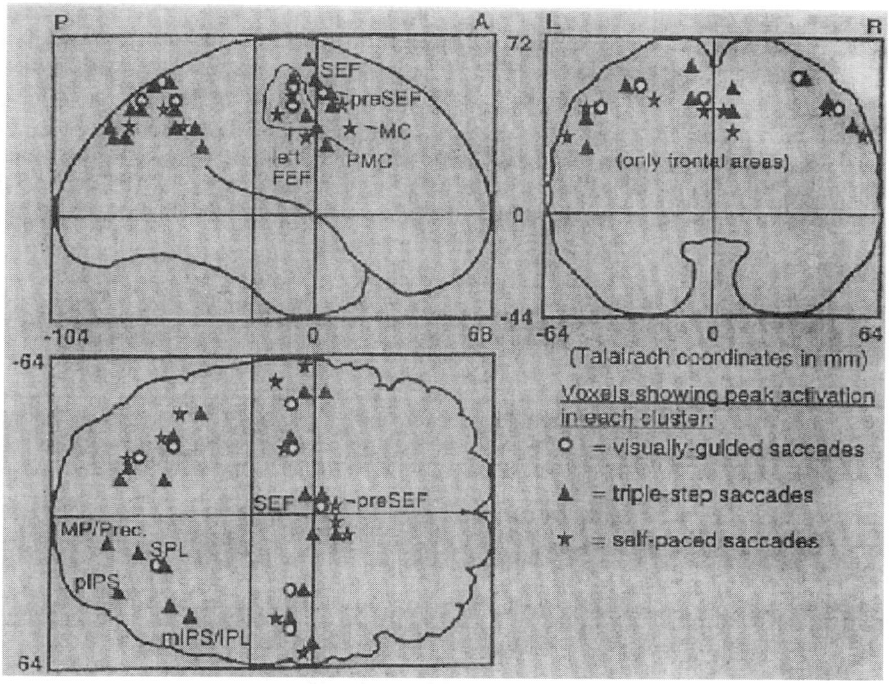

Figure 2. Task-dependent variability of cortical foci activated during the performance of visually guided saccades (vs. central fixation), triple-step saccades (vs. central fixation) and self-paced saccades (vs. rest in darkness). The respective voxels of peak activation (Z > 4.0) are plotted in Talairach space for each cluster of the population data. Compared with the coordinates in Tables 1 and 4, z-coordinates of the parietal foci were corrected for incongruency between Montreal Neurological Institute (MNI) templates and Talairach coordinates (Stephan et al., 1997) by subtracting 13 mm. For the sake of clarity, the right FEF is not shown in the saggital plot (top left) and parietal foci not in the coronal plot (top right). The symbols for the different tasks are explained in the figure (lower right). (From Heide et al., 2001.)

Grasping is usually visually guided and therefore employs eye movements as well. Directing eye movements to the target and focussing attention on it is accomplished by neural aggregates that closely overlap with those controlling limb movement. The activation patterns during visually-guided eye movements (Heide et al., 2001) depend on the experimental conditions but is slightly different from that seen during limb activity. Clusters of activation during self-paced, visually guided and memorised saccades are seen in the frontal eye fields (FEF), supplementary eye fields (SEF), ventral premotor cortex (vPMC), anterior cingulate, the precuneus and several superior posterior parietal areas. Part of this circuitry is also involved during covert shifts of attention without eye movements so that attentive subfunctions are also represented in partially overlapping areas.

Active Touch

Active touch illustrates the close interdependence between somatic sensation and movement. It elaborates the micro- and macrogeometry of objects on the basis of the finely tuned scanning movements of the fingers during the sequential sampling of mechanoreceptive information. The breakdown of digital palpation along with disturbed object exploration and recognition typically seen in patients with damage of the superior parietal lobule (SPL) but normal elementary sensation emphasises the significance of that parietal area for this integrative somatomotor function. It is this ability to engage the hand in the motor performance required to collect the sensory information that is specifically disturbed. This unimodal sensorimotor disturbance, tactile apraxia, is confined to the somatosensory modality and affects only the contralesional hand (Pause et al., 1989; Binkofski et al., 2000).

Functional neuroimaging during object exploration and manipulation shows parietal, premotor and primary sensorimotor activations similar to those identified in the monkey. Comparing the manipulation of complex objects versus manipulation of a simple sphere showed similar patterns of parietal and premotor activations. The subtraction between the two conditions shows bilateral activation of the cortex lining the anterior part of the IPS, the SPL and vPMC. The bilateral activations in the parietal-premotor network during unimanual tasks raises the question about the role of the ipsilateral activations for functional recovery.

Modelling Objects

One of the most proficient human hand skills including the cognitive dimension is the shaping and construction of objects. This formative motor act requires the mental representation of the intended 'Gestalt' and its transformation into the hand-finger trajectories. Asking subjects to shape and construct objects out of amorphous plastiline lumps with both hands evoked similar activations along the IPS as seen during exploration. However, there were striking differences between the two conditions in that SPL and medial wall motor activations were much more pronounced during manual modelling. Part of the activations of the medial wall motor areas may be due to the well known strong activations of the SMA and cingulate motor areas during bilateral hand movements (Stephan et al., 1999). Imagination of the model that underlies the constructive concept activated the left SPL and the left angular and ventral premotor cortex - a pattern emphasising the close vicinity of the circuitry for cognitive manipulative motor behaviour and language (Jäncke et al., 2001).

The consistent and strong concomitant activation of vPMC during object exploration, manipulation and modelling is not due to covert naming (Binkofski et al., 1999). When the explored objects had to be named additional activations in more anterior vPMC and in the triangular part of the inferior frontal gyrus corresponding to Brodmann area 45 were associated with naming. The robust activation of the parietal-premotor circuitry during active touch, manipulation, and modelling is therefore specific for the involvement of our hand in a wide range of sensorimotor behaviours.

Correspondingly, lesions of these areas interfere with the respective functions. These somatomotor disturbances following parietal lesions signify an important and principal difference as compared to visuomotor information flow: they show, that the functional architecture of cognitive and action related aspects of hand use is closely intertwined.

There is no analogue of the dorsal-ventral stream dichotomy of visual processing for somatosensation. Consequently, parietal damage compromises perception and action. Temporal lobe damage that interferes with visual cognitive functions leaves tactile cognition unaffected.

Movement Observation

There is ample evidence that movement observation and imagery activate the same networks that are recruited during performance (Grezes and Decety, 2001). The mirror neurone concept elaborated for a subpopulation of neurones in the ventral premotor area F5 in the monkey (Rizzolatti et al., 1996) has gained new facets by observations in humans. Buccino et al. (2001) demonstrated that action observation activates not only ventral premotor cortex possibly homologous to the monkey's area F5 but also dorsolateral premotor cortex in a somatotopic manner. Another new finding was that whenever the observed movement was object-directed, parietal cortex was strongly activated as well. Again, this activation was bilateral and somatotopically organised. The object related parietal activations are in agreement with the assumed pragmatic role of parietal cortex, specifying how to do a motor task - in contrast to the specification what to do - that are commonly regarded as a frontal lobe function.

Buccino et al. (2001) therefore concluded that during action observation there is a recruitment of the same neural structures which would be normally involved in the actual execution of the observed motor behaviour. When individuals observe an action, they code that action in terms of the related voluntary movements. The "seen" actions are mapped onto the corresponding motor representations of the frontal lobe and, in the case of object-related actions, the "seen" objects on the effector-related, pragmatic representations in the parietal lobe.

Adding the Fourth Dimension: Time

The considerable overlap of the functional maps elaborated for different aspects of sensorimotor integration shows the limitations of the approach to segregate brain functions into distinct modules. Too many variables such as the imagination and observation of movement, attention, eye movements, generating motor plans, spatial and temporal coding, coordinate transformations, sensorimotor transformations, memory processes all activate parietal cortex. If such sensorimotor behaviours partially share the same cerebral areas at least at the level of resolution that can presently be accomplished by fMRI, the question arises whether the added dimension of the time domain will provide new means to disentangle the overlapping functional maps. This is even more at issue as not only strong overlap between coextensive functional networks is increasingly shown but there are also seemingly conflicting results.

The advantage of adding high temporal resolution is illustrated by a recent study on single word reading in developmental stutterers using magnetoencephalography (MEG). Salmelin et al. (2000) identified source areas in different brain regions and defined their temporal course. The flow of activity throughout the activated brain areas specified the relation between sensory stimuli and the temporal course of the magnetic field potentials as related to the stimulus or to speech onset or other relational variables.

Nine developmental stutterers and 10 fluent speakers were asked to read a word presented on a screen for 300 ms. After a blank interval of 500 ms a question mark appeared for 2000 ms prompting the subject to read the word aloud. Figure 3 gives the standard error of the mean source wave forms in each of 11 regions of interest (ROI) averaged over fluent subjects and stutterers. The responses can be classified in three groups: within the first 200 ms after word onset occipital cortex (area 1) and left and right occipito-temporal cortices (areas 2 + 3) showed strong initial signals and subsequent smaller responses after the question mark. A later response group ranging between 200 and 600 ms responses occurred in both inferior frontal cortices (areas 4 + 5), left superior temporal area (6) and both inferior parietal regions (areas 7 + 8). Finally, both fronto-parietal cortices (areas 9 + 10) and the vertex. (area 11) became involved and remained active throughout the vocalisation prompt (question mark at 800 ms and mouth movement and speech onsets (at about 960 ms).

Although the overt performance was basically identical in the two groups the cortical activation patterns differed significantly. Processing in fluent speakers advanced from left inferior frontal cortex (articulatory programming) to the left motor and premotor cortex within the first 400 ms after seeing the word. This sequence was reversed in the stutterers where motor cortex activation was followed by a delayed left inferior frontal signal.

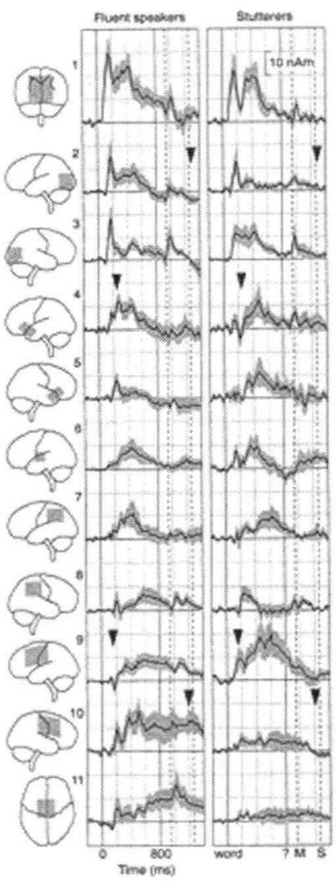

Figure 3. Mean ± standard error of the mean (black curve and shading) source strengths as a function of time in fluent subjects (left) and stutterers (right). The word and question mark onsets are indicated with solid vertical lines and the mouth movement (M) and speech (S) onsets with dashed lines. The black arrowheads denote the ROIs (regions of interest) and TOIs (time windows of interest) where the responses of stutterers and fluent speakers differed significantly from each other. The studied ROIs are illustrated on the schematic drawings of brains on the left. (From Salmelin et al., 2000.)

These data exemplify the advantage of monitoring sequential processing stages in real time by means of the superimposition of MEG activity on 3D-MRI-data sets allocating the sequential processing stages to distinct cortical areas. This method therefore provides a new powerful tool to study sensorimotor processing in normals and to elucidate pathophysiology as shown for a temporal disorder like stuttering.

The two approaches to functional neuroimaging, fMRI and MEG have obtained a new link because a recent landmark article by Logothetis et al. (2001) has shown that the physiological correlate of the Bold signal is the local field potential (LFP) that can be measured by macroelectrodes in the brain or by EEG and MEG. It shows that the Bold effect does not reflect spike activity as previously assumed (Rees et al., 2000) but measures the synaptic potential fields. It is the gamma band component of the LFP that mimics the Bold contrast changes. Against this background the comparison of the two different approaches opens new avenues for future research.

REFERENCES

Binkofski, F., Dohle, S., Posse, S., Stephan, K. M., Hefter, H., Seitz, R. J., and Freund, H.-J., 1998, Human anterior intraparietal area subserves prehension: a combined lesion and functional MRI activation study, *Neurology*, 50, 1253-1259.

Binkofski, F., Buccino, G., Posse, S., Seitz, R. J., Rizzolatti, G., and Freund, H.-J., 1999, A fronto-parietal circuit for object manipulation in man: evidence from an fMRI-study, *European Journal of Neuroscience*, 11, 3276-3286.

Binkofski, F., Amunts, K., Stephan, K. M., Posse, S., Schormann, T., Freund, H.-J., Zilles, K., and Seitz, R. J., 2000, Broca's region subserves imagery of motion: a combined cytoarchitectonic and fMRI study, *Human Brain Mapping*, 11, 273-285.

Buccino, G., Binkofski, F., Fink, G. R., Fadiga, L., Fogassi, L., Gallese, V., Seitz, R. J., Zilles, K., Rizzolatti, G., and Freund, H.-J., 2001, Action observation activates premotor and parietal areas in a somatotopic manner: an fMRI study, *European Journal of Neuroscience*, 13, 400-404.

Grezes, J., and Decety J., 2001, Functional anatomy of execution, mental simulation, observation, and verb generation of actions: a meta-analysis, *Human Brain Mappng*, 12, 1-19.

Heide, W., Binkofski, F., Seitz, R. J., Posse, S., Nitschke, M. F., Freund, H.-J., and Kömpf, D., 2001, Activation of frontoparietal cortices during memorized triple-step sequences of saccadic eye movement: an fMRI study, *European Journal of Neuroscience*, 13, 1177-1189.

Jäncke, L., Kleinschmidt, A., Mirzazade, S., Shah, N. J., and Freund, H.-J., 2001, The role of the inferior parietal cortex in linking the tactile perception and manual construction of object shapes, *Cerebral Cortex*, 11, 114-121.

Logothetis, N. K., Pauls, J., Augath, M., Trinath, T., and Oeltermann, A., 2001, Neurophysiological investigation of the basis of the fMRI signal, *Nature*, 412, 150-157.

Pause, M., Kunesch, E., Binkofski, F., and Freund, H.-J., 1989, Sensorimotor disturbances in patients with lesions of the parietal cortex, *Brain*, 112, 1599-1625.

Rees, G., Friston, K., and Koch, C., 2000, A direct quantitative relationship between the functional properties of human and macaque V5, *Nature Neuroscience*, 3, 716-723.

Rizzolatti, G., Fadagia, L., Gallese, V., and Fogassi, L., 1996, Premotor cortex and the recognition of motor actions, *Brain Research, Cognitive Brain Research*, 3, 131-141.

Salmelin, R., Schnitzler, A., Schmitz, F., and Freund, H.-J., 2000, Single word reading in developmental stutterers and fluent speakers, *Brain*, 123, 1184-1202.

Stephan, K. M., Binkofski, F., Halsband, U., Dohle, C., Wunderlich, G., Schnitzler, A., Tass, P., Posse, S., Herzog, H., Sturm, V., Zilles, K., Seitz, R. J., and Freund, H.-J., 1999, The role of ventral medial wall motor areas in bimanual co-ordination. A combined lesion and activation study, *Brain*, 122, 351-368.

DYNAMIC USE OF TACTILE AFFERENT SIGNALS IN CONTROL OF DEXTEROUS MANIPULATION

Roland S. Johansson*

ABSTRACT

During object manipulation, humans select and activate neural action programs acquired during ontogenetic development. A basic issue in understanding the control of dexterous manipulation is to learn how people use sensory information to adapt the output of these neural programs such that the fingertip actions matches the requirements imposed by the physical properties of the manipulated object, e.g., weight (mass), slipperiness, shape, and mass distribution. Although visually based identification processes contribute to predictions of required fingertip actions, the digital tactile sensors provide critical information for the control of fingertip forces. The present account deals with the tactile afferent signals from the digits during manipulation and focuses on some specific issues that the neural controller has to deal with to make use of tactile information.

INTRODUCTION

Successful use of the hands in manipulatory tasks requires the selection of motor commands tailored to the task at hand, the manipulative intent, and the relevant physical properties of the object in hand. A basic issue in understanding the control in manipulation is to learn how people use sensory information to adapt their fingertip forces to the constraints imposed by various object properties such as mass (weight), slipperiness, shape, and mass distribution. These factors, in turn, impose constraints on the feasible fingertip forces – including their magnitudes, directions, and sites of application. Central to the control of fingertip forces is to maintain grasp stability. Most manipulatory tasks require that we apply both linear forces and torques tangential to the grasped surfaces as we move the object or use it as a tool. These self generated tangential fingertip loads destabilize the grasp. To counteract this, we therefore apply adequate forces normal to the grasped surfaces (grip forces) in relation to the destabilizing

* Physiology Section, Department of Integrative Medical Biology, Umeå University, Umeå, Sweden.
Email: roland.s.johansson@physiol.umu.se

tangential loads. Grasp stability thus implies that the grip forces are controlled such that they are adequate to prevent accidental slips but not so large as to cause unnecessary fatigue or damage to the object or hand.

Vision normally provides critical sensory information for the control of kinematic aspects of manipulation, i.e., planning and control of hand motion prior to grasping and motion of the objects held in the hands (Jeannerod, 1986; Ballard et al., 1992; Land et al., 1999; Johansson et al., 2001). Although visual object identification processes contribute to prediction of required fingertip forces, the digital tactile sensors provide the critical information for the control of fingertip forces in dexterous manipulation. Indeed, individuals with impaired digital sensibility have great difficulty performing manipulation tasks even under visual guidance. For instance, they often drop objects, may crush fragile objects, and have difficulties in otherwise simple tasks such as buttoning a shirt.

The present account addresses the use of tactile afferent signals from the digits in manipulation and focuses on some specific issues that the neural controller has to deal with. One such issue is the long time delays between mechanical events at the fingertip and when they can be expressed as motor consequences. Closed feedback loops based on digital sensory input do not work for moment-to-moment adjustment of fingertip forces. The time delays in such loops are simply too long in relation to the temporal characteristics of most manipulative actions (cf. Rack, 1981). Due to nerve conduction and processing delays and the sluggishness of the muscles, the effective time delays in such loops are 100 ms or longer. Another problem is that the tactile input arises from principally two causes: as a result of the self-generated movements as such, and as a result of events related to the specific properties of the object, both central for the control. Yet another related issue is that various fingertip parameters, critical for the control of fingertip forces, are likely to show interaction effects on the afferent responses. Given that the tactile afferents eventually are driven by fingertip deformations that cause strain/stress changes at the nerve endings, interactions can be expected between, for instance, the local shape of the grasped surfaces and the direction of fingertip forces. Knowledge of how these and other parameters of stimuli on the fingertips may interact is crucial for understanding the neural computations that occur in sensorimotor systems that support dexterous manipulation.

‘ANTICIPATORY PARAMETER CONTROL’ (APC) AND ‘DISCRETE EVENT, SENSORY-DRIVEN CONTROL’ (DESC)

As for neural central pattern generators implied in other self-paced motor behaviors, the coordination patterns generated by the sensorimotor programs activated in manipulation show a clever design. Similar to the vocal language, the brain appears to use a ‘vocabulary’ for manipulation with sequential ‘syllables’, i.e., delineated phases that are characterized by a distinctive use of tactile sensory information. Furthermore, the programs express coordinative constraints that provide a basis for the control of grasp stability (Johansson and Westling, 1984). Rather than driven by sensory feedback related to the tangential load, the grip force is constrained by the programs to change in parallel with the applied load. Thus, the grip force increases and decreases in phase with – and thus predicts – changes in the tangential load whether it is linear force (Fig. 1A) and tangential torque (Fig. 1B) and combinations thereof (Kinoshita et al., 1997). This control strategy is not specific to any particular task, grip configuration and mode of object transport (Flanagan and Tresilian, 1994; Burstedt et al., 1997; Flanagan et al., 1999; Johansson et al., 1999).

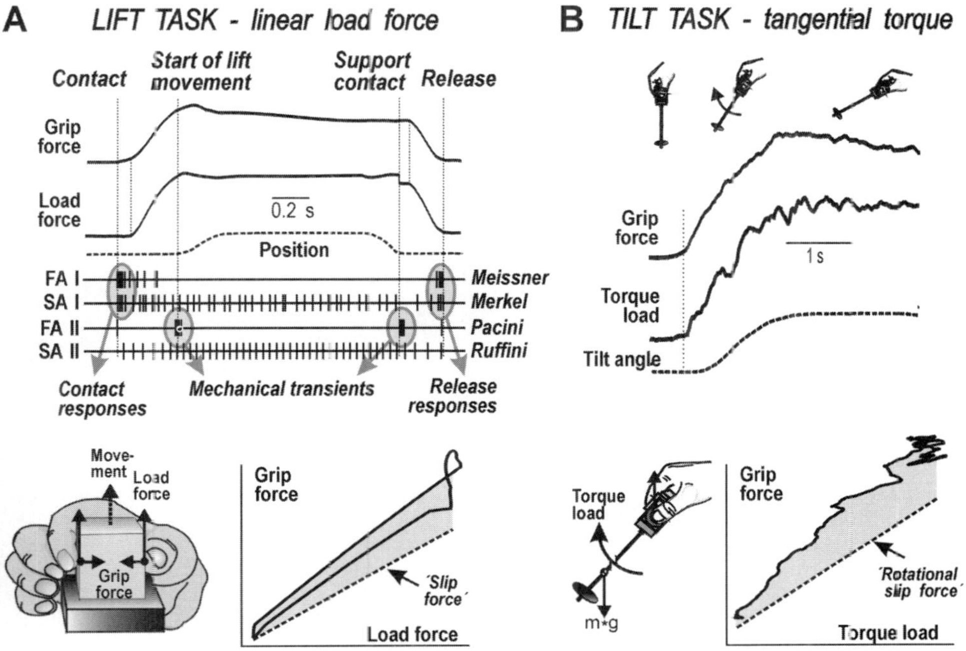

Figure 1. Fingertip forces during two prototypical manipulatory tasks. *A,* Subject lifts an instrumented test object from a table, holds it in the air and then replaces it, using a precision grip. Top graph: horizontally oriented grip force, vertically oriented load force ('lift force') and object's vertical position as a function of time. After contact with the object, demarcated by the left-most vertical line in the top graph, the grip force increases by a short period while the grip is established before the command is released for a parallel increase in grip and load force during isometric conditions (2nd vertical line). This increase continues until the start of object movement when the load force overcomes the force of gravity; the object lifts off at the 3rd vertical line. After the replacement of the object and table contact occurs (4th line) there is a short delay before the two forces decline in parallel (5th line) until the object is released (6th line). Note that these forces are constrained by the active sensorimotor program to change in parallel with no time delay. This emphasized by the phase plane plot in the bottom graph showing the grip force against the load force. Dashed line indicates the minimum grip-to-load force ratio required to prevent slips and the safety margin against slips is indicated by hatching. Middle graph shows schematically obligatory tactile afferent responses during a lift. Indicated events are: (i) contact responses occurring preferentially in FA-I (Meissner) and SA-I (Merkel) tactile afferents; (ii) burst discharges in FA II-afferents (Pacinian) related to the mechanical transients that occur at lift-off and at the moment of table contact. Also note (iii) the release responses when the receptor bearing digit breaks the contact with the object. During most phases there was a tonic input from tactile units dominated by a fairly regular impulse activity in the SA II (Ruffini) afferents. SA I afferents also fire but at lower and more irregular rates (Adapted from Johansson and Westling, 1990). *B,* Subjects tilted an already lifted object by 65° around its grip axis, which caused tangential torques at each grasp surface. The top graph show the grip force, tangential torque and tilt angle against time. Note the parallel change in grip force and the tangential load; the bottom graph emphasizes this coordinative constraint by plotting the grip force against tangential torque load for the same data. The shading indicates the safety margin against rotational slips. (Modified from Goodwin et al., 1998.)

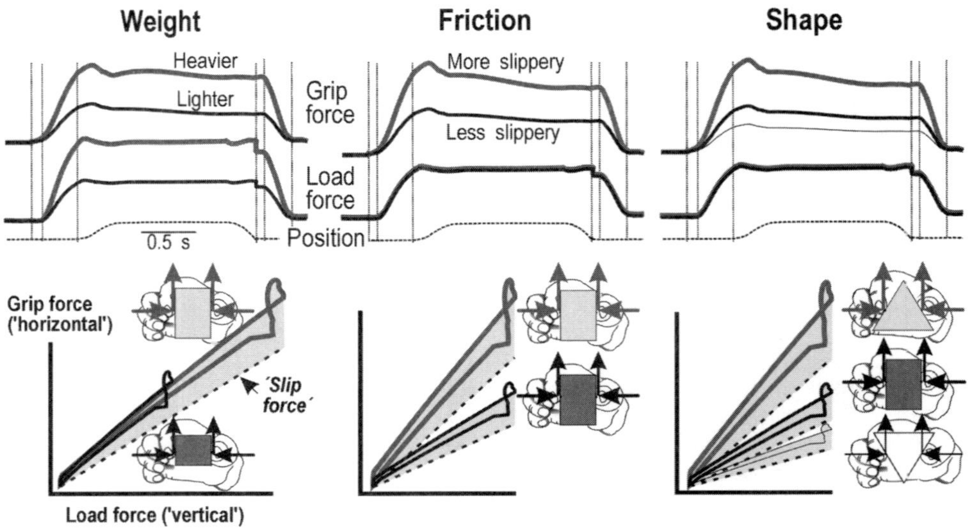

Figure 2. Parametric adjustments of motor output to object weight, friction between the object and skin, and shape of contact surface. Subject lifts an instrumented test object from a table, holds it in the air and then replaces it, using a precision grip. Upper graphs show the horizontally oriented grip force, the vertically oriented load force ('lift force') and the object's vertical position as a function of time for superimposed trials, indicated by differently hatched curves. The lower graphs show the grip force as a function of the load for the same trails and dashed line indicates the minimum grip force to prevent slips; the safety margin against slips is indicated by hatching. With weight variations the parallel change in grip and lift forces produced by the sensorimotor program ensures grasp stability when lifting objects of different weights. To deal with changes in friction, the sensorimotor system adjusts the balance between grip force and load force. A similar scaling of the grip-to-load force ratio is observed when object shape is varied. A greater ratio is used when the grip surfaces are tapered upwards compared to downwards. In either instance, the coordination of grip and load force ensures an adequate safety margin against slips. (Modified from Johansson and Westling, 1984, Johansson and Westling, 1988a and Jenmalm and Johansson, 1997.)

The coupling between changes in grip force and tangential load provided by the sensorimotor programs engaged likewise offers neat solutions for parametric adaptation of fingertip forces to the physical properties of the object at hand, such as its as mass, distribution of mass, shape and friction against the fingertips (Fig. 2). Again, rather than depending on moment-to-moment somatosensory feedback, the neural controller operates primarily in a feedforward manner. According to a control policy termed "anticipatory parameter control" (APC) motor command parameters are specified in advance by internal models (memory representations) pertaining to the object properties (Johansson and Cole, 1992, 1994). Thus, through APC subjects tailor fingertip forces for the properties of the object to be manipulated prior to the execution of the motor commands. For familiar objects, visual and haptic information can be used to identify and select the appropriate internal models. People may also use geometric information (e.g., size and shape) for anticipatory control, relying on internal forward models capturing relationships between geometry and force requirements. In general terms, there is now ample evidence from various areas of motor control that the brain makes use of internal models to form and adapt motor commands, i.e., neural circuits that mimic the behavior of the motor system and environment and capture the mapping between motor outputs and sensory inputs (Lacquaniti, 1992; Prochazka, 1993; Miall and Wolpert, 1996; Blakemore et al., 1998a; Kawato, 1999; Ohki et al., 2002).

How then are these internal models related to object properties are acquired, maintained and updated? Tactile sensory signals from the digits provide critical information by intervening intermittently according to a control policy called "discrete event, sensory-driven control" (DESC) (Johansson and Cole, 1992, 1994). The DESC policy is based on a continuous comparison of the actual and a predicted sensory inflow – an internal sensory signal often referred to as corollary discharge (Sperry, 1950). Thus, when we lift and further manipulate an object the active sensorimotor programs generate both efferent motor commands to accomplish the task as well as this internal time-varying estimate of the expected sensory input in a task and phase of task specific manner. The CNS continuously monitors the tactile input and if signals don't appear at the expected times and at expected strengths, control actions are taken: Disturbances in task execution due to erroneous parameter specification of the sensorimotor program give rise to a mismatch between predicted and actual sensory input. Detection of such a mismatch triggers pre-programmed patterns of corrective responses along with an updating of the relevant internal models used to predict sensory events and estimate the motor commands required. In manipulation, this updating typically takes place within a single trial. As such, the DESC policy principally solves the problem that the tactile signals arise both from the self-generated movements and from the specific properties of the object.

Interplay between the APC and DESC Policies

Our experimental evidence indicates that APC and DESC can explain adaptation to frictional condition between a fingertips and objects (Johansson and Westling, 1984, 1987; Edin et al., 1992), to object weight (Johansson and Westling, 1988a-b; Gordon et al., 1991), to mass distribution (Johansson et al., 1999) and to object shape (Jenmalm and Johansson, 1997; Goodwin et al., 1998; Jenmalm et al., 1998, 2000). Figure 3A exemplifies the interplay between the APC and DESC control polices in adaptation to object shape without vision of the grasped surfaces; visual cues pertaining to object geometry can be used in a feedforward manner to parametrically adapt the finger force coordination to object shape in anticipation of the upcoming force requirements (Jenmalm and Johansson, 1997). Three consecutive lifts are shown, taken from a lift series in which the angle of the grasped surfaces was changed between trials in an unpredictable order. The sequence is 30°, –30° and –30° and thus includes a transition from an upward tapered object (30°) to a downward tapered object (–30°). In the trials preceding this sequence, a 30° object was lifted. On the first trial after the switch to the –30° object, grip force develops initially according to the force requirements in the previous trial. Thus, memory of the previous surface angle determines the default force coordination in a feedforward manner, i.e., APC. However, about 100 ms after the digits contacted the object the grip force is modified and tuned for the actual surface angle (see first trial with the –30°). This time is required to obtain tactile information pertaining to surface angle, detect a mismatch and translate that into an updating of the motor commands. By the second trial after the switch, the force output is appropriately adapted to the –30° surface angle right from the onset of force application. Thus, an internal model related to object shape determines the force coordination in a feedforward fashion and a mismatch between the expected and actual tactile sensory information obtained at initial contact with the object mediate an updating of this model to changes in object shape. Furthermore, a single trial is enough to update the relevant internal model. With digital anesthesia, leaving neither visual nor somatosensory cues about shape, the force output is no longer adjusted to object shape; people adapt to the loss of tactile sensibility by applying strong grip forces regardless of the surface angle (Jenmalm and Johansson, 1997).

The curvature of spherically curved grasp surfaces is another aspect of object shape and becomes acute in tasks involving torsional loads (Fig. 3B) (Goodwin et al., 1998). Again, the relationship between the grip force and tangential load is scaled parametrically by curvature: for a given torque load, people increase grip force when curvature increases and maintain a small but adequate safety margin against rotational slip within a wide range of curvatures. The principal manner by which tactile information influences the force coordination resembles that illustrated for tapered surfaces in Fig. 3A (Jenmalm et al., 2000). Likewise, visual cues about surface curvature can be used for feedforward control of force.

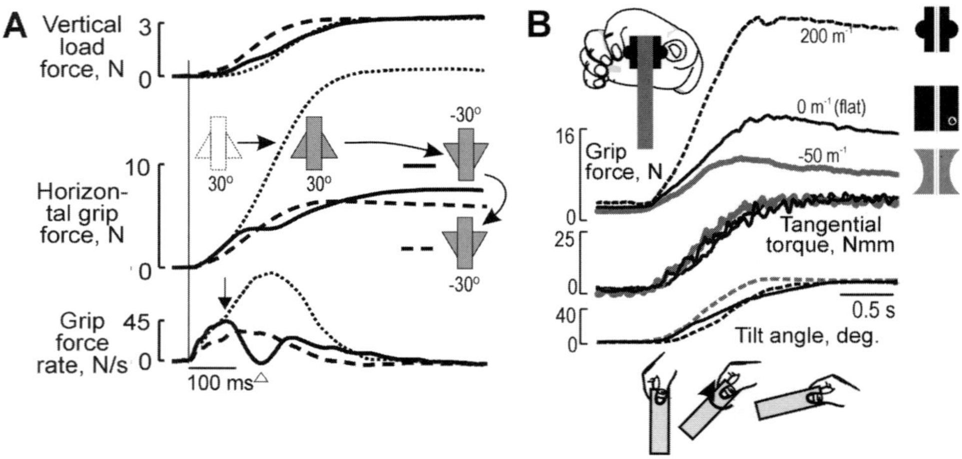

Figure 3. Adaptation of fingertip forces to object shape. *A*, Adjustments to changes in surface angle during lift series in which surface angle was unpredictably varied between trials. Vertical load force, horizontal grip force and grip force rate is shown as a function of time for trials without vision of the grasped surfaces. Object shape in current and previous trial is illustrated by the inset figures. Adjustment to a smaller angle is illustrated by a trial with -30° surface angle (solid line) that was preceded by a trial with 30°. Trials with -30° (dashed line) and 30° (dotted line) not preceded by a change in surface angle are shown for comparison. Arrow indicates the point in time where the new surface angle was expressed in the motor output. Modified from Jenmalm and Johansson, 1997. *B*, As in Fig. 1B subject tilt an object by 65° around its grip axis, which causes tangential torques at each grasp surface. Three superimposed trials are shown; one by each of the following surface curvatures: -50 m^{-1} (concave with a radius of 20 mm), 0 m^{-1} (flat) and 200 m^{-1} (convex with a radius of 5 mm). (Modified from Goodwin et al., 1998.)

NATURE OF TACTILE AFFERENT SIGNALS IN MANIPULATORY TASKS

Each fingertip is equipped with some 2000 tactile sensors (Johansson and Vallbo, 1979). They inform about goal completion of sequential phases of manipulative tasks and provide checkpoints for task progress. Figure 1A schematically indicates obligatory responses in the four types of tactile afferents innervating the fingertips (Johansson and Vallbo, 1983) during a prototypical lifting-task (Westling and Johansson, 1987). Note that, at four points corresponding to phase transitions there are distinct discharges in tactile afferents (Fig. 1A, middle panel).

Contact and release responses. Contact responses appear in primarily FA-I (Meissner) and SA-I (Merkel) afferents when a digit contacts an object and convey with high acuity both spatial and temporal information about the contact event; already at a contact force of 0.5 N some 400 tactile afferents are engaged (Westling and Johansson, 1987). As predicted from DESC, if these signals are not available due to topical anesthesia of the fingertips the application of tangential action forces in dexterous tasks is delayed and typically takes place at abnormally high grip forces (Johansson and Westling, 1984). Furthermore, subjects cannot control appropriately the direction of the fingertip forces when accurate spatial contact information is unavailable during finger numbness. Small items typically escape the grip making, for instance, buttoning impossible. The release responses that occur in primarily FA-I and SA-I afferents when a digit breaks the contact with an object verify the digit becomes disengaged. The significance of spatio-temporally accurate contact and release information is obvious for stereognostic tasks.

The contact responses also provide early information about object shape (see Fig. 3A) and the frictional condition between a fingertip and the grasped surface. However, unlike shape there is no evidence that subjects can use visual cues for frictional adaptation. The adjustment after an unexpected change in friction takes place about 100 ms after contact with the object; the frictional condition modulates the contact responses in subpopulations of FA-I afferents. The detection of a mismatch between the actual and expected frictional related tactile input results in a change in the grip-to-load force ratio. Likewise, according to the DESC-policy this mismatch results in an updating of the internal model so as to capture the new frictional condition for predictive control of the grip-to-load force ratio in further interactions with the object. Occasionally, accidental slip may occur at a later point, often at one digit only. Slip-evoked unexpected burst responses generated in dynamically sensitive tactile afferents promptly trigger – by means of the DESC-policy – an automatic upgrading of the grip-to-load force ratio along with an updating the relevant internal model. The tactile encoding of object shape will be dealt with in some detail further down, with emphasize on the curvature of grasped surfaces.

Responses to mechanical transients. The vibration sensitive FA-II (Pacinian) afferents efficiently detects mechanical transients that occur when making and breaking contact between hand-held object and other objects. Such mechanical transient regularly occur in natural manipulation, including tool use. Conservatively estimated, in a dexterous lifting task about 500 - 1,000 FA-IIs inject a nearly synchronous impulse volley into the CNS at liftoff and at support contact when the object is replaced (Fig. 1A, middle panel) (Westling and Johansson, 1987). The other three types of tactile afferents are virtually indifferent to such transient mechanical events. Certain musculotendinous receptors are known to respond to mechanical transients as well, but, microneurography recordings indicate that their sensing of mechanical events at the fingertips in manipulation is very low in comparison to that of cutaneous receptors. Instead, muscle spindles and the Golgi tendon organs are primarily concerned with events in the muscle itself (Macefield and Johansson, 1996; see also Evarts, 1981; Vallbo, 1985).

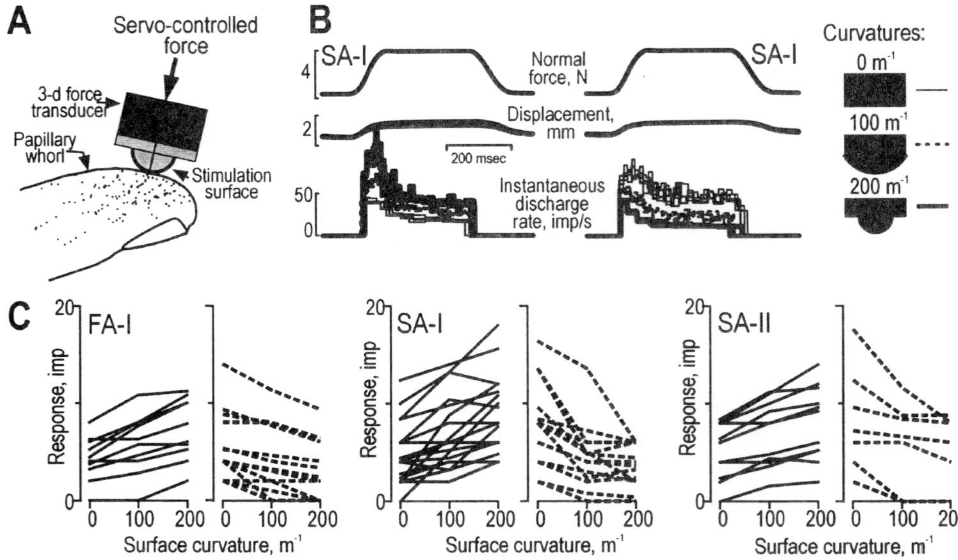

Figure 4. Encoding of object curvature by human tactile afferents assessed by force stimulation in the direction normal to the skin at a standard site on distal phalanx. *A*, the surface was advanced, under position control to contact the skin with a force of 0.2 N and force stimuli were superimposed on this background contact force. *B*, temporal profile of the force and probe displacement together with examples of typical responses obtained from two different SA-I afferents (microneurography). A flat surface and two spherically curved surfaces (curvatures 0, 100 and 200 m^{-1}) were applied and data from 5 trials by each curvature are superimposed. Each stimulus consisted of a protraction phase (125 ms), a plateau phase (250 ms; 4N amplitude) and a retraction phase (125 ms). *C*, FA-I , SA-I and SA-II afferents grouped according to the effect of surface curvature on the response intensity during the protraction phase. Afferents with responses positively and negatively correlated with surface curvature are shown in the matching left and right panels, respectively. The fine dots in *A* indicate the receptive field centers of all afferents SA-I, SA-II and FA-I afferents recorded from; the side view of the fingertip includes afferents located on either side of the finger. (Adapted from Jenmalm et al., 1999, paper under review.)

When grasping and lifting familiar objects that we can identify either visually or haptically, the force development is via APC tailored to the weight of the object before sensory information related to weight becomes available at lift-off. As we have all experienced, however, our predictions of objects' weight may sometimes be erroneous. In such cases, the lifting movement may be either jerky or slow. If the object is lighter than anticipated, the force drive will be too strong when the lift-off takes place. Although burst responses in FA-II afferents evoked by the unexpectedly early lift-off, trigger an abrupt termination of the force drive, this occurs too late (due to control loop delays) to avoid an excessively high lift. Conversely, if the object is heavier than expected, people will initially increase load force to a level that is not sufficient to produce lift-off and no sensory event will be evoked to confirm lift-off. Importantly, this absence of a sensory event at the expected lift-off now causes the release of a 'new' set of motor commands. These generate a slow discontinuous force increase, until terminated by a neural event at the true lift-off. Thus, corroborating the DESC policy, control actions are taken as soon there is a mismatch between an expected sensory event and the actual sensory input. Moreover, once an error occurs, the internal model of the object is updated to capture the new weight for use in subsequent interactions with the object, i.e., single trial learning.

Tactile Encoding of Object Curvature and Interaction Effects between Surface Curvature and the Direction of the Fingertip Force

We have recently investigated the encoding of object curvature by human FA-I, SA-I and SA-II tactile fingertip sensors and interaction effects between surface curvature and the direction of the fingertip force on the afferent's responses (Jenmalm et al., 1999; submitted). To get an overall view of the afferent input, we applied forces to a standard site at the fingertip (Fig. 4A) while recording from 172 afferents whose receptive field centers were distributed over the distal phalanx. A flat surface and two spherically curved surfaces (curvatures 0, 100 and 200 m^{-1}) were applied under force control in one of five directions: normal force, and forces at a 20° angle from the normal in the radial, distal, ulnar or proximal directions (cf. Fig. 5A). Nearly all afferents responded and the response intensity was correlated with curvature in at least one force direction for 86%, 87% and 83% of the responding SA-I, SA-II and FA-I afferents, respectively. Thus, in contrast to FA-I afferents observed in monkeys, our data show a clear representation of curvature in the human FA-I afferent responses (Goodwin et al., 1995; Khalsa et al., 1998).

Figure 4B illustrates responses in a two different SA-I afferents stimulated in the direction normal to the skin at the primary contact site. The afferent in the left panel behaved as previously described in monkeys (Goodwin et al., 1995) and in a human study (Goodwin et al., 1997). That is, this and many of the SA-I afferents that we recorded from showed stronger responses with more curved surfaces. However, nearly as many showed a negative correlation with increased curvature. These were often located more peripheral in the fingertip. The right panel of Fig. 4B shows responses in one such afferent. Furthermore, we found that curvature sensitivity characteristics were similar for all three classes of afferents. That is, there were afferents of each type for which response intensity correlated either positively or negatively with curvature (Fig. 4C). The occurrence of two groups of afferents, in which responses are correlated either positively or negatively with curvature, respectively, results in the presence of a curvature contrast signal across the two groups. This signal is likely to be tolerant against factors that may influence the overall discharge rates of the afferents, e.g., changes in the magnitude of the contact force, skin temperature, and variations in fingertip compliance.

Like most neurons in the sensorimotor systems, tactile fingertip afferents also show spatial directional tuning, i.e., they are broadly tuned to a preferred direction of force (Birznieks et al., 2001). For the SA-II afferent illustrated in Fig. 5A the responses were strongest for stimuli with tangential force components in-between the proximal and ulnar directions. By vector summation of the responses obtained with different directions of the tangential force component, we could establish a statistically reliable preferred direction of fingertip force for the large majority of the nearly 200 afferents that we recorded from (Fig. 5B). One important role of tactile directional information is presumably to signal critical sensory consequences of performed manipulatory actions necessary for the acquisition and maintenance the internal models used for predictive control. Interestingly, tactile directional information obtained during the first 50 ms after target contact can also be used to update the position of the fingertip relative to the body in target pointing movements (Lackner and DiZio, 2000; see also Chapter 9).

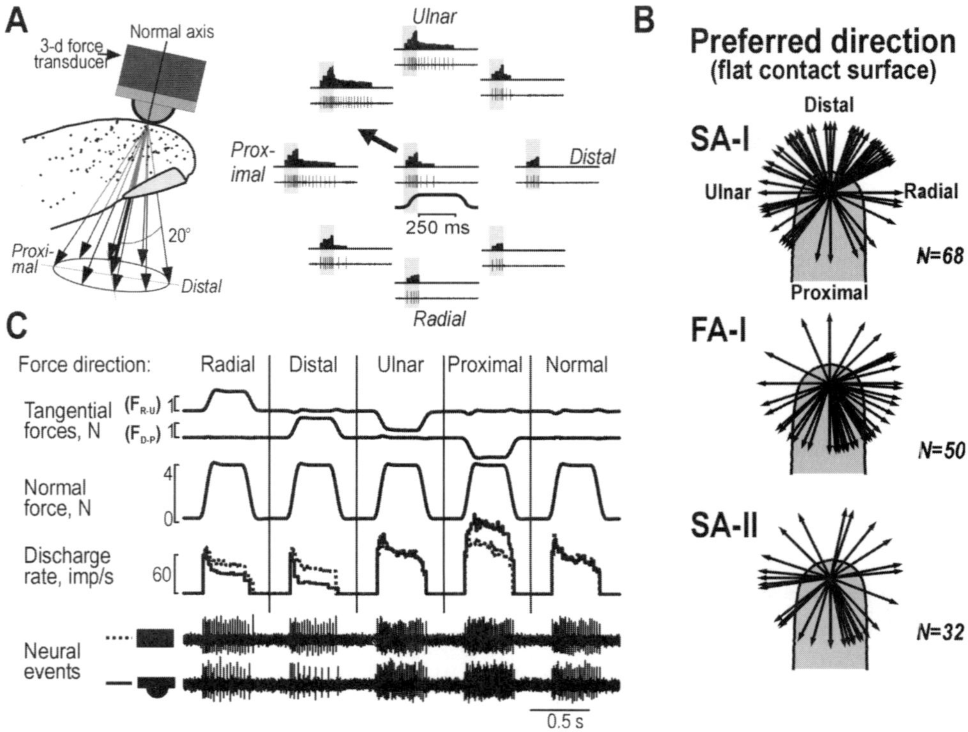

Figure 5. Encoding of direction of fingertip forces by human tactile afferents and interaction effects between the force direction and surface curvature on afferent responses. *A*, Responses in a single SA-II afferent during stimulation in each of nine force directions. Force stimuli were superimposed on a 0.2 N background contact force and delivered in the normal direction and at an angle 20° to the normal with tangential components in eight directions 45° apart; temporal profile of the applied forces as in Fig. 4B. The arrow indicates the preferred direction of force stimuli computed as the vector sum of the responses during the protraction phases of the eight stimuli with tangential force components. *B*, The arrows (unit vectors) show the estimated preferred directions of 68 SA-I, 50 FA-I and 32 SA-II afferents that were directionally sensitive (modified from Birznieks et al., 2001). *C*, Responses in one SA-I afferent together with records of instantaneous frequency and normal and tangential forces (FD-P and FR-U) for the two extreme curvatures (0 and 200 m^{-1}) during various directions of force stimulation. 'Radial', 'Distal', 'Ulnar' and 'Proximal' refer to the direction of the tangential force component and dashed and solid curves refer to the 0 m^{-1} and 200 m^{-1} curvature, respectively. (Adapted from Jenmalm et al., 1999.)

Interaction effects between surface curvature and direction of fingertip force. For nearly all afferents, changing the direction of force affected the afferent's sensitivity to curvature. Figure 5C shows the responses to the normal direction and four directions with tangential components for a SA-I afferent, stimulated with the flat and a curved surface. Surface curvature influenced modestly the direction preference of this afferent; the strongest responses occurred in the proximal direction and weakest in the distal. However, the sensitivity to surface curvature was dramatically influenced by force direction. This afferent could reverse its sensitivity to curvature depending on force direction. With proximal stimulation, the afferent responded stronger with a more curved surface. In contrast, in the radial and distal directions the flat surface caused the strongest response. With stimuli with only normal and with ulnar tangential force components the afferent response was indifferent to curvature. About 40 percent of the SA-I afferents showed reversal of curvature sensitivity with changes in direction; this number was smaller for the other two types. Conversely, the preferred force direction of afferents could vary with different curvatures (not illustrated); the SA-I and FA-I afferents showed least correspondence in preferred direction for the different curvatures. For many afferents changes in curvature changed the preferred direction by more than 45 degrees. For the SA-I afferents we often observed changes by some 180 degree, i.e., a shift to an opposite direction.

Thus, there are marked interactions between effects of curvature and direction of force on tactile afferent responses. We don't know how the brain resolves these and many other possible interactions between various fingertip parameters critical for manipulation, e.g., force magnitude, force direction, torques tangential to the contact surfaces, object shape, object position relative to the hand and frictional characteristics of the surfaces. However, through to the DESC policy the CNS would dramatically reduce the effective degrees of freedom in the tactile domain by analyzing the sensory inflow from the fingertips with reference to predicted sensory outcomes produced by internal forward models in conjunction with a copy of the motor command, i.e., efference copy (e.g., Miall and Wolpert, 1996). There is evidence that processes underlying such forward models engage cerebellar mechanisms while interacting with cortical networks (Wolpert and Miall R C Kawato, 1998; Tamada et al., 1999; Imamizu et al., 2000) and that the cerebellum contribute signals that are used to cancel the sensory response to self-generated stimulation (Blakemore et al., 1998b, 2001; see also Roberts and Bell, 2000). Furthermore, it is likely that such signals impinge on the somatosensory ascending pathways already at subcortical levels. Various descending, largely inhibitory, projections indeed control the signal processing in these pathways (Harris et al., 1965; Adkins et al., 1966; Abdelmoumene et al., 1970; Ergenzinger et al., 1998). A cancellation of predicted sensory inflow in the somatosensory pathways may also explain the puzzling phenomenon of so-called central "gating" of irrelevant somatosensory afferent input during motor actions (e.g., Cohen and Starr, 1987; Williams et al., 1998). Indeed, the CNS quickly identifies and disregards experimentally imposed stimuli that are inappropriate for the control of the motor task (see Johansson and Westling, 1987).

Furthermore, it is likely that the somatosensory pathways have an important role in resolving interactions between various fingertip parameters critical for control. These programmable pathways could model functionally the mechanics of the fingertip as represented in the tactile domain. The abundant structural convergence/divergence in these pathways combined with the somatotopic arrangements and feed-forward and recurrent surround inhibitory mechanisms at each relay level (e.g., Dykes and Craig, 1998) would provide for effective processing of the spatiotemporal impulse pattern in the populations of afferents from each fingertip. Finally, appropriate neural computation could be learned and maintained by activity dependent short and long term plasticity known to be distributed in the somatosensory pathways (Buonomano and Merzenich, 1998; Jones, 2000).

REFERENCES

Abdelmoumene, M., Besson, J.M., and Aleonard, P., 1970, Cortical areas exerting presynaptic inhibitory action on the spinal cord in cat and monkey, *Brain Research,* 20, 327-329.

Adkins, R.J., Morse, R.W. and Towe, A.L., 1966, Control of somatosensory input by cerebral cortex, *Science,* 153, 1020-1022.

Ballard, D.H., Hayhoe, M.M., Li, F., and Whitehead, S.D.,1992, Hand-eye coordination during sequential tasks, *Philosophical Transactions of the Royal Society of London. Series B: Biological Sciences,* 337, 331-338.

Birznieks, I., Jenmalm, P., Goodwin, A.W., and Johansson, R.S.,2001, Encoding of direction of fingertip forces by human tactile afferents, *Journal of Neuroscience,* 21, 8222-8237.

Blakemore, S.J., Frith, C.D., and Wolpert, D.M., 2001, The cerebellum is involved in predicting the sensory consequences of action, *Neuroreport,* 12, 1879-1884.

Blakemore, S.J., Goodbody, S.J., and Wolpert, D.M., 1998a, Predicting the consequences of our own actions: The role of sensorimotor context estimation, *Journal of Neuroscience,*18, 7511-7518.

Blakemore, S.J., Wolpert, D.M., and Frith, C.D., 1998b, Central cancellation of self-produced tickle sensation, *Naure Neurosciece,* 1, 635-640.

Buonomano, D.V., and Merzenich, M.M., 1998, Cortical plasticity: from synapses to maps, *Annual Review of Neuroscience,* 21, 149-186.

Burstedt, M.K.O., Edin, B.B., and Johansson, R.S., 1997, Coordination of fingertip forces during human manipulation can emerge from independent neural networks controlling each engaged digit, *Experimental Brain Research,* 117, 67-79.

Cohen, L.G., and Starr, A., 1987, Localization, timing and specificity of gating of somatosensory evoked potentials during active movement in man, *Brain,* 110, 451-467.

Dykes, R.W., and Craig, A.D., 1998, Control of size and excitability of mechanosensory receptive fields in dorsal column nuclei by homolateral dorsal horn neurons, *Journal of Neurophysiology,* 80, 120-129.

Edin, B.B., Westling, G., and Johansson, R.S., 1992, Independent control of fingertip forces at individual digits during precision lifting in humans, *Journal of Physiology,* 450, 547-564.

Ergenzinger, E.R., Glasier, M.M., Hahm, J.O., and Pons, T.P., 1998, Cortically induced thalamic plasticity in the primate somatosensory system, *Nature Neuroscience* 1, 226-229.

Evarts, E.V., 1981, Sherrington's concepts of proprioception, *Trends in Neuroscience,* 4, 44-46.

Flanagan, J.R., Burstedt, M.K.O., and Johansson, R.S., 1999, Control of fingertip forces in multi-digit manipulation, *Journal of Neurophysiology,* 81, 1706-1717.

Flanagan, J.R., and Tresilian, J.R., 1994, Grip load force coupling: A general control strategy for transporting objects, *Journal of Experimental Psychology: Human Perception and Performance,* 20, 944-957.

Goodwin, A.W., Browning, A.S., and Wheat, H.E., 1995, Representation of curved surfaces in responses of mechanoreceptive afferent-fibers innervating the monkeys fingerpad, *Journal of Neuroscience,*15, 798-810.

Goodwin, A.W., Jenmalm, P., and Johansson, R.S., 1998, Control of grip force when tilting objects: effect of curvature of grasped surfaces and of applied tangential torque, *Journal of Neuroscience,* 18, 10724-10734.

Goodwin, A.W., Macefield, V.G., and Bisley, J.W., 1997, Encoding of object curvature by tactile afferents from human fingers, *Journal of Neurophysiology,* 78, 2881-2888.

Gordon, A.M., Forssberg, H., Johansson, R.S., and Westling, G., 1991, Visual size cues in the programming of manipulative forces during precision grip, *Experimental Brain Research,* 83, 477-482.

Harris, F., Jabbur, S.J., Morse, R.W., and Tow, A.L., 1965, Influence of the cerebral cortex on the cuneate nucleus of the monkey, *Nature,* 208, 1215-1216.

Imamizu, H., Miyauchi, S., Tamada, T., Sasaki, Y., Takino, R., Putz, B., Yoshioka, T., and Kawato, M., 2000, Human cerebellar activity reflecting an acquired internal model of a new tool, *Nature,* 403, 192-195.

Jeannerod, M., 1986, The formation of finger grip during prehension. A cortically mediated visuomotor pattern, *Behavioural Brain Research,* 19, 99-116.

Jenmalm, P., Birznieks, I., Goodwin, A., and Johansson, R., 1999, Differential responses in populations of fingertip tactile afferents to objects' surface curvatures, *Acta Physiologica Scandinavica,* 167, A24-A25.

Jenmalm, P., Dahlstedt, S., and Johansson, R.S., 2000, Visual and tactile information about object curvature control fingertip forces and grasp kinematics in human dexterous manipulation, *Journal of Neurophysiology,* 84, 2984–2997.

Jenmalm, P., Goodwin, A.W., and Johansson, R.S., 1998, Control of grasp stability when humans lift objects with different surface curvatures, *Journal of Neurophysiology,* 79, 1643-1652.

Jenmalm, P., and Johansson, R.S., 1997, Visual and somatosensory information about object shape control manipulative finger tip forces, *Journal of Neuroscience,* 17, 4486-4499.

Johansson, R.S., Backlin, J.L., and Burstedt, M.K.O., 1999, Control of grasp stability during pronation and supination movements, *Experimental Brain Research,* 128, 20-30.

Johansson, R.S., and Cole, K.J., 1992, Sensory-motor coordination during grasping and manipulative actions, *Current Opinion in Neurobiology,* 2, 815-823.

Johansson, R.S., and Cole, K.J., 1994, Grasp stability during manipulative actions. *Canadian Journal of Physioliogy and Pharmacology,* 72, 511-524.

Johansson, R.S., and Vallbo, A.B., 1979, Tactile sensibility in the human hand: relative and absolute densities of four types of mechanoreceptive units in glabrous skin, *Journal of Physiolog,* 286, 283-300.

Johansson, R.S., and Vallbo, Å.B., 1983, Tactile sensory coding in the glabrous skin of the human hand, *Trends in Neuroscience,* 6, 27-31.

Johansson, R.S., and Westling, G., 1984, Roles of glabrous skin receptors and sensorimotor memory in automatic control of precision grip when lifting rougher or more slippery objects, *Experimental Brain Research,* 56, 550-564.

Johansson, R.S., and Westling, G., 1987, Signals in tactile afferents from the fingers eliciting adaptive motor responses during precision grip, *Experimental Brain Research,* 66, 141-154.

Johansson, R.S., and Westling, G., 1988a, Coordinated isometric muscle commands adequately and erroneously programmed for the weight during lifting task with precision grip, *Experimental Brain Research,* 71, 59-71.

Johansson, R.S., and Westling, G., 1988b, Programmed and triggered actions to rapid load changes during precision grip, *Experimental Brain Research,* 71, 72-86.

Johansson, R.S., and Westling, G., 1990, Tactile afferent signals in the control of precision grip, in: *Attention and Performance, vol XIII,* Jeannerod M. Erlbaum, ed., Hilldale, New Jersey, pp. 677-713.

Johansson, R.S., Westling, G., Bäckström, A., and Flanagan, J.R., 2001, Eye-hand coordination in object manipulation, *Journal of Neuroscience,* 21, 6917-6932.

Jones, E.G., 2000, Cortical and subcortical contributions to activity-dependent plasticity in primate somatosensory cortex, *Annual Review of Neuroscience,* 23, 1-37.

Kawato, M., 1999, Internal models for motor control and trajectory planning, *Current Opinion in Neurobiology,* 9, 718–727.

Khalsa, P.S., Friedman, R.M., Srinivasan, M.A., and Lamotte, R.H., 1998, Encoding of shape and orientation of objects indented into the monkey fingerpad by populations of slowly and rapidly adapting mechanoreceptors, *Journal of Neurophysiology,* 79, 3238-3251.

Kinoshita, H., Bäckström, L., Flanagan, J.R., and Johansson, R.S., 1997, Tangential torque effects on the control of grip forces when holding objects with a precision grip, *Journal of Neurophysiology,* 78, 1619-1630.

Lackner, J.R., and DiZio, P.A., 2000, Aspects of body self-calibration, *Trends in Cognitive Science, Regular Edition,* 4, 279-288.

Lacquaniti, F., 1992, Automatic control of limb movement and posture, *Current Opinion in Neurobiology,* 2, 807-814.

Land, M., Mennie, N., and Rusted, J., 1999, The roles of vision and eye movements in the control of activities of daily living, *Perception,* 28, 1311-1328.

Macefield, V.G., and Johansson, R.S., 1996, Control of grip force during restraint of an object held between finger and thumb: responses of muscle and joint afferents from the digits, *Experimental Brain Research,* 108, 172-184.

Miall, R.C., and Wolpert, D.M., 1996, Forward models for physiological motor control, *Neural Networks* 9, 1265-1279.

Ohki, Y., Edin, B.B., and Johansson, R.S., 2002, Predictions specify reactive control of individual digits in manipulation, *Journal of Neuroscience,* 22, 600-610.

Prochazka, A., 1993, Comparison of natural and artificial control of movement, *IEEE Transactions on Rehabilitation Engineering,* 1, 7-17.

Rack, P.M.H., 1981, Limitations of somatosensory feedback in control of posture and movement, in: *Handbook of Physiology. Sect. 1: The Nervous System,* Brookhart, J.M. and Mountcastle, V.B., eds., American Physiological Society, Bethesda, Maryland, pp. 229-256.

Roberts, P.D., and Bell, C.C., 2000, Computational consequences of temporally asymmetric learning rules: II. Sensory image cancellation, *Journal of Computational Neuroscience,* 9, 67-83.

Sperry, R.W., 1950, Neural basis of the spontaneous optokinetic response produced by visual inversion, *Journal of Comparative and Physiological Psychology,* 43, 482-489.

Tamada, T., Miyauchi, S., Imamizu, H., Yoshioka, T., and Kawato, M., 1999, Cerebro-cerebellar functional connectivity revealed by the laterality index in tool-use learning, *Neuroreport* 10, 325-331.

Vallbo, Å.B., 1985, Proprioceptive activity from human finger muscles, in: *Feedback and motor control in invertebrates and vertebrates,* Barnes, W.J.P. and Gladden, M.H., eds., Croom Helm Ltd, London, pp. 411-430.

Westling, G., and Johansson, R.S., 1987, Responses in glabrous skin mechanoreceptors during precision grip in humans, *Experimental Brain Research,* 66, 128-140.

Williams, S.R., Shenasa, J., and Chapman, C.E., 1998, Time course and magnitude of movement-related gating of tactile detection in humans. I. Importance of stimulus location, *Journal of Neurophysiolgy,* 79, 947-963.

Wolpert, D.M., and Miall R C Kawato, M., 1998, Internal models in the cerebellum, *Trends in Cognitive Science,* 2, 338-347.

46

MOTOR CORTEX AND THE DISTRIBUTED ANATOMY OF FINGER MOVEMENTS

Marc H. Schieber*

ABSTRACT

Voluntary movements are thought to be controlled via a well-ordered, spatially discrete, somatotopic map in the primary motor cortex (M1). We examined this hypothesis in monkeys trained to perform visually-cued, individuated flexion and extension movements of each digit and of the wrist. Single neurone recordings in M1 during such finger movements revealed two unexpected features. First, single M1 neurones often discharge during instructed movements of multiple digits. Second, neurones active during any particular instructed movement are distributed widely throughout the same M1 territory as neurones active during any other movement. Reversible, partial inactivation of the M1 hand representation produced by injection of 5 -10 μg muscimol at one site impaired the monkeys' ability to perform finger movements, but no relationship was evident between the particular finger movements that were affected and the mediolateral location of the injection site along the central sulcus. Thus each finger movement is represented by activity distributed widely in the M1 upper extremity representation. If not controlled from spatially segregated M1 regions, movements of different fingers might be controlled by groups of spatially scattered but physiologically similar neurones. Cluster analysis of M1 neurones demonstrated a large group that discharged during most finger movements, and a small group that paused during most movements. Distinct functional groups of M1 neurones that might control particular finger movements were identified inconsistently. We therefore hypothesize that M1 neurones are a very diverse network controlling finger movements.

INTRODUCTION

The primary motor cortex, via its corticospinal projection, plays a major role in controlling fine finger movements. After lesions of these structures, gross opening and

* Departments of Neurology and of Neurobiology & Anatomy, University of Rochester School of Medicine and Dentistry Rochester, New York, 14642 USA. Email: mhs@cvs.rochester.edu

Sensorimotor Control of Movement and Posture
Edited by Gandevia *et al.*, Kluwer Academic/Plenum Publishers, 2002

closing movements of all the fingers together may recover, but fine movements of individual fingers are lost, suggesting that M1 acts on a rudimentary foundation of whole hand movement to individuate movements of particular fingers. Control of individuated finger movements long has been thought to be accomplished through a somatotopic map of the fingers in M1. Studies over the past quarter century have demonstrated two major features, however, that necessarily constrain the degree to which M1 can control movements via a somatotopic map: i) corticospinal inputs to the motoneurone pool of any particular muscle converge from a relatively large cortical territory that overlaps extensively with the cortical territory controlling other nearby muscles, and ii) outputs of many single corticospinal neurones diverge to innervate the motoneurone pools of multiple muscles. Nevertheless, given that any voluntary finger movement entails the simultaneous contraction of multiple muscles, somatotopically organized M1 regions could be activated differentially during individuated movements of each finger. Here I summarize our work examining the hypothesis of somatotopic control, which provides a more complex, but more realistic, view of the "anatomy" of finger movement control by M1.

REVISITING THE MOTOR APPARATUS OF THE HAND

Although movements of different fingers often are assumed to be independent of one another, careful examination reveals that such is not the case. In the vast majority of everyday uses of the hand, such as picking up your coffee cup or writing with a pen, multiple digits move simultaneously. Even in tasks such as typing or piano playing, the epitomes of independent finger movements, multiple digits move with each keystroke (Engel et al., 1997; Soechting and Flanders, 1997). These elegant performances do not require that only one digit move, only that no unintended digits strike keys. Even when asked to move just one digit, normal human subjects typically produce motion simultaneously in other digits as well (Häger-Ross and Schieber, 2000). Monkeys likewise produce motion in non-instructed digits (more than humans) when they are instructed to flex or extend a particular digit (Schieber, 1991).

In both monkeys and humans, the lack of complete independence of the fingers arises in part from the biomechanical properties of the tissues of the hand and forearm. The skin and connective tissues of the web spaces between the digits produce some degree of passive coupling between the digits. The active contraction of multitendoned muscles, flexor digitorum profundus (FDP), flexor digitorum superficialis (FDS), and extensor digitorum communis (EDC) in humans; plus extensor digiti secundi et tertii (ED23) and extensor digiti quarti et quinti (ED45) in macaques, also may act simultaneously on multiple digits for a number of reasons. First, EDC has marked interconnections between the tendons to different fingers, so that even if only a part of the muscle belly serving a particular tendon contracts, tension will be distributed to the extensor hood of more than one digit. The tendons of FDP also are interconnected to some degree in humans, and are interconnected markedly in macaques. Second, some multitendoned muscles may lack separate neuromuscular compartments for each digit they serve. The macaque FDP has 4 compartments, each of which act on multiple digits, and the macaque ED45 contains many motor units that exert similar tension on each of its two tendons, one to digit 4, the other to digit 5 (Schieber et al., 1997; Schieber et al., 2001). By comparison, human multitendoned muscles may be more compartmentalized for each digit. Third, the activity in different compartments of the same muscle, or in different muscles, may be linked by common inputs as evidenced by short-term synchronization between pairs of motor units (Farmer et al., 1997). Because of these passive, active, and central interconnections between the elements of different digits, an active movement of one digit is likely to draw other digits along as well. To move one digit alone insofar as possible, the nervous system therefore often contracts additional muscles to prevent unwanted motion of other digits (Schieber, 1995). Active stabilization of the posture of other digits thus becomes part and parcel of "isolated" movement of a single digit.

M1 ACTIVITY DURING INDIVIDUATED FINGER MOVEMENTS

Single neurone recordings in the M1 of monkeys trained to perform individuated flexion and extension movements of each digit of the right hand and of the right wrist showed two overt features that were inconsistent with the hypothesis that somatotopically organized M1 regions are activated differentially during individuated movements of each finger (Schieber and Hibbard, 1993). First, the majority of M1 neurones were active during movements of multiple digits. Often a given neurone was active during movements of non-adjacent digits. Second, neurones active during any particular individuated finger or wrist movement were distributed throughout a large region of M1 (extending 8-9 mm along the central sulcus), and were extensively intermingled with the neurones active during any other individuated finger or wrist movement. Finding the centroid of activity for each movement separately showed that all 12 centroids were clustered together in the middle of the field of active neurones, with little somatotopic separation.

We further tested the hypothesis that somatotopically organized regions of M1 control individuated movements of each finger by reversibly inactivating parts of the M1 hand representation (Schieber and Poliakov, 1998). Reversible inactivation of different portions of the M1 hand representation was performed in different sessions by injecting 5-10 μg of muscimol (a long acting GABA agonist) in a 1 μl volume at a single site in each session. Such injections in the physiologically identified M1 hand representation consistently impaired performance of individuated finger movements within 5-15 minutes, whereas similar injections in the leg representation or premotor cortex had no effect (Fig. 1). In some cases the response time for particular finger movements became progressively longer until finally the monkey failed to close the switch within the 700 ms required by the task. In other cases particular finger movements became progressively less individuated until finally non-instructed digits moved enough to close the wrong switch. Given that a single muscimol injection inactivated cortex over an estimated 3 mm radius, injections at different locations along the central sulcus might have been expected to affect different fingers in a somatotopic fashion. Yet no relationship between the injection location along the central sulcus and the sequence of finger movement failures (Fig. 1), or the ultimately impaired movements, could be identified. Like the findings of single neurone recording, these inactivation findings suggest that movement of each finger is controlled by a network of neurones widely distributed in the M1 hand representation.

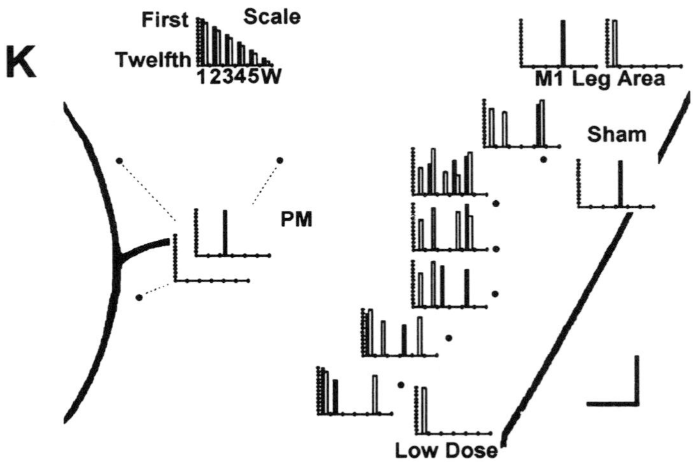

Figure 1. A bargraph for each muscimol injection session in one monkey shows the instructed movements for which the monkey failed 8 of 10 consecutive trials in what order. Each bargraph has been positioned on an enlarged map of monkey K's left hemispheric surface close to the point at which the injection was made (small dots), or else connected to it with a fine dashed line. (Three dashed lines indicate the three different locations injected during each PM session.) Each instructed movement that failed is shown as a bar positioned along the abscissa to indicate the instructed digit, and shaded to indicate the instructed direction: filled - flexion; open-extension. The height of each bar indicates the serial order in which different instructed movements failed within each session from first (tallest) to twelfth (shortest). The scale at top thus would illustrate an idealized result in which an injection placed laterally in the hand area impaired instructed movements starting with those of the thumb and spreading somatotopically to those of the little finger and wrist, with instructed flexion of each digit failing before instructed extension. Injections in the M1 hand representation along the central sulcus (thick line at right), impaired instructed movements of several fingers, though which finger movements were impaired in a given session was unrelated to the location of the injection along the central sulcus. In comparison, after injections in the M1 leg area (top right), premotor cortex (PM, left), and low dose or sham injections in M1, only one movement failed, since normally upon satiation the monkey stopped working and let trials time out.

If not controlled from spatially segregated M1 regions, how can one finger be moved at one time, and another finger at another time? Movements of different fingers might be controlled by groups of spatially scattered but physiologically similar neurones. To search for such groupings, cluster analysis was performed on the activity of M1 neurones recorded during finger movements (Poliakov and Schieber, 1999). Two groups of physiologically similar neurones were identified in all 3 monkeys studied: a large group that discharged during most or all finger movements (Fig. 2, BFE), and a small group that paused during most or all movements (Fig. 2, BFI). Groups of M1 neurones that discharged specifically during a particular finger movement (e.g. 1f or 1e groups in Fig. 2), or during a small subset of finger movements (e.g. the 3e,5e group in Fig. 2), were few in number and varied from monkey to monkey. These findings suggest that execution of different finger movements is not controlled by activity in distinct functional classes of M1 neurones.

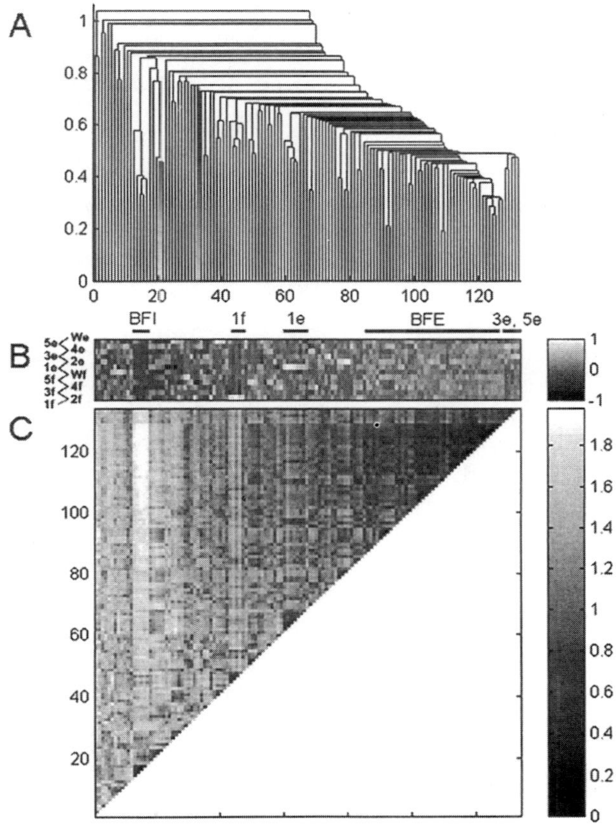

Figure 2. Cluster analysis of physiologic activity in 133 M1 neurones in monkey K is shown in 3 complementary displays. In (B) each neurone is represented by a column and each instructed movement by a row (from 1f , thumb flexion, at bottom, to We, wrist extension, at top). The grayscale value in each cell represents the change in that neurone's discharge during that movement, normalized such that the values for all 12 movements constitute a unit vector in 12-dimensional space. The columns representing different neurones have been ordered by the clustering process, such that the column representing each neurone is close to the columns of other neurones with similar activity patterns. Groups of similar neurones therefore appear in (B) as contiguous columns with similar shades in each row, forming horizontal bands of relatively constant shade. In the corresponding dendrogram (A), neurones are represented by vertical lines rising from the abscissa in the same left-to-right order as in (B). Horizontal lines join the vertical lines for two neurones at the ordinate value representing the distance between them in the 12-dimensional space. In the corresponding similarity matrix (C), each recording is represented in the same order as in B along both the abscissa and the ordinate. The distance in 12-dimensional space between each possible pair of neurones is displayed in the appropriate cell using a grayscale [different from the scale used in (B)] to represent distances from 0 (black) to 2 (white). In the similarity matrix (C), groups of similar neurones therefore appear as black triangles with hypotenuses along the main diagonal. Horizontal black bars between (A) and (B) indicate groups of neurones identified on the basis of repeating such analysis using different discharge measures and clustering algorithms. These groups include: a large group of neurones whose discharge increased (light grays) with many or all instructed movements (BFE), a small groups of neurones whose tonic discharge decreased (dark grays) with many or all movements (BFI), a small group that discharged more for movement 1f than for others (1f), a small group that discharged more for 1e than for others (1e), and a small group that discharged more for movements 3e and 5e than for others (3e,5e). Note, however, that none of these groups are particularly distinct from other members of the population. Only the BFE and BFI groups were present consistently in all three monkeys studied.

CONCLUSION

We hypothesize that M1 neurones act as a very diverse network in controlling finger movements. We are exploring this hypothesis using spike-triggered averaging to study M1 neurones in this network that have relatively direct connections to spinal motoneurone pools. Preliminary results indicate that neurones which produce post-spike effects (PSpikeEs) in EMG activity are as diverse as the M1 population in general. Such neurones may i) be active during movements of several different fingers, ii) be active during movements for which target muscles are inactive, and iii) facilitate (or suppress) both flexor and extensor muscles. Finally, synchrony is evident in the PSpikeEs of many M1 neurones, particularly in monkeys whose skill has developed over several years. Individuated finger movements thus are controlled by overlapping subpopulations of M1 neurones distributed throughout the M1 hand representation, each neurone active during some movements and not others, each output neurone having a different distribution of connections to the spinal motoneurone pools. This complex, physiologically diverse, anatomically distributed network generates apparently simple finger movements.

REFERENCES

Engel, K. C., Flanders, M., and Soechting, J. F., 1997, Anticipatory and sequential motor control in piano playing, *Experimental Brain Research*, 113, 189-199.

Farmer S. F., Halliday, D. M., Conway, B. A., Stephens, J. A., and Rosenberg, J. R., 1997, A review of recent applications of cross-correlation methodologies to human motor unit recording, *Journal of Neuroscience Methods*, 74, 175-187.

Häger-Ross, C., and Schieber, M. H., 2000, Quantifying the independence of human finger movements: Comparisons of digits, hands, and movement frequencies, *Journal of Neuroscience*, 20, 8542-8550.

Poliakov, A. V., and Schieber, M. H., 1999, Limited functional grouping of neurons in the motor cortex hand area during individuated finger movements: A cluster analysis, *Journal of Neurophysiology*, 82, 3488-3505.

Schieber, M. H., 1991, Individuated finger movements of rhesus monkeys: a means of quantifying the independence of the digits, *Journal of Neurophysiology*, 65, 1381-1391.

Schieber, M. H., 1995, Muscular production of individuated finger movements: The roles of extrinsic finger muscles, *Journal of Neuroscience*, 15, 284-297.

Schieber, M. H., Chua, M., Petit, J., and Hunt, C. C., 1997, Tension distribution of single motor units in multitendoned muscles: comparison of a homologous digit muscle in cats and monkeys, *Journal of Neuroscience*, 17, 1734-1747.

Schieber, M. H., Gardinier, J., and Liu, J., 2001, Tension distribution to the five digits of the hand by neuromuscular compartments in the macaque flexor digitorum profundus, *Journal of Neuroscience*, 21, 2150-2158.

Schieber, M. H., and Hibbard, L. S., 1993, How somatotopic is the motor cortex hand area? *Science*, 261, 489-492.

Schieber, M. H., and Poliakov, A. V., 1998, Partial inactivation of the primary motor cortex hand area: effects on individuated finger movements, *Journal of Neuroscience*, 18, 9038-9054.

Soechting, J.F., and Flanders, M., 1997, Flexibility and repeatability of finger movements during typing: analysis of multiple degrees of freedom, *Journal of Computational Neuroscience*, 4, 29-46.

47

REWARD-BASED PLANNING OF MOTOR SELECTION IN THE ROSTRAL CINGULATE MOTOR AREA

Jun Tanji[1,2], Keisetu Shima[1], and Yoshiya Matsuzaka[1]

ABSTRACT

The cingulate motor areas, located in the banks of the cingulate sulcus, constitute a portion of the cingulate cortex of primates. We here present experimental evidence showing that the rostral cingulate motor area (CMAr), but not the caudal one (CMAc) is crucial for the selection of future movements based on reward information. After muscimol injection into the CMAr, monkeys were impaired in selecting movements appropriately on the basis of the amount of reward obtained by performing correct movements. Furthermore, four types of cells in the CMAr were found to reflect a process intervening between detection of reward alteration and selection of a future movement. Each type of cell seems to be involved in responding to the quality of the reward, and to relay that information to change planned movements, and prepare a new movement.

INTRODUCTION

The cingulate motor areas (CMAs) have been defined in the banks of the cingulate sulcus in the medial surface of the cerebral hemisphere of primate brain (Vogt and Gabriel, 1993). Because of their afferent inputs from the limbic structures and the prefrontal cortex (Morecraft et al., 1993; Morecraft and Van-Hoesen, 1993), and their efferent projections to cortical and subcortical motor areas (Matelli et al., 1991; Picard and Strick, 1996), CMAs are generally thought to be situated at a nodal point in the neuronal circuit that processes information about motivation and the internal state of subjects to select voluntary actions, in accordance with their own need (Vogt et al., 1987; Vogt and Gabriel, 1993). However, it was not known how exactly the CMAs were used for the motor selection when confronted with an actual need to do so. Moreover, no previous studies have examined differences in the use of subareas in the CMA, the rostral and caudal parts of the CMA (CMAr and CMAc), in subjects performing motor selection.

[1] Department of Physiology, Tohoku University School of Medicine, Sendai, 980-8575, Japan.
[2] CREST, Kawaguchi, Japan. Email: tanjij@mail.cc.tohoku.ac.jp

Therefore, we performed a series of experiments to study neuronal activity in the CMAr and CMAc, together with the study in the two neighboring motor areas, the supplementary and pre-supplementary motor areas (SMA and pre-SMA).

REWARD-BASED PLANNING OF MOTOR SELECTION: AN EXPERIMENTAL MODEL

Motor selection can be made when following instructions given by signals in the outside world. On the other hand, the selection of what motor act to do is often determined by individual's own decision. When selecting self-actions, the most crucial factor that governs the decision process is the estimation of the outcome of that action, i.e., whether the selected action is likely to be rewarding or non-rewarding to himself. To investigate where and how in the brain the process of motor selection based on reward is achieved, we need an experimental model. For this purpose we set up a simple model of reward-based motor selection to study the role of CMAs in that process (Shima and Tanji, 1998).

We used three monkeys to select one out of two different arm movements, either pushing or turning a handle, in response to a visual trigger signal. The essence of the motor task was that the animals voluntarily selected a movement based on the amount of reward. When the monkey held the handle in a neutral position for 3 to 8 s, an LED was illuminated as a signal to start the correct movement. Initially, the subject had to guess which of the two choices was correct. Performing the correct movement was rewarded with a drop of fruit juice, and the correct movement remained unchanged in a block of trials, so that the monkey was required to keep selecting the same movement. The amount of the reward remained constant for four to 12 successive trials, unless the subject made a mistake and selected the wrong movement, in which case a warning tone signal replaced the reward. Subsequently, the amount of the reward decreased by 30% for each correct trial. At this stage, monkeys were free to select the alternate movement. They usually did so after the first to the third decrement (30 to 65.7% decrease in reward). If they did, the alternate movement was then redefined as the correct movement, the reward reverted to the full amount, and a new series of constant-reward trials began, with the redefined correct movement. After training, monkeys performed the behavioral task with an error rate of less than 2%.

ROSTRAL CMA IS CRUCIALLY INVOLVED IN THE REWARD-BASED MOTOR SELECTION

To study the effects of transiently inactivating each of CMAr and CMAc on the performance of the behavioral task described above, we applied GABA agonist, muscimol, topically into a small portion of each area (Fig. 1). We inserted small-caliber (300 μm) injection canulae bilaterally into the CMAs and injected a small amount (2 to 4 μl, 1 to 10μg/μl) of muscimol, using a motor-driven microinjector (1 μl in 10 minutes). When 3 to 4μl of muscimol was injected bilaterally in the forelimb part of the CMAr, the monkey began to fail in selecting an alternative, more-rewarding movement, 10 to 15 min after injection. Even if the reward was reduced considerably, the monkey kept selecting the previously performed movement and failed to select the alternative movement. At other times, the animal made a mistake and prematurely selected the other movement before the reward was reduced. These effects, observed when muscimol was injected into the forelimb part of the CMAr, were dose-dependent, and not observed with a concentration of less than 5 μg/μl. The animal's movement time was not altered even if the animal's motor selection was inappropriate. The effects were not observed when we injected muscimol into the hindlimb representation area of the CMAr or into the forelimb part of the CMAc. Was the effect selectively related to the reward-based selection or on the motor selection in general? To answer this question, we performed additional injection experiments in which the animals were required to select an alternative movement under an auditory instruction. We found that the animal had no problems in selecting the alternative movements when the alteration was cued with the tone signal, with normal reaction time. These observations led us to conclude that the CMAr, but not CMAc, is crucially involved in selecting an appropriate movement, based on the amount of reward. The next question is *how* the CMAr is involved in that process. To answer that question, we analyzed cellular activity in the CMAr and adjacent areas of task-performing monkeys.

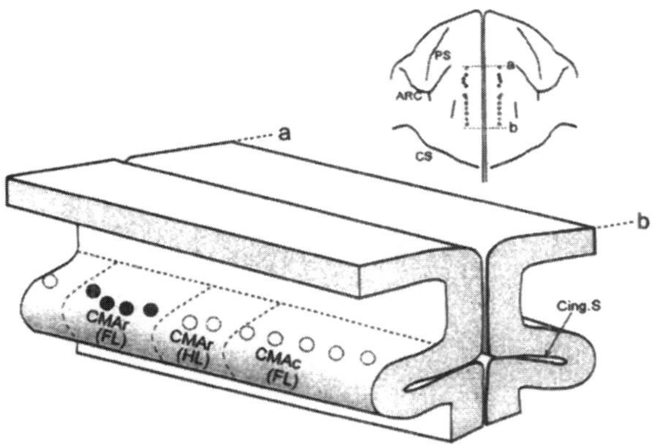

Figure 1. A schematic drawing of muscimol injection sites. Muscimol was effective when injected into the forelimb part (FL) of the CMAr. The labels a and b correspond to rostral and caudal limits of injection sites shown in the top view of the cortex (inset at top right). ARC, arcuate sulcus; CS, central sulcus; PS, principal sulcus, Cing. S, cingulate sulcus

CELLULAR ACTIVTY IN THE CMAr and CMAc

We confirmed our recording sites as being in the CMAr and CMAc, or their adjacent areas based on histological and physiological criteria (Shima et al., 1991). As reported previously, we found cellular activity related to execution of movements in both CMAr and CMAc (Shima et al., 1991). However, we here focus on the activity during the most critical period for motor selection in the present behavioural task. We found that four types of cells in the CMAr exhibit changes in activity during the interval between the occurrence of the reduced reward and the initiation of a movement that was selected as an alternate movement (Table 1). The first type of cell had short-lasting activity that began

Table 1. Distribution of cells in multiple motor areas that were preferentially active in different phases of the behavioral task.

	Reduced Reward /Movement Alteration				Ordinary Reward	Preparatory	Pre-movement	Total
	Type 1	Type 2	Type 3	Type 4				
CMAr	35	23	16	10	16	98	54	231
CMAc	0	1	0	3	2	39	97	132
Pre-SMA	10	3	6	4	5	118	38	166
SMA	0	3	0	3	1	41	71	109
MI	0	0	0	1	0	11	78	82

200 - 600 ms after the occurrence of the reduced reward, and well before the monkey initiated the alternate movement for the next trial, as shown in Figure 2A (left panel). The second type showed a long-lasting activity (>1 s) following the reduced reward (Fig. 2B, left) that also built up rapidly but decayed before initiation of the next movement. The third type had a rapid build-up in activity that lasted continuously until the initiation of the next alternative movement (C). The fourth type showed a gradual increase in activity that built up toward a peak at the initiation of the next movement (D). It is important to note that none of these types of cellular activity was observed unless the monkey selected the alternative movement, even when the reward was reduced. These four types of cells are interpreted as useful in relaying information about the reward reduction to the process of selecting an alternative movement. A prominent property found in these cells was that a majority (68%) of activity was specific to the alternative selection of one movement but not to another. For instance, a selective response to the reduced reward was observed in a CMAr cell when the animal selected an alternative movement TURN, but not when selecting PUSH.

An important question that should be asked was whether the activity described above is specific to the motor selection based on reward or the selection of movement in general. To answer that question, we performed a control experiment. For that control study, we used a tone signal (1 KHz, 300 ms) to tell the animals that they should change the currently performing movement to a different movement in the future. In a great majority of cases (89%), the cells did not respond to the tone signal, although responding to the reduced reward, thus exhibiting selectivity to the reward-based motor selection.

A striking contrast was found between the activity in the CMAr and CMAc. In the CMAr, 36% of 231 task-related cells exhibited the activity occurring between the reduced reward and the choice of an alternative movement. In contrast, in the CMAc, only 3% showed such activity. A majority of cells in the CMAc were active in close time relation to the initiation of movements (66 %), or during preparation of a forthcoming movement or movements (29%), irrespective of whether the movement was the same as or different from the previous movement.

An interesting question is whether there exists an activity responding preferentially to the constant-amount reward. Indeed we found such activity in 8 CMAr cells, one of which is shown in Figure 3. In that example, the responses to ordinary, constant-amount reward were bigger than to the reduced reward.

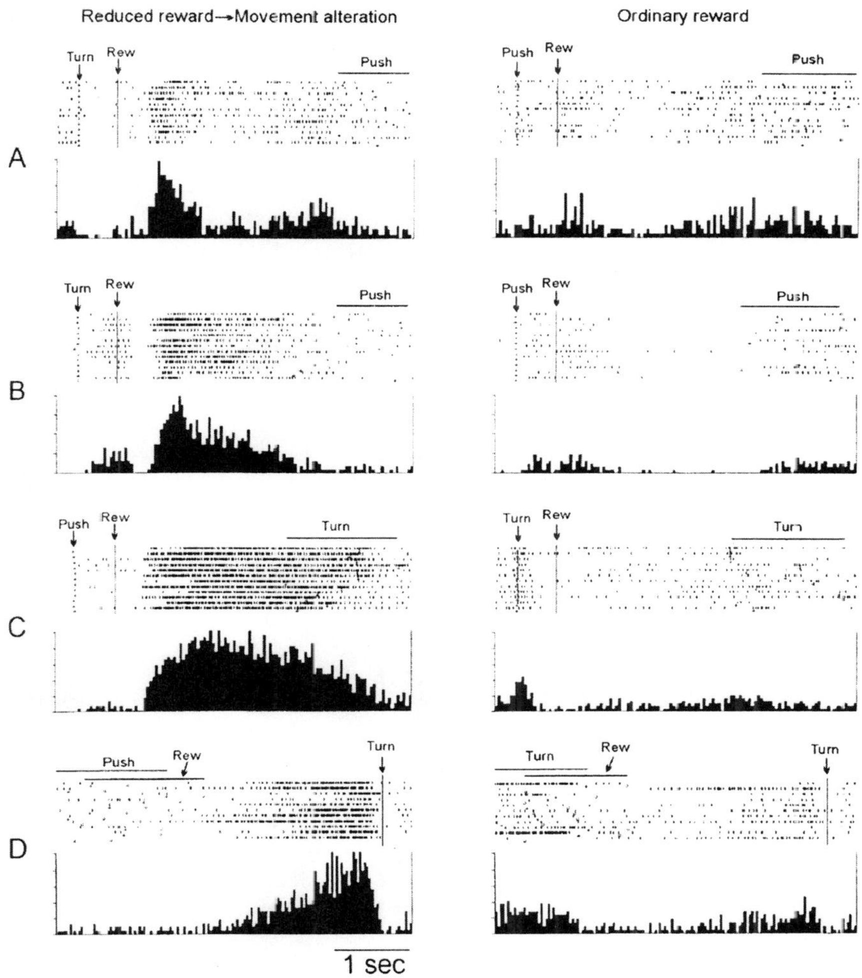

Figure 2. Activity of four types of CMAr cells showing increased discharges after the reward was reduced and before the monkey initiated a newly selected movement (left panels). *A*, Short-lasting activity after the reduced reward. *B*. Long-lasting but decaying activity. *C*. Continuous activity. *D*. Activity that increases progressively before the next newly-selected movement. Little or no activity is seen if the reward is constant amount (right panels). (Adapted from Shima and Tanji, 1998.)

ACTIVITY IN THE SMA AND PRE-SMA

We found that 6% of SMA and 14% of pre-SMA cells exhibited the type of activity selective for the reduced reward, suggesting more participation of the pre-SMA than SMA in the selection of alternate movement (Table 1). A previous report from our laboratory showed that pre-SMA cells take part in changing plans for forthcoming movements (Matsuzaka and Tanji, 1996). Therefore, we interpret the present finding as suggesting participation of pre-SMA cells in modifying plans for future movements, not necessarily based on reward information.

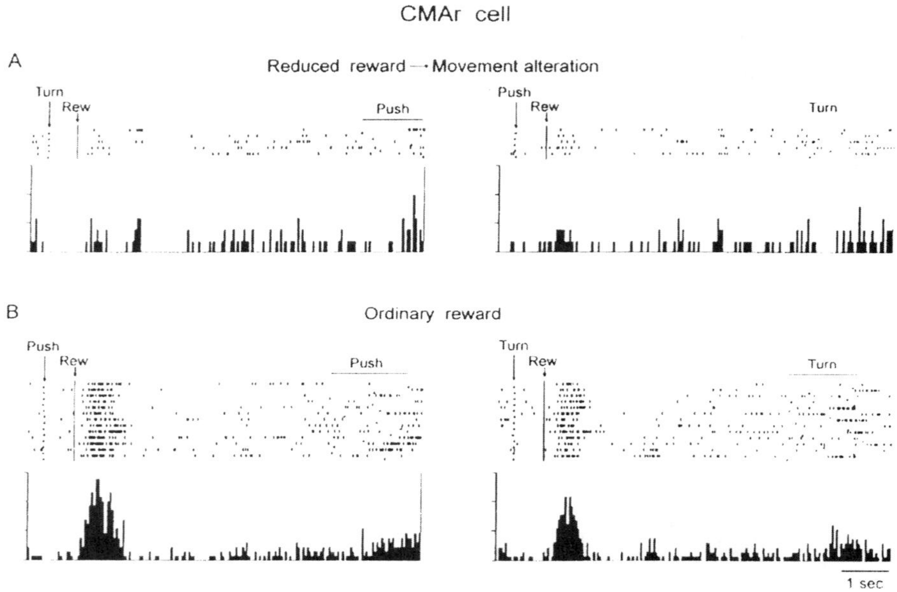

Figure 3. Activity of a CMAr cell. This cell exhibits preferential responses to the constant-amount reward that was ordinarily given to the monkey. In contrast, responses to the reduced reward is meagre.

CONCLUSION

The cingulate motor areas occupy a specific portion of the cingulate cortex that is connected to primary motor cortex and other motor areas, and should be considered separate from other portions of the cingulate cortex. Although the CMAs should be viewed as playing a broad range of functional roles covering various aspects of behaviour, our findings indicate the involvement of the CMAr, in particular, in selecting movements based on reward. The present findings indicate that, on one hand, the first type of CMAr cells monitor the nature of reward (constant-amount or reduced). On the other hand, the fourth type of CMAr cells are involved in preparing a newly-selected movement. Other two types of activity appeared intermediate between the first and fourth (Fig. 2). It is tempting to speculate that information is transmitted through the four types of cells reflecting the process of motor selection based on the amount of reward. In future studies, it will be interesting to look for functional connectivity between different types of cells. On the other hand, CMAc cells are hardly involved in the selection process. We still do not know what characterizes activity in that area, except to find similarities of CMAc cells to MI cells (Shima et al., 1991). As for cells in the pre-SMA, they seem to be involved in selecting a movement that is different from a previous movement, regardless of the source of information that induced or instructed the movement shift (Matsuzaka and Tanji, 1996; Tanji, 1996).

REFERENCES

Matelli, M., Luppino, G., and Rizzolatti, G., 1991, Architecture of superior and mesial area 6 and the adjacent Cingulate cortex in the macaque monkey, *Journal of Comparative Neurology,* 311, 445-462.

Matsuzaka, Y., and Tanji, J., 1996, Changing directions of forthcoming arm movements: neuronal activity in the presupplementary and supplementary motor area of monkey cerebral cortex. *Journal of Neurophysiology,* 76, 2327-2342.

Morecraft, R. J., Geula, C., and Mesulam, M. M., 1993, Architecture of connectivity within a cingulo-fronto-parietal Neurocognitive network for directed attention, *Archives of Neurolog,* 50, 279-284.

Morecraft, R. J., and Van-Hoesen, G.W., 1993, Frontal granular cortex input to the cingulate (M3), supplementary (M2) and primary (M1) motor corteices in the rhesus monkey, *Journal of Comparative Neurology,* 337, 669-689.

Picard, N., and Strick, P. L., 1996, Motor areas of the medial wall: a review of their location and functional activation, *Cerebral Cortex,* 6, 342-353.

Shima, K., Aya, K., Mushiake, H., Inase, M., Aizawa, H., and Tanji, J., 1991, Two movement-related foci in the primate cingulate cortex observed in signal-triggered and self-paced forelimb movements, *Journal of Neurophysiology,* 65, 188-202.

Shima, K., and Tanji, J., 1998, Role for cingulate motor area cells in voluntary movement selection based on Reward, *Science,* 282, 1335-1338.

Tanji, J., 1996, New concepts of the supplementary motor area, Current Opinion in Neurobiology. 6, 782-787.

Vogt, B. A., and Gabriel, M., 1993, *Neurobiology of Cingulate Cortex and Limbic Thalamus,* Birkhauser, Boston, USA.

Vogt, B. A, Pandya, D. N., and Rosene, D. L., 1987, Cingulate cortex of the rhesus monkey: I. Cytoarchitecture and Thalamic afferents, *Journal of Comparative Neurology,* 262, 256-270.

FUNCTIONAL DIFFERENCES IN CORTICOSPINAL PROJECTIONS FROM MACAQUE PRIMARY MOTOR CORTEX AND SUPPLEMENTARY MOTOR AREA

Roger N. Lemon[1], M. A. Maier[2], J. Armand[3], Peter A. Kirkwood[1] and H-W. Yang[4]

ABSTRACT

We made a quantitative comparison of the density of macaque corticospinal projections from primary motor cortex (M1) and supplementary motor area (SMA) to spinal motor nuclei supplying hand and finger muscles. We also compared the action of corticospinal outputs from these two areas on 84 upper limb (mostly hand) motoneurones in chloralose-anaesthetised macaques. The hand representations of M1 and SMA were first identified using MRI and intracortical microstimulation. We made focal injections of WGA-HRP into these representations. Densitometric analysis showed that corticospinal projections from M1 were far denser and occupied a much greater proportion of the hand muscle motor nuclei than did SMA projections. Stimulation of M1 and SMA with bipolar intracortical pulses evoked monosynaptic EPSPs. These were significantly larger and more common from M1 (88% of motoneurons) than from SMA (48%). The results demonstrate cortico-motoneuronal connections from both M1 and SMA, some converging upon single motoneurons. Both areas give rise to CM projections but that those from M1 are far more numerous and exert stronger excitatory effects than those from the SMA.

INTRODUCTION

Many different areas of the cerebral cortex contribute to the corticospinal projection. In the frontal lobe, projections arise from primary motor cortex (M1), dorsal and ventral premotor areas, from a number of cingulate motor areas and from the supplementary motor area (SMA) (Dum and Strick, 1991; 1996; Galea and Darian-Smith, 1994). These different

[1] Sobell Department of Neurophysiology Institute of Neurology University College London Queen Square London WC1N 3BG UK. Email: rlemon@ion.ucl.ac.uk
[2] INSERM U.483, Université Pierre et Marie Curie, 75005 Paris, France.
[3] CNRS & Université de la Mediterranée, Faculté des Sciences du Sport, Marseille, France.
[4] Department of Life Sciences, Chung Shan Medical and Dental College, Taichung 402, Taiwan.

areas show characteristic functional differences, but share others in a distributed fashion (Fetz, 1992; Tanji, 1994; Dum and Strick, 1996). Insight into the functional specificity of these cortical areas can be gained by comparing their patterns of corticospinal output, since these connections ultimately determine the influence a cortical motor area can exert on the spinal machinery for movement. For upper limb movements, it is of particular interest to compare the direct cortico-motoneuronal (CM) connections from different cortical areas. CM connections are a distinctive feature of the primate motor system and are important for the capacity to perform independent finger movements (Porter and Lemon 1993; Bortoff and Strick, 1993; Lemon, 1993; Maier et al., 1997; Nakajima et al., 2000).

We made a quantitative comparison of the density of anterogradely labelled corticospinal *projections* from the hand representations of M1 and SMA to the hand muscle motor nuclei in the lower cervical cord. Then, to assess the strength of functional CM *connections* from M1 and SMA, we made intracellular recordings from motoneurones innervating hand and arm muscles while stimulating these cortical areas. A fuller account can be found in Maier et al. (2001).

Identifying the Hand Representation in M1 and SMA

The experiments involved seven purpose bred adult macaques. All animal care and use was in accordance with the U.K. Animals (Scientific Procedures) Act 1986. It was of critical importance to identify accurately the hand representation in M1 and SMA. MRI was used to determine the sulcal geometry of the relevant cortical regions and to allow accurate placement of anatomical tracers and cortical stimulating electrodes (Baker et al., 1999). Mapping with repetitive intracortical microstimulation (ICMS; twenty 0.2 ms pulses at 300 Hz delivered at rates of 0.5-1 Hz) was then carried out under light, sedative i.m. doses of ketamine. Mapping included the full extent of the SMA motor representation, including face, hand/arm, trunk, leg and tail areas. In all animals, the low threshold (<15 μA) zone giving rise to hand and digit movements was clearly identified; this zone was intermingled with other regions giving rise to movements at the elbow and shoulder (cf. Mitz and Wise, 1987; Luppino et al., 1991)

Corticospinal Projections to Hand Muscle Motor Nuclei

We were able to make a direct comparison of corticospinal projections to the hand motor nuclei from M1 hand area (injected on the left side; Fig. 1A) and SMA (injected on the right side; Fig. 1B) *in the same animal* because there are few, if any, ipsilateral projections to these motor nuclei from either SMA or M1 (Kuypers 1981; Armand, 1982; Dum and Strick, 1996; Armand et al., 1997). Any labelling within lamina IX must have arisen from the contralateral injection site. This experiment, performed in the same animal, avoided any possible errors due to variations in survival time, anterograde transport, histochemical reaction etc.

Projections from both areas included widespread terminations in the cervical enlargement, and these were particularly dense in the intermediate zone. Labelling was much stronger in the hemi-spinal grey matter contralateral to M1 than to SMA. We used a densitometric method (Armand et al., 1997) to make a quantitative comparison of the extent and density of anterograde labelling within the motor nuclei of the C8 and Th1 segments which supply the hand and finger muscles. The proportion of the motoneuronal area that was occupied by labelled projections in different density ranges was determined

(Fig. 1C). Irrespective of whether all labelling was considered (open/filled columns for M1 and SMA, respectively), or the 40% densest labelling (hatched columns), it is clear that the projection from M1 is much the heavier: in caudal Th1, which received the heaviest projections from both areas, there was a 12-13 fold difference in the area of the hand muscle motor nuclei occupied by the densest labelling from M1 compared to that from SMA. Notice also the pronounced rostro-caudal gradient in the strength of the labelling, which probably reflects the particularly heavy projection to motor nuclei supplying intrinsic hand muscles (Armand et al., 1997). These results were confirmed by densitometric analysis of three other cases (Maier et al., 2001).

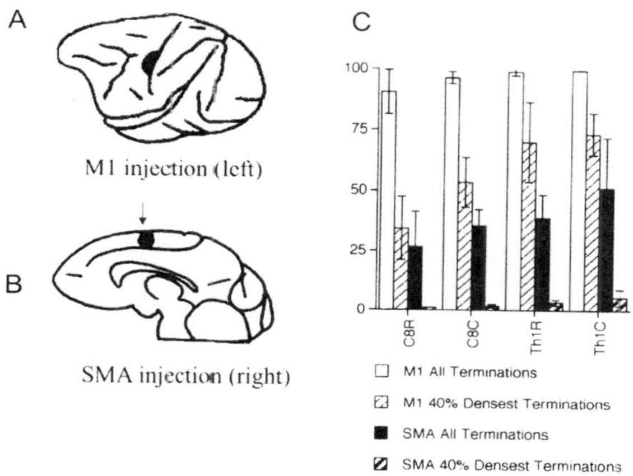

Figure 1. Densitometric analysis of corticospinal projections to hand muscle motor nuclei in the monkey. *A* and *B*, WGA-RF injection sites in M1 (A; left hemisphere) and SMA (B; left hemisphere) in a single monkey (case CS8). The M1 injection site consisted of three needle tracks in the anterior bank of the central sulcus and centred on the M1 hand area. WGA-HRP (total volume 1.2 μL) was deposited along these tracks between 2.5 and 6.5 mm from the surface. The SMA injection site consisted of four needle tracks 1 mm apart in the medial wall of the hemisphere, and centred on the hand representation of SMA as defined by previous ICMS. 10% WGA-HRP (total volume 1.2 μL) was deposited along these tracks between 2 and 4 mm from the pial surface. The level of the genu of the arcuate sulcus is indicated by the arrow. *C*, Comparison in a single monkey (CS8) of the proportion of the selected area of lamina IX, containing the motor nuclei of the hand and finger muscles at C8-Th1, that received corticospinal terminations ('all terminations') and the densest 40% of these terminations, arising from M1 of one hemisphere and from SMA of the other hemisphere For each half-segment (R, rostral; C, caudal), the mean ± SD was computed from data gathered from 6-12 different sections.

Corticospinal outputs generated by stimulation of M1 and SMA

The anatomical study indicated large quantitative differences in the number and density of corticospinal *projections* to lamina IX from SMA and M1. To explore the significance of these differences, we compared the effects of activating these projections on upper limb motoneurones in three adult macaques. The hand representation in SMA and M1 were identified as before with MRI and ICMS, and the terminal experiment was carried out under chloralose anaesthesia (50-80 mg.kg^{-1} i.v.) with all preparatory surgery being carried out under isoflurane anaesthesia (see above and Maier et al., 1998).

We found that bipolar intracortical stimulation of either M1 or SMA evoked a series of descending corticospinal volleys (Fig. 2). The first volley represents a direct or D-wave, produced by direct excitation of corticospinal neurons, while the later volleys were I-waves, resulting from indirect stimulation of these neurons (Patton and Amassian, 1954; Edgley et al., 1990, 1997; Baker et al., 1995). The longer latency of the D volley, compared with that evoked from stimulation of the corticospinal projection in the medullary pyramidal tract (PT), is consistent with the extra conduction distance from the cortex. The interval between the D-wave and the first I-wave (labelled I_1 in Figs. 2,3) was typically 1.0–1.4 ms, and a similar interval was observed between I_1 and I_2, and between I_2 and I_3 (Fig. 2B). Stimulation of the SMA also elicited D- and I-waves (Fig. 2); these had longer latencies and smaller amplitudes than those elicited from M1, probably reflecting both the lower density and maximum conduction velocity of corticospinal neurons in SMA vs M1 (Macpherson et al., 1982; Dum and Strick, 1991; Wise, 1996). The stimulation was focal, evoking descending volleys that were 1-9% of the large volley evoked by supramaximal stimulation of the PT (Figs. 2, 3); occlusion tests established independent and non-overlapping activation of the corticospinal projections from M1 and SMA.

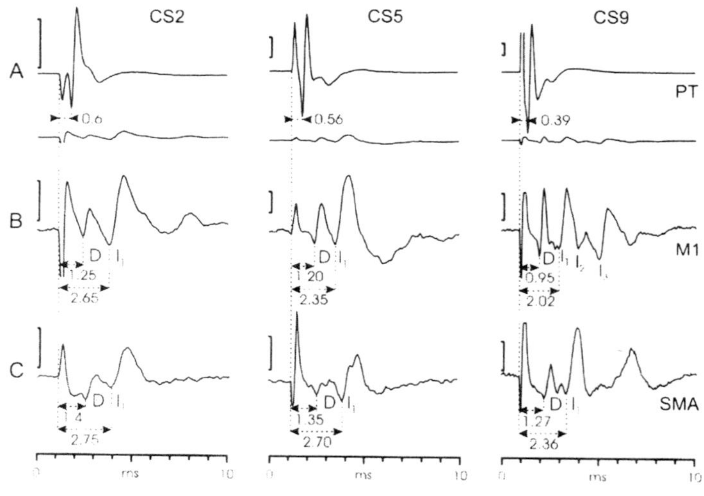

Figure 2. Corticospinal surface volleys evoked by stimulation of M1 and SMA. Row *A*, *B*, and *C* show recordings of corticospinal volleys from the dorsolateral surface of the spinal cord at the C3 or C4 level that were elicited by stimulation of the PT, M1 and SMA respectively, for the three monkeys investigated. Cortical stimulation involved bipolar low-impedance electrodes, with an interelectrode separation of ~ 2mm. The anode was targeted at the upper cortical layers, while the cathode was located either in layer V or in the immediately adjacent white matter. In M1 the electrodes were located in the anterior bank of the central sulcus; in SMA, they were located in the medial wall of the hemisphere. In both cases, they were within the hand representation previously identified by ICMS. In all three monkeys, M1 as well as SMA stimulation evoked a direct (D) volley and at least one indirect (I) volley. Absolute latencies given in ms from stimulus onset to arrival of volley. Clearly, the M1 evoked D-volley and the I-volley arrive earlier than the volleys evoked by SMA stimulation, indicating slower conduction velocity for SMA fibres. Second trace in *A* corresponds to the M1 volley scaled down to a similar magnification as the pyramidal volley, to show its relative size. . Traces are averages of 100 sweeps. Calibration bars: 10 μV for cortical stimulation, 100 μV for PT stimulation. Chloralose anaesthesia.

Postsynaptic Responses of Upper Limb Motoneurons to Corticospinal Outputs.

The effects of these descending corticospinal volleys were investigated in intracellular recordings from 84 upper limb motoneurones: most of these were identified as supplying intrinsic hand muscles (16), wrist or finger flexors (37) or extensors (14). We recorded a large variety of different post-synaptic responses. Maier et al., (2001) categorised these effects according to their segmental delays: the time between arrival of descending corticospinal activity and onset of the post-synaptic potential.

EPSPs

Three types of EPSP were recognised by Maier et al. (2001). First, *direct-wave EPSPs* were unequivocally identified as being due to monosynaptic action of corticospinal volleys in the D-wave because of the short delay (0.6-1.3 ms) between the D-wave and EPSP onset. *Indirect-wave EPSPs* were attributed in the same way to monosynaptic action of the first I-wave; the segmental delays of these EPSPs were 0.5-1.4 ms after the I_1 wave. The latencies of these EPSPs were too long to have been due to monosynaptic D-wave action. They are unlikely to have been due to oligosynaptic corticospinal excitation, because this is rare in the macaque monkey, at least under these experimental conditions (see Maier et al., 1998; Alstermark et al., 1999; Nakajima et al., 2000). Finally *late EPSPs* which were probably evoked by later I wave activity.

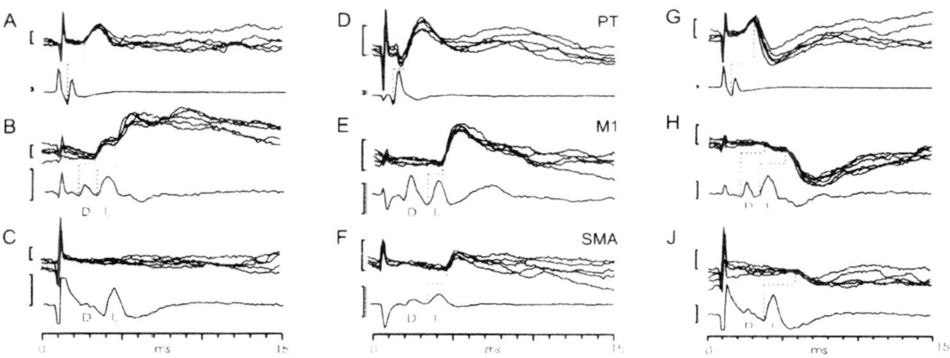

Figure 3. Excitation and inhibition of motoneurones by M1 and SMA stimulation. Examples of responses in three motoneurones (*A-C*, *D-F*, *G-J*). In these specimen records the upper traces display intracellular records of the motoneurone (5 sweeps superimposed), the lower traces display the averaged surface volley (labelled D, I_1 or I_2 the case of cortical stimulation) recorded at the spinal level of the motoneurone. For each motoneurone responses to PT stimulation at 200 μA are shown in *A*, *D*, *G*, while responses to M1 and SMA stimulation at 400 μA are shown in *B*, *E*, *H* and *C*, *F*, *J*, respectively. Stippled lines indicate segmental latency as measured from the D- or I_1-volley. Note presence in *B* of an early EPSP from M1 with a monosynaptic segmental delay after the D volley, followed by a later EPSP with a similar delay after the I_1 volley; only this later EPSP was seen in *E* and *F*. Traces *G-J* show IPSPs. M1 stimulation (H) evoked a small early IPSP with a disynaptic latency after the D volley (left dotted line), with a later larger IPSP after the I_1 volley (right dotted line). SMA stimulation (*J*) evoked only the later IPSP. *A-C*: radial motoneurone (action potential amplitude 58 mV); D-F radial motoneurone (action potential 70 mV); *G-J* median motoneurone (action potential 55 mV). Calibration bars 1 mV for intracellular records, 10 μV for volleys.

Examples of EPSPs are shown in Figure 3. In each case responses from M1 and SMA (using a single 400 μA shock) are compared with those from stimulating the PT (200 μA). In the radial motoneurones shown in Figure 3A-C, PT stimulation evoked an early EPSP, followed by an IPSP. The brief segmental latency of the EPSP (1.0 ms; dotted line in Fig. 3A) identifies the EPSP as monosynaptic, i.e. a direct CM effect (Jankowska et al., 1975; Maier et al., 1998). A single 400 μA shock to M1 (Fig. 3B) evoked two EPSPs, the second following shortly after the first. Their segmental latencies indicate that they are respectively direct-wave and indirect-wave EPSPs. The former had a delay of 1.1 ms from the D-wave and the latter the same delay from the I_1 wave (see dotted lines in Fig. 3B). This motoneurone did not respond to SMA stimulation (Fig. 3C). In the second motoneurone (Fig. 3D-F) M1 stimulation produced a clear EPSP (Fig. 3E) with a segmental latency of 2.3 ms after the D-wave, but only 1.1 ms after the I_1 wave. This EPSP was therefore also classified as an indirect-wave EPSP. Note that the form and size of the EPSP is practically identical to that from the PT (Fig. 3D), and that there was very little jitter in EPSP onset latency. SMA stimulation evoked an EPSP that was smaller than that from M1, but with a similar form (Fig. 3F); it had a segmental latency of 1.2 ms from the I_1 wave. This demonstrates convergence of excitation from both cortical areas upon the same motoneurone.

In many intrinsic hand muscle motoneurones, successive direct- and indirect-wave EPSPs showed temporal summation leading to motoneuronal discharge. This presumably reflects the well-known augmenting responses of CM synapses to successive corticospinal inputs (Phillips and Porter, 1964; Kernell and Wu, 1967; Muir and Porter, 1973).

IPSPs

Figure 3G-J shows a median motoneurone with a monosyaptic EPSP followed by a large IPSP from the PT (Fig. 3G). The IPSP had a disynaptic latency (1.5 ms) from the corticospinal volley (Maier et al., 1998). Stimulation of M1 evoked only IPSPs: a small IPSP after the D-volley and a much larger one after the I-volley (Fig. 3H). In both cases, the segmental delays (1.6 and 1.7 ms) were within the disynaptic range. A disynaptic IPSP after the I-volley was also seen after SMA stimulation (Fig. 3J).

Comparison of CM Effects from M1 and SMA

M1 stimulation (400 μA) produced a response in all 84 tested motoneurones. Direct-wave EPSPs were found in 16 cases (19%), all 16 also showed later I- wave related effects. Indirect-wave EPSPs were found in 74 motoneurones (16 + 58; 88%); the remainder (12%) had only IPSPs. In general, I-wave EPSPs had the lower threshold. In any given motoneurone, there was a close parallel between excitatory effects from PT and M1: thus 15/16 motoneurons with a 'pure' EPSP from the PT showed a similar response from M1. However, of the 63 motoneurones with an EPSP/IPSP sequence from the PT, only 24 (38%) showed an EPSP/IPSP from M1, while 34 (54%) showed a pure EPSP, and the remainder (8%), a pure IPSP. Thus cortical and PT stimulation activate a different balance of corticospinal inputs to a given motoneurone, with the latter giving rise to more pronounced signs of inhibition. .

SMA stimulation (400 μA) did not produce any direct-wave EPSPs (Fig. 3C, F). Indirect-wave EPSPs were recorded in 36 motoneurones (48%) and IPSPs in 10 (13%). Thus all of these received convergent effects from both SMA and M1.

CONCLUSIONS

Both the anatomical and electrophysiological results from this study indicate that corticospinal projections from both M1 and SMA can exert a direct excitatory influence on hand and arm muscles, and this includes convergence of CM connections upon single motoneurones. This study provides the first direct electrophysiological proof of CM effects in upper limb motoneurones from the SMA-proper (Tanji, 1994), or area F3 (Luppino et al., 1991). We have shown that corticospinal projections from M1 occupy a larger area of lamina IX than do those from SMA and the projections are much denser from M1 than SMA, at least for the hand and finger motor nuclei (Fig. 1). These anatomical findings are in keeping with the electrophysiological results. CM connections from M1 to the hand muscle motor nuclei were more numerous and stronger than those from SMA, and the amplitudes of EPSPs from M1 were consistently larger than those from SMA.

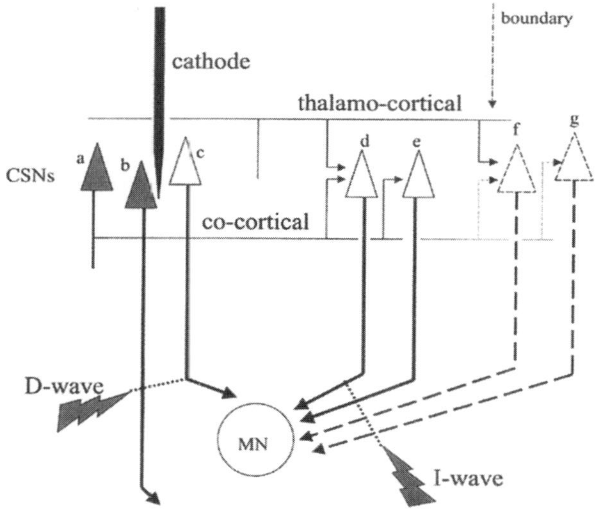

Figure 4. Highly schematic diagram of the possible actions of intracortical stimulation on layer V cortical pyramidal neurons (Δ,σ). The open triangles (Δ) are corticospinal neurons (CSNs) which all project monosynaptically to a single motoneuron (MN) from which intracellular records are taken (i.e. they are the cortico-motoneuronal 'colony' for that motoneuron). Filled triangles (σ) indicate pyramidal/corticospinal neurons with other targets. Stimulation directly excites a restricted population of cell bodies or axons (a,b,c) close to the tip of the cathode. This gives rise to D-wave activity, only some of which is made up of CSNs terminating on the sampled MN; impulses in these axons give rise to the early, direct-wave EPSP. Stimulation also excites intracortical axons; for simplicity, only a single cortico-cortical and thalamo-cortical axon are shown. Some of these axons represent a functionally discrete set of inputs to specific sets of CSNs (d,e,f,g). Activation of these axons leads to indirect, trans-synaptic activation of other members of the 'colony' in cortical areas not excited directly by the stimulus (d,e), and possibly including CSNs (f,g) in other cortical motor areas (dashed symbols and projections). Activation of this network of corticospinal neurons contributes to the I-wave, and leads to the generation of later indirect-wave EPSPs.

Our results could be most parsimoniously explained by bipolar stimulation activating more CM cells indirectly than directly. It is known that, because of temporal variation in the responses times of corticospinal neurons to indirect activation, there can be considerable cancellation of activity making up the I waves (Edgley et al., 1997; Deletis et al., 2000); the greater dispersion of the I- compared to the D-waves is evident from Figure 2. Thus the indirect activation we recorded may actually have been much larger than the direct activation. Cortical stimulation may be particularly effective at evoking these indirect effects because it excites intracortical axonal systems which are presynaptic to CM cells and which are functionally organised to engage large numbers of output neurons. These outputs are known to be distributed over a relatively large cortical area (Andersen et al., 1975; Lemon et al., 1987; Lemon, 1988; Schieber and Hibbard, 1993), and thus it is possible that a much larger number of outputs belonging to the same cortical 'colony' (the group of CM cells all terminating on the sampled motoneurone) were recruited to the I-wave than to the D-wave; the latter is restricted by the physical extent of current flow at the stimulation site (Fig. 4). The recruitment, through intracortical synaptic excitation, of a convergent network of corticospinal neurons might also explain why indirect-wave EPSPs could be as large as those from the whole PT (Fig. 3). The precise extent of this recruitment is unknown: it probably depends primarily on the cortico-cortical connections that are excited at the stimulation site, and these can be extensive (Huntley and Jones, 1991); it may include members of the cortical 'colony' in other cytoarchitectonic areas (see Fig. 4).

We have here concentrated mainly on CM excitation of motoneurones; however, it is well-established that corticospinal tract terminates widely throughout the spinal grey matter, with the densest projection to the intermediate zone (Dum and Strick, 1991; 1996; Armand et al., 1997; Maier et al., 2001). IPSPs were frequently evoked from both SMA and M1, and these are presumably mediated, at least in part, by last-order inhibitory interneurons within the intermediate zone (Perlmutter et al., 1998), which receive convergent peripheral afferent and corticospinal inputs (Jankowska et al., 1976).

Thus although there are similarities in the corticospinal projections from the two motor areas, our studies highlight the special importance of the CM projection to upper limb motoneurons from M1, and make it doubtful that SMA, with a much weaker CM input, could substitute M1 in this respect (cf. Roullier et al., 1998). The SMA, although often referred to as having a motor 'executive' role may have other important functions concerned with the preparation and modulation of intrinsic spinal circuitry. The contrasting functions of M1 and SMA appear to be reflected in the differences in CM action.

ACKNOWLEDGEMENTS

This work was funded by grants from the Wellcome Trust and International Spinal Research Trust. We thank Dr. D.G. MacManus, NMR Unit, Institute of Neurology for help with MRI, and C. Seers, N. Philbin, R. Spinks, K. Sunner and N. Ognjenovic for expert technical assistance.

REFERENCES

Alstermark, B., Isa, T., Ohki, Y., and Saito, Y., 1999, isynaptic pyramidal excitation in forelimb motoneurons mediated via C3-C4 propriospinal neurons in the *Macaca fuscata*, *Journal of Neurophysiology*, 82, 3580-3585.

Andersen, P., Hagan, P. J., Phillips, C. G., and Powell, T. P. S., 1975, Mapping by microstimulation of overlapping projections from area 4 to motor units of the baboon's hand, *Proceedings of the Royal Society*, 188, 31-60

Armand, J., 1982, The origin, course and terminations of corticospinal fibers in various mammals, *Progress in Brain Research*, 57, 330-360.

Armand, J., Olivier, E., Edgley, S. A., and Lemon, R. N., 1997, The postnatal development of corticospinal projections from motor cortex to the cervical enlargement in the macaque monkey, *Journal of Neuroscience*, 17, 251-266.

Baker, S. N., Olivier, E., amd Lemon, R. N., 1995, Task-related variation in corticospinal output evoked by transcranial magnetic stimulation in the macaque monkey, *Journal of Physiology*, 488, 795-801.

Baker, S. N., Philbin, N., Spinks, R., Pinches, E. M., Pauluis, Q., McManus, D., Lemon, R. N., and Wolpert, D. M., 1999, Multiple single unit recording in the cortex of monkeys using independently moveable microelectrodes, *Journal of Neuroscience Methods*, 94, 5-17.

Bortoff, G. A., Strick, P.L., 1993, Corticospinal terminations in two New-World primates: Further evidence that corticomotoneuronal connections provide part of the neural substrate for manual dexterity, *Journal of Neuroscience*, 13, 5105-5118.

Deletis, V., Rodi, Z., and Amassian, V, 2001, Neurophysiological mechanisms underlying motor evoked potentials in anaesthetized humans. Part 2. Relationships between epidural recordings and muscle responses, *Clinical Neurophysiology*, 112, 445-452.

Dum, R. P., and Strick, P. L., 1991, The origin of corticospinal projections from the premotor areas in the frontal lobe, *Journal of Neuroscience*, 11, 667-689.

Dum, R. P., and Strick, P. L., 1996, Spinal cord terminations of the medial wall motor areas in macaque monkeys, *Journal of Neuroscience*, 16, 6513-6525.

Edgley, S. A., Eyre, J. A., Lemon, R. N., and Miller, S., 1990, Excitation of the corticospinal tract by electromagnetic and electrical stimulation of the scalp in the macaque monkey, *Journal of Physiology*, 425, 301-320.

Edgley, S. A., Eyre, J. A., Lemon, R. N., and Miller, S., 1997, Comparison of activation of corticospinal neurones and spinal motoneurones by magnetic and electrical stimulation in the monkey, *Brain*, 129, 839-853.

Fetz, EE., 1992, Are movement parameters recognizably coded in activity of single neurons? *Behavioural and Brain Sciences*, 15, 679-690.

Galea, M. P., and Darian-Smith, I., 1994, Multiple corticospinal neuron populations in the Macaque monkey are specified by their unique cortical orgins, spinal terminations, and connections, *Cerebral Cortex*, 4, 166-194.

Huntley, G. W., and Jones, E.G., 1991, Relationship of intrinsic connections to forelimb movement representations in monkey motor cortex: A correlative anatomic and physiological study, *Journal of Neurophysiology*, 66, 390-413.

Jankowska, E., Padel, Y., and Tanaka, R., 1975, Projections of pyramidal tract cells to þ-motoneurones innervating hind-limb muscles in monkey, *Journal of Physiology*, 249, 637-667.

Jankowska, E., Padel, Y., and Tanaka, R., 1976, Disynaptic inhibition of spinal motoneurones from the motor cortex in the monkey, *Journal of Physiology*, 258, 467-487

Kernell, D., and Wu, C-P., 1967, Post-synaptic effects of cortical stimulation on forelimb motoneurones in the baboon, *Journal of Physiology*, 191, 673-690.

Kuypers, H. G. J. M., 1981, Anatomy of the descending pathways, in: *Handbook of Physiology - The Nervous System II*, Brookhart, J. M., and Mountcastle, V. B., ed., American Physiological Society, Bethesda, pp. 597-666.

Lemon, R. N., 1988, The output map of the primate motor cortex, *Trends in Neuroscience*, 11, 501-506

Lemon, R. N., 1993, Cortical control of the primate hand, *Experimental Physiology*, 78, 263-301.

Lemon, R. N., Muir, R. B., and Mantel, G. W. H., 1987, The effects upon the activity of hand and forearm muscles of intracortical stimulation in the vicinity of corticomotor neurones in the conscious monkey, *Experimental Brain Research*, 66, 621-637.

Luppino, G., Matelli, M., Camarda, R. M., Gallese, V., and Rizzolatti, G., 1991, Multiple representations of body movements in mesial area 6 and the adjacent cingulate cortex: An intracortical microstimulation study in the Macaque monkey, *Journal of Comparative Neurology*, 311, 463-482.

Macpherson, J. M., Wiesendanger, M., Marangoz, C., and Miles, TS., 1982, Corticospinal neurones of the supplementary motor area of the monkey, *Experimental Brain Research*, 48, 81-88.
Maier, M. A., Armand, J., Kirkwood, P. A., Yang, H-W., Davis, J. N., and Lemon, R. N., 2001, Differences in the corticospinal projection from primary motor cortex and supplementary motor area to macaque upper limb motoneurons: an anatomical and electrophysiological study, *Cerebral Cortex*, in press.
Maier, M. A., Illert, M., Kirkwood, P. A., Nielsen, J., and Lemon, R. N., 1998, Does a C3-C4 propriospinal system transmit corticospinal excitation in the primate? An investigation in the macaque monkey, *Journal of Physiology*, 511, 191-212.
Maier, M. A., Olivier, E., Baker, S. N., Kirkwood, P. A., Morris, T., and Lemon, RN., 1997, Direct and indirect corticospinal control of arm and hand motoneurons in the squirrel monkey (Saimiri sciureus), *Journal of Neurophysiology*, 78, 721-733.
Mitz, A. R., and Wise, S. P., 1987, The somatotopic organization of the supplementary motor area: intracortical microstimulation mapping, *Journal of Neuroscience*, 7, 1010-1021.
Muir, R. B., and Porter, R., 1973, The effect of a preceding stimulus on temporal summation at corticomotoneuronal synapses, *Journal of Physiology*, 228, 749-763.
Nakajima, K., Maier, M. A., Kirkwood, P. A., and Lemon, R. N., 2000, Striking differences in the transmission of corticospinal excitation to upper limb motoneurons in two primate species, *Journal of Neurophysiology*, 84, 698-709.
Patton, H. D., Amassian, VE., 1954, Single- and multiple-unit analysis of cortical stage of pyramidal tract activation, *Journal of Neurophysiology*, 17, 345-363.
Perlmutter, S. I., Maier, M. A., and Fetz, E. E., 1998, Activity of spinal interneurons and their effects on forearm muscles during voluntary wrist movements in the monkey, *Journal of Neurophysiology*, 80, 2475-2494.
Phillips, C. G., and Porter, R., 1964, The pyramidal projection to motoneurones of some muscle groups of the baboon's forelimb, *Progress in Brain Research*, 12, 222-245.
Porter, R., and Lemon, R. N., 1993, *Corticospinal Function and Voluntary Movement,* Clarendon Press, Oxford.
Rouiller, E. M., Yu, X. H., Moret, V., Tempini, A., Wiesendanger, M., and Liang, F., 1998, Dexterity in adult monkeys following early lesion of the motor cortical hand area: the role of cortex adjacent to the lesion, *European Journal of Neuroscience*, 10, 729-740.
Schieber, M. H., Hibbard, L. S., 1993, How somatotopic is the motor cortex hand area? *Science*, 261, 489-492.
Tanji, J., 1994, The supplementary motor area in the cerebral cortex, *Neuroscience Research*, 19, 251-268.
Wise, SP., 1996, Corticospinal efferents of the supplementary sensorimotor area in relation to the primary motor area, in: *Advances in Neurology*, Luders, H. O., ed., Lippincott-Raven, Philadelphia, pp. 57-69.

CORTICOSPINAL TRANSMISSION AFTER VOLUNTARY CONTRACTIONS

Janet L. Taylor, Nicolas T. Petersen, Jane E. Butler and Simon Gandevia*

ABSTRACT

Transmastoid stimulation in human subjects evokes a single descending volley in corticospinal axons. Electromyographic (EMG) responses to transmastoid stimuli measured in the relaxed elbow flexor muscles after a maximal voluntary contraction (MVC) are depressed by approximately 50% compared to prior to the MVC. The depression recovers over 2 min. Responses can also be depressed in the relaxed muscle after submaximal contractions and when measured during weak contractions after an MVC. The depression may reflect a decrease in efficacy at the corticospinal-motoneuronal synapse. If so, an activity-dependent decrease in the effectiveness of cortical drive to the motoneurones could affect the control of many voluntary movements.

INTRODUCTION

The corticomotoneuronal system is very important in control of the upper limb in human subjects. It is particularly important in the control of fine movements of the fingers and hand. Although the firing of cells in the cortex has been examined during various tasks in animals including non-human primates (see Porter & Lemon, 1993), it is not feasible to examine directly the activity of corticomotoneurones in human subjects. Thus the precise way in which cortical output translates into movement remains unclear.

In awake human subjects, axons in the descending tracts can be activated at the level of the cervicomedullary junction by passing an electrical pulse between electrodes fixed over each mastoid process (Ugawa et al., 1991, Gandevia et al., 1999). A transmastoid stimulus generates a single descending volley which can evoke a response from the motoneurones and this can be recorded as an EMG response in the muscle (cervicomedullary evoked potential; CMEP). It is likely that many of the descending axons activated by transmastoid stimulation are in the corticospinal tract. This means that the

* Prince of Wales Medical Research Institute, Randwick, Sydney, NSW, 2031, Australia.
Email: jl.taylor@unsw.edu.au

behaviour of CMEPs under different conditions should reflect changes in the effective muscle response produced by a constant cortical output.

Evidence that transmastoid stimulation activates corticospinal axons comes from experiments in which transmastoid stimulation and electrical cortical stimulation were given at short interstimulus intervals (Ugawa et al., 1991). The response to cortical stimulation measured in an intrinsic muscle of the hand could be occluded by an appropriately timed transmastoid stimulus. The simplest explanation for this finding is that the volley descending from the cortex collided with the antidromic volley initiated by transmastoid stimulation. This collision would only occur if the two stimuli activated the same neurones. As cortical stimulation generates descending volleys in corticospinal axons, transmastoid stimulation must also activate these axons. Figure 1 shows similar results from the biceps brachii. Here, responses to magnetic cortical stimulation are largely occluded by transmastoid stimulation given 1 ms later (Gandevia et al., 1999). This occurs despite the multiple descending volleys produced by transcranial magnetic stimulation. Thus, CMEPs in proximal arm muscles, as well as those in the muscles of the hand, are likely to be produced through activation of the corticospinal tract.

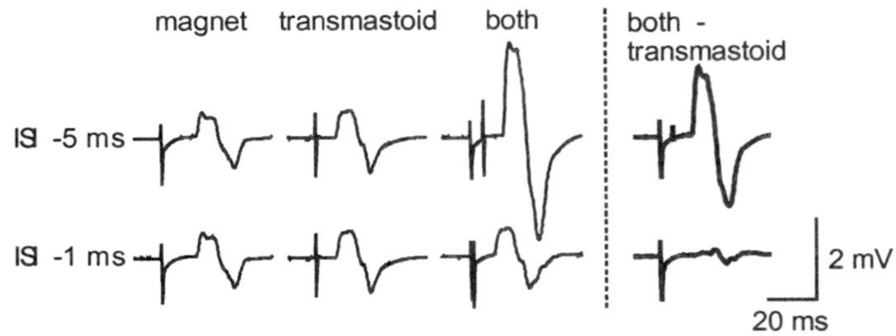

Figure 1. Transmastoid stimulation occludes the response to transcranial magnetic stimulation. EMG potentials recorded from biceps brachii in one subject. Each trace is an average of three responses. Responses to transcranial magnetic stimulation alone (magnet), transmastoid stimulation alone (transmastoid) and both stimuli at 2 different interstimulus intervals (ISIs) were elicited in the relaxed muscle in random order. When the magnetic cortical stimulus was given 5 ms before the transmastoid stimulus (ISI -5 ms), the response to both stimuli together was large. Subtraction of the response to transmastoid stimulation from this response to both stimuli shows a remainder that is larger than the response to cortical stimulation alone. The response was facilitated. In contrast, when the magnetic cortical stimulus was given 1 ms before the transmastoid stimulus (ISI -1 ms), the response was small and the subtraction shows substantial occlusion of the response to cortical stimulation

DEPRESSED CORTICOSPINAL TRANSMISSION AFTER VOLUNTARY ACTIVITY

We have found that CMEPs in the elbow flexor muscles can be altered after relatively small amounts of voluntary activity. Subjects performed isometric elbow flexions and EMG was recorded from biceps brachii and brachioradialis. Transmastoid stimuli (100□s pulse,~500 V, Digitimer D180 stimulator) were delivered to evoke CMEPs of an amplitude of 40-50 % of the maximal M-wave in biceps brachii and of 20-30 % of the maximal M-wave in brachioradialis. When subjects relaxed after performing a maximal voluntary contraction (MVC), CMEPs in both muscles were immediately depressed by about 50% compared to their size measured in the resting muscles before the MVC. This depression recovered over approximately 2 minutes. Surprisingly, the depression was similar in depth and duration after MVCs as short as 5 s and as long as 2 min (Fig. 2). Thus, the depression was not related to fatigue. The size of the CMEP was also not critical. Depression after an MVC was seen in CMEPs of amplitudes of 5% to 50% of the maximal M-wave.

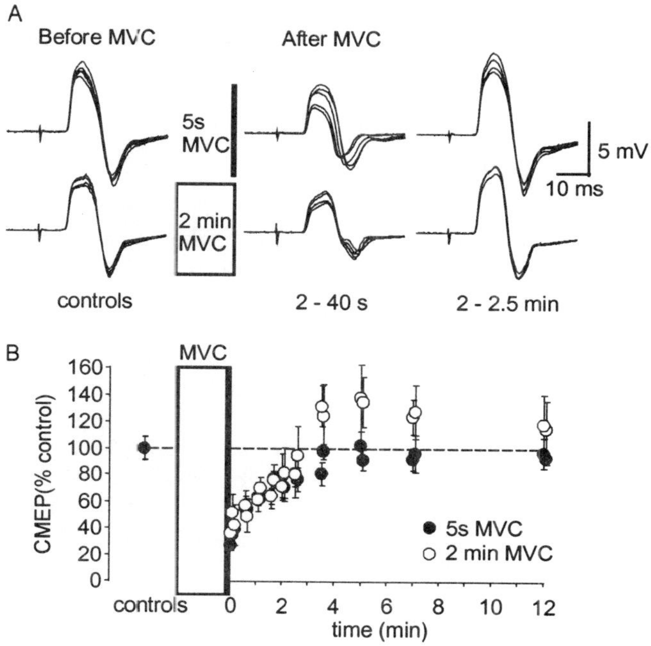

Figure 2. Depression of CMEP after 5 s or 2 min maximal voluntary contraction (MVC). *A*, EMG responses to transmastoid stimulation recorded from biceps brachii in one subject on 2 days. The upper traces show CMEPs evoked before and after an MVC of 5-s duration. The lower traces were recorded before and after a 2-min MVC. A similar reduction in the size of the CMEP occurs after both MVCs. The smallest CMEP in each middle set of traces (2-40 s post-MVC) was the first one evoked after each contraction. CMEPs have recovered by 2 mins post-MVC. *B*, Group data for 7 subjects (mean ± SEM) show the reduction in area of CMEPs recorded in the relaxed biceps brachii after a 5 s MVC (filled circles) or a 2-min MVC (open circles). For each subject, the area of CMEPs is expressed as a percentage of the area of CMEPs elicited prior to each contraction.

With stimuli delivered to the corticospinal axons and a response measured in the muscle there are a number of potential sites at which changes in function might affect the size of the CMEP. These include: the site of stimulation where changes in axonal excitability might reduce the descending volley; any interneurones which might have altered excitability; the corticospinal-motoneuronal synapse where the efficacy of synaptic transmission could be altered; the motoneurones which might have decreased responsiveness through inhibition, disfacilitation or alterations in intrinsic properties; and the muscle where the muscle fibre action potential can be altered by activity. The last of these sites is easily excluded. Maximal M-waves evoked in biceps brachii and brachioradialis by supramaximal stimulation over the brachial plexus are not depressed after MVCs. They tend to be slightly enlarged immediately after the contraction (see Gandevia et al., 1999, Figs. 2 & 3). Thus, changes in the muscle fibre action potential cannot account for changes in the CMEP.

Other sites are more difficult to exclude. However, single motor unit studies which have examined responses to transcranial magnetic stimulation of the motor cortex suggest that in human subjects biceps brachii has strong monosynaptic connections from the cortex (Palmer & Ashby, 1992). The responses of single motor units to transmastoid stimulation also suggest that there is a large excitatory monosynaptic component in the CMEP (Petersen et al., 2001b). Therefore, the influence on the CMEP of excitatory interneurones in the corticospinal-motoneuronal pathway is probably limited.

There is also evidence that argues against activity-dependent changes in axonal excitability at the site of stimulation causing the depression of the CMEP. In peripheral motor axons, repetitive activation during voluntary contractions results in hyper-polarisation of the cell membrane and an increased threshold to electrical stimulation. After a 1 min MVC, the compound muscle action potential evoked by a constant stimulus can decrease by about 30%. However, after a 15 s MVC this decrease is only 15% (Vagg et al., 1998). The strength and duration of the changes in axonal excitability depend directly on the extent of activity. In contrast, the depression of the CMEPs does not depend on the duration of the contraction but is similar following short-duration (5 s) and sustained MVCs (Fig. 2). This behaviour is not consistent with reported activity-dependent changes in axons. A second argument follows from the observation that if the CMEP is evoked during brief maximal efforts rather than at rest, it is not depressed after an MVC (Taylor et al., 1996; Butler et al., 1999). A reduction in the descending volley through a change in axonal thresholds should act to decrease the CMEP during activity as well as at rest.

Finally, activation of the motoneurones through a route other than via the corticospinal tract does not produce depression of the CMEPs. The musculocutaneous nerve was stimulated at 30 Hz for 10 s with an intensity greater than that needed to elicit a maximal M-wave in biceps brachii. Such stimulation should activate the axons of all α-motoneurones as well as many of the afferents in the nerve. Thus, the muscle contracts tetanically and the motoneurones should be activated either antidromically or trans-synaptically through reflex pathways at a rate comparable to the mean rates seen during MVCs. Tetanic stimulation did not depress the responses evoked by transmastoid stimuli, whereas an MVC of the same duration did (Fig. 3). This result suggests that the depression of the CMEP after a voluntary contraction is not due to an intrinsic property of the motoneurone dependent on its activity, nor is it a consequence of increased afferent firing during the contraction or altered feedback subsequent to the contraction.

Experiments which demonstrate that the depression of the CMEP can be seen not only in the resting muscle but also during weak voluntary contractions offer further evidence that this depression is not the result of changes in motoneurone excitability. When CMEPs were recorded while subjects matched EMG in biceps brachi (5% MVC) before and after a 10 s maximal effort, CMEPs after the MVC were smaller than those recorded in the control period. As maintaining a constant level of EMG in a muscle should hold the "excitability" of the motoneurone pool steady, it is unlikely that inhibition of the motoneurones could cause this reduction of the CMEP.

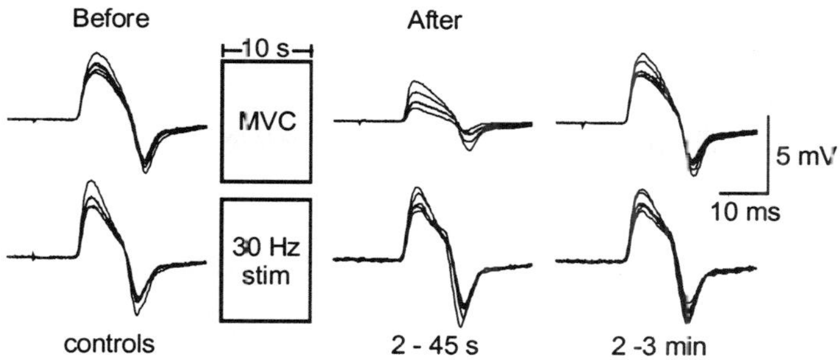

Figure 3. Depression of CMEP after a voluntary contraction but not after tetanic peripheral nerve stimulation. Responses to transmastoid stimulation recorded from biceps brachii in one subject. Upper traces were recorded before and after a 10 s MVC and show a reduction in the size of the CMEP immediately after the contraction. Lower traces were recorded before and after 10 s supramaximal peripheral nerve stimulation at 30 Hz. No depression of the CMEP occurs.

If depression of the CMEP is not the result of changes in axonal excitability in the corticospinal axons or a post-synaptic change in the motoneurones, then it may reflect a change in the efficacy of transmission at the corticospinal-motoneuronal synapse. Corticomotoneuronal neurones are thought not to have classical presynaptic inhibition such as that which provides afferent and descending control of Ia afferent input to the motoneurones (Nielsen & Petersen, 1994). However, repetitive activation alters the release of transmitter at many synapses in the central nervous system. Some synapses show augmentation or facilitation of release whereas others show depression. A number of different mechanisms with different time courses have been postulated for short-term depression. These include release site refractoriness which can result in the depression of the response to the second pulse in a paired-pulse paradigm, and depletion of the immediately and readily releasable pools of synaptic vesicles. Such depletion results from repetitive activation and is frequency-dependent (for review see Thomson, 2000). Post-activation depression of the H-reflex probably also represents an example of altered efficacy of a synapse onto motoneurones in human subjects (Hultborn et al., 1996; see also Li & Burke, 2001). Its time-course (recovery in ~10 s) is much shorter than the depression of the CMEP but activity-dependent changes differ widely between synapses.

FUNCTIONAL IMPLICATIONS

There is a surprising contrast between the behaviour of the responses to transmastoid stimulation and that of the responses to transcranial magnetic stimulation recorded from biceps brachii after a sustained MVC. While the CMEPs are initially depressed, the MEPs are initially facilitated (Samii et al., 1996). As the CMEPs recover over 2 min and then show a long-lasting facilitation, the MEPs become less facilitated and, with fatiguing contractions, become depressed (Brasil-Neto et al., 1993; Samii et al., 1997). The two responses follow quite different time courses (Gandevia et al., 1999). These differences could occur because the responses to magnetic cortical stimulation are the result of multiple descending volleys whereas responses to transmastoid stimulation are produced by a single descending volley (Edgley et al., 1997). A second explanation is that the excitability of the motor cortex is altered during and after voluntary contractions such that the output evoked by the magnetic stimulus varies greatly. The degree of change that must occur in the cortex can only be appreciated when the behaviour of the MEP is contrasted with that of the response to a constant corticospinal volley.

The depth and duration of the depression of the CMEP can be altered both by the strength of the voluntary contraction which causes it and by whether the CMEPs are elicited during rest or during voluntary activity. In the resting muscle, CMEPs can be depressed after a 10 s submaximal voluntary contraction of 25 to 50% of maximal force. However, depression varies between subjects and is graded, with less depression occurring after weaker contractions (Petersen et al., 2001a). When the CMEP is elicited *during* submaximal voluntary contractions, the depression seen after a 10 s MVC is smaller than in the resting muscle but can occur during contractions of up to 50% of maximal force. During MVCs and strong submaximal contractions, the CMEP is not reduced.

The occurrence of depression of the CMEP during voluntary contractions suggests that this decreased effectiveness of corticospinal activity may have functional consequences. For example, a weak voluntary contraction performed *after* a strong contraction might require more output from the motor cortex than the same weak contraction made *before* the strong contraction. Such an effect might be apparent through an increase in the effort needed to make the contraction.

The absence of depression of the CMEP during strong voluntary contractions at a time after an MVC when the CMEP in the resting muscle would be depressed may have interesting implications. It could indicate that the reduced synaptic efficacy might be overcome by sufficient activity. On the contrary, it might indicate that the efficacy of corticospinal output is reduced after relatively few action potentials. If synaptic transmission was reduced after only a few impulses, CMEPs evoked during the brief control MVCs after approximately 1 s of activity might already be depressed and so no further depression would be apparent after the extra activity of the 10 s MVC. Some support for this possibility is given by a report of depression of responses to transcranial electrical stimulation after just 1 s of 10 Hz rapid-rate transcranial magnetic stimulation. MEPs elicited at the same time were facilitated (Modugno et al., 1998). As electrical cortical stimulation is thought to generate corticospinal output primarily through direct activation of corticospinal axons, this contrasting behaviour is similar to that of CMEPs and MEPs after voluntary contractions.

CONCLUSIONS

For the elbow flexor muscles in human subjects, the response elicited by a single corticospinal volley can be depressed by moderate voluntary activity. This decreased effectiveness of corticospinal output may result from activity-dependent changes in synaptic function. Given that the depression can occur during a voluntary contraction, it is likely to influence the descending drive required to activate motoneurones and thus affect the control of movement. The decrease in the response to a constant corticospinal input to the motoneurones may be significant after only a few action potentials. It may therefore be a factor in most voluntary contractions.

ACKNOWLEDGEMENTS

This work was supported by the National Health and Medical Research Council of Australia. Nicolas Petersen is supported by the Danish Sports Research Council and Team Danmark.

REFERENCES

Butler, J. E., Taylor, J. L., and Gandevia, S. C., 1999, Reduction of responses to stimulation of descending tracts during sustained maximal voluntary contractions in human subjects, *Proceedings of the Australian Neuroscience Society,* 10, 196.

Brasil-Neto, J. P., Pascual-Leone, A., Valls-Solé, J., Cammarota, A., Cohen, L. G., and Hallett, M., 1993, Postexercise depression of motor evoked potentials: a measure of central nervous system fatigue, *Experimental Brain Research,* 93, 181-184.

Edgley, S. A., Eyre, J. A., Lemon, R. N., and Miller, S., 1997, Comparison of activation of corticospinal neurons and spinal motor neurons by magnetic and electrical transcranial stimulation in the lumbosacral cord of the anaesthetized monkey, *Brain,* 120, 839-853.

Gandevia, S. C., Petersen, N., Butler, J. E., and Taylor, J. L., 1999, Impaired response of human motorneurones to corticospinal stimulation after voluntary exercise, *Journal of Physiology,* 521, 749-759.

Hultborn, H., Illert, M., Nielsen, J., Paul, A., Ballegaard, M., and Wiese, H., 1996, On the mechanism of the post-activation depression of the H-reflex in human subjects, *Experimental Brain Research,* 108, 450-462.

Li, Y., and Burke, R. E., 2001, Short-term synaptic depression in the neonatal mouse spinal cord: effects of calcium and temperature, *Journal of Neurophysiology,* 85, 2047-2062.

Modugno, N., Priori, A., Gilio, F., Lorenzano, C., and Berardelli, A., 1998, Repetitive transcrainal magnetic stimulation, rTMS) of the motor cortex elicits opposite excitability changes at cortical and subcortical level in man, *Journal of Physiology,* 513.P, 34P.

Nielsen, J., and Petersen, N., 1994, Is presynaptic inhibition distributed to corticospinal fibres in man? *Journal of Physiology,* 477, 47-58.

Palmer, E., and Ashby, P., 1992, Corticospinal projections to upper limb motoneurones in humans, *Journal of Physiology,* 448, 397-412.

Petersen, N. T., Taylor, J. L., Butler, J. E., and Gandevia, S. C., 2001a, Reduced corticospinal transmission to motoneurones after strong voluntary contraction persists during a weak contraction, *Proceedings of the Australian Neuroscience Society,* 12, 226.

Petersen, N. T, Taylor, J. L., and Gandevia, S. C., 2001b, Stimulation of the corticospinal tract activates descending neurones with direct connections to human motoneurones, *XXXIV International Congress of Physiological Sciences Abstract, CD-Rom,* 1853.

Porter, R., and Lemon, R., 1993, *Corticospinal Function and Voluntary Movement,* Oxford University Press, Oxford, New York.

Samii, A., Wasserman, E. M., Ikoma, K., Mercuri, B., and Hallett, M., 1996, Characterization of postexercise facilitation of motor evoked potentials to transcranial magnetic stimulation, *Neurology,* 46, 1376-1382.

Samii, A., Wasserman, E. M., and Hallett, M., 1997, Post-exercise depression of motor evoked potentials as a function of exercise duration, *Electroencephalography and Clinical Neurophysiology,* 105, 352-356.

Taylor, J. L., Butler, J. E., Allen, G. M., and Gandevia, S. C., 1996, Changes in motor cortical excitability during human muscle fatigue, *Journal of Physiology,* 490, 519-528.

Thomson, A.M., 2000, Molecular frequency filters at central synapses, *Progress in Neurobiology,* 62, 159-196.

Ugawa, Y., Rothwell, J. C., Day, B. L., Thompson, P. D., and Marsden, C. D., 1991, Percutaneous electrical stimulation of corticospinal pathways at the level of the pyramidal decussation in humans, *Annals of Neurology,* 29, 418-427.

Vagg, R., Mogyoros, I., Kiernan, M. C., and Burke, D., 1998, Activity-dependent hyperpolarization of motor axons produced by natural activity, *Journal of Physiology,* 507, 919-925.

50

AFFERENT AND CORTICAL CONTROL OF HUMAN MASTICATORY MUSCLES

Timothy S. Miles and Michael A. Nordstrom*

ABSTRACT

Like most other muscles, the human masticatory muscles are controlled by descending signals from the cortex and other supraspinal structures, as well as afferent signals arising in receptors in muscles, skin and other tissues. However, the special functional roles of the masticatory system, and in particular the fact that the muscles on both sides are usually used together, has led to some special adaptations of function.

INTRODUCTION

The human trigeminal motor system is adapted for several purposes. Its primary function is to bite off food which may vary from the soft (fruit) to the brittle (nuts), and prepare it for swallowing. It also contributes to respiration and speech. The latter involves rapid, precise movements to shape the mouth for forming phonemes. Three pairs of jaw-closing muscles and two main pairs of jaw-opening muscles operate bilaterally on the mandible to position it appropriately *via* its sliding-hinge joint to exert the forces required to achieve these diverse functions.

Because of the complexity of the roles of the masticatory system, the high forces that it can exert over short distances, its potential for damaging other tissues (e.g., the tongue and cheeks), the fixed end-point to jaw closing when the teeth come together, and the speed with which the jaw must move in speech, the muscles and the mechanisms that control them are highly specialized (Miles and Nordstrom, 1995).

Normal chewing movements are based on a basic rhythm driven by a pattern generator in the brainstem reticular formation, which is subject to modulation by volition and by afferent feedback. Fine, voluntary control of jaw muscles in primates is enhanced by projections from motor cortex to the trigeminal motor nuclei in the pons. In the first

* Department of Physiology, Adelaide University, Adelaide, SA 5005 Australia.
Email: timothy.miles@adelaide.edu.au

part of this review we briefly summarise the descending control of trigeminal motoneurones that allows fine voluntary control of jaw muscle activity. In the second section, we describe the reflexes which modulate masticatory activity with particular emphasis on the role of stretch reflexes.

Cortical Control

The motor cortex initiates and controls movements of the mandible through corticobulbar projections to the trigeminal motor nuclei and brainstem reticular formation (Luschei & Goldberg 1981). Unilateral repetitive electrical stimulation of specific regions of cortex elicits rhythmic jaw movements in a number of species, including primates. The bilateral, rhythmic, alternating activity in jaw-opening and closing muscles resembles the pattern seen in natural mastication, and is influenced by sensory feedback from the jaw, tongue and face. The alternating pattern of muscle activity is produced by a central pattern generator (CPG) in the brainstem reticular formation, which receives descending projections from the cortex (Nakamura and Katakura, 1995).

Human corticobulbar fibres project bilaterally to the pons near the trigeminal motor nuclei (Kuypers, 1958). This is consistent with the observation that bilateral responses can be elicited in jaw muscles by unilateral electrical stimulation of the human and primate motor cortex (Penfield and Boldrey, 1937), and with the preservation of masticatory function bilaterally following unilateral lesions of the primary motor cortex in humans and monkey (Willoughby and Anderson, 1984). The corticotrigeminal projection provides a direct path for voluntary activation of jaw muscles and, by analogy with the corticospinal system, is likely to be important for their fine voluntary control.

Transcranial magnetic stimulation (TMS) elicits muscle evoked potentials (MEPs) in human jaw-closing and opening muscles bilaterally. Focal TMS using a figure-of-eight coil over the face area of motor cortex on one side has shown that corticotrigeminal projection from one hemisphere excites the masseter and anterior digastric motoneurons bilaterally (Gooden et al., 1999; Nordstrom et al., 1999; Butler et al., 2001). TMS data suggest that corticobulbar neurons make direct monosynaptic excitatory connections with trigeminal motor neurons. Masseter motorneurons may receive direct excitatory projections from either hemisphere (Nordstrom et al., 1999) while direct projections to digastric motoneurons appear to be mostly ipsilateral (Gooden et al., 1999). Some single corticotrigeminal neurons appear to branch to innervate the masseter motoneuron pools on both sides (Carr et al., 1994). This raises questions about the functional role of the corticotrigeminal projection in humans. Does this projection provide the capacity for fractionated activation of individual jaw muscles, including homologous pairs across the midline? Are there distinct roles for the contralateral and ipsilateral hemispheres in the control of masticatory muscles?

The projections from the two hemispheres have a distinct role in control of the jaw muscles. The corticobulbar projection to the trigeminal motor pools is bilateral but not symmetrical. The contralateral projection providing short-latency excitation is more effective than the ipsilateral for both masseter and anterior digastric muscles. MEPs evoked in active jaw muscles by focal TMS are 30-40% larger in the muscles contralateral to the hemisphere stimulated (Gooden et al., 1999; Butler et al., 2001). TMS of the motor cortex of one hemisphere may excite a masseter or digastric motor unit (MU) while the opposite hemisphere inhibits it (Gooden et al., 1999; Nordstrom et al., 1999).

The asymmetry is marked for masseter, in which MUs were most commonly excited at short latency with contralateral TMS and inhibited with no short-latency excitation by ipsilateral TMS. Most digastric MUs were excited by TMS applied to either hemisphere, however one of the 14 MUs tested was inhibited by ipsilateral TMS and excited by contralateral TMS. The corticotrigeminal projections from each hemisphere could contribute to differential activation of homologous jaw muscles across the midline, but more for masseter than digastric. The greater hemispheric asymmetry of motor cortical control of MUs in masseter than digastric is consistent with the normal activation patterns of these muscles. Masseter muscles are commonly activated independently on each side during mastication (Luschei & Goldberg, 1981) whereas the digastric muscles are said to be co-activated during all jaw movements (Widmalm et al., 1988).

The contralateral hemisphere dominates the descending commands mediating selective activation of jaw-closing muscles on one side of the body required for a unilateral bite. We compared MEPs evoked by unilateral focal TMS while subjects attempted unilateral activation of one masseter, with those obtained in the same muscle during a bilateral bite at an equivalent EMG level (Butler et al., 2001). MEPs in the masseter contralateral to the stimulated hemisphere were significantly smaller during unilateral compared with bilateral biting. The ipsilateral masseter MEPs were not different during ipsilateral and bilateral biting. Hence the command for unilateral biting is associated with reduced excitability of corticotrigeminal neurons in the contralateral, but not the ipsilateral motor cortex. This may be accomplished by reduced activity in the contralateral hemisphere of a population of corticotrigeminal neurons which branch to innervate both masseter motoneuron pools. The observations that the contralateral masseter MEP is larger than the ipsilateral response, and that only the contralateral MEP is modulated during unilateral biting suggest that there is a separate population of corticotrigeminal neurons with exclusively contralateral projections to masseter. The ipsilateral MEP could arise from corticotrigeminal neurons which branch to innervate both masseter motoneuron pools, and/or a separate population of cells with exclusively ipsilateral projections.

In summary,

1. corticotrigeminal projections from either hemisphere contribute to the command mediating voluntary activation of masseter and anterior digastric muscles;
2. there are direct corticomotoneuronal projections to masseter and digastric motoneurons;
3. the contralateral hemisphere provides relatively more of the excitatory input via the fast corticotrigeminal pathway to both masseter and digastric motoneurons;
4. projections from contralateral and ipsilateral hemispheres can have opposing effects on a single trigeminal motoneuron; and
5. modulation of excitatory drive from the contralateral, but not ipsilateral, hemisphere is involved in selective activation of one masseter muscle during a unilateral bite.

Reflex Control

The masticatory muscles are subject to feedback control via a number of reflexes. The jaw-closing muscles have muscle spindles and probably Golgi tendon organs (Lund et al., 1978). There are the usual sensory receptors in the peri-oral skin and oral mucosa which exert primarily inhibitory reflex control over the jaw-closing muscles (e.g, Miles et al., 1987). There are also mechanoreceptors in the connective tissue ligament in which the teeth are suspended, the so-called periodontal receptors. These are activated by low forces applied to the tooth surface and are saturated by forces of about 2 N (Trulsson and Johansson, 1996). Only slowly-adapting periodontal receptors have been described in humans (Trulsson, 1992). Sharp taps on teeth elicit a primarily inhibitory response at segmental latency, whereas pressing slowly on teeth elicits a longer-latency excitatory response (Türker et al., 1994). Perhaps the sustained afferent activity from pressing on a tooth gives rise to sustained excitation of jaw-closing motoneurones, while the brisk afferent volley from tapping on teeth opens an inhibitory pathway that over-rides this.

The periodontal mechanoreceptors are optimally placed to register the forces applied to the teeth in a manner analogous to the Golgi tendon organs in muscles. The pattern of periodontal reflex responses suggests their functional roles. For example, the reflexly-induced facilitation evoked by the initial contact of the teeth onto the food bolus during chewing may help to maintain the bolus between the teeth so that it may then be crushed. However, when one bites unexpectedly on a rigid object such as a bone while chewing, the resulting brisk forces on the teeth will evoke a strong afferent volley from the periodontal receptors and possibly others in the periosteum, which reflexly inhibits the jaw-closing muscles (van der Glas et al., 1985; Brodin et al., 1993).

The stretch reflex in the jaw-closing muscles is used clinically to test trigeminal sensory and motor pathways; however, jaw-opening muscles have few if any spindles. "Jaw-jerk" reflexes have been evoked by tapping on the chin with a tendon hammer (Godaux and Desmedt, 1975) and by controlled stretch (Lamarre and Lund 1975). These studies concluded that, in contrast to most limb muscles in which stretch evokes a short-latency segmental response followed by a long-latency, presumably transcortical response, the human jaw-closing muscles respond to stretch with only a short-latency excitatory reflex. We subsequently found that rapid stretches evoke only the monosynaptic excitatory response reported earlier. However, slower rates of stretch evoke an additional, long-latency excitatory response which appears analogous to that which exists in the limbs (Poliakov and Miles, 1994). The dependency on stretch velocity was explained by recording responses of single masseter MUs to stretch (Miles et al., 1995). Rapid stretches reflexly excite low-threshold masseter motoneurones with high probability. Having discharged, the motoneurones are then hyperpolarised for up to 100 ms and are therefore less likely to respond to the second excitatory volley that arrives during this period. However, with slower stretches, the probability that motoneurones will be discharged at short latency is reduced: hence, there is a greater probability that they will discharge when the second volley arrives about 25 ms later.

The stretch reflex acts as a load-compensating mechanism during mastication, automatically adding excitation to the jaw-closing motoneurones when the texture of the food adds unexpected resistance during jaw closing. When additional loads are added to the jaw-closing muscles at unpredictable times during simulated chewing, the activity of the jaw-closing muscle increased at a latency of about 20 ms (Ottenhof et al., 1992a, b). While this response overcomes only rather small increases in force, it is also adaptive. When a larger resistive force is first encountered, the additional muscle activity required to overcome the resistance occurs at reflex latency. In subsequent chewing cycles, additional muscle activity occurs before the extra resistance appears. That is, the first chewing stroke is used to assess the resistance of the food, and the force of subsequent chewing strokes is automatically increased to deal with this.

The stretch reflexes in the masticatory muscles also unload the jaw-closing muscles when the resistance to closing suddenly falls unexpectedly. This is particularly important when high forces are being used, for instance, to crack a hard nutshell between the teeth (Hannam et al., 1968; Miles and Wilkinson, 1982). A second, non-reflex event also acts to arrest the upward movement of the mandible. During forceful, isometric contractions when rapid unloading of the jaw-closing muscles is anticipated, the jaw-opening muscles are co-activated. While the downward force that they exert on the mandible is weak, contracting muscles are stiff therefore act as a retaining strap that limits the rapid upward movement on the mandible on unloading, and the resulting forceful crashing together of the teeth (Miles and Madigan, 1983).

The understanding of the functional roles of stretch reflexes in the limbs has been hindered by the lack of a good model for the operation of the reflex under entirely physiological conditions, and without the complication of muscle excitation for other purposes. The control of mandibular position during locomotion offers an unique opportunity to observe a stretch reflex operating in an entirely physiological mode. In a person sitting or standing quietly upright, the mandible is suspended from the maxilla with a 4-8 mm separation between upper and lower anterior teeth: this is the "rest" position of the mandible. During running, the jaw moves up and down as the result of gravity and inertia.

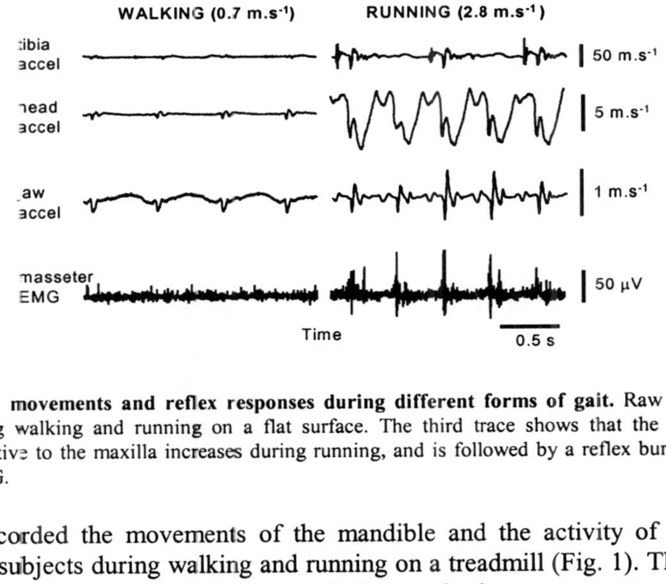

Figure1. Jaw movements and reflex responses during different forms of gait. Raw data from a single subject during walking and running on a flat surface. The third trace shows that the acceleration of the mandible relative to the maxilla increases during running, and is followed by a reflex burst of activity in the masseter EMG.

We recorded the movements of the mandible and the activity of the jaw-closing muscles in subjects during walking and running on a treadmill (Fig. 1). The head and jaw moves up and down with each step at both speeds, but movements of the mandible relative to the maxilla are slow and small during walking. These movements increase in amplitude when subjects run, and the mandible moves downwards relative to the maxilla following landing. This stretch of the jaw-closing muscles is followed by a reflex excitation of masseter and a subsequent upward movement of the mandible. The distance moved by the mandible relative to the maxilla and the amplitude of the reflex response evoked depends on the gait, being greatest in subjects running "downhill" when heel-landing is obligatory and the resulting deceleration of the skull is greatest. These data indicate that the relative downward movement of the mandible during running stretches the masseter, evoking a short-latency stretch reflex which causes the mandible to move upwards towards its normal resting position.

Thus, the mechanisms controlling the masticatory system share many of the features of spinal control mechanisms, but with some unique features which are the result of its adaptations to its functional role, and particularly to the bilateral activation of its muscles which occurs during normal function.

ACKNOWLEDGEMENTS

Our research is funded by the NHMRC of Australia. Stanley Flavel conducted the experiments on the reflex control of mandibular position during locomotion.

REFERENCES

Brodin, P., Türker, K. S., and Miles, T. S., 1993, Mechanoreceptors around the tooth evoke inhibitory and excitatory reflexes in the human masseter muscle, *Journal of Physiology,* 464, 711-723.

Butler, S. L., Miles, T. S., Thompson, P. D. and Nordstrom, M. A., 2001, Task-dependent control of human masseter muscles from ipsilateral and contralateral motor cortex, *Experimental Brain Research,* 137, 65-70.

Carr, L. J., Harrison, L. M., and Stephens, J. A., 1994, Evidence for bilateral innervation of certain homologous motoneurone pools in man, *Journal of Physiology,* 475, 217-227.

Godaux, E., and Desmedt, J.E., 1975, Human masseter muscle: H- and tendon reflexes, *Archives in Neurology,* 32, 229-238.

Gooden, B. R., Ridding, M. C., Miles, T. S., Nordstrom, M. A. and Thompson, P. D., 1999, Bilateral cortical control of the human anterior digastric muscles, *Experimental Brain Research,* 129, 582-591.

Hannam, A. G., Matthews, B., and Yemm, R., 1968, The unloading reflex in masticatory muscles of man, *Archives in Oral Biology,* 13, 361-364.

Kuypers, H. G. J. M., 1958, Corticobulbar connexions to the pons and lower brain-stem in man, *Brain,* 81, 364-388.

Lamarre, Y., and Lund, J. P., 1975, Load compensation in human masseter muscle, *Journal of Physiology,* 253, 21-35.

Luschei, E., and Goldberg L. J., 1981, Neural mechanisms of mandibular control: mastication and voluntary biting, in: *Handbook of Physiology* Sect. 1, Vol. 2, American Physiological Society, Bethesda, pp. 1237-1274.

Lund, J. P., Richmond, F. J., Touloumis, C., Patry, Y., and Lamarre, Y., 1978, The distribution of Golgi tendon organs and muscle spindles in masseter and temporalis muscles of the cat, *Neuroscience* 3, 259-270.

Miles, T. S., and Madigan, M. L., 1983, Programming of antagonist muscle stiffness during masticatory muscle unloading in man, *Archives in Oral Biology,* 28, 947-951.

Miles, T. S., and Wilkinson, T. M., 1982, Limitation of jaw movement by antagonist muscle stiffness during unloading of human jaw closing muscles, *Experimental Brain Research,* 46, 305-310.

Miles, T. S., and Nordstrom, M. A., 1995, Fatigue of jaw muscles and speech mechanisms, *Advances in Experimental Medicine and Biology,* 384, 415-426.

Miles, T. S., Poliakov, A. V., and Nordstrom, M.A., 1995, Responses of human masseter motor units to stretch, *Journal of Physiology,* 483, 251-264.

Miles, T. S., Türker, K. S., and Nordstrom, M. A., 1987, Reflex responses of motor units in human masseter muscle to electrical stimulation of the lip, *Experimental Brain Research,* 65, 331-336.

Nakamura, Y. and Katakura, N., 1995, Generation of masticatory rhythm in the brainstem, *Neuroscience Research,* 23, 1-19.

Nordstrom, M. A., Miles, T. S, Gooden, B. R., Butler, S. L., Ridding, M. C., and Thompson, P. D., 1999, Motor cortical control of human masticatory muscles, *Progress in Brain Research,* 123, 203-214.

Ottenhoff, F. A. M., Van der Bilt, A., van der Glas, H. W., and Bosman, F., 1992a, Peripherally induced and anticipating elevator muscle activity during simulated chewing in humans, *Journal of Neurophysiology,* 67, 75-83.

Ottenhoff, F. A. M., Van der Bilt, A., van der Glas, H. W., and Bosman, F., 1992b, Control of elevator muscle activity during simulated chewing with varying food resistance in humans, *Journal of Neurophysiology,* 68, 933-944.

Penfield, W., and Boldrey, E., 1937, Somatic motor and sensory representation in the cerebral cortex of man as studied by electrical stimulation, *Brain,* 60, 389-443.

Poliakov, A.V., and Miles, T.S., 1994, Stretch reflexes in human masseter, *Journal of Physiology,* 476, 323-331.

Trulsson, M., and Johansson, R. S., 1996, Encoding of tooth loads by human periodontal afferents and their role in jaw motor control, *Progress in Neurobiology,* 49, 267-284.

Trulsson, M., Johansson, R. S., and Olsson, K. A., 1992, Directional sensitivity of human periodontal mechanoreceptive afferents to forces applied to the teeth, *Journal of Physiology,* 447, 373-389.

Türker, K. S., Brodin, P., and Miles, T. S., 1994, Reflex responses of motor units in human masseter muscle to mechanical stimulation of a tooth, *Experimental Brain Research,* 100, 307-315.

Van der Glas, H. W., de Laat, A., and van Steenberghe, D., 1985, Oral pressure receptors mediate a series of inhibitory and excitatory periods in the masseteric poststimulus EMG complex following tapping of a tooth in man, *Brain Research,* 337, 117-125.

Widmalm, S.-E., Lillie, J. H., and Ash Jr., M. M., 1988, Anatomical and electromyographic studies of the digastric muscle, *Journal of Oral Rehabilitation,* 15, 3-21.

Willoughby, E. W., and Anderson, N. E., 1984, Lower cranial nerve motor function in unilateral vascular lesions of the cerebral hemisphere, *British Medical Journal,* 289, 791-794.

MECHANISMS FOR ACUTE CHANGES IN SENSORY MAPS

Mike B. Calford*

ABSTRACT

Many studies have examined changes in the topographic representations of the special senses in cerebral cortex following partial peripheral deafferentations. This approach has demonstrated the short- medium- and long-term aspects of plasticity. However, the extensive capacity for immediate plasticity, while first demonstrated more than 15 years ago, still challenges explanation. What such studies indicate is that each locus in sensory cortex receives viable input from a far wider area of the sensory epithelium than is represented in the normal receptive field, with the implication that much of this input is normally inhibited. Consideration of the geometric and temporal aspects of receptive field plasticity suggests that this inhibition must be tonic and must derive its driving input from a tonically active periphery.

INTRODUCTION

Any hopes for therapy after major brain or spinal trauma ultimately depend on the capacity for the central nervous system to encode and integrate a changed input into its working systems. Whereas we know that nearly every synapse in the nervous system is capable of some form of plasticity, be that long-term potentiation or depression, or ultrastructural changes or control by a neuromodulator, we can also point to many cases of inflexibility of neural systems. As a simple example, consider phantom limb experiences. Even decades after a limb amputation, and despite the altered perceptual relationships and new neural discharge correlations which ensue, a perceptual presence of the missing limb remains (Melzack, 1990; Ramachandran et al., 1992; Halligan et al., 1993).

As a way of studying the adaptive capacity necessary for neural therapies, the sensory representations of the cerebral cortex have been examined: the body surface representation of the somatosensory cortex, the tonal-frequency representation of the

* Discipline of Human Physiology, School of Biomedical Science & Hunter Medical Research Institute, University of Newcastle, Newcastle, NSW, 2308, Australia. Email: mike.calford@newcastle.edu.au

Sensorimotor Control of Movement and Posture
Edited by Gandevia *et al.*, Kluwer Academic/Plenum Publishers, 2002

auditory cortex, and the visual-projection representation of the visual cortex. These representations serve as metrics for measuring the capacity for plasticity in the adult brain.

A productive approach has been to examine the plasticity of the topographic representations resulting from partial loss of afferent input. In the somatosensory cortex, these losses come from amputations or nerve sections, peripheral local anaesthesia, or from interruption of the central or spinal pathways. Determinations have been based primarily upon two parameters: the response areas (receptive fields) of individual neurons and the topographies which these neural responses form. Such work has demonstrated 3 phases in the response to a partial loss of input.

1. Short term: immediate to hours
   - unmasking of new responsiveness
2. Medium term: hours to days
   - new filtering of the unmasked responsiveness
     - expansion of area of new responsiveness
     - dependence upon synaptic plasticity
3. Long term: weeks to months
   - structural changes
     - local axonal sprouting
     - dendritic remodelling

The initial event in each case is that partial peripheral lesion or deactivation can lead to short-term changes in the topographic representations which are greater than the mere loss of the affected body part, frequency range or visual projection. This indicates an inherent capacity for functional plasticity within the circuitry. Studies of the somatotopic map of the primary somatosensory area (area 3b) in New World monkeys revealed plasticity in the form of expansion of the neighbouring representations shortly after median (Merzenich et al., 1983) or combined median and ulnar nerve section (Kolarik et al., 1994). In these cases substantial areas remained unresponsive, but the expansions of the remaining radial, or radial plus ulnar, nerve representations were dramatic. Similar rapid plasticity of the topographic map has been described in macaque monkey area 3b following digit amputation (Calford and Tweedale, 1991b).

Examination of the response profile of a single neuron (or a group of neurons recorded at one site) following a peripheral manipulation, has revealed that the expression of new or expanded receptive fields is extremely rapid. A robust result is the finding that local anaesthesia of a small body area leads immediately to the expression of large receptive fields on the adjoining body surface in the affected region of Area 3b. During the period of local anaesthesia, and for a short time after the return of sensitivity, single- and multi-neuron recordings from the digit or toe representations in macaque monkeys (Calford and Tweedale, 1990) flying foxes (Calford and Tweedale, 1991a) and rats (Byrne and Calford, 1991; Doetsch et al., 1996) revealed greatly expanded receptive fields.

To examine the capacity for immediate expansion of frequency tuning of neurons in cat primary auditory cortex, temporary threshold shifts in cochlear sensitivity were induced with loud pure tones (Calford et al., 1993). The procedure allowed a period of about 30 minutes, after recovery from the initial dramatic effects of the loud tone, in which notch-like losses in the peripheral sensitivity of 20 dB to 50 dB were centred on the tuning range of previously studied neurons. In around one-third of cases there was an expansion of the range of frequency sensitivity. While pointing to an inherent capacity for expansion of the responsive range of some neurons, the broad width of the losses compared with the frequency-tuning range of many cells precluded population estimates of this capacity.

Recently Snyder and Sinex have developed the technique of lesioning the cell bodies of auditory nerve afferents in the tonotopically arranged spiral ganglion. As this technique does not disrupt cochlear mechanics it has allowed the effects of acute frequency-limited lesions to be examined. In these experiments, post-lesion recordings of single neurons in primary auditory cortex (Snyder and Sinex, 1998) and inferior colliculus (Snyder et al., 2000) of the anaesthetized cat were compared with pre-lesion recordings of neurons near the same sites. Extensive topographic plasticity was evident as a shift in frequency tuning at affected sites. Furthermore, the threshold sensitivities of the newly tuned neurons did not differ from normal at these frequencies, indicating an unmasking of previously unexpressed inputs rather than responses reflecting a residual portion of the previously expressed frequency tuning curve of the neurons.

A capacity for rapid representational plasticity has also been demonstrated in visual cortex in monkey (Fiorani et al., 1992) and cat (Pettet and Gilbert, 1992) using a non-invasive paradigm. In independent studies it was shown that, after defining the classical receptive field of a neuron, a mask could be placed over the field area and the neuron would become responsive to large stimuli outside the original field limits. Such masking resulted in the rapid 'filling in' of these areas. This was apparent by responses to stimuli which extended beyond the boundaries of the mask and an equivalent five-fold expansion of receptive field size with a topographic organization of the interpolated centres. These demonstrations have been criticised for the incompleteness of the classical receptive field in defining the response area of a cortical neuron (see DeAngelis et al., 1995; and for a counterpoint, Chapman and Stone, 1996) but they, nevertheless, show a dynamic aspect to the spatial limits of responsiveness under the two stimulus conditions.

Short-term plasticity of the topographic representation in adult cat primary visual cortex has been demonstrated within a few hours of a discrete laser lesion of one retina (Calford et al., 1999). However the method of laser lesioning does not allow the study of early effects in visual cortex as there are widespread deactivating effects of the laser heating procedure at the retinal level (Tassignon et al., 1991; Yamamoto et al., 1996; Humphrey et al., 1997; Chu et al., 1998; Calford et al., 1999). Using a strategy which induces a retinal detachment, Schmid et al. (1995) were able to examine the capacity for representational plasticity within an hour of retinal deactivation. In this study, a lower power laser light was used to make a small lesion in animals which had been preloaded with excess fluids, by intravenous infusion. Haemodynamic forces induced a retinal detachment 4 to 5 times larger than the initial lesion. Separation of the neural retina from the pigment epithelial layer rendered the detached portion inactive and, in most cases, the effect was stable after 1 hour. With this paradigm some cortical neurons within the projection zone of the detached part of the retina were immediately responsive to stimulation of ectopic receptive fields, displaced to unaffected parts of the retina. Beyond the first few hours, the proportion of neurons showing such ectopic responses was similar to that after laser lesions (Calford et al., 1999). In both short-term paradigms the final proportions of recording sites at which neural responses could be evoked by photic stimulation of ectopic receptive fields (75 % at 12 hours post-lesion) were similar to those obtained, in the same laboratory, for cases with post-lesion survival times of two weeks to three months (Schmid et al., 1996).

With the short-term monocular lesion and detachment paradigms it is possible to compare directly the location of an ectopic receptive field for the affected eye with the normal receptive field for the unaffected eye (Schmid et al., 1995; Calford et al., 1999). The latter were not found to change from those recorded prior to the lesion. Converting these to cortical-representational dimensions (Albus, 1975) revealed that the maximal ectopic displacement was in the range 6 to 9 mm.

This review is concentrated on the first stage of plasticity – the unmasking of new responsiveness. However, it is worth quickly summarizing some of the critical studies that have identified the later stages. In studies of somatosensory (Calford and Tweedale, 1988) and visual cortex (Calford et al., 2000; Dreher et al., 2001) the size of the ectopic receptive fields revealed by partial peripheral denervations shrinks dramatically over the first 1 to 2 weeks. In many studies of somatosensory cortex, the extent of denervated cortex which shows this new responsiveness increases (e.g. Merzenich et al., 1983; Wall and Cusick, 1986; Turnbull and Rasmusson, 1991). There is evidence that this second stage depends upon some form of synaptic plasticity. The precise nature, and the number, of such mechanisms is not yet established. These second stage events have been variously shown to be dependent upon cholinergic modulation via basal forebrain inputs (Dykes, 1990; Webster et al., 1991a,b; Rasmusson, 1993) or upon NMDA-receptor mediated long-term potentiation (Kano et al., 1991; Garraghty and Muja, 1996; Glazewski and Fox, 1996). In addition, there have been direct demonstrations of potentiation at corticocortical synapses (Smits et al., 1991).

Local structural changes and expression of growth-associated factors (Dunn-Meynell et al., 1992; Laskawi et al., 1997) have been described and represent a third stage of peripheral-denervation-induced plasticity. An increase in the density (but probably not extent) of within field corticocortical connections resulting from terminal sprouting of axons follows retinal lesions (Darian-Smith and Gilbert, 1994) and vibrissectomy (Kossut and Juliano, 1999) in the respective cortices.

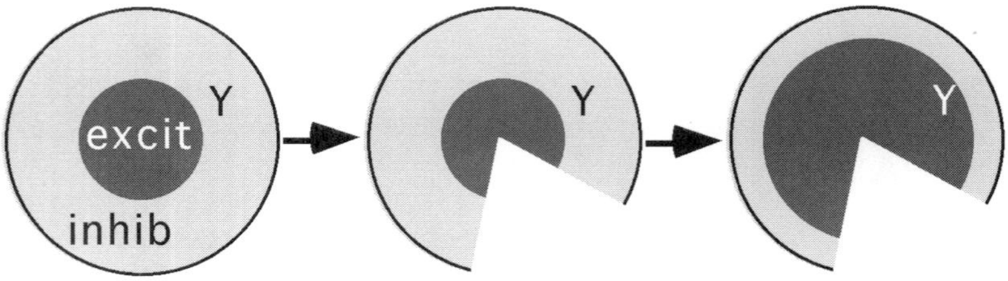

Figure 1. Geometric representation of a receptive field illustrating the failure of disruption of stimulus-driven lateral inhibition to account for expansion of the excitatory receptive field. The missing segment represents a partial peripheral denervation. Without postulating disruption of a tonic output from the denervated segment there is no reason why the outcome of a stimulus at position **Y** should change from net suppression to net excitation.

Independent of the changes seen in later stages, the capacity for immediate expression of new responsiveness raises questions about the nature of the inhibition which normally stops an extensive input being expressed in the receptive field of a cortical neuron. Stimulus-driven lateral inhibition has been well demonstrated in cortical neurons representing each of the special senses (e.g. Mouncastle and Powell, 1959; Henry et al., 1969; Laskin and Spencer, 1979; Calford and Semple, 1995). It was therefore tempting to consider that much of the array of extensive inputs to a cortical neuron is suppressed by this mechanism such that only a subset of the input is reflected in the receptive field. Disturbance of this balance by a peripheral denervation, which removes some of the input, would therefore result in an expansion of the receptive field. Consider the simple geometric representation of a receptive field with lateral inhibition in Figure 1A. If some of the input is lost by partial peripheral denervation (Fig. 1B), then can disruption of lateral inhibition account for an expansion of the manifested excitatory area (Fig. 1C)? Now consider a point (Y) within the original inhibitory surround and also within the final expanded receptive field. Initially stimulation of point Y results in a net suppression of activity. It is clear that if all of the inhibition which normally suppresses the expression of point Y in the original receptive field comes from stimulus-driven inhibition then the outcome for a single stimulus applied to point Y would be the same (suppression) irrespective of the denervation affecting another part of the receptive field. In order for the loss of input to affect the outcome of a stimulus delivered to point Y in this way there must be some input from the denervated area that normally contributes to the suppression of excitation and which is lost as a result of the denervation. In the context of sensory afferents not contributing directly to inhibition, this requirement can be thought of as a loss of a tonic signal from afferents to the denervated area removing drive that normally contributes to the inhibition at point Y.

Irrespective of the place in a sensory pathway where the contribution of a tonic excitatory input has its effect by activating a local interneuron-based inhibition, there is ultimately a need for tonic activity in primary afferents. Numerous studies have reported that myelininated A-fibre mechanoreceptive afferents have no spontaneous activity. Hence unmyelinated C-fibres were examined as a source of tonic input to central inihibition. In an exposed nerve C-fibres can be selectively blocked with the toxin capsaicin. The receptive fields of single neurons responsive to cutaneous stimulation in cortex of cat (Calford and Tweedale, 1991b), flying fox (Calford and Tweedale, 1991b), rat (Katz et al., 1999; Calford and Clarey, unpublished) and mouse (Nussbaumer and Wall, 1985), the thalamus of raccoon (Rasmusson et al., 1993) and rat (Katz et al., 1999), and the cuneate nucleus of cat (Pettit and Schwark, 1996) have been shown to expand with capsaicin blockade of C-fibres. Since A-fibres are unaffected, the expanded receptive fields include the original receptive field. It is unclear at which level C-fibre activity affects inhibition. Dykes and Craig (1998) have shown that, within matched body part representations, receptive fields of neurons in the dorsal column nuclei expand after blockade of ipsilateral dorsal horn cellular responsiveness. This suggests that the pathway for propagation of the peripheral tonic activity is via the ipsilateral finiculus; termination may be in the border zones of the dorsal column nuclei which contain inhibitory neurons projecting into the lemniscal relay regions (Fyffe et al., 1986). While the capsaicin deactivation result appears robust, the interpretation is not as simple as reasoning that some C-fibres provide a source of activity which ultimately drives an inhibition that contributes to shaping the receptive fields of central pathway neurons. Fundamentally, this is because C-fibres have little or no spontaneous activity. A few years ago this disparity may have been dismissed by arguing that no study had appropriately and specifically looked for such activity (which may be at very low rates). However a recent study – aimed primarily at examining spontaneous activity induced by nerve section –

reported no spontaneous activity from 68 rat sural nerve cutaneous C-fibres (Michaelis et al., 2000).

Turning to the visual and auditory systems, there are no problems identifying sources of tonic activity in primary afferents but is there any evidence that these contribute to central inhibition which limits receptive field extent? Auditory nerve afferents can be separated into high- and low-spontaneously active classes which are further differentiated by their central projections (Fekete et al., 1984; Liberman, 1991; Liberman, 1993). Acoustic noise damage has a greater effect on afferents with high rates of spontaneous activity (Liberman and Dodds, 1984). Rajan (1998) has shown that noise-induced mild unilateral chronic hearing losses over well defined frequency ranges reveal a marked reduction of the monaural inhibitory component in the responses of cortical neurons. This was expressed as broadening of the receptive field (response area in frequency by level space) and an increase in the proportion of neurons with the unusual property of multiple-peaked frequency tuning curves. Both of these changes reflect losses in inhibitory components which normally contribute to the frequency tuning of cortical neurons, but which do not have the capacity to operate at near threshold levels. Hence the threshold-defined topographic representation is preserved. The effect on auditory cortical responses parallels that produced in the somatosensory cortex with peripheral-capsaicin-induced loss of C-fibre-mediated inhibition: normal sensitivity but expanded response areas.

In normal vision there is a constant output from all areas of the retina. However, adaptation is rapid, such that an area viewing a uniform colour and intensity for more than a few seconds can be considered as equivalent to a scotoma (Kuffler, 1953). Psychophysical investigation has shown that such adaptation can rapidly affect perception of objects close to the masked area (Kapadia et al., 1994). As a result, it could be argued that the visual-field-masking-induced receptive field expansion experiments discussed above (Fiorani et al., 1992; Pettet and Gilbert, 1992) depend upon such adaptation and the ensuing loss of drive to inhibitory circuits. It is therefore interesting to examine a neural network simulation of the Pettet and Gilbert experiment (receptive field expansion with optical masking paradigm) by Xing and Gerstein (1994). This was a scale-model visual cortex simulation with an input layer, an output layer and excitatory and inhibitory *hidden* layers equivalent to horizontal connections. Successful simulation depended upon giving the unmasked input layer spontaneous activity.

Work of the type reviewed in this chapter raises questions concerning the value of the concept of receptive fields and the topographic maps they form. A detailed analysis of the fine structure of the cortical somatotopic representations undertaken by Favorov and colleagues (Favorov et al. 1987; Favorov and Whitsel, 1988a,b; Favorov and Diamond, 1990) considered the relationship between the near-threshold-stimulation receptive field (minimal RF), the maximal area of effective stimulation using a neuron's optimal stimulus (maximal RF), and the topographic map. For a given cortical column, the minimal RF was interpreted as a measure of the central tendency of the distribution of neural response properties (firing rate, threshold, feature filtering) toward the common intersection of the individual maximal RFs. Thus, minimal RFs, and the threshold-field maps constructed from them, should be considered not as physiological measures of neural response properties in the spatial dimension but as statistical constructs. This is emphasized by the few studies which have examined topographies in terms of the spatial distribution of responses to a given stimulus rather than as a compilation of threshold-response properties. For example, Phillips and colleagues (1994) used multiple placement of an extracellular recording electrode throughout primary auditory cortex and found that, even for pure-tone stimulation, there was little relationship between the isofrequency contours of the threshold-field map and the activation contours formed by plotting the discharge rates of the recorded neurons, except at the lowest sound levels examined. Nevertheless, the feature detection filtering properties of neurons – of which the receptive field is the most fundamental – stands as a valid description for predicting neural discharge in most circumstances. The capacity for short-term plasticity can be thought of as extending the definition of the feature detection properties. As with the conclusion arrived at from many areas of sensory physiology, it is clear that activation of a given a neuron can not be interpreted as an indication that a particular stimulation (which satisfies the conditions of a simple receptive field) has occurred. How the brain interprets some of the plasticity inducible in topographies is not clear. Many arm amputees, when appropriately asked, describe secondary sensations referred to their phantom limb when they are stimulated on the cheeks, lips or chin (Ramachandran et al., 1992; Halligan et al., 1993), implying that, while the face representation may have expanded, the identity of the remapped cortex as 'hand' was not entirely lost.

ACKNOWLEDGEMENTS

Work in the author's laboratory was carried out with support from the National Health and Medical Research Council and the Australian Research Council.

REFERENCES

Albus, K., 1975, A quantitative study of the projection area of the central and the paracentral visual field in area 17 of the cat I, The precision of the topography, *Experimental Brain Research,* 24, 159-179.

Byrne, J. A., and Calford, M. B., 1991, Short-term expansion of receptive fields in rat primary somatosensory cortex after hindpaw digit denervation, *Brain Research,* 565, 218-224.

Calford, M. B., Rajan, R., and Irvine, D. R. F., 1993, Rapid changes in the frequency tuning of neurons in cat auditory cortex resulting from pure-tone-induced temporal threshold shift, *Neuroscience*, 55, 953-964.

Calford, M. B., Schmid, L. M., and Rosa, M. G. P., 1999, Monocular focal retinal lesions induce short-term topographic plasticity in adult visual cortex, *Proceedings of the Royal Society of London. Series B: Biological Sciences,* 266, 499-507.

Calford, M. B., and Semple, M. N., 1995, Monaural inhibition in cat auditory-cortex, *Journal of Neurophysiology,* 73, 1876-1891.

Calford, M. B., and Tweedale, R., 1988, Immediate and chronic changes in responses of somatosensory cortex in adult flying-fox after digit amputation, *Nature*, 332, 446-448.

Calford, M. B., and Tweedale, R., 1990, Interhemispheric transfer of plasticity in the cerebral cortex, *Science*, 249, 805-807.

Calford, M. B., and Tweedale, R., 1991a, Acute changes in cutaneous receptive fields in primary somatosensory cortex after digit denervation in adult flying fox, *Journal of Neurophysiology,* 65, 178-187.

Calford, M. B., and Tweedale, R., 1991b, Immediate expansion of receptive fields of neurons in area 3b of macaque monkeys after digit denervation, *Somatosensory and Motor Research*, 8, 249-260.

Calford, M. B., Wang, C., Taglianetti, V., Waleszczyk, W. J., Burke, W., and Dreher, B., 2000, Plasticity in adult cat visual cortex, area 17) following circumscribed monocular lesions of all retinal layers, *Journal of Physiology,* 524, 587-602.

Chapman, B., and Stone, L. S., 1996, Turning a blind eye to cortical receptive fields, *Neuron*, 16, 9-12.

Chu, Y., Humphrey, M. F., Alder, V. V., and Constable, I. J., 1998, Immunocytochemical localization of basic fibroblast growth factor and glial fibrillary acidic protein after laser photocoagulation in the Royal College of Surgeons rat, *Australian and New Zealand Journal of Ophthalmology.*, 26, 87-96.

Darian-Smith, C., and Gilbert, C. D., 1994, Axonal sprouting accompanies functional reorganization in adult cat striate cortex, *Nature*, 368, 737-40.

DeAngelis, G. C., Anzai, A., Ohzawa, I., and Freeman, R. D., 1995, Receptive field structure in the visual cortex: does selective stimulation induce plasticity? *Proceedings of the National Academy of Sciences of the United States of America,* 92, 9682-9686.

Doetsch, G. S., Harrison, T. A., MacDonald, A. C., and Litaker, M. S., 1996, Short-term plasticity in primary somatosensory cortex of the rat: rapid changes in magnitudes and latencies of neuronal responses following digit denervation, *Experimental Brain Research*, 112, 505-512.

Dreher, B., Burke, W., and Calford, M. B., 2001, Cortical plasticity revealed by circumscribed retinal lesions or artificial scotomas, *Progress in Brain Research*, 134, in press.

Dunn-Meynell, A. A., Benowitz, L. I., and Levin, B. E., 1992, Vibrissectomy induced changes in GAP-43 immunoreactivity in the adult rat barrel cortex, *Journal of Comparative Neurology*, 315, 160-170.

Dykes, R. W., 1990, Acetylcholine and neuronal plasticity in somatosensory cortex, in: *Brain Cholinergic Systems*, Steriade, M., Biesold, D., eds., Oxford University Press, New York, pp. 294-313.

Dykes, R. W., and Craig, A. D., 1998, Control of size and excitability of mechanosensory receptive fields in dorsal column nuclei by homolateral dorsal horn neurons, *Journal of Neurophysiology,* 80, 120-129.

Favorov, O., and Whitsel, B. L., 1988a, Spatial organization of the peripheral input to area 1 cell columns. I. the detection of 'segregates', *Brain Research Reviews.* 113, 25-42.

Favorov, O., and Whitsel, B. L., 1988b, Spatial organization of the peripheral input to area 1 cell columns. II. the forelimb representation achieved by a mosaic of segregates, *Brain Research Reviews*, 113, 43-56.

Favorov, O. V., and Diamond, M. E., 1990, Demonstration of discrete place-defined columns - segregates - in the cat SI, *Journal of Comparative Neurology*, 298, 97-112.

Favorov O. V., Diamond M. E., Whitsel B. L., 1987, Evidence for a mosaic representation of the body surface in area 3b of the somatic cortex of cat, *Proceedings of the National Academy of Sciences of the United States of America*, 84, 6606-6610.

Fekete, D. M., Rouiller, E. M., Liberman, M. C., and Ryugo, D. K., 1984, The central projections of intracellularly labeled auditory nerve fibers in cats, *Journal of Comparative Neurology*, 229, 432-450.

Fiorani, M., Rosa, M. G. P., Gattass, R., and Rocha-Miranda, C. E., 1992, Dynamic surrounds of receptive fields in primate striate cortex: a physiological basis for perceptual completion? *Proceedings of the National Academy of Sciences of the United States of America*, 89, 8547-8551.

Fyffe, R. E., Cheema, S. S., and Rustioni, A., 1986, Intracellular staining study of the feline cuneate nucleus. I. Terminal patterns of primary afferent fibers, *Journal of Neurophysiology,* 56, 1268-1283.

Garraghty, P. E., and Muja, N., 1996, NMDA receptors and plasticity in adult primate somatosensory cortex, *Journal of Comparative Neurology*, 367, 319-326.

Glazewski, S., and Fox, K., 1996, Time course of experience-dependent synaptic potentiation and depression in barrel cortex of adolescent rats, *Journal of Neurophysiology,* 75, 1714-1729.

Halligan, P. W., Marshall, J. C., Wade, D. T., Davey, J., and Morrison, D., 1993, Thumb in cheek? Sensory reorganization and perceptual plasticity after limb amputation, *Neuroreport*, 4, 233-236.

Henry, G. H., Bishop, P. O., and Coombs, J. S., 1969, Inhibitory and sub-liminal excitatory receptive fields of simple units in cat striate cortex, *Vision Research*, 9, 1289-1296.

Humphrey, M. F., Chu, Y., Mann, K., and Rakoczy, P., 1997, Retinal GFAP and bFGF expression after multiple argon laser photocoagulation injuries assessed by both immunoreactivity and mRNA levels, *Experimental Eye Research*, 64, 361-369.

Kano, M., Lino, K., and Kano, M., 1991, Functional reorganization of adult cat somatosensory cortex is dependent upon NMDA receptors, *Neuroreport*, 2, 77-80.

Kapadia, M. K., Gilbert, C. D., and Westheimer, G., 1994, A quantitative measure for short-term cortical plasticity in human vision, *Journal of Neuroscience*, 14, 451-457.

Katz, D. B., Simon, S. A., Moody, A., and Nicolelis, M. A. L., 1999, Simultaneous reorganization in thalamocortical ensembles evolves over several hours after perioral capsaicin injections, *Journal of Neurophysiology,* 82, 963-977.

Kolarik, R. C., Rasey, S. K., and Wall, J. T., 1994, The consistency, extent, and locations of early-onset changes in cortical nerve dominance aggregates following injury of nerves to primate hands, *Journal Neuroscience*, 14, 4269-4288.

Kossut, M., and Juliano, S. L., 1999, Anatomical correlates of representational map reorganization induced by partial vibrissectomy in the barrel cortex of adult mice, *Neuroscience*, 92, 807-817.

Kuffler S. W., 1953, Discharge patterns and functional organization of the mammalian retina, *Journal of Neurophysiology,* 16, 37-68.

Laskawi, R., Rohlmann, A., Landgrebe, M., and Wolff, J. R., 1997, Rapid astroglial reactions in the motor cortex of adult rats following peripheral facial nerve lesions, *European Archives of Oto-Rhino-Laryngology*, 254, 81-85.

Laskin, S. E., and Spencer, W. A., 1979, Cutaneous masking. II. Geometry of excitatory and inhibitory receptive fields of single units in somatosensory cortex of the cat, *Journal of Neurophysiology*, 42, 1061-1082.

Liberman, M. C., 1991, Central projections of auditory-nerve fibers of differing spontaneous rate. I. Anteroventral cochlear nucleus, *Journal Comparative Neurology*, 313, 240-258.

Liberman, M. C , 1993, Central projections of auditory nerve fibers of differing spontaneous rate, II: Posteroventral and dorsal cochlear nuclei, *Journal of Comparative Neurology*, 327, 17-36.

Liberman, M. C., and Dodds, L. W., 1984, Single-neuron labeling and chronic cochlear pathology. II. Stereocilia damage and alterations of spontaneous discharge rates, *Hearing Research*, 16, 43-53.

Melzack, R., 1990, Phantom limbs and the concept of a neuromatrix, *Trends in Neuroscience*, 13, 88-92.

Merzenich, M. M., Kaas, J. H., Wall, J. T., Sur, M., Nelson, R. J., and Felleman, D. J., 1983, Progression of change following median nerve section in the cortical representation of the hand in areas 3b and 1 in adult owl and squirrel monkeys, *Neuroscience*, 10, 639-665.

Michaelis, M., Liu, X., and Jänig, W., 2000, Axotomized and intact muscle afferents but no skin afferents develop ongoing discharges of dorsal root ganglion origin after peripheral nerve lesion, *Journal of Neuroscience*, 20, 2742-2748.

Mountcastle, V. B., and Powell, T. P., 1959, Neural mechanisms subserving cutaneous sensibility, with special reference to the role of afferent inhibition in sensory perception and discrimination, *Bulletin of the Johns Hopkins Hospital*, 105, 201-232.

Nussbaumer, J. C., and Wall, P. D., 1985, Expansion of receptive fields in the mouse cortical barrelfield after administration of capsaicin to neonates or local application on the infraorbital nerve in adults, *Brain Research*, 360, 1-9.

Pettet, M. W., and Gilbert, C. D., 1992, Dynamic changes in receptive-field size in cat primary visual cortex, *Proceedings of the National Academy of Sciences of the United States of America*, 89, 8366-8379.

Pettit, M. J., and Schwark, H. D., 1996, Capsaicin-induced rapid receptive field reorganization in cuneate neurons, *Journal of Neurophysiology*, 75, 1117-1125.

Phillips D. P., Semple M. N., Calford M. B., Kitzes L. M., 1994, Level-dependent representation of stimulus frequency in cat primary auditory cortex, *Experimental Brain Research*, 102, 210-226.

Rajan, R., 1998, Receptor organ damage causes loss of cortical surround inhibition without topographic map plasticity, *Nature Neuroscience*, 1, 138-143.

Ramachandran, V. S., Stewart, M., and Rogers-Ramachandran, D. C., 1992, Perceptual correlates of massive cortical reorganization, *Neuroreport*, 3, 583-586.

Rasmusson, D. D., Louw, D. F., and Northgrave, S. A., 1993, The immediate effects of peripheral denervation on inhibitory mechanisms in the somatosensory thalamus, *Somatosensory and Motor Research*, 10, 69-80.

Schmid, L. M., Rosa, M. G. P., and Calford, M. B., 1995, Retinal-detachment induces massive immediate reorganization in visual cortex, *Neuroreport*, 6, 1349-1353.

Schmid, L. M., Rosa, M. G. P., Calford, M. B., and Ambler, J. S., 1996, Visuotopic reorganization in the primary visual cortex of adult cats following monocular and binocular retinal lesions, *Cerebral Cortex*, 6, 388-405.

Smits, E., Gordon, D. C., Witte, S., Rasmusson, D. D., and Zarzecki, P., 1991, Synaptic potentials evoked by convergent somatosensory and corticocortical inputs in raccoon somatosensory cortex: substrates for plasticity, *Journal of Neurophysiology*, 66, 688-695.

Snyder, R. L., and Sinex, D., 1998, Tonotopic reorganization of cat primary auditory cortex, AI) after acute lesions of restricted sectors of the spiral ganglion, *Society for Neurosciences Abstract*, 24, 904.

Snyder, R. L., Sinex, D. G., McGee, J. D., and Walsh, E. W., 2000, Acute spiral ganglion lesions change the tuning and tonotopic organization of cat inferior colliculus neurons, *Hearing Research*, 147, 200-220.

Tassignon, M. J., Stempels, N., Nguyen, L. J., De Wilde, F., and Brihaye, M., 1991, The effect of wavelength on glial fibrillary acidic protein immunoreactivity in laser-induced lesions in rabbit retina, *Graefes Archive for Clinical and Experimental Ophthalmolog*, 229, 380-388.

Turnbull, B. G., and Rasmusson, D. D., 1991, Chronic effects of total or partial digit denervation on raccoon somatosensory cortex, *Somatosensory and Motor Research*, 8, 201-213.

Wall J. T., Cusick C. G., 1986, The representation of peripheral nerve inputs in the S-1 hindpaw cortex of rats raised with incompletely innervated hindpaws, *Journal of Neuroscience*, 6, 1129-1147.

Webster, H. H., Hanisch, U. K., Dykes, R. W., and Biesold, D., 1991a, Basal forebrain lesions with or without reserpine injection inhibit cortical reorganization in rat hindpaw primary somatosensory cortex following sciatic nerve section, *Somatosensory and Motor Research*, 8, 327-346.

Webster, H. H., Rasmusson, D. D., Dykes, R. W., Schliebs, R., Schober, W., Bruckner, G., and Biesold, D., 1991b, Long-term enhancement of evoked potentials in raccoon somatosensory cortex following co-activation of the nucleus basalis of Meynert complex and cutaneous receptors, *Brain Research*, 545, 292-296.

Xing, J., and Gerstein, G. L., 1994, Simulation of dynamic receptive-fields in primary visual-cortex, *Vision Research*, 34, 1901-1911.
Yamamoto, C., Ogata, N., Yi, X., Takahashi, K., Miyashiro, M., Yamada, H., Uyama, M., and Matsuzaki, K., 1996, Immunolocalization of basic fibroblast growth factor during wound repair in rat retina after laser photocoagulation, *Graefes Archive for Clinical and Experimental Ophthalmology*, 234, 695-702.

52

VISION AS MOTIVATION: INTERHEMISPHERIC OSCILLATION ALTERS PERCEPTION

Jack D. Pettigrew and Olivia Carter*

ABSTRACT

The well-recognised proclivity of primates for things visual can sometimes mislead us into thinking that a high order visual process is taking place at a much earlier level of processing than it is. We illustrate this with a new illusion, Bonneh's "motion-induced-blindness", that involves the intermittent disappearance and reappearance of simple bright targets under the influence of a moving background. This perceptual oscillation is strongly affected by the mood and impulsivity of the subject, in line with other evidence that this simple disappearance and reappearance must be taking place at a high level of processing, perhaps at the level of the cerebral hemsipheres themselves.

INTRODUCTION

In preparing for this meeting, we were conscious of the heavy predominance of motor systems participants as opposed to those with a "more" sensory background. We therefore present some work on the visual system that reverses the usual "bottom-up" approach and instead looks at a visual phenomenon that can only be understood in a "top-down" context. This reversal of thinking is becoming fashionable again, as in the continuing debates about whether the physiology of the primary visual cortex can explain aspects of visual perception or whether it is necessary to resort to "higher" levels of analysis. Even phenomena as basic as negative after-images, which are given as examples of retinal processes in text books, have recently been shown to have a high-level central component (Shimojo et al., 2001). Hochstein and colleagues (e.g. Ahissar and Hochstein, 1997) are also exponents of the emerging view that late visual processes may intuitively seem like an early processes (such as a brightness estimation, or the blinking of simple target on and off, as in the example we present below). Apparently more complex processes may be much earlier. The visual phenomenon we present seems to be

* Vision Touch and Hearing Research Centre, School of Biomedical Sciences, University of Queensland, 4072, Australia. Email: j.pettigrew@vthrc.uq.edu.au

Sensorimotor Control of Movement and Posture
Edited by Gandevia *et al.*, Kluwer Academic/Plenum Publishers, 2002

connected to the overall motivation of the subject, even though it concerns the simple appearance and disappearance of a bright target.

The phenomenon in question is Bonneh's illusion, which he and his colleagues call Motion-induced-blindness (Bonneh et al., 2001). Three, stationary, bright yellow discs are presented against a swirling cloud of smaller blue dots. As bright and clear as the yellow discs are, they disappear and reappear in a quasi-regular but unpredictable cycle. This illusion was presented at the meeting and it can be viewed on the web at http://www.uq.edu.au/nuq/jack/rivalry.html or on Bonneh's site http://www.weizmann.ac.il/~masagi/MIB/mib.html. When it was first presented, some reacted angrily, claiming that Bonneh had switches under the table that he was using to turn off the yellow discs! Although there are a number of related illusions such as Troxler fading (Troxler, 1804) and the visual disappearance phenomenon reported by Ramachandran and colleagues (1993), Bonneh and colleagues deserve some credit for taking this one to the point where all doubts can be eliminated over the possibility of its early origin. Because the disappearance of a yellow disc, in relation to its fellow discs, follows complex Gestalt rules of grouping, closure, good continuation etc, it must be occurring at a high level of analysis where such forms of pattern recognition can take place, and not at early levels in the visual system. There are many experimental examples on Bonneh's website to illustrate the influence of Gestalt principles. Even a brief look makes it easy to convince oneself that this is a high level phenomenon that "knows" a lot about image grouping and segmentation based on form and colour. We will give one example here: If each disc is filled with parallel lines, its disappearance becomes subject to Gestalt rules of grouping by orientation; e.g. discs filled with lines of the same orientation disappear together, whereas discs oriented with their respective lines aligned in an orthogonal manner disappear independently. In other experiments one can show similar effects of grouping by colour or proximity.

We emphasise that the effect is taking place at a very high level by taking a further step. We show that the Bonneh illusion has all the hallmarks of perceptual rivalry (even though the rival alternatives are not obvious from inspection of the pattern before a disappearance takes place). In view of the raging controversy about whether binocular rivalry takes place in the primary visual cortex (V1) or much higher in the visual pathway, it might be argued that aligning the Bonneh illusion with the perceptual rivalries in general, does not constitute a compelling argument that Bonneh's illusion is "high". We will ignore this argument in view of the fact that V1 is already a relatively high level and, like perceptual rivalry, there is mounting evidence that Bonneh's illusion reflects interhemispheric oscillations (Pettigrew and Funk, 2001). Moreover, in keeping with the proposed interhemispheric involvement, the precise details of the illusion are influenced by two factors that show hemispheric interaction, an individual's mood (Heller, 1993; Robinson, 1995; Baker et al., 1997) and personality (Caplan and Shechter, 1990; Sutton and Davidson, 1997). In particular, the proportion of time spent with one of the yellow discs disappeared is directly related both to the degree to which the observer has a positive mood and an impulsive nature, with euphoric and impulsive individuals reporting the highest proportion of time spent with a disc absent. In contrast, no disappearance at all was experienced by a highly stressed individual. This surprising finding emphasises the general point that apparently early visual processes may be influenced at the highest level. The result can be explained in terms of the growing evidence for hemispheric involvement in both rivalry and mood, given that the perceptual oscillations of Bonneh's illusion are a form of rivalry (Carter and Pettigrew, 2002). We discuss this result in the light of the oscillatory nature of Bonneh's illusion, with particular emphasis on the possible role of neural oscillations in general.

Methods

Twenty-nine subjects with normal or corrected to normal vision, were presented with Bonneh's illusion. They were unaware of the object of the experiment and were naïve with regard to the illusion itself. The illusion consisted of three stationary yellow dots and a fixation cross, overlaid on a global moving pattern of 150 blue dots set on a black background. The yellow dots subtended 0.5° of visual angle arranged along a 4° radius circle forming a triangular arrangement around a yellow fixation cross (0.5°) situated in the middle of the display. Subjects were instructed to fixate on the cross while attending to the yellow dots and report whether (i) they could see all of the yellow dots in the display or (ii) if any of the dots had disappeared. Responses were indicated *via* key-press on a standard computer keyboard. Data was collected over a 10 minute period divided into 4x100 second trials, with a 30 second break between each trial. The display was presented to subjects on a standard Macintosh (iMac) computer monitor, in a dimly lit, quiet room and viewed from a distance of 60 cm.

To assess an individuals mood state, we used a rating system developed by Watson and colleagues (1988), which has been shown to be a reliable, valid and efficient means of measuring positive affect (Watson et al., 1988). Prior to the presentation of the MIB stimulus, each subject was asked to rate ten adjectives *(interested, distressed, excited, upset, strong, guilty, scared, hostile, enthusiastic* and *proud)* in relation to how they felt "right now, that is at this moment." The adjectives were rated from 1 to 5 where 1 = Very slight or not at all, 2 = A little, 3 = Moderately, 4 = Quite a bit and 5 = Very much. The ratings were then summed to give a score between 10 and 50.

As a link has been shown between positive affect and impulsivity (Gray, 1987), twenty-one subjects were additionally administered the Eysenck impulsivity sub-scale (Eysenck, 1973). Subjects were asked to answer "Yes" or "No" to nine short questions such as "Do you often do things on the spur of the moment?" Each response corresponding to an impulsive tendency was given a score of 1, from which an overall impulsivity score out of 9 was calculated for each subject.

Results

The proportion of disappearance experienced in Bonneh's illusion, calculated as the percentage of the total testing time, that any of the dots were reported to have disappeared, was found to increase with an increase in the subject's self-reported level of positive affectivity ($R = 0.46$). Through an analysis of variance, this effect was found to be significant ($F_{(1,27)}=7.30$, $p = 0.012$). An even greater effect on the proportion of disappearance was observed with the degree to which the subject reported impulsive tendencies ($R = 0.62$). Again this effect was found to be significant ($F_{(1,19)}=13.04$, $p < 0.01$).

A)

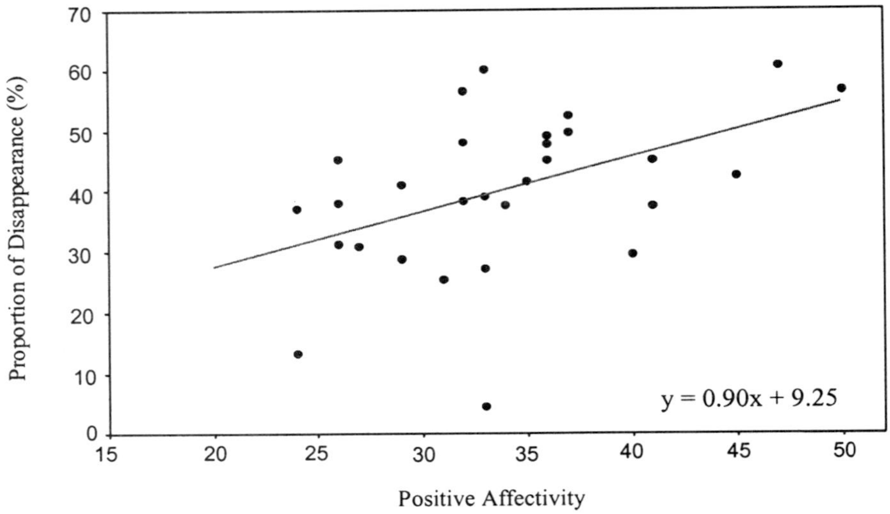

B)

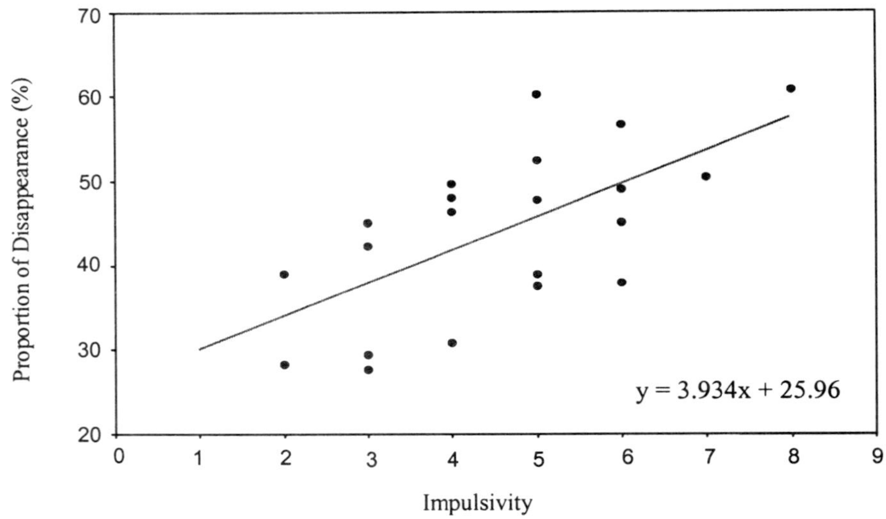

Figure 1. Proportion of disappearance induced by Bonneh's illusion in relation to (A) positive affect measured by the PANAS (Watson et al., 1988) and (B) impulsivity measure by Eysenck's impulsivity sub-scale (Eysenck, 1973). Note the surprising correlation between these measures of mood and personality and the details of Bonneh's illusion experiences by each individual. A high index of impulsivity or positive affect is associated with a greater illusion, with the yellow discs disappearing for a greater proportion of time. This result is consistent with our hypothesis that the basis of this perceptual oscillation is an interhemispheric oscillation, with the left hemisphere responsible for the disappearance phase and hence for the longer disappearance phase in subjects with high positive affect and impulsivity, both of which are left hemisphere attributes.

Discussion

Bonneh's illusion is a perceptual oscillation that shares many similarities with more familiar perceptual rivalries such as binocular rivalry, the Necker cube, Shroeder's staircase etc. There is much debate about the origin of these rivalries and whether it is appropriate to place them under the same rubric (e.g. Andrews and Purves, 1997) or whether they represent completely independent phenomena with separate neural mechanisms and loci (Tong, 2001). We take the former view, arguing that the timing of these oscillations at least share common elements that explain the remarkable coupling of the rhythms of different perceptual oscillations (Carter and Pettigrew, 2002). The current controversy surrounding the location and exact nature of the neuronal interactions responsible for perceptual rivalry is too extensive to be covered here, so the interested reader is referred to recent position papers on perceptual rivalries by Blake (2001), Tong (2001) and Pettigrew (2001). In this discussion we focus on the remarkable fact that a basic visual event, the presence or absence of a yellow disc, is dependent upon the mood state and personality type of the observer! While mood and visual perception are intuitively unrelated, hemispheric asymmetries, like those implicated in perceptual rivalries (Miller et al., 2000), have been associated with positive and negative mood states (Heller, 1993). Dominant activation of anterior regions of the left hemisphere have been linked to positive emotions (Baker et al., 1997; Grisaru et al., 1998; Robinson, 1995) while dominant activation of the right hemisphere appears to correspond to negative emotions (Martinot et al., 1990; Pascual-Leone et al., 1996; Soares and Mann, 1997). If one accepts that perceptual oscillations are linked to oscillations of the hemispheres, it is therefore possible to see that perception itself will be mood-related, an outcome that might be more familiar to artists than to vision scientists, but which is nevertheless robustly supported by our results. If we accept for the moment that perceptual oscillations can be linked to mood *via* interhemispheric oscillation (skeptical readers are referred to the position papers on rivalry cited above), then a burning question concerns the oscillations themselves. Why does perception oscillate?

Oscillations are a well-recognised feature of motor systems, if only because of the common locomotor requirement for rhythmic alternation of antagonists that is seen in both vertebrates and invertebrates. But antagonistic muscle action is not a satisfactory basis for the explanation of many neuronal oscillations. Even single cells oscillate, as in chemotaxis, where bacteria have a clock-like alternation (every 1.5 sec) between the opposite directions of rotation of the flagellar bundle that produce, respectively, forward motion ("go") and tumbling ("stop"). In the case of the phenomenon that we have just described, interhemispheric oscillation, and its impact on mood and visual perception, the role of the oscillation is mysterious and generally elicits a strident "why?" along with the general discomfort that accompanies new, counterintuitive viewpoints. We would like to suggest that the oscillation in perception reflects high-level perceptual decision making. This would place visual perception closer to motor output than is generally recognized, but is consonant with emerging evidence that what may seem superficially like an early process is actually late (Ahissar and Hochstein, 1997). Brightness estimations, for example, are usually thought of as "early vision" with explanations based in retinal terms, whereas the surprising new illusions of Dale Purves resurrect Helmholtz' notion of "unconscious inference" where complex deductions precede apparently simple estimations such as that of brightness (Lotto and Purves, 1999; Williams et al., 1998). The key problem emphasised by Purves is that visual inputs are inherently ambiguous and therefore require additional information to help resolve ambiguities, such as the relative contributions of the illuminant and the intrinsic brightness of the object. In the Bonneh illusion, we think that this additional information is provided by the coordinated activity of the two hemispheres. Each hemisphere adopts a different cognitive style, with regular oscillations between the hemispheres leading to alternations between the two styles. The pattern seeking left hemisphere, in one interpretation, rejects the stationary yellow disk on the basis of its discrepancy, a form of concrete "denial" in keeping with Ramachandran's (1994, 1995) formulation of the cognitive style of the left hemisphere. An alternative view is based on the superiority of the left hemisphere in "form-from-motion" tasks that would emphasise the moving cloud of blue dots. Both these interpretations are quite compatible with each other and with the Purves thesis that global empirical information is brought to bear to make perceptual decisions in the face of ambiguity. Our addition to this thesis is that the coordinated activity of a whole hemisphere with a particular cognitive style could incorporate a number of related viewpoints of this kind and relay the sum to the decision-making apparatus. In the case of the right hemisphere, the style is veridical with discrepancies highlighted for investigation instead of being denied for the sake of a particular interpretation. An oscillation between right and left would ensure that neither of the complementary styles of the two hemispheres would prevail indefinitely. In this case one can see that the oscillation might also be connected to the resolution of the ambiguity (as in a variety of other areas that deal with ambiguity such as quantum physics and Taoist philosophy). Other explanations for the oscillation are possible.

We would therefore like to propose a number of possible roles for neuronal oscillation. These are not mutually exclusive, as evolutionary stable strategies, of which oscillations are likely to be one, are often syndromes with many related features. Nor would we suggest that they are exhaustive, as we would welcome any further suggestions as to the role of oscillation. As a general principle, we think it important to keep in mind that current information about the encoding of timing within the nervous system is very sketchy compared with the information about the spatial location of different functions within the brain.

1. Coupled, Nested Clocks: Richard Feynman Thesis

In the years before his untimely death, Richard Feynman was very concerned about how the brain represented time and visited JDP's lab at Caltech to discuss this problem. Feynman asked whether it was possible that the brain had a central clock that was responsible for ordering all processes in time, in the same way that modern computers have a central timekeeper. The answer to Feynman's question is still uncertain nearly thirty years later, but much more has been learned about neural clocks and pacemaker neurons, with intriguing evidence from flies, worms and humans that neural oscillators with vastly different scales, from milliseconds to hours, are coupled, in keeping with Feynman's notion of a master oscillator. According to this thesis, neural oscillations would be pervasive manifestations of timekeeping operations including perception and we would expect robustness and high heritability of the period for perceptual oscillation in an individual, as is observed.

2. Adaptation to Ambiguity: Dale Purves Empirical Vision Thesis

In an exciting new series of experiments that extends earlier work on visual illusions, Purves and his colleagues have elaborated a new thesis on the essential ambiguity of visual objects. It is impossible to decide, on the basis of the light emanating from an object, how much of this light is a function of the illuminant conditions and how much is a function of the intrinsic colour properties and reflectance of the object. Resolving the ambiguity must involve extra information apart from the patch of light reaching the retina from the object. This inescapable ambiguity requires that the brain resort to all sorts of empirical data on the probabilities of different illuminant and object conditions. Taking this thesis on ambiguity a step further, we propose that ambiguity may require that perception have an inbuilt oscillation, to avoid getting "stuck." This proposal is supported by new experiments that combine the Purves thesis about ambiguity with the prediction that they would be accompanied by perceptual oscillation (Carter and Pettigrew, unpublished). One important aspect of these perceptual oscillations is that one phase may be unconscious (e.g. the disappearance, or "denial" phase in the Bonneh illusion) and so they may not be obvious to inspection without special manoeuvres to reveal their presence.

3. Phylogenetic Extension of a CPG Mechanism

Rhythmic alternations between antagonistic muscle groups and opposite sides of the body are familiar to students of the motor system who are studying central pattern generators. Perhaps at high levels of processing in the CNS there is a duplication of these interhemispheric oscillators for increasingly more rostral (and therefore more metaphorical) functions. Alternation between complementary cognitive styles in the cerebral hemispheres may thus be a phylogenetic extension of earlier patterns of neural oscillation for opposing muscle groups.

4. Binding Mechanism: Wolf Singer Thesis

Much has been written about this alternative thesis to explain the ubiquitous occurrence of oscillation in the nervous system. Dispersed neurons that are all part of a combined percept are thought to be linked (the "binding" phenomenon) by synchronous oscillations. This explanation for 40 Hz oscillations in the visual system has received some support from experiments on cats, monkeys, and even humans, but not from some monkey experiments where one would have expected it.

5. Improved Predictability and Precision of Timing: "Chorus Girl Problem"

When a line of chorus girls achieves perfect synchrony of their kicks, a feed-forward strategy involving internal oscillators is a more likely explanation than feedback from one chorus girl to another which would introduce small phase delays associated with latencies and reaction time. Instead of relying on moment-to-moment communication between chorus girls, each chorus girl could have a tunable internal oscillator whose phase and frequency could be set in relation to external cues. An array of oscillators, the line of chorus girls, could then be set up with perfect synchrony.

CONCLUSION

The blatant perceptual oscillations of Bonneh's Illusion are somewhat unnerving considering our apparently stable view of the world. We think that the disturbing quality comes from the unique awareness that the illusion gives us into the inherently oscillatory nature of visual perception. This is in contrast to more classical examples of perceptual rivalry such as ambiguous figures where such oscillations can be easily rationalised away. The disappearance of bright targets also has a second feature that is interesting:- viz. It seems connected with a process akin to denial which eliminates component(s) of sensory information, if that component is discrepant with the individual's current belief system. If one accepts our interhemispheric interpretation for the moment, with the left hemisphere involved in a process like denial that ignores discrepant stimuli, one can see how visual perception can be influenced by seemingly irrelevant variables such as the mood or personality type of the observer. In this way our everyday experience of perceptual stability and objectivity may be even more illusory than the great mystics would have us believe. One phase of the perceptual oscillation, involving a denial-like process in the left hemisphere, that may be quite unconscious of the discrepancy with the other phase of the oscillation! Exploring this new view of perception should be an interesting pursuit.

ACKNOWLEDGEMENT

This work was supported by the Stanley Foundation and the National Health and Medical Research Council of Australia.

REFERENCES

Ahissar, M., and Hochstein, S., 1997, Task difficulty and the specificity of perceptual learning, Nature, 387, 401-406.

Andrews, T. J., and Purves, D., 1997, Similarities in normal and binocularly rivalrous viewing, *Proceedings of the National Academy of Sciences of the United States of America,* 94, 9905-9908.

Baker, S. C., Frith, C. D., and Dolan, R. J., 1997, The interaction between mood and cognitive function studied with PET, *Psychological Medicine,* 27, 565-578.

Blake, R., 2001, A primer on binocular rivalry, including current contraversies, *Brain and Mind* 2, 5-38.

Bonneh, Y., Cooperman, A., and Sagi, D., 2001, Motion induced blindness in normal observers, *Nature,* 411, 798-801.

Caplan, B., and Shechter, J., 1990, Clinical applications of the Matching Familiar Figures Test: impulsivity vs. unilateral neglect, *Journal of Clinical Psychology,* 46, 60-67.

Carter, O. L., and Pettigrew, J. D., 2002, Motion induced blindness is a form of perceptual rivalry with implications for bipolar disorder, *Australian Society for Neuroscience, Abstract,* in press.

Eysenck, H. J., 1973, *Eysenck on extraversion*, Crosby Lockwood Staples, London.

Gray, J. A., 1987, Perspectives on anxiety and impulsivity: A commentary, *Journal of Research in Personality,* 21, 493-509.

Grisaru, N., Chudakov, B., Yaroslavsky, Y., and Belmaker, R. H., 1998, Transcranial magnetic stimulation in mania: a controlled study, *American Journal of Psychiatry,* 155, 1608-1610.

Heller, W., 1993, Neuropsychological mechanisms of individual differences in emotion, personality and arousal, *Neuropsychology,* 7, 476.

Lotto, R. B., and Purves, D., 1999, The effects of color on brightness, *Nature Neuroscience,* 2(11), 1010-4.

Martinot, J. L., Hardy, P., Feline, A., Huret, J. D., Mazoyer, B., Attar-Levy, D., Pappata, S., and Syrota, A., 1990, Left prefrontal glucose hypometabolism in the depressed state: a confirmation, *American Journal of Psychiatry,* 147(10), 1313-1317.

Miller, S. M., Liu, G. B., Ngo, T. T., Hooper, G., Riek, S., Carson, R. G., and Pettigrew, J. D., 2000, Interhemispheric switching mediates perceptual rivalry, *Current Biology,* 10, 383-392.

Pascual-Leone, A., Rubio, B., Pallardo, F., and Catala, M. D., 1996, Rapid-rate transcranial magnetic stimulation of left dorsolateral prefrontal cortex in drug-resistant depression, *Lancet,* 348, 233-237.

Pettigrew, J. D., 2001, Searching for the switch: Neural bases for perceptual rivalry alternations, *Brain and Mind,* 2, 85-118.

Pettigrew, J. D., and Funk, A. P., 2001, Opposing effects on perceptual rivalry caused by right vs left TMS, *Society for Neuroscience Abstracts.*

Ramachandran, V. S., 1995, Anosognosia in parietal lobe syndrome, *Consciousness and Cognition,* 4, 22-51.

Ramachandran, V. S., 1994, Phantom limbs, neglect syndromes, repressed memories, and Freudian psychology, *International Review of Neurobiology,* 37, 291-333.

Ramachandran, V. S, Gregory, R. L., and Aiken, W., 1993, Perceptual fading of visual texture borders, *Vision Research,* 33, 717-721.

Robinson, R. G., 1995, Mapping brain activity associated with emotion, *American Journal of Psychiatry,* 152, 327-329.

Shimojo, S., Kamitani, Y., and Nishida, S., 2001, Afterimage of perceptually filled-in surface, *Science,* 293, 1677-1680.

Soares, J. C., and Mann, J. J., 1997, The Anatomy of Mood Disorders - Review of Structural Neuroimaging Studies, *Biological Psychiatry,* 41, 86-106.

Sutton, S. K., and Davidson, R. J., 1997, Prefrontal Brain Asymmetry: A biological substrate of the behavioral approach and inhibition system, *Psychological Science,* 8, 204-210.

Tong, F., 2001, Competing Theories of Binocular Rivalry: A possible Resolution, *Brain and Mind,* 2, 55-83.

Troxler, D., 1804, In *Opthalmologische Bibliothek,* Himly, K., and Schmidt, J. A., eds., Fromman, Jena, pp. 51-53.

Watson, D., Clark, L. A., and Tellegen, A., 1988, Development and validation of brief measures of positive and negative affect: the PANAS scales, *Journal of Personality and Social Psychology,* 54, 1063-1070.

Williams, S. M., McCoy, A. N., and Purves, D., 1998, The influence of depicted illumination on brightness, *Proceedings of the National Academy of Sciences of the United States of America,* 95, 13296-13300.

SECTION VIII

Mechanics and Movement

Not all symposia on sensori-motor control include consideration of muscle mechanics and its integration into models of motor behaviour. A common view is that most of the fundamental issues relating to muscle properties were resolved long ago. T.R. Nichols (Chapter 53) points to resurgent interest in recent years in the musculoskeletal system, in the search for a predictive model of motor behaviour. An obvious reason is that biomechanics gives meaning to neuronal mechanism in an expression of the final response, action or inaction. There are also other emerging influences, such as new imaging methods that allow visualisation of motion of muscle fascicles and tendons during motor tasks. This comes against a background of recognition that several basic issues in muscle mechanics still need to be resolved (see Proske and Morgan, 1999).

Nichols emphasises the co-operative interaction between neuronal and biomechanical components of the motor system, citing as an example the combination of muscle spindle non-linearity and extrafusal properties to extend the range of linear behaviour of the muscle in stiffness regulation. He concludes that three issues to be addressed in the future are, (i) the mechanical variables that are encoded by the nervous system, (ii) the degrees of freedom available with respect to, for example, flexibility in recruitment order of motor units and (iii) redundancy, in terms of the number of joint positions defined by the skeleton.

G.E. Loeb (Chapter 54) reminds us that one important goal in striving to improve our understanding of the musculoskeletal system is to apply the knowledge in a rehabilitation setting. The miniaturisation of computer components, and sophistication of available software, means that portable controller systems can now be considered in FES (functional electrical stimulation) applications. Considerations for modelling of the musculoskeletal system include prioritisation of components, a computational platform and successful application when interfaced with real biomechanical variables. Two important components of biomechanical models that had been overlooked or ignored were co-operativity in both cross-bridge action and Ca^{2+} binding at the sarcomere level, and changes in size and direction of moment arms during flexion-extension movements. Multi-component models can now be generated, making use of some recently available commercial software. These can be incorporated into a controller which acted through micro-stimulators and feed-back sensors injected into paralysed muscle.

D.L. Morgan (Chapter 55) raises questions based on the proposition that a muscle can be considered a scaled sarcomere plus tendinous attachments. More sarcomeres in series increases shortening velocity but is energetically expensive. More sarcomeres in parallel increase force output and is similarly expensive. One further consideration is that for muscles which routinely undergo eccentric contractions, their working range would need to avoid the descending limb of the length-tension relation, the region of sarcomere instability and damage (Morgan, 1990). So prime movers, which typically undergo shortening contractions need fewer sarcomeres in series than postural muscles that regularly undergo eccentric contractions. The lower limit in numbers of sarcomeres in series would be set by the need to maintain a level of active tension over the muscle's full working range.

Morgan emphasises the ability of muscle to adapt to changes in activity patterns. So rats which had been trained to run downhill, had larger numbers of sarcomeres in series in their knee extensor muscles than rats trained to run uphill. It is concluded that muscles are plastic structures, constantly adjusting their properties according to the needs of the moment.

P. Sheard (Chapter 56) reminds the reader that what happens in the muscle is just as important for the final outcome as what happened in the brain. It is now clear that the traditional, simplified view of force transmission in muscle was no longer tenable. This applies particularly to muscles in which fibres are arranged in series with one another. Furthermore, the existence of lateral attachments means that fibres would not be able to undergo length changes independently of one another. It means that the motor unit may not always be the functionally single unit. It may be a group of units so placed to maximise force transmission. Also the effective series compliance faced by a contracting motor unit will depend on its connections with adjacent inactive fibres. Is the longer optimum length for active tension of some units, a reflection of their greater series compliance? (See Chapter 55)

The power of the new imaging methods is demonstrated in a presentation by T. Fukunaga (Chapter 57). Muscle fibre lengths in human gastrocnemius were visualised with ultrasound, while ground reaction force, EMG and ankle angle were measured. Subjects were asked to execute a jump from a standing start or preceded by a dorsiflexion (counter movement). It is shown that after muscle fibres had been passively lengthened at the onset of dorsiflexion, they contract isometrically, lengthening the series elastic element. The catapult action of the recoiling series elastic element plus the force of the actively shortening fibres provide more push, leading to a greater jump height with a countermovement than without it.

A. De Troyer (Chapter 58) describes the distribution of airway opening pressures obtained by calculation and the distribution of EMG recorded in intercostal muscles during resting breathing. It leads to the conclusion that the central nervous system is able to selectively recruit the intercostals with greater mechanical advantage for inspiration and expiration. This appears to be regulated centrally since deafferentation did not change the pattern. It raises questions about how such motor programs are generated, are they examples of an adaptive controller principle and what is their significance for lung inflation?

REFERENCES

Morgan, D. L., 1990, New insights into the behaviour of muscle during active lengthening, *Biophysics Journal,* 57, 209-221.

Proske, U., and Morgan, D. L., 1999, Do cross-bridges contribute to the tension during stretch of passive muscle? *Journal of Muscle Research and Cell Motility,* 20, 433-442.

53

MUSCULOSKELETAL MECHANICS: A FOUNDATION OF MOTOR PHYSIOLOGY

T. Richard Nichols*

ABSTRACT

The design of the musculoskeletal system has always been a major consideration in the interpretation of experiments on the motor system. However, as motor physiology progresses toward a more comprehensive picture of motor behaviour, the study of the musculoskeletal system has of necessity, and of interest, come to depend more and more on the quantitative methods of biomechanics. Biomechanical studies have led to new hypotheses about the design of the motor system and biomechanical considerations have provided important tests of existing hypotheses concerning the neural control of movement. These hypotheses include global issues such as redundancy and encoded variables as well as specific hypotheses such as Stiffness Regulation, Selective Recruitment and the concept of Flexor Reflex Afferents.

INTRODUCTION

A major challenge for research on the motor system is to integrate the various mechanisms associated with the nervous and musculoskeletal systems into a predictive model of motor behavior. Neurophysiological mechanisms have received the greater emphasis of study in the past, but attention to the biomechanics of the motor system has been steadily increasing among researchers. The musculoskeletal system possesses intrinsic mechanisms that contribute to motor behavior and that provide important constraints on the resulting dynamics of motor coordination. These constraints translate the output of motoneurons into movement and determine the sensory feedback that is returned to the central nervous system. Therefore, the results of biomechanical investigations of the musculoskeletal system help to give physiological meaning to neuronal mechanisms with regard to motor coordination.

---

* Department of Physiology and Center of Spinal Cord Research, Emory University, Atlanta, GA 30322. Email: trn@physio.emory.edu

A number of researchers have contributed to this more integrated view by making original contributions in biomechanics from the perspective of motor control and in some cases by studying the two systems together, as exemplified by the late Dr. Peter Rack (Joyce and Rack, 1969; Joyce et al., 1969; Rack, 1981). This more global approach to the motor system is a growing trend (Young et al., 1993; Kirsch et al., 1994; Gielen et al., 1997; Macpherson and Fung, 1999; Bawa et al., 2000; Lin and Rymer, 2000; Ting et al., 2000). The study of musculoskeletal mechanics from the perspective of motor control has led to progress in this area by contributing new design principles of the musculoskeletal system and insights into putative neuronal mechanisms of motor coordination and more global issues relating to the motor system. Some examples of these contributions are given in this brief overview.

MUSCULOSKELETAL DESIGN

A major benefit of the increased attention to the musculoskeletal system has been the discovery of additional principles of design that have implications for theories of motor control. For example, new facts are emerging concerning the routes of force transmission within muscles and between muscles and skeleton. It has been shown that many muscles do not possess a purely in parallel organization of muscle fibers. Instead, muscle fibers are short relative to the distance between aponeurosis of origin and tendon of insertion, and so are arranged at least partially in-series (Loeb et al., 1987; Young et al., 2000). This organizational feature promotes rapid distribution of activation to all parts of the muscle, since activation is limited by the conduction velocity of nerve rather than muscle. However, this feature appears to impose the constraint that motor units associated with the in-series fibers contract simultaneously (Sokoloff et al., 1998). Otherwise, the series compliance of inactive fibers in series would substantially reduce force production by the entire in-series unit.

Beyond the confines of the muscle, forces may be transmitted to the skeleton through structures other than tendons and aponeuroses. The hamstring muscles of felines, for example, make strong connections to the calcaneus through thickened portions of the crural fasciae. The transmission of forces through these structures as well as through the tendons of origin and insertion lead to torque production at the hip, knee and ankle with the directions appropriate to enable the animal to spring toward its prey (Ingen Schenau, 1994). This feature may help to explain the ambiguity in the measured activation patterns of the biceps femoris posterior muscle (Rossignol, 1996), and the actions of this muscle seem difficult to classify across behaviors. The designation "knee flexor" might be appropriate in some cases, but "retractor of the leg" in others.

Besides the identity of the joints across which a given muscle exerts torque through tendons or fasciae, the direction of torque generation about a given joint is equally important to motor function. Joints and axes of rotation both represent degrees of freedom of motion. The contributions of a given muscle to torque about all the axes at a given joint are important to consider in terms of muscle action. The forces in the calcaneal tendon of the cat developed by the triceps surae muscles are transmitted remarkably independently, so that each muscle has a unique direction of action at the ankle (Nichols et al., 1993). Correspondingly, each muscle has a unique pattern of force output (Fowler et al., 1993).

MECHANISMS OF POSTURE AND STABILITY

An accurate representation of the architecture of the musculoskeletal system is clearly important to interpret patterns of muscle activation. The dynamic mechanical properties of muscles and of the musculoskeletal system are critical as well. Knowledge of these properties has allowed existing theories of motor control to be tested, new hypotheses to be proposed, and new functions to be proposed for known neural circuits.

The Stiffness Regulation Hypothesis (Nichols and Houk, 1976; Houk et al., 1981) was based on a consideration of the integrated dynamics of muscles and reflexes. The complementary nonlinear properties of muscle spindle receptors and extrafusal muscle combined in such a way as to improve the extent of linear behavior of the muscle. In addition, the stretch reflex was found to substantially reduce the dependence of muscular stiffness on force, so that the muscle could respond vigorously to mechanical disturbances even at low forces (Hoffer and Andreassen, 1981). This hypothesis has been criticized on various grounds, including a presumed lack of applicability to different muscles and different behavioral states. Newer evidence, however, justifies reconsideration of this idea. Ongoing movement can result in an increase in the linearity of intrinsic muscle properties, but the extent of this increase is contingent on the nature of movement history (Huyghues-Despointes, 1998). Reflex action is apparently automatically adjusted so as to maintain spring-like behavior of the muscle. The history dependence of the compensation is thought to be due to complementary nonlinear properties of muscle and spindle (Proske et al., 1993; Nichols et al., 1999b). The regulatory actions of the stretch reflex may also be subject to descending control through neuromodulatory pathways. Newer pharmacological evidence suggests that the action of the stretch reflex to reduce the dependence of muscular stiffness on background force is strongly dependent upon the modulation of serotonergic receptors and perhaps the presence of repetitive motoneuron firing (Machacek et al., 2000). Finally, the importance of reflex action in governing stiffness during natural behavior has been supported by behavioral studies in cats during ramp walking (Abelew et al., 1997). One of the functions of a pathway closely related to the stretch reflex, reciprocal inhibition, has also been clarified by biomechanical arguments. This pathway works synergistically with the stretch reflex and intrinsic elastic properties of muscle to regulate the stiffness of joint systems, and may be enhanced during cocontraction to increase joint stiffness (Nichols et al., 1999a).

A system of pathways whose functions are much less clear than those of the stretch reflex mediates the clasp-knife response. This stretch-evoked response is released during spinal injury, causes widespread effects on muscles, and arises from small diameter muscle afferents (Cleland et al., 1990; Cleland and Rymer, 1990). A consideration of the mechanical conditions under which this response is expressed have shown that, contrary to previous accounts, this response is evoked by forces and stretches that occur during normal movements, does not require abnormally high stresses within the muscle, and has a unique intermuscular distribution that further distinguishes it from a flexion-withdrawal reflex (Nichols and Cope, 2001). This finding indicates that the corresponding pathways may well have important functions during normal activities that are not accompanied by muscle damage. If these pathways function to minimize damage, then they seem capable of responding before rather than after damage actually occurs. In addition, the unique intermuscular distribution of this response provides further evidence that the neuronal circuits arising from high-threshold afferents do not constitute a functionally monolithic interneuronal system mediating a generalized flexion withdrawal reflex, but instead form a heterogeneous collection of individual circuits that have unique adequate stimuli, distinct muscular distributions (Schouenborg and Kalliomaki, 1990) and distinct mechanical effects.

The foregoing examples illustrate the value of biomechanical data to interpreting the actions of specific neuronal circuits. Knowledge of musculoskeletal design will also be critical for filling even larger gaps in our understanding of the basis of motor coordination. Although research on mechanisms of central pattern generation has advanced rapidly (Rossignol, 1996), investigations of integrated spinal mechanisms of posture and stability have not. The strategy for postural regulation in three dimensions during standing in cats, the force constraint strategy, have been described in terms of ground reaction forces and electromyographic patterns (Macpherson, 1988a,b) and appears to be largely proprioceptive in origin (Thomson et al., 1991). There is a substantial and growing literature on mechanisms of posture (Allum et al., 1998), but there has not been a concerted attempt to determine the role of known spinal pathways in the force constraint strategy, even though the simplest of spinal reflexes has long been implicated in postural regulation. Although more recent evidence suggests that the force constraint strategy is not supported following spinal injury (Macpherson and Fung, 1999), the extent to which purely spinal mechanisms can mediate the strategy has not been thoroughly tested. Rapid spinal pathways are often dismissed on the basis of arguments about latency, but these arguments are less than compelling due to the lack of information about the mechanical conditions of the muscles during the execution of this strategy. An investigation that combines a study of musculoskeletal mechanics in three dimensions and the distribution of rapid feedback in the spinal cord would test whether the proprioceptive component of postural responses could be supported by purely spinal pathways.

GLOBAL ISSUES OF MOTOR CONTROL

Among the more global issues of motor control that can be addressed using the results of biomechanical studies are three that can be mentioned here briefly. First and perhaps most obviously, attempts have been made to determine the mechanical variables that are encoded by signals in receptors and motor circuits. Recently, studies that combine neural recordings with kinematic measurements of the feline hindlimb indicate that global limb position is encoded in the discharges of cells of the dorsal spinocerebellar tract (Bosco and Poppele, 2001). Second, the number of degrees of freedom possessed by the motor system has been a matter of considerable argument. For example, flexibility in recruitment order of motor units has been the subject of a large number of experimental studies. Evidence for selective recruitment (Kanda et al., 1977; Nardone et al., 1989) has not been supported by a number of more recent studies (Cope and Clark, 1995; Bawa and Jones, 1999), but the issue has not been finally resolved for all parties concerned and remains contentious. Previous arguments concerning the functional utility of selective recruitment have been largely qualitative (Nardone et al., 1989) and few reports exist (Hutton and Enoka, 1986) that actually test the hypothesis that recruitment of slow twitch muscle is disadvantageous for any movement. Quantitative arguments that support a role for selective recruitment would go a long way toward resolving this issue. A third major issue has been that of redundancy in the motor system. Considerable redundancy exists in the mammalian motor system if one considers the possible combinations of joint positions defined by the skeleton. However, the musculature and associated tissues impose additional constraints on movement. Muscle attachments and fasciae tend to couple joints and axes of rotation. Limitations on joint rotations by connective tissue, length-dependent properties of muscle, and changing moment arms can also limit the targets available to a multi-segmented limb. In cases where movements are constrained within the available skeletal degrees of freedom, the constraints imposed by the musculature must also be considered. An interesting new approach that provides a measure of the degrees of freedom that are controlled in a movement also quantifies redundancy in terms of variability (Scholz and Schoner, 1999). Further biomechanical and neurophysiological studies will be needed in order to determine which constraints are due to properties of the musculoskeletal system and which are due to constraints imposed by the nervous system.

ACKNOWLEDGEMENTS

This work was supported by NS20855, HD32571 and NS40405.

REFERENCES

Abelew, T. A., Miller, M. D., Huyghues-Despointes, C. M. J. I., and Nichols, T. R., 1997, Disruption of interjoint coordination during ramp locomotion in self-reinnervated cats, *Society for Neuroscience Abstract*, 23, 760.

Allum, J. H., Bloem, B. R., Carpenter, M. G., Hulliger, M., and Hadders-Algra, M., 1998, Proprioceptive control of posture: a review of new concepts, *Gait Posture*, 8, 214-242.

Bawa, P., Chalmers, G. R., Jones, K.E., Soggard, K. and Walsh, M.L., 2000, Control of the wrist joint in humans, *European Journal of Applied Physiology*, 83, 116-127.

Bawa, P., and Jones, K. E., 1999, Do lengthening contractions represent a case of reversal in recruitment order? In: *Peripheral and Spinal Mechanisms in the Neural Control of Movement*. M. D. Binder, ed., Elsevier, Amsterdam, 123, 215-221.

Bosco, G., and Poppele, R. E., 2001, Proprioception from a spinocerebellar perspective, *Physiological Reviews*, 81, 539-568.

Cleland, C. L., Hayward, L., and Rymer, W. Z., 1990, Neural mechanisms underlying the clasp-knife reflex in the cat II: stretch-sensitive muscular-free nerve endings, *Journal of Neurophysiology*, 64, 1319-1330.

Cleland, C. L., and Rymer, W. Z., 1990, Neural mechanisms underlying the clasp knife reflex in the cat I characteristics of the reflex, *Journal of Neurophysiology*, 64, 1303-1318.

Cope, T. C., and Clark, B. D., 1995, Are there important exceptions to the Size Principle of α-motoneuron recruitment? in: *Alpha and Gamma Motor Systems*, Taylor, A., Gladden, M. H., and Durbaba, R., eds., Plenum, New York, pp. 71-78.

Fowler, E. G., Gregor, R. J., Hodgson, J. A., and Roy, R. R., 1993, Relationship between ankle muscle and joint kinetics during the stance phase of locomotion in the cat, *Journal of Biomechanics*, 26, 465-483.

Gielen, C. C. A. M., Vrijenhoek, E. J., Flash, T., and Neggers, S. F., 1997, Arm Position constraints during pointing and reaching in 3-D space, *Journal of Neurophysiology*, 78, 660-673.

Hoffer, J. A., and Andreassen, S., 1981, Regulation of soleus muscle stiffness in premammillary cat intrinsic and reflex components, *Journal of Neurophysiology*, 45, 267-285.

Houk, J. C., Crago, P. E., and Rymer, W. Z., 1981, Function of the spindle dynamic response in stiffness regulation - a predictive mechanism provided by non-linear feedback, in: *Muscle Receptors and Movement*. Taylor, A., and Prochazka, A., eds., Macmillan, London, pp. 299-309.

Hutton, R. S., and Enoka, R. M., 1986, Kinematic assessment of a functional role for recurrent inhibition and selective recruitment, *Experimental Neurology*, 93, 369-379.

Huyghues-Despointes, C. M. J. I., 1998, *Effects of Movement History on the Intrinsic Properties and the Neural Regulation of Feline Skeletal Muscle*, doctoral dissertation, Emory University, Atlanta.

Ingen Schenau, G. J. v., 1994, Proposed actions of bi-articular muscles and the design of hindlimbs of bi- and quadrupeds, *Human Movement Sciences* 13, 665-681.

Joyce, G. C., and Rack, P. M. H., 1969, Isotonic lengthening and shortening movements of cat soleus muscle, *Journal of Physiology,* 204, 475-491.

Joyce, G. C., Rack, P. M. H., and Westbury, D. R., 1969, The mechanical properties of cat soleus muscle during controlled lengthening and shortening movements, *Journal of Physiology,* 204, 461-474.

Kanda, K., Burke, R. E., and Walmsley, B., 1977, Differential control of fast and slow twitch motor units in the decerebrate cat, *Experimental Brain Research* 29, 57-74.

Kirsch, R. F., Boskov, D., and Rymer, W. Z., 1994, Muscle stiffness during transient and continuous movements of cat muscle: perturbation characteristics and physiological relevance, *IEEE Transactions on Biomedical Engineering,* 41, 758-770.

Lin, D. C., and Rymer, W. Z., 2000, Damping actions of the neuromuscular system with inertial loads: soleus muscle of the decerebrate cat, *Journal of Neurophysiology,* 83, 652-658.

Loeb, G. E., Pratt, C.A., Chanaud, C. M., and Richmond, F. J. R., 1987, Distribution and innervation of short, interdigitated muscle fibres in parallel-fibered muscles of the cat hindlimb, *Journal of Morphology,* 191, 1-15.

Machacek, D. W., Livingston, B. P., Hochman, S., and Nichols, T. R., 2000, Spinal 5-HT2 receptor activation produces long-lasting modulation of flexor-withdrawal reflex and changes in stiffness regulation, *Society Neuroscience Abstract,* 26, 57.16.

Macpherson, J. M., 1988, Strategies that simplify the control of quadrupedal stance I forces at the ground, *J. Neurophysiology,* 6, 204-217.

Macpherson, J. M., 1988, Strategies that simplify the control of quadrupedal stance II.Electromyographic activity, *Journal of Neurophysiology,* 60, 218-231.

Macpherson, J. M., and Fung, J., 1999, Weight support and balance during perturbed stance in the chronic spinal cat, *Journal of Neurophysiology,* 82, 3066-3681.

Nardone, A., Romano, C., and Schieppati, M., 1989, Selective recruitment of high -threshold human motor units during voluntary isotonic lengthening of active muscles, *Journal of Physiology,* 409, 451-471.

Nichols, T. R., and Cope, T. C., 2001, The organization of distributed proprioceptive feedback in the chronic spinal cat, in: *Motor Neurobiology of the Spinal Cord,* T. C. Cope, ed., CRC Press, Boca Raton, pp: 305-326.

Nichols, T. R., Cope, T. C., and Abelew, T. A., 1999, Rapid spinal mechanisms of motor coordination, *Exercise and Sport Sciences Reviews,* 27, 255-284.

Nichols, T. R., and Houk, J. C., 1976, The improvement in linearity and regulation of stiffness that results from action of the stretch reflex, *Journal of Neurophysiology,* 39, 119-142.

Nichols, T. R., Lawrence, J. H. III, and Bonasera, S. J., 1993, Control of torque direction by spinal pathways at the cat ankle joint, *Experimental Brain Research,* 97, 366-371.

Nichols, T. R., Lin, D. C., and Huyghues-Despointes, C. M. J. I., 1999, The role of musculoskeletal mechanics in motor coordination, in: *Peripheral and Spinal Mechanisms in the Neural Control of Movement,* M. D. Binder, ed., Elsevier, Amsterdam, *Progress in Brain Research,* 123, 369-378.

Proske, U., Morgan, D. L., and Gregory, J. E., Thixotropy in skeletal muscle and in muscle spindles: A review, *Progress in Neurobiology,* 41, 705-721.

Rack, P. M. H., 1981, Limitations of somatosensory feedback in control of posture and movement, *Handbook of Physiology, Section 1: The Nervous System; Volume II, Motor Control, Part 1.* American Physiological Society, Bethesda, pp. 229-256.

Rossignol, S., 1996, Neural control of stereotypic limb movements, in: *Handbook of Physiology. Section 12: Excercise: Regulation and Integration of Multiple Systems,* Rowell, L. B., and Shepherd, J. T., Oxford, New York, pp. 173-216.

Scholz, J. P., and Schoner, G., 1999, The uncontrolled manifold concept: identifying control variables for a functional task, *Experimental Brain Research,* 126, 289-306.

Schouenborg, J., and Kalliomaki, J. 1990, Functional organization of the nociceptive withdrawal reflexes. I. Activation of hindlimb muscles in the rat, *Experimental Brain Research,* 83, 67-78.

Sokoloff, A. J., Ryan, J. M., Valerie, E., Wilson, D. S., and Goslow, G. E., 1998, Neuromuscular organization of avian flight muscle: morphology and contractile properties of motor units in the pectoralis (pars thoracicus) of pigeon (columba livia), *Journal of Morphology,* 236, 179-208.

Thomson, D. B., Inglis, J. T., Schor, R. H., and Macpherson, J. M., 1991, Bilateral labyrinthectomy in the cat: motor behaviour and quiet stance parameters, *Experimental Brain Research,* 85, 364-372.

Ting, L. H., Kautz, S. H., Brown, D. A., and Zajac, F. E., 2000, Contralateral movement and extensor force generation alter flexion phase muscle coordination in pedaling, *Journal of Neurophysiology,* 83, 3351-3365.

Young, M., Paul, A., Rodda, J., Duxson, M., and Sheard, P., 2000, Examination of intrafascicular muscle fiber terminations: implications for tension delivery in series-fibered muscles, *Journal of Morphology,* 245, 130-145.

Young, R. P., Scott, S. H., and Loeb, G. E., 1993, The distal hindlimb musculature of the cat: multiaxis moment arms at the ankle joint, *Experimental Brain Research,* 96, 141-151.

THE IMPORTANCE OF BIOMECHANICS

Gerald E. Loeb, Ian E. Brown, Ning Lan and Rahman Davoodi*

ABSTRACT

When neuroscientists gather to discuss "Movement and Sensation", they tend to discuss neurons rather than muscles and bones. Neurons may be more interesting, but their roles in motor control depend on the mechanical properties of the system to be controlled. Understanding of those properties has been surprisingly elusive, despite the well-developed disciplines of biomechanics and muscle physiology. Each experimental field has its favorite, often unique preparation. Mathematical models range in scale from individual cross-bridges to articulated limbs, usually written in different computer languages. The shortcomings of such fragmented knowledge become particularly apparent when biomedical engineers must design safe and effective control systems for real limbs, such as for functional electrical stimulation (FES) of reach and grasp in quadriplegic patients. We are addressing the question of how to model neuromusculoskeletal systems so that they are sufficiently complete, valid and accessible to be useful in both basic and applied sensorimotor research.

INTRODUCTION

This essay focuses on three elements that are required to appreciate fully the role of biomechanics in sensorimotor control:

1. **Prioritization** according to the importance of various attributes of the component parts. Models inevitably simplify reality because they are based on limited experimental data and because they must be computationally tractable. This makes it necessary to determine which attributes can be simplified or omitted from the model and which attributes require additional experimental data to define their models.
2. **A computational platform** in which to bring the component parts together. This is required because models of complete neuromusculoskeletal systems are inherently complex and subject to changes and upgrades that are likely to be derived from many

* A.E. Mann Institute for Biomedical Engineering, University of Southern California, Los Angeles, USA. Email: gloeb@usc.edu

different sources. It is not practical for one researcher to build and maintain a complete model and it would be difficult for other researchers to benefit from such a personal model system.

3. **A real-world application** to serve as a test of the adequacy of our understanding. This is required because artificial scenarios may obscure the shortcomings of model controllers (not always coincidentally). If we truly understand what the nervous system must compute to control a limb, then we should be able to build a prosthetic controller that works successfully when interfaced with the real biomechanical system.

PRIORITIZATION OF ATTRIBUTES

Most models incorporate various terms that represent well-known aspects of biomechanical systems that seem to be important, such as the active force-length relationship and the sluggish response of skeletal muscle in response to dynamically changing neural activation. Other attributes tend to be omitted because they are less well-known or less well-described experimentally or because their mathematical descriptions appear to be too complex; the relative importance of the omitted attributes often is neither obvious nor tested. The space available here does not permit an exhaustive presentation of model components, so we will focus on two that are often omitted or over-simplified but seem to be particularly important.

Cross-Bridge Cooperativity

In most muscle models, the cycle of cross-bridge attachment/detachment is described in terms of Ca^{++} regulation alone. However, there is a substantial body of evidence that attached, force-producing cross-bridges can themselves facilitate the formation of nearby cross-bridges (see Gordon et al., 2000 for review). This is in addition to the cooperativity that arises from the second order kinetics of calcium binding to troponin and from the distributed reconfiguration of tropomyosin along the thin filament. For muscle models that use a black-box approach to describe activation dynamics (e.g. Brown and Loeb, 2000), understanding cooperativity is not necessary. Such models tend to become arbitrarily complex, however, as they are extended to account for important properties such as the frequency dependence of relaxation time and the energy consumption of muscle under varying conditions of work. By incorporating both types of cooperativity explicitly in our newer model, we have found that the overall structure of the model actually becomes simpler even as the fit to experimental data improves.

The various forms of cooperative binding for both Ca^{++} and cross-bridges all contribute importantly to the steepness of the force-frequency relationship. In Figure 1 we plot two force-frequency relationships, one for a 'realistic' fibre and one for a 'hypothetical' fibre that is lacking only the cross-bridge cooperativity term. The steeper curves are for the realistic fibre, with model coefficients adjusted to fit the measured force-frequency relationships. The equation used for this plot is the biochemical Hill equation, with a Hill co-efficient (n_H) of 5. Typical values of n_H for single fibres range from 4-6 (Gordon et al., 2000). The shallower force-frequency relationship is an estimate for our 'hypothetical' fibre in which there is no cross-bridge cooperativity. We used n_H=2 for this fibre based upon the range of measured Hill co-efficients for Ca^{++} binding to troponin (these range from 1.2-2.0; Gordon et al., 2000). In order to compare the force-frequency relationships of these two fibres we had to assume relative values for cross-bridge attachment/detachment and Ca^{++} sensitivity. We have matched the two simulations to produce identical rise and fall times for force and to consume similarly low energy in the relaxed state. The timing constraint leads us to conclude that cross-bridge attachment/detachment rates should be identical as should Ca^{++} release/uptake. Minimizing basal energy consumption requires us to equate Ca^{++} sensitivities at some small value of [Ca]. We chose arbitrarily to equate these at the level necessary to produce a force of 0.05 F_0. This approximation probably overestimates the hypothetical Ca^{++} sensitivity, thus underestimating the differences between realistic and hypothetical fibres. Note that our choice of n_H for the hypothetical fibre at the high end of the range for Ca^{++} binding alone will also tend to underestimate the differences between the fibres.

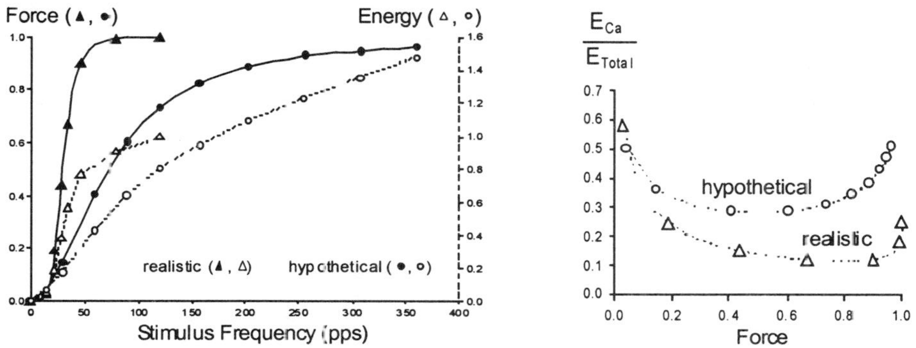

Figure 1. Isometric active force and energy consumption. Data for models of mammalian skeletal muscle with realistic cross-bridge cooperativity (triangles) and no cooperativity (circles). Force and energy are each normalized to their peak realistic values. See text for model details. On the right we show the fraction of energy consumption related to calcium kinetics at equivalent force levels for the two models.

As would be expected, the hypothetical fibre has a much shallower force-frequency relationship. We have also plotted the total energy consumption vs. frequency relationships for the two fibres on the same plot (open symbols, dashed lines). To calculate energy consumption from the Ca^{++} pump, we assumed that under tetanic conditions (120 pps for the realistic fibre) ~25% of the total energy consumption is due to Ca^{++} uptake (Homsher, 1987). For a given stimulus frequency the plots show that the hypothetical fibre consumes less energy. This result is expected given that the hypothetical fibre is producing less force (i.e. fewer cross-bridges are cycling). A more relevant comparison is shown in Figure 1 (right) in which we plot the fraction of Ca^{++} pump energy consumption vs. total energy. At optimal force levels, the realistic fibre uses only ~15% of its energy consumption on Ca^{++} uptake. In order to achieve the same force levels, the hypothetical fibre must operate at a higher frequency, forcing it to use at least ~30% of its energy on Ca^{++} uptake. This result suggests that one of the reasons that cross-bridge cooperativity evolved may have been to minimize energy wasted by the Ca^{++} pump so as to maximize the force and work production per unit energy consumption.

Omission of effects due to cross-bridge cooperativity has large effects on both force and energy predictions in skeletal muscle models.

Dynamic Moment Arms

Muscles generate forces along their effective lines of pull. These must be multiplied by moment arms to compute the joint torques that actually move a limb. Moment arms in biological systems are more complex than they might seem because muscles often cross joints obliquely to one or more axes of rotation and because the paths that their tendons follow across those joints are often constrained by bony protuberances, ligamentous retinacula and the bulging of adjacent muscles. Thus, the actual moment arm in any one axis tends to depend on the angular position of the joints in all axes. These changes in moment arm, and hence muscle torque, may be quite large. Furthermore, the contractile force of the muscle cannot be computed accurately without knowing the length and velocity of the muscle itself. This is usually computed from the same features of the musculotendon path on the skeleton that are used to define the moment arms.

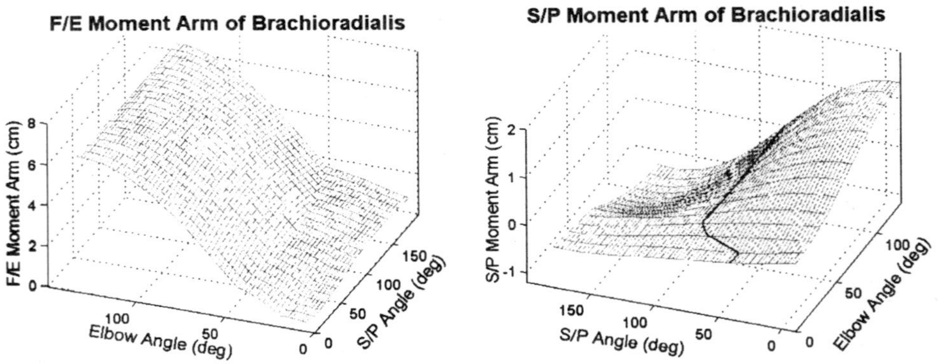

Figure 2. Moment arms of brachioradialis muscle. Data shown as functions of elbow flexion/extension angle (left: F/E, 0°, full extension) and forearm supination/pronation (right: S/P; 0°, max. pronation) posture. The S/P angle for which the S/P moment arm reverses sign (heavy line) changes with elbow angle.

Moment arms may actually change sign as joint angles change. In at least two systems, these zero-crossings effectively define "equilibrium joint angles"; any contractile force in the muscle will tend to resist motion away from the neutral angle at which moment arms (and torques) go through zero. Young et al. (1993) described the tendency of several feline muscles to stabilize the ankle in neutral inversion/eversion and abduction/adduction postures regardless of flexion/extension angle or torque. Figure 2 shows one example from an even more complex system in which the human elbow flexors tend to pull the forearm toward equilibrium angles in pronation/supination. The surface plots were generated by operating an anatomically accurate model of the musculotendon path. This was created in the graphical environment of the SIMM™ commercial software package (see Fig. 3) and adjusted to match the available data on flexion/extension moment arms (Murray, 2000). The slopes and zero-crossings of the pronation/supination moment arms change as flexion/extension angle changes; biceps has a similar pattern with different zero-crossings. It seems likely, but remains unproven, that the central nervous system selects kinematic trajectories and muscle activation patterns that take advantage of these biomechanical features.

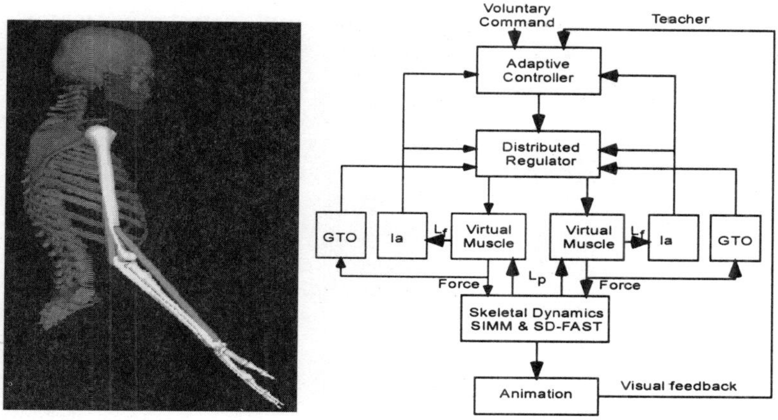

Figure 3. Model of neuromuscular system. *Left*, model of human arm in SIMM with brachicradialis (elbow flexor) and triceps lateralis (elbow extensor). *Right*, Simulink blocks corresponding to this system with sensors (Ia, spindle primary; GTO, Golgi tendon organ) and a hierarchical controller based on a spinal-cord like distributed regulator and a brain-like adaptive controller. L_p, musculotendon path length, L_f, fascicle length.

COMPUTATIONAL PLATFORM

The two attributes described above belong to different levels of the musculoskeletal system. Cross-bridge cooperativity is a feature of muscle physiology that is probably universal to all mammalian skeletal muscles, while dynamically changing moment arms are a consequence of the anatomical details of a given musculotendon unit and its attachments to the skeleton. A set of sensors and a control system constitute still other levels of the system, all of which must exchange data about state variables with the other levels. The neural control system itself is organized into hierarchical levels such as spinal cord and cerebral cortex (Loeb et al., 1999).

We have selected the commercially supported Simulink™ software application in which to develop and link the components into complete systems. We are developing a set of modeling tools to facilitate the construction and manipulation of the component models. The first two tools are largely complete:

- MMS™ (Musculoskeletal Modeling in Simulink) is a C program that operates on SIMM musculoskeletal models (see Fig. 3), converting them to Simulink blocks.
- Virtual Muscle™ (Cheng et al., 2000) is a Matlab program that creates Simulink blocks representing the tension-generating properties of mixed fibre-type muscles under the full dynamic range of length, velocity and recruitment history.

Each Virtual Muscle block is driven by a neural command signal that produces recruitment and frequency modulation of its various motor units. Muscle path length (Lp) information from the SIMM block is apportioned among muscle fascicles and tendon-aponeurosis to provide the fascicle length (L_f) and velocity information to compute contractile force and to drive a model of a spindle primary afferent. The forces of all muscles produce torques on the limb segments according to the musculoskeletal model; the limb moves according to Newtonian mechanics incorporated in a kinetic model generated by SD-FAST™ (another commercial software package). Figure 3 shows the SIMM anatomical model and corresponding Simulink blocks required to represent and simulate its operation and control by a simple, hierarchical sensorimotor nervous system.

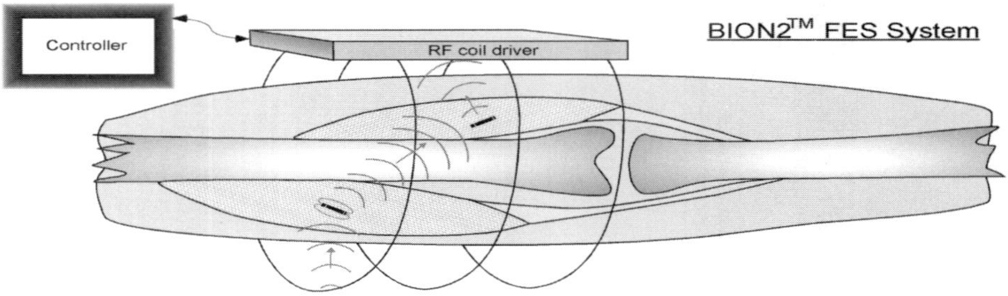

Figure 4. BION™ (BIOnic Neuron) modules for stimulation and sensing of muscle activity. Each injectable module receives power and digital command signals from a common external RF inductive coil. Each module can stimulate a muscle (now in clinical trials to treat disuse atrophy in patients with stroke and osteoarthritis); future modules will also transmit data from internal sensors, such as the range-finding function depicted here between two such implants to create a BIONic muscle spindle for joint angle sensing.

TEST APPLICATION

We have chosen the neural prosthetic control of functional reach and grasp movements in quadriplegic patients in order to apply and test biomechanically realistic strategies for sensorimotor control:

- Quadriplegia as a result of spinal cord injury is a fairly common and particularly devastating disability because it leaves the patient entirely dependent on an attendant for even simple self-care activities like eating.
- Reach and grasp tasks require the simultaneous coordination of distal hand muscles (normally controlled mostly directly by sensorimotor cortex) as well as proximal arm muscles (normally controlled mostly indirectly through propriospinal and other interneuronal circuits of the spinal cord).
- A new neural prosthetic interface technology is available that facilitates implantation of large numbers of independent stimulation and sensing channels in many arm muscles without requiring extensive surgery (Loeb et al., 2001; Fig. 4 above).
- Many quadriplegic patients retain voluntary control of shoulder motion, which can be used to infer the location, orientation and grasp state desired for the hand.
- Biomimetic design principles will be used in an intermediate "regulator" that integrates the output of an adaptive controller with sensory feedback from the periphery, much as the spinal cord does normally (Loeb et al., 1999).

REFERENCES

Brown, I. E., and Loeb, G. E., 2000, Measured and modeled properties of mammalian skeletal muscle: IV. dynamics of activation and deactivation, *Journal of Muscle Research and Cell Motility,* 21, 33-47.

Gordon, A. M., Homsher, E., and Regnier, M., 2000, Regulation of contraction in striated muscle, *Physiological Reviews,* 80, 853-924.

Homsher, E., 1987, Muscle enthalpy production and its relationship to actomyosin ATPase, *Annual Review of Physiology,* 49, 673-690.

Loeb, G. E., Brown, I. E., and Cheng, E. J., 1999, A hierarchical foundation for models of sensorimotor control, *Experimental Brain Research,* 126, 1-18.

Loeb, G. E., Peck, R. A., Moore, W. H., and Hood, K., 2001, BION™ system for distributed neural prosthetic interfaces, *Medical Engineering & Physics,* 23, 9-18.

Murray, W. M., Buchanan, T. S., and Delp S. L., 2000, The isometric functional capacity of muscles that cross the elbow, *Journal of Biomechanics,* 33, 943-952.

Young, R. P., Scott, S. H., and Loeb, G. E., 1993, The distal hindlimb musculature of the cat: multiaxis moment arms at the ankle joint, *Experimental Brain Research,* 96, 141-151.

55

THE ROLE OF THE LENGTH-TENSION CURVE IN THE CONTROL OF MOVEMENT

D. L. Morgan[1], C. L. Brockett[2], J. E. Gregory[2], and U. Proske[2]

ABSTRACT

The length-tension curve of muscle is one of the important descriptors of mechanical performance, and also a direct reflection of the underlying structure, particularly the number of sarcomeres connected in series in muscle fibres. This number is one of the most plastic properties of muscle, changing within days after changes in activity patterns. We propose that this adaptation is to prevent eccentric contractions from occurring beyond the optimum length for tension generation, since this is the region of sarcomere instability and muscle damage. Evidence for this is presented for muscles from rats trained on a treadmill, and from motor units of the gastrocnemius muscle of the cat.

INTRODUCTION

The maximally activated length tension curve of a sarcomere is believed to be determined only by the overlap of thick and thin filaments, and hence by the sarcomere length. For any given (uniform) sarcomere length, the muscle length can be determined as sarcomere length multiplied by the number of sarcomeres in series, added to the length of the tendon (Proske and Morgan, 2001). Thus a wide range of combinations of parameter values can give the same optimum length. The question then becomes "which combination should be chosen?" A real muscle will contain fibres with a variety of sarcomere numbers and tendon lengths, giving a smoothly rounded length tension curve. However it is useful to think of an ideal muscle as shown in Figure 1.

Clearly too many sarcomeres in series is costly. They introduce increased muscle mass without increasing muscle tension at optimum length. The extra mass impedes the organism, and the metabolic costs of tension production and maintenance are increased. It would seem then that the optimum number is the minimum number required.

[1] Department of Electrical and Computer Systems Engineering, Monash University, Melbourne, VIC 3800, Australia. Email: david.morgan@eng.monash.edu.au.
[2] Department of Physiology, Monash University, Melbourne, VIC 3800, Australia.

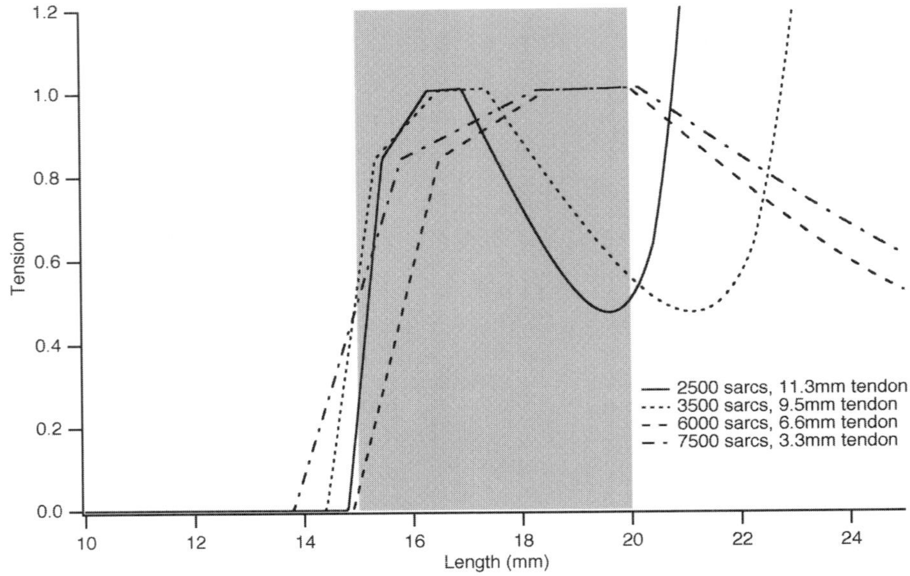

Figure 1. Ideal length tension curves for a variety of muscle geometries. The grey area indicates the working range of muscle lengths. The curve has an active component taken from Gordon et al. (1966), and an exponential passive component has been added.

The minimum number depends on the criterion applied. In Figure 1, the working range of the muscle is taken to be from 15 to 20 mm. If the requirement is the ability to generate some active tension over the whole working range, then the fibre with 2500 sarcomeres is adequate. If the requirement is an ability to generate at least half of optimum active tension over the whole working range, then about 3500 sarcomeres are needed.

Another factor that may increase the number of sarcomeres needed is a requirement for rapid shortening. The unloaded shortening velocity of muscle, and hence the shortening velocity at a given fraction of optimum tension, is determined on a sarcomere basis by more sarcomeres in series, increasing the speed at which a fibre can shorten. As power is the product of force and velocity, this will also increase the rate at which it can do work. This means that power output at the optimum velocity is proportional to muscle mass, whether muscle mass is determined by the number of sarcomeres in series, increasing optimum velocity, or the number of filaments in parallel, increasing tension at optimum velocity. If the need for speed were the important criterion, then muscle fibres that are required to shorten rapidly would be predicted to have more sarcomeres than muscle fibres with a postural role.

Our work with eccentric contractions, stretching muscle while it is generating active tension, has suggested that eccentric contractions beyond optimum length cause microscopic damage to the muscle. While areas of damage are initially confined to a few sarcomeres within a single fibre, they can lead to eventual cell death. If this were the limiting factor on the reduction of the number of sarcomeres, then fibres that naturally undergo eccentric contractions would adapt to keep the working range confined to muscle lengths below optimum. In Figure 1, this would lead to the need for 6000 sarcomeres to put the whole working range on the ascending limb and plateau of the length-tension relation or about 7500 sarcomeres to do this while maintaining half optimum tension over the working range. Clearly, postural muscles are more likely to undergo eccentric contractions, for example, while walking downhill, than are jumping and running muscles. Hence, if this were the criterion determining the number of sarcomeres, the slow postural fibres would be predicted to have more sarcomeres than the fast fibres.

ADAPTATION

These ideas have been examined in two ways. The first involves choosing a particular muscle, and examining optimum length variation between individuals exposed to different kinds of exercise. Rats were given mild exercise, walking on an inclined or declined treadmill. Decline training causes an increase in the number of sarcomeres in vastus intermedius, the deep red portion of the knee extensors, comprising mainly slow fibres and thought to be the principal muscle active during posture. Sarcomere number was measured by fixing the muscles, acid digesting and dispersing fibres, measuring the lengths of isolated complete fibres under the microscope, and estimating sarcomere length by laser diffraction. Sarcomere number was then found as the ratio of fibre length to sarcomere length (Lynn and Morgan, 1994).

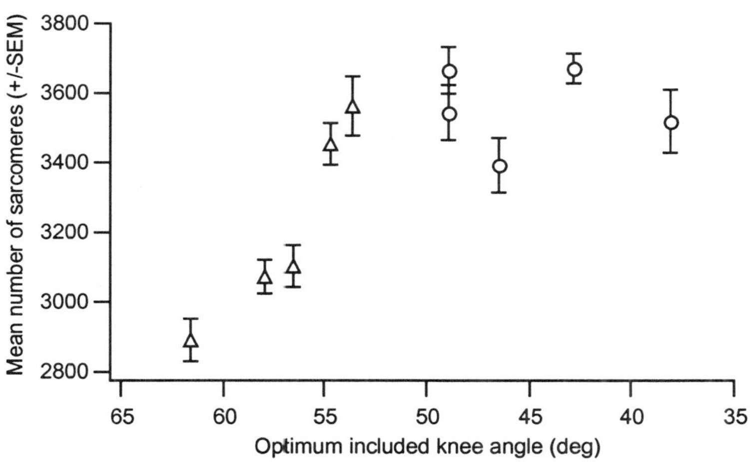

Figure 2. Relation between sarcomere number and optimum knee angle. Data for rat vastus intermedius muscle. Smaller included knee angles to the right represent longer muscle length. Circles are from rats trained by decline running, and triangles for rats trained by incline running.

Figure 2 shows experiments that measured both sarcomere numbers and optimum knee angles for active tension generation (Lynn et al., 1998) in the same muscles. It is clear that these parameters were correlated, with eccentrically exercised muscles generally having fibres with longer optimal lengths and more sarcomeres. The correlation is also apparent within the incline group.

The second test of our predictions measured the optimum lengths of different types of motor units in the medial gastrocnemius muscle of the cat (Whitehead et al., 2001). The whole muscle was dissected free connected to a length controlling device, and a whole muscle length-tension curve was constructed. A laminectomy was performed, and the ventral roots supplying the hindlimb exposed. Functionally single motor units were prepared by splitting cut ventral roots, and classified as fast or slow based on their twitch time to peak at optimum length. Typically six units were obtained at a time. Length tension curves were constructed for each individual motor unit, and the optimum length determined for each unit.

Figure 3 shows the results. All of the slow units, determined as having times to the peak of a twitch contraction at optimal length greater than 60 ms and shown as circles, had optimal lengths longer than the whole muscle optimum. The fast motor units generally had optima less than the whole muscle, but there were some exceptions to this. This suggests that most fast units were not adapted to avoid operating on the descending limb of the whole muscle length-tension curve, but all slow units, and some fast units were. It may well be that the fast units with longer optima were fatigue resistant units, but we did not test fatiguability.

In order to check that optimum angles were related to susceptibility to damage from eccentric exercise, the motor units were then stimulated while the muscle was stretched by 6 mm, beginning 3 mm short of whole muscle optimum. Thus the range of muscle lengths involved in the stretch was the same for all units, but the range of sarcomere lengths was different, fibres with shorter optima being stretched further beyond their optimum. This was repeated for 20 contractions, and then the individual motor unit length-tension curves were again measured. The shift in optimum length has been shown to be a useful measure of the degree of damage done to the muscle, and to correlate well with other measures, such as the fall in tension (Talbot and Morgan, 1998). Fall in tension was also measured, and showed a similar trend, but with somewhat more scatter of values.

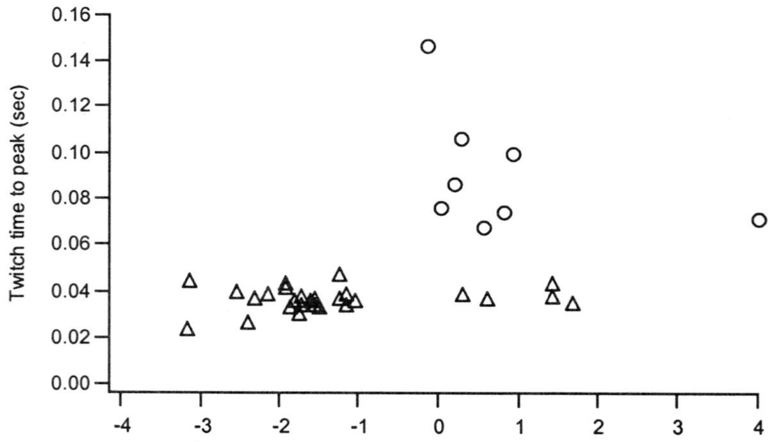

Figure 3. Optimum length for active tension generation of motor units. Data expressed relative to the whole muscle optimum compared to the time-to-peak tension in a twitch at optimum length (triangles, fast twitch units). Units with times to peak greater than 60 ms (circles) were classified as slow. Measurements of tetanic fusion frequencies and twitch:tetanus ratios confirmed this classification.

Figure 4 shows that damage, as measured by shift in optimum, was strongly correlated with the initial optimum length. Shift in optimum was also correlated with fibre type, with fast fibres being, on average, more damaged. However, the graph makes it clear, and statistical analysis confirmed, that after taking the effect of optimum length into account, fibre type was no longer a significant predictor of damage. That is, both fast and slow motor units lie on the same line in Figure 4. This suggests that in maximally activated mixed muscles, the previously reported differences in susceptibility of different fibres types to damage from eccentric exercise are principally due to different optimum lengths of the different fibre types (Lieber and Fridén, 1988). If other factors, such as structural proteins or tension per unit area were responsible, then the fast and slow fibres would be expected to fall on different lines, that is fibre type would still be significant after accounting for the effect of optimum length.

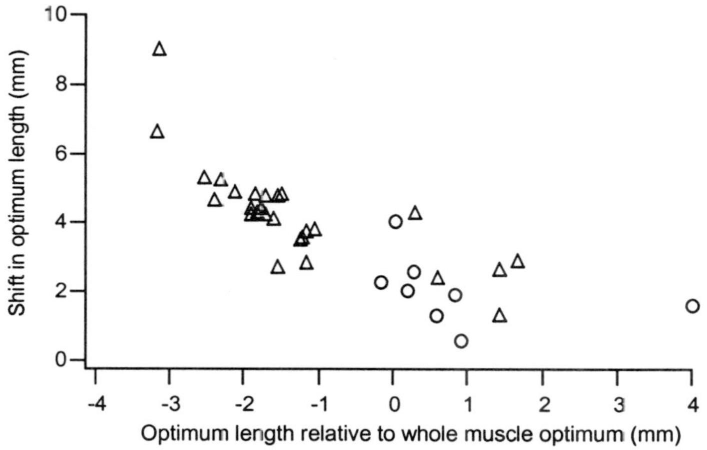

Figure 4. The shift in optimum length for active tension generation is strongly correlated with optimum length, regardless of fibre type. Circles are slow, triangles are fast motor units.

CONCLUSION

The experiments described here support the idea that muscles adapt to particular activities, with the number of sarcomeres matching the activity patterns. Muscles or motor units regularly used in eccentric contractions adapt to ensure that contractions occur at lengths less than optimum. Muscles or motor units with short optima and fewer sarcomeres, when used outside of their normal activity patterns, concerned primarily with shortening contractions, become damaged and adaptation may result.

ACKNOWLEDGEMENTS

Supported by NHMRC and Monash University grants.

REFERENCES

Brockett, C. L., Morgan, D. L., Gregory, J. E., and Proske, U., 2002, Damage to different motor units from active lengthening of the medial gastrocnemius muscle of the cat, *Journal of Applied Physiology*, 92, 1104-1110

Lieber, R. L., and Fridén, J., 1988, Selective damage of fast glycolytic fibres with eccentric contraction of the rabbit tibialis anterior, *Acta Physiologica Scandinavica*, 133, 587-588.

Lynn, R., and Morgan, D. L., 1994, Decline running produces more sarcomeres in rat vastus intermedius muscle fibres than incline running, *Journal of Applied Physiology*, 77, 1439-1444.

Lynn, R., Talbot, J. A., and Morgan, D. L., 1998, Differences in rat skeletal muscles after incline and decline running, *Journal of Applied Physiology*, 85, 98-104.

Proske, U., and Morgan, D. L., 2001, Muscle damage from eccentric exercise: mechanism, mechanical signs, adaptation and clinical applications, *Journal of Physiology*, 537, 333-345.

Talbot, J. A., and Morgan, D. L., 1998, The effects of stretch parameters on eccentric exercise induced damage to toad skeletal muscle, *Journal of Muscle Research and Cell Motility*, 19, 237-245.

Whitehead, N. P., Weerakkody, N. S., Gregory, J. E., Morgan, D. L., and Proske, U., 2001, Changes in passive tension in humans and animals after eccentric exercise, *Journal of Physiology*, 533, 593-604.

56

INTRAMUSCULAR FORCE TRANSMISSION

Philip Sheard,[1] Angelika Paul,[2] and Marilyn Duxson[1,2]

ABSTRACT

The architectural form of skeletal muscle, the pattern of activity/usage between neighbouring fibres, and the pathways for lateral and lengthwise tension delivery are all of interest in understanding muscle function and dysfunction. We have attempted to contribute to understanding of intramuscular force transmission by investigating the functional relationships between coactive motor units, and by examining the detailed molecular and morphological features at sites of tension transfer. We found that tension delivery is modulated by interaction between active and inactive fibres, that many muscle fibre terminations feature structural coupling between fibres, and that sites of tension delivery feature a variety of proteins including acetylcholinesterase, NCAM, dystrophin and two splice variants of the $\alpha 7$ integrins. We conclude that structural and molecular pathways exist to deliver force within, along, and between muscle fibres, and that the quality/quantity of tension delivered from any single fibre is at least partly a consequence of whether its neighbouring fibres are synchronously coactive.

INTRODUCTION

Much of this volume is devoted to recent findings regarding the neurophysiological underpinnings of voluntary movement. The notion of voluntary movement incorporates our expectation that when activated by the sophisticated output from the motor cortex, our skeletal muscles will respond to develop the force required to execute the planned action. By devoting a whole section to the mechanics of force production and delivery from the sarcomere to the joint, the editors have acknowledged that what happens *within the muscle* is as significant as what happens in the brain.

Division of mammalian skeletal muscles into functional types on the basis of gross morphological form has been common practice for many years. What the various

[1] Department of Physiology, [2] Dept. of Anatomy and Structural Biology, Otago School of Medical Sciences, University of Otago, PO Box 913, Dunedin, New Zealand. Email: phil.sheard@stonebow.otago.ac.nz

architectural forms have in common (especially in non-primate mammals) is the numerical dominance of short fibres. Even when fascicles are long (tens of cm) individual fibres are seldom longer than 40mm (although primates are rather exceptional in this regard (Paul, 2001). This conspicuous dominance of short fibres arises by a variety of architectural devices including pennate arrays, tendinous inscriptions, and intrafascicularly terminating fibres. With the use of modern microanatomical techniques we have recently seen a rapid increase in our knowledge of the functional microanatomy of skeletal muscles, and this has fuelled a growing interest in the field, as evidenced by the number of recent review articles dealing broadly with the subject of intramuscular force delivery (Tidball, 1991; Patel and Lieber, 1997; Huijing, 1999; Monti et al., 1999; Sheard, 2000; Monti et al., 2001). What is now plainly evident is that the pathway and mechanism of force delivery from the sarcomere to the tendon is complex, and its potential impact on muscle function justifies its careful examination.

FUNCTIONAL MICROANATOMY

The basic morphological forms of mammalian skeletal muscles each bring with them their own special functional costs and benefits. Pennate muscles, for instance have many fibres packed into a small cross-sectional area and are therefore very strong, but can effect only small changes in length. Parallel fibred (strap) muscles, by contrast, tend to be less powerful but produce much greater changes in length. Consideration of these aspects of muscle function is beyond the scope of this article but interested readers should see Gans and Gaunt (1991). Here, we wish to concentrate on functional aspects of muscle microanatomy, and especially how this must alter when muscles are long.

Individual muscle fibres are not simple independent cylindrical arrays of contractile elements that insert onto tendon at each end. A wide variety of morphological variations in fibre shape exists in mammalian and non-mammalian species alike (Loeb et al., 1987; Gaunt and Gans, 1992; Young et al., 2000). Irrespective of the particular details of individual fibre microanatomy, it is a non-trivial task to understand how force developed within the sarcomere may be efficiently delivered to the tendon (perhaps more than 30 cm away) where it is required to do effective work.

To contribute to our understanding of intramuscular force delivery, we recently published a detailed account of the morphological variations seen at myotendinous junctions (MTJs), intrafascicular terminations (IFTs), and sites of lateral adhesion (Young et al., 2000) in a series-fibred muscle. The shape of the MTJ is very consistent, the fibres were at maximal calibre and inserted with complex deep infoldings of the sarcolemma. We described sites of lateral adhesion between adjacent fibres in which their sarcolemmae came to lie very close together (the narrow space being spanned by a short bridge of transverse collagen fibrils) and we proposed that these structures are involved with lateral transmission of tension between fibres. In addition to providing a potential route for direct lateral tension transfer between adjacent fibres, such adhesions render impossible independent length changes by individual fibres. Finally, we made a detailed survey of the morphology of IFTs. These typically feature long tapers with tiny steplike endings that can either associate directly with a collagen bridge or with a complementary ending on another fibre. Blunt terminations and coarse stepwise endings with complex infoldings of the terminal membrane and a structural association with either the extracellular matrix or a neighbouring fibre feature at IFTs. These and other observations show that the likely pathways and mechanisms for intramuscular force delivery are more varied than is commonly recognised.

MOLECULAR PATHWAYS

Having investigated the fine structure of many types of fibre terminations we then wondered what molecules might be specifically localised at these putative sites of tension transfer, and how they might vary in muscles of different types. In particular, we wished to know whether the structural proteins implicated in tension delivery from the myofibre vary as a function of the type of ending and its location within the muscle. Two major protein systems are thought to be involved in transfer of tension from the filamentous actin at the fibre's periphery to the connective tissue matrix of the muscle. The first involves the integrins with their associated binding proteins vinculin and talin, the second involves dystrophin with its associated dystroglycans (see Fig. 6 in Monti et al., 2001).

We used a panel of antibodies to examine the protein composition of muscle-muscle and muscle-extracellular matrix junctions in a variety of non-primate mammals (Paul, Sheard, Kaufman, Duxson, unpublished observations). We found that MTJs and IFTs were molecularly distinct, and furthermore that the molecular profile of the MTJ varied with muscle architectural type. Dystrophin and α7B integrins were present at both IFTs and MTJs, while tenascin (the sole ligand for $\alpha 9\beta 1$ integrins) was present at MTJs but not IFTs. α7B integrin and dystrophin are also present on the sarcolemma away from overtly specialised sites of tension delivery, but while dystrophin levels are further upregulated at the MTJs of simple muscles they are not significantly elevated above non-junctional regions in series-fibred muscles. These interesting observations show that putative tension transmitting proteins vary at different sites of tension delivery and at the corresponding sites in muscles of different architectural type. In simple muscles where all fibres span the full length of the fascicle, we propose that the MTJ is the major site of tension delivery (and therefore the site requiring greatest structural reinforcement) with lateral tension delivery playing a relatively minor role. For muscles featuring non-spanning fibres, diffuse tension delivery is likely to be a feature of this structural form so one would expect more uniform reinforcement of the sarcolemma, with less overt specialisation at the MTJ. Our findings support this hypothetical view.

FUNCTIONAL CONSEQUENCES

While these recent morphological and anatomical findings add to our understanding of muscle biology, they beg the obvious question "What is the significance of this information to our understanding of muscle function?" If tension produced in the sarcomere were transmitted in series along the fibre to be delivered to the tendon at the MTJ in every case, then the answer to that question would likely be "not very much". However, we have ample direct experimental evidence to show that simple linear transmission of force along the series of sarcomeres is far from an accurate description of events. There have been many demonstrations that tension is delivered, at least in part, laterally away from the fibre of origin (Street, 1983; Proske and Morgan, 1984; Clamann and Schelhorn, 1988; Emonet-Dénand et al., 1990; Sheard et al., 1999; Troiani et al., 1999) in a variety of systems. These papers show, collectively, that individual motor units/fibres do not act independently when delivering force to the tendon. Obviously, some force is delivered directly to the MTJ (where these exist) but force is also transmitted laterally to the extracellular matrix and to neighbouring fibres, the latter being particularly important in muscles with a series-fibred architecture.

An important determinant of the functional properties of active muscle is compliance, and this is equally significant when considering tension delivery from individual fibres or motor units. During submaximal contractions, the delivery of tension from active to inactive fibres is inevitable. Particularly in series-fibred muscles, these inactive fibres will contribute a significant amount of elasticity to the tension delivery pathway. The total elasticity (or compliance) of any motor unit will therefore be at least partly a consequence of its spatial relationship with other active and inactive fibres. This is important because series elasticity impacts on the magnitude of twitch tension and the rate at which that tension can be delivered (Hill, 1951). Thus, any motor unit's performance profile may vary as a function of the precise way it is used in combination with other units. Perhaps we need to begin to consider the motor unit's performance profile as a much more dynamic entity than we have in the past, recognising that force production and delivery from any unit will be influenced by the activity patterns of its neighbours. While it has been relatively easy and undoubtedly useful to define and investigate the isolated motor unit on the basis of the connectivity pattern of the motoneuron, perhaps it is time now to coin a more functional definition of the motor unit (the functional unit) that includes the intramuscular pathway through which tension is delivered. In series fibred muscles this must include fibres that would normally not have been considered part of the motor unit.

The simplest way of overcoming the theoretical difficulties that arise when considering the issue of variable compliance would be to propose that, for any given motor unit, it should always be recruited in synchrony with the same set of partner units for a contraction of any particular magnitude. For series-fibred muscles this might comprise, at best, units whose fibres are aligned in series, thereby creating a functional unit whose fibres run full length and that are routinely coactivated. A small amount of experimental evidence exists to suggest that such "functional units" may exist (Pratt and Loeb, 1991). However we have recently shown (using whole mount immunohistochemistry) that in the series-fibred guinea pig sternomastoid muscle slow-myosin containing fibres may terminate in mid belly surrounded entirely by fast fibres (Young et al., 2000). In situations such as this, coactivation (and efficient tension delivery) can only occur when the slow unit is recruited along with fast units, only likely with moderate to high level contractions.

CONCLUSIONS

Our knowledge of muscle microanatomy is growing quickly, and with it an increasing need to re-evaluate our assumptions about muscle function. Whereas the isolated motor unit has for many years been a valuable and standard model of the smallest functional muscle unit, evidence now suggests that real motor units do not work in isolation, even if they can be activated experimentally in isolation. We are becoming increasingly aware of the close structural and functional relationships that neighbouring fibres have with one another, whether or not they share common neural drive. It is apparent that delivery of tension/force across the sarcolemma is a generalised property of the whole sarcolemma, not just the myotendinous junction. Our new data suggests that the amount (and possibly the significance) of lateral tension delivery differs as a function of muscle architectural type, being more important in series fibred muscles than in those lacking non-spanning fibres. Muscles with unusually long fibres, such as those of primates, are also likely to rely heavily on lateral transmission of tension, although information in this area is currently sparse. The data briefly discussed here should allow new ways of thinking about how force output from muscles is modulated during real behaviours, give new insights into the pathology of muscle diseases, and contribute to a greater understanding of the way in which muscles should be driven when being artificially reanimated.

REFERENCES

Clamann, H. P., and Schelhorn, T. B., 1988, Nonlinear force addition of newly recruited motor units in the cat hindlimb, *Muscle and Nerve,* 11, 1079-1089.

Emonet-Dénand, F., Laporte, Y. and Proske, U., 1990, Summation of tension in motor units of the soleus muscle of the cat, *Neuroscience Letters,* 116,112-117.

Gans, C., and Gaunt, A. S., 1991, Muscle architecture in relation to function, *Journal of Biomechanics,* 42, 53-65.

Gaunt, A. S., and Gans, C., 1992, Serially arranged myofibres: An unappreciated variant in muscle architecture, *Experientia,* 48, 864-868.

Hill, A. V., 1951, The effect of series compliance on the tension developed in a muscle twitch, *Proceedings of the Royal Society of London, Series B: Biological Sciences,* 138, 325-329.

Huijing, P. A., 1999, Muscle as a collagen fiber reinforced composite: a review of force transmission in muscle and whole limb, *Journal of Biomechanics,* 32, 329-345.

Loeb, G., Pratt, C., Chanaud, C., and Richmond, F., 1987, Distribution and innervation of short, interdigitated muscle fibres in parallel-fibred muscles of the cat hindlimb, *Journal of Morphology,* 191, 1-15.

Monti, R. J., Roy, R. R., and Edgerton, V. R., 2001, Role of motor unit structure in defining function, *Muscle and. Nerve,* 24, 848-866.

Monti, R. J., Roy, R. R., Hodgson, J. A., and Edgerton, V. R., 1999, Transmission of forces within mammalian skeletal muscles, *Journal of Biomechanics,* 32, 371-380.

Patel, T. J., and Lieber, R. L., 1997, Force transmission in skeletal muscle: from actomyosin to external tendons, *Exercise and Sport Sciences Reviews,* 25, 321-363.

Paul, A. C., 2001, Muscle length affects the architecture and pattern of innervation differently in leg muscles of mouse, guinea pig, and rabbit compared to those of human and monkey muscles, *Anatomical Record,* 262, 301-309.

Pratt, C. A., and Loeb, G. E., 1991, Functionally complex muscles of the cat hindlimb. I. Patterns of activation across sartorius, *Experimental Brain Research,* 85, 243-256.

Proske, U., and Morgan, D. L., 1984, Stiffness of cat soleus muscle and tendon during activation of part of muscle, *Journal of Neurophysiology,* 52, 459-468.

Sheard, P. W., 2000, Tension delivery from short fibers in long muscles, *Exercise and Sport Sciences Reviews,* 28, 51-56.

Sheard, P.W., McHannigan, P., and Duxson, M.J., 1999, Single and paired motor unit performance in skeletal muscles: Comparison between simple and series-fibred muscles from the rat and the guinea pig, *Basic and Applied Myology,* 9, 79-87.

Street, S.F., 1983, Lateral transmission of tension in frog myofibers: A myofibrillar network and transverse cytoskeletal connections are possible transmitters, *Journal of Cellular Physiology,* 114, 346-364.

Tidball, J.G., 1991, Force transmission across muscle cell membranes, *Journal of Biomechanics,* 24S1, 43-52.

Troiani, D., Filippi, G.M., and Bassi, F.A., 1999, Nonlinear tension summation of different combinations of motor units in the anesthetized cat peroneus longus muscle, *Journal of Neurophysiology,* 81, 771-780.

Young, M., Paul, A., Rodda, J., Duxson, M., and Sheard, P., 2000, Examination of intrafascicular muscle fiber terminations: Implications for tension delivery in series-fibered muscles, *Journal of Morphology,* 245, 130-145.

MUSCLE AND TENDON RELATIONS IN HUMANS

Power enhancement in counter-movement exercise

Tetsuo Fukunaga, Yasuo Kawakami, Tetsuro Muraoka and Hiroaki Kanehisa*

ABSTRACT

To clarify the mechanisms of power enhancement during counter-movement exercise, in vivo muscle fibre behavior during plantar flexion exercise was estimated by real time ultrasonography. Six healthy male subjects were requested to perform ankle plantar flexion exercise with counter-movement (CM, plantar flexion preceded by dorsiflexion) and without counter-movement (noCM, plantar flexion only) on a specially designed dynamometer. In CM, in the dorsiflexion phase, muscle fascicle length was initially lengthened, following which its length remained unchanged while the whole muscle-tendon unit was still lengthened, and decreased in the plantar flexion phase. In noCM, fascicle length decreased throughout the movement and it was longer at the onset of movement than in CM. During dorsiflexion phase in CM, muscle fascicles were not actively lengthened, but contracted isometrically at near optimum length of fibre; thus the increase in muscle-tendon unit length was taken up by elongation of the tendinous tissues. These results demonstrate that power enhancement during CM is due to higher force production by isometric contraction at optimal fibre length.

INTRODUCTION

Movement performance in humans is affected by mechanical properties of tendinous tissues as well as by muscle fibre. Shortening of muscle preceded by active lengthening has shown to be more powerful than that resulting from shortening alone (Cavagna et al., 1965). Mechanisms for the enhancement of performance of the final action have been proposed, including the potentiation of force exerted by pre-stretched muscle fibres and the additional power provided by elastic potential of muscle-tendon unit (Huijing, 1992, Komi et al., 1992). In the latter case, tendinous tissues are thought to play an important

* Department of Life Sciences (Sports Sciences), University of Tokyo, Komaba, 3-8-9, Meguro, Tokyo, Japan. Email: fukunaga@idaten.c.u-tokyo.ac.jp

role as a spring (Alexander and Bennet-Clark, 1977; Hof et al., 1983). However, it has not been possible to observe separately the behaviour of muscle fibre and tendon. In the present study, to estimate the interaction between muscle fibres and tendinous tissues, we measured in vivo muscle fibre behavior during plantar-flexion exercise with and without counter movement by means of real-time ultrasonography.

METHOD

Six healthy men (24-45 yrs, 171 ± 5 cm, 72 ± 6 kg, mean ± SD) participated in this study as subjects. The nature and possible consequences of the study were explained to each subject before informed consent was obtained.

The subject took a spine position on a specially designed machine, with the trunk secured onto a sliding table which was designed to slide with a minimum friction with a constant load through a steel cable connected adjustable weight. A force plate (Kistler, Switzerland) was mounted onto the footplate of machine. The joint angle of the right ankle was measured with an electric goniometer. EMG was recorded from the midbellies of medial (MG) and lateral (LG) gastrocnemii and soleus (Sol) muscles in the right leg using a surface electrode.

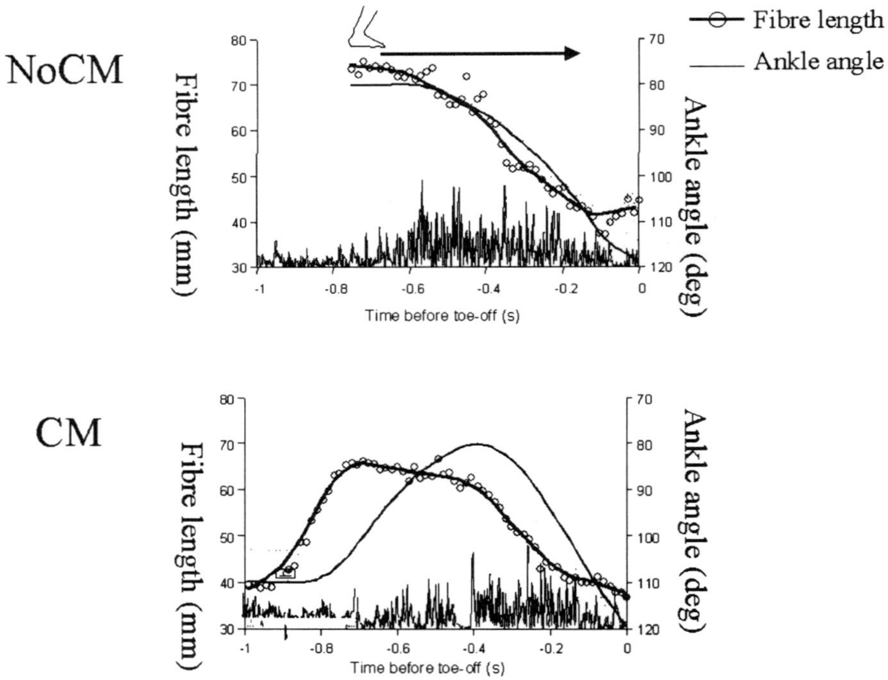

Figure 1. Changes in muscle fascicle length and angle of ankle joint before toe-off for counter movement (CM) and no counter-movement exercises (noCM).

Longitudinal images of the MG were obtained using a real-time ultrasound apparatus (SSD-5000, Aloka, Japan). The images were obtained at a rate of 65 images/sec. For each image, MG fascicles were traced along their length from the superficial and deep aponeuroses (Kawakami et al., 1998). The reliability and reproducibility of the present technique for measuring muscle fascicle length have been confirmed elsewhere with the coefficient of variation of 0-2% (Fukunaga et al., 1997, Kawakami et al., 1998). We regarded the fascicle length equivalent to muscle fibre length, based on a previous observation of tendon-to-tendon fibre arrangement within a fascicle in human MG (Kawakami et al., 2000).

The subject was requested to perform a single plantar flexion exercise with the load equivalent to 40% of the maximum isometric strength until the toe took off from the force plate, with and without a counter-movement (CM and noCM conditions, respectively).

RESULTS

Changes in joint angle, force at foot, EMG activities and length of muscle fascicle with CM and noCM are illustrated in Figure 1. There was no significant difference in the ankle angle between CM and noCM before toe-off. The maximal force was larger for CM than noCM, especially at the transition phase from dorsi- to plantar flexion. There was no significant effect of conditions or muscles on EMG amplitudes.

The fascicle length decreased throughout exercise in noCM. In CM on the other hand, the fascicle length increased at the onset of joint movement (dorsiflexion), then remained in constant as the ankle was dorsiflexed. When the ankle was finally plantar flexing, the fascicle length decreased. In CM, the EMG amplitude in all muscles decreased when the fascicle length increased during dorsiflexion of the ankle. In the transition phase from dorsi- to plantar flexion EMG activities increased.

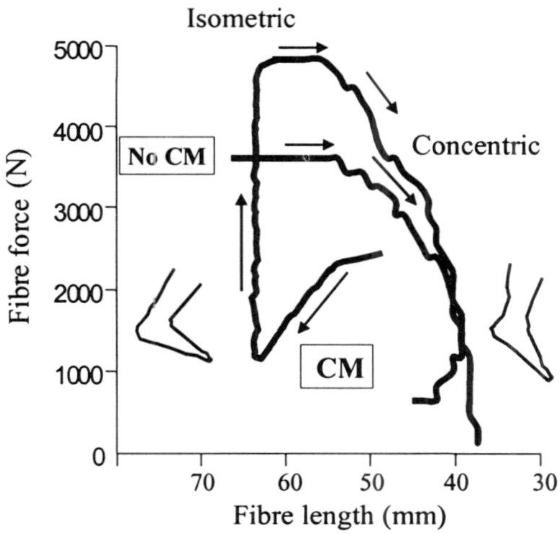

Figure 2. Changes in force developed at Achilles tendon for CM and noCM, at a given muscle fibre length.

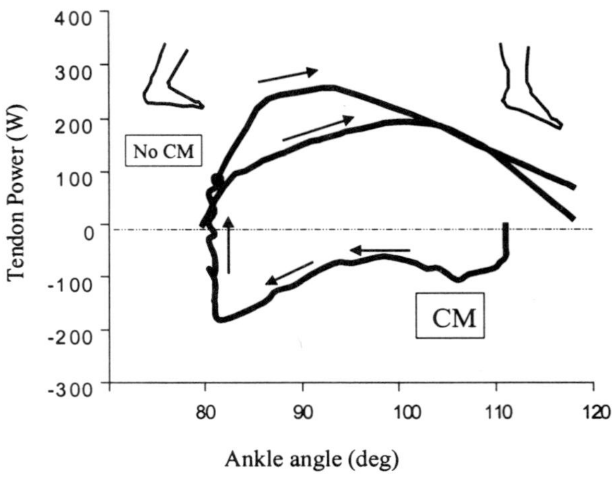

Figure 3. Changes in mechanical power developed at Achilles tendon for CM and noCM, at a given angle of ankle joint.

Figures 2 and 3 illustrate changes in force and power developed at the Achilles tendon, plotted against the fascicle length and ankle joint angle, respectively. In CM Achilles tendon force was greatest at the onset of the plantar flexion phase, compared with noCM. It was obvious that in CM fascicles exerted the force isometrically at the transition phase. The power at the Achilles tendon showed a similar pattern with CM, demonstrating greater values when the ankle started plantar flexion.

DISCUSSION

One of the major findings of the present study was that changes in joint angle and fascicle length did not match in CM and noCM. A number of studies on the stretch-shortening cycle in humans have measured only joint angles and reaction forces, and to distinguish the eccentric and concentric phases from those parameters (e.g. Bosco et al., 1982, Komi, 1992). The present study showed that in an "eccentric action" where the ankle joint was being dorsiflexed, fascicle length was almost constant when the muscle was activated. Mismatching of muscle and muscle fibre length during movements has already been observed in animals, e.g., during the stance phase of walking in cats when MG length increased, muscle fascicles kept an isometric contraction (Hoffer et al., 1989).

Based on the present findings, kinetics of contractile component (CC) and series elastic component (SEC) during CM and noCM can be considered as follows. For CM initially CC is contracting isometrically with the ankle most plantar flexed, being stretched by the CC tension. Then the ankle begins to be dorsiflexed when the torque around the ankle decreases without CC activation, accompanying CC is stretched passively by the tension from elongated SEC. When the subject start to activate the muscle to resist to the inertial load until dorsiflexion finally stopped, CC activates isometrically with lengthened SEC. After the transition from dorsi- to plantar flexion, both CC and SEC started to shorten until the toe takes off the block. This made powerful plantar flexion like a "catapult action" (Hof et al., 1983). In noCM, CC contract isometrically to sustain the load at initial dorsiflexed position. When CC started to develop greater force to launch the ankle at the onset of plantar flexion, CC stretched SEC, resulting greater shortening velocity of CC than that of muscle-tendon unit. Towards the take-off of the toes, both CC and SEC shortened.

However, the question is why more power is generated in CM than noCM. In case of CM, at the transition phase from dorsi- to plantar flexion CC contract isometrically. Given force-velocity characteristics, a muscle fibre can produce the highest force with isometric contraction. On the contrary in noCM CC contacts accompanied by some shortening through whole exercise phase, resulting in lower force production than with isometric contraction. Other possibility for force enhancement in CM might reside in the difference in CC length between CM and noCM. Our recent study showed the MG fascicle length was ~50mm, corresponding to the optimal sarcomere length (Kawakami et al., 2000). At the onset of noCM fascicle length was 64mm, compared to 55mm at the transition from dorsi- to plantar flexion for CM. It is, therefore, considered that in noCM the sarcomers are on the descending limb of the length-tension curve and contract in this disadvantageous condition, while in CM the sarcomeres would be closer to the optimal length for force development.

In conclusion, MG fibre behaviour during counter-movement revealed isometric, not lengthening contraction, in the "eccentric" phase of the joint. The increase in muscle-tendon-unit length is taken up by the elongation of tendinous tissues.

REFERENCES

Alexander, R. McN., and Bebbet-Clark, H. C., 1977, Storage of elastic strain energy in muscle and other tissues, *Nature*, 265, 114-117.

Bosco, C., Vitasalo, J. A., Komi, P. V., and Luhtanen, P., 1982, Combined effect of elastic energy and myoelectrical potentiation during stretch-shortening cycle exercise, *Acta Physiologica Scandinavica*, 114, 557-567.

Cavagna, G. A., Saibene, F. P., and Margaria, R., 1965, Effect of negative wqork on the amount of positive work performed by an isolated muscle, *Journal of Applied Physiology*, 20, 157-158.

Fukunaga, T., Ichinose, Y., Ito, M., Kawakami, Y. and Fukashiro, S., 1997, Determination of fascicle length and pennation in a contracting human muscle in vivo, *Journal Applied Physiology*, 82, 354-358.

Hof, A. F., Geelen, B. A., and van den Berg, J. W., 1983, Calf muscle moment, work and efficiency in level walking: role of series elasticity, *Journal of Biomechanics*, 16, 523-537.

Hoffer, J. A., Acputi, A. A., Pose, I. E., and Griffiths, R. I., 1989, Roles of muscle activity and load on the relationship between muscle spindle length and whole muscle length in the freely walking cat, in: *Progress in Brain Research*, Allum, J. H. J. and Hulliger, M., eds., pp. 80, 75-85.

Huijing, P. A., 1992, Elastic potential of muscle, in: *Strength and Power in Sport*, Komi, P.V., ed., pp. 151-168.

Kawakami, Y., Ichinose, Y., and Fukunaga, T., 1998, Architectural and functional feastures of human triceps surae muscle during contraction, *Journal of Applied Physiology*, 85, 398-404

Kawakami,Y., Kumagai,K., Huijing, P. A., Hijikata, T. and Fukunaga,T., 2000, The length-force characteristics of human gastrocnemius and soleus muscles in vivo, in: *Skeletal Muscle Mechanics*, Herzog, W., ed., pp. 327-341.

Komi, P. V., 1991, Stretch shortening cycle, in: *Strength and Power in Sport*, Komi, P. V., ed., pp. 169-179.

58

RELATIONSHIP BETWEEN NEURAL DRIVE AND MECHANICAL EFFECT IN THE RESPIRATORY SYSTEM

André De Troyer*

ABSTRACT

The actions of the canine external and internal interosseous intercostal muscles on the lung were assessed by applying the Maxwell reciprocity theorem. The external intercostals in the dorsal part of the cranial interspaces were found to have a large inspiratory effect. However, this effect decreases continuously in the caudal and the ventral direction, such that the muscles in the ventral part of the caudal interspaces have an expiratory effect. The internal intercostals also show marked gradients, such that the muscles in the dorsal part of the caudal interspaces have a large expiratory effect and those in the ventral part of the most cranial interspaces have a small inspiratory effect. During breathing, however, inspiratory activity is found only in the external intercostals with an inspiratory effect, and expiratory activity is confined to the internal intercostals with an expiratory effect. The spatial distribution of inspiratory activity among the canine external intercostals closely mirrors the distribution of inspiratory effect, and the distribution of expiratory activity among the internal intercostals closely mirrors the distribution of expiratory effect. Therefore, the external intercostals have a clear-cut inspiratory action on the lung during breathing, whereas the internal intercostals have a definite expiratory action. The distribution of neural drive among these muscles appears to be equally well matched to the distribution of respiratory effect in humans.

INTRODUCTION

The intercostal muscles are two thin layers of muscle fibres occupying each of the intercostal spaces. The external intercostals form the superficial layer and their fibres slope obliquely caudal and ventrally from the rib above to the rib below. In contrast, the internal intercostals, which form the deeper muscle layer, slope caudal and dorsally. The external intercostals, however, do not extend ventrally beyond the costochondral

* Laboratory of Cardiorespiratory Physiology, Brussels School of Medicine, and Chest Service, Erasme University Hospital, Brussels 1070, Belgium. Email: a_detroyer@yahoo.fr

junctions. From these junctions to the sternum, the only fibres are those of the internal intercostals; because of their location, these muscle areas are usually called the "parasternal intercostals".

The actions of these muscles are conventionally regarded according to the theory proposed 250 years ago by Hamberger (1749). This theory predicts that when an intercostal muscle contracts in a single interspace, it pulls the upper rib caudally and the lower rib cranially. However, as the fibres of the external intercostal slope caudal and ventrally, their lower insertion is further from the centre of rotation of the ribs (i.e., the costovertebral articulations) than their upper insertion. When this muscle contracts, therefore, the torque acting on the lower rib is greater than that acting on the upper rib. Consequently, the net effect of the muscle should be to raise the ribs and to inflate the lung. On the other hand, the fibres of the internal intercostal slope caudal and dorsally, so their lower insertion is closer to the centre of rotation of the ribs than their upper insertion. As a result, the torque acting on the lower rib is smaller than that acting on the upper rib, and hence, the net effect of the muscle should be to lower the ribs and to deflate the lung. Hamberger (1749) also concluded that the action of the parasternal intercostals should be referred to the sternum, rather than the costovertebral articulations. Therefore, even though these muscle fibres are part of the internal intercostal layer, they exert a greater torque on the lower rib than on the upper rib, and their net effect should be to raise the ribs and to inflate the lung.

A number of studies in dogs and in humans have clearly established, in agreement with the theory, that the parasternal intercostals do inflate the lung when they contract (De Troyer and Kelly, 1982; De Troyer et al., 1996; De Troyer et al., 1998). However, the actions of the external and internal (interosseous) intercostals remain uncertain.

RESPIRATORY EFFECTS OF INTERCOSTAL MUSCLES IN DOGS

As these actions cannot be determined by external stimulation, they were evaluated by using an indirect approach, based on the Maxwell reciprocity theorem. When applied to the respiratory system (Wilson and De Troyer, 1992), this standard theorem of mechanics predicts that the change in airway opening pressure (ΔPao) produced by a particular muscle during a maximal isolated contraction against a closed airway is related to the mass (m) of the muscle, the maximal active muscle tension per unit cross-sectional area (σ), and the fractional change in muscle length ($\Delta L/L$) per unit volume increase of the relaxed chest wall $(\Delta V_L)_{Rel}$, such that

$$\Delta Pao = m\sigma[\Delta L/(L\Delta V_L)]_{Rel} \quad (1)$$

This equation has been validated experimentally in dogs on a number of inspiratory and expiratory muscles (De Troyer et al., 1996; De Troyer and Legrand, 1998), and the coefficient of proportionality (σ) between $\Delta Pao/m$ and $[\Delta L / (L \Delta V_L)]_{Rel}$ for all the muscles was found to be 3.0 kg/cm^2, in close agreement with values of maximal active muscle tension measured in vitro (Close, 1972). Based on these results, we have therefore measured the masses of the canine external and internal intercostals throughout the rib cage and the changes in muscle length during passive inflation, and the maximum ΔPao for each muscle area was calculated by multiplying ($\Delta L/L$) by m and by 3.0 (De Troyer et al., 1999).

As shown in Figure 1A, the external intercostals in the dorsal third of the second and fourth interspace produce a significant fall in Pao during maximal contraction, thus indicating that these muscle areas have a clear-cut inspiratory effect on the lung. However, this inspiratory effect decreases rapidly and continuously toward the base of the rib cage. This inspiratory effect also decreases toward the costochondral junctions, such that the muscles in the middle third and the ventral third of the eighth and tenth interspace have an expiratory, rather than inspiratory effect. The internal intercostals also show marked rostrocaudal and dorsoventral gradients (Fig. 1B): the muscles in the dorsal third of the eighth and tenth interspace have a large expiratory effect, but this effect decreases in both the rostral and the ventral direction. As a result, the internal intercostal in the ventral third of the second interspace has a small inspiratory, rather than expiratory effect.

The observation that the external intercostals in the dorsal third of the second through sixth interspaces have an inspiratory effect and that the internal intercostals over a large fraction of the rib cage have an expiratory effect is in agreement with the theory of Hamberger (1749) and confirms that the orientation of the muscle fibres is an important determinant of the actions of these muscles. However, this theory cannot explain the dorsoventral or the rostrocaudal gradients, and this is due to the fact that it is based on a model that contains a number of simplifications (De Troyer et al., 1999). More importantly, these observations imply that the actions of the muscles during breathing are largely determined by the spatial distribution of neural drive among them, and this prompted us to re-examine the pattern of external and internal intercostal activity in anesthetized, spontaneously breathing dogs (Legrand and De Troyer, 1999).

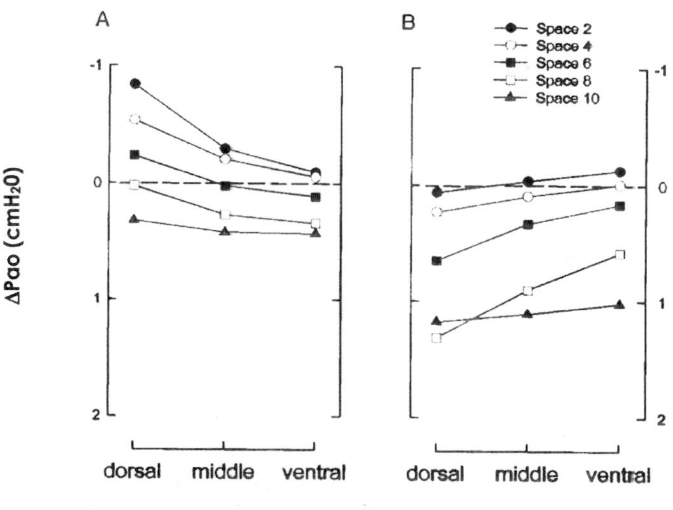

Figure 1. Maximal changes in airway opening pressure (ΔPao) generated by the canine external (*A*) and internal interosseous (*B*) intercostal muscles in the dorsal third, the middle third, and the ventral third of the second to tenth interspace. A negative ΔPao indicates an inspiratory effect, and a positive ΔPao indicates an expiratory effect. (Reproduced, with permission from De Troyer et al., 1999.)

INTERCOSTAL MUSCLE ACTIVITY IN DOGS

We first assessed the pattern of activity of the external intercostals with an expiratory effect and of the internal intercostals with an inspiratory effect. As Le Bars and Duron (1984) have reported in the cat, phasic expiratory discharges were recorded from the external intercostals situated in the ventral portion of the caudal interspaces (Legrand and De Troyer, 1999). Inspiratory discharges were also recorded from the internal intercostal in the ventral portion of the second interspace, in particular when the inspiratory mechanical load was augmented. However, selective denervation procedures demonstrated that these inspiratory discharges were, in fact, arising in the adjacent external intercostals and that the expiratory discharges recorded from the caudal external intercostals originated from the underlying internal intercostals (Legrand and De Troyer, 1999). Thus, in agreement with the previous conclusion by Sears (1964) and Kirkwood and colleagues (1982), the external intercostals in our animals, when active, were active only during inspiration, and the internal intercostals, when active, were active only during expiration.

We then evaluated in a quantitative manner the spatial distribution of external intercostal inspiratory activity and internal intercostal expiratory activity. This quantitative assessment was made by normalizing the activity recorded from each muscle area during breathing relative to the activity induced by tetanic, supramaximal stimulation of the corresponding intercostal nerve, and the results obtained during resting breathing are shown in Figure 2. External intercostal inspiratory activity in the dorsal part of the rib cage was consistently greatest in the second interspace and smallest in the sixth interspace (Fig. 2A); the muscle in the dorsal part of the eighth interspace showed no activity at all. External intercostal inspiratory activity in the third interspace also decreased from the dorsal third to the ventral third (Fig. 2B). In contrast, although

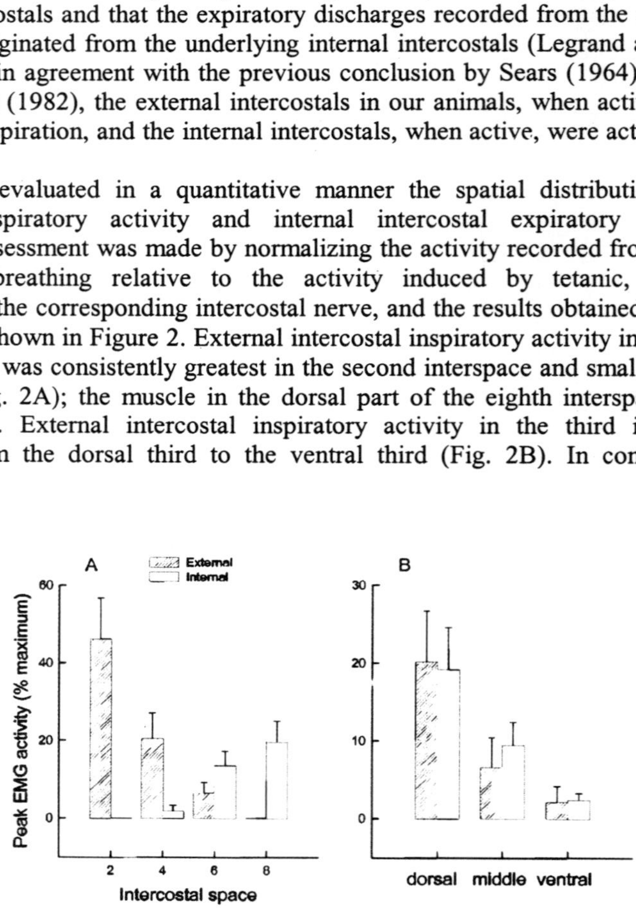

Figure 2. Spatial distribution of external intercostal inspiratory activity (hatched bars) and internal intercostal expiratory activity (open bars) during resting breathing in dogs. The data in panel *A* were obtained in the dorsal third of interspaces 2, 4, 6 and 8; those in panel *B* were obtained in the dorsal third, the middle third, and the ventral third of the external intercostal in the third interspace and of the internal intercostal in the eighth interspace. Each value is expressed as a percentage of the activity recorded during tetanic, supramaximal stimulation of the corresponding external or internal intercostal nerve (maximal activity). (Reproduced with permission from Legrand and De Troyer, 1999.)

internal intercostal expiratory activity was also greatest in the dorsal part of the rib cage (Fig. 2B), this activity was greatest in the eighth interspace and declined continuously to the second interspace (Fig. 2A).

In view of the spatial distributions of respiratory effects (Fig. 1), these observations thus confirm the conventional idea (Hamberger, 1749) that the external intercostal muscles have an inspiratory action on the lung during breathing and that the internal intercostals have an expiratory action. However, these observations also indicate that these opposite actions are largely the result of selective regional activation of the muscles. In addition, they suggest that the distribution of neural drive among the canine respiratory muscles is closely matched to the distribution of respiratory effect, such that the muscle areas with the greatest effect are also those that receive the greatest drive. The parasternal intercostals provide other examples of this matching.

THE PARASTERNAL INTERCOSTALS

As previously pointed out, it is clearly established that the parasternal intercostals have an inspiratory effect on the lung and contract only during the inspiratory phase of the breathing cycle. In the dog, however, this inspiratory effect is not uniform. In a given segment of the rib cage, the effect is greatest in the muscle bundles situated in the immediate vicinity of the sternum, and it decreases rapidly in the lateral direction such that the muscle bundles situated near the costochondral junction have no inspiratory effect at all (De Troyer and Legrand, 1995). Similarly, during breathing, inspiratory activity in a given parasternal intercostal is greatest in the muscle bundles near the sternum and decreases toward the costochondral junction (De Troyer and Legrand, 1995). The bundles near the junction, in fact, remain consistently silent, including when the respiratory drive is augmented by CO_2-enriched gas mixtures or by elevated inspiratory airflow resistance. Also, inspiratory activity in the medial parasternal bundles decreases progressively from the third to the seventh interspace (Legrand et al., 1996), as does the inspiratory effect (De Troyer et al., 1996). It is worth emphasizing that these mediolateral and rostrocaudal gradients of parasternal activity persist after complete deafferentation of the rib cage, thus indicating that they are primarily determined by central mechanisms (Legrand et al., 1996).

THE INTERCOSTAL MUSCLES IN HUMANS

The respiratory effects of the external and internal interosseous intercostals in humans have also been evaluated by using the principle defined in Eqn (1) (Wilson et al., 2001). The muscles were first dissected in cadavers to measure their masses and the orientations of the muscle fibres relative to the ribs. Five healthy individuals were then placed in a computed tomographic (CT) scanner to determine the shape of the ribs and their precise transformation during passive inflation, and these data, combined with the orientations of the muscle fibres, were used to compute the fractional changes in muscle length. The values thus obtained were finally multiplied by the values of muscle mass measured in cadavers and by the value of σ (3.0 kg/cm^2) to calculate the maximal ΔPao's for the muscles in the ventral and dorsal portions of the even-numbered interspaces.

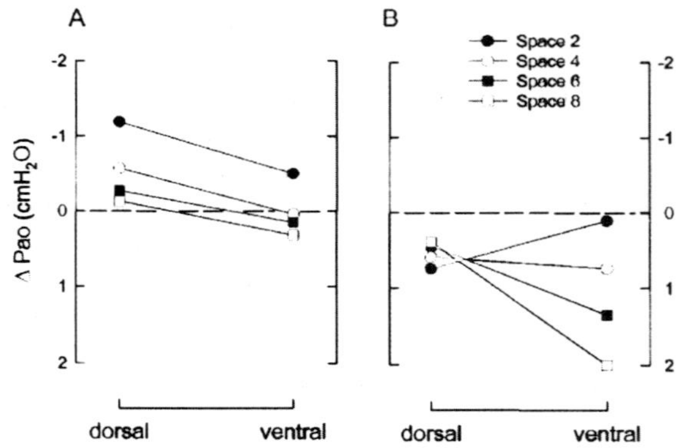

Figure 3. Computed maximal changes in airway opening pressure (ΔPao) generated by the human external (*A*) and internal interosseous (*B*) intercostal muscles situated in the dorsal half of the second to eighth interspaces. (Reproduced with permission from Wilson et al., 2001.)

The topographic distribution of respiratory effect among the external intercostals in humans is qualitatively similar to that observed in the dog (Fig. 3A). Thus the inspiratory effect of the muscle in humans is greatest in the dorsal portion of the rostral interspaces and is reversed into a small expiratory effect in the ventral portion of the caudal interspaces. On the other hand, the human internal intercostals have an expiratory effect throughout the rib cage, and this effect is greatest in the ventral portion of the caudal segments (Fig.3B). If the distribution of respiratory activity among these muscles were closely matched to the distribution of respiratory effect, one would therefore predict that the external intercostals in humans contract during inspiration, in particular in the dorsal portion of the rostral segments. One would also predict that the internal intercostals contract during expiration and display greater activity in the ventral portion of the caudal segments.

In agreement with these predictions, the extensive electrical recordings performed by Taylor (1960) have shown that as in the dog, the external intercostals in normal humans are active only during inspiration and that the internal intercostals are active only during expiration. Also, expiratory activity during resting breathing was detected only in the internal intercostals of the ventrolateral aspect of the caudal interspaces, thus indicating that neural drive is greatest in these areas. Taylor (1960) hardly investigated the dorsal aspect of the rib cage in his subjects, but we have recently inserted concentric needle electrodes in this area in six healthy individuals (A. De Troyer, R. Gorman and S.C. Gandevia, unpublished observations). The external intercostal in the dorsal portion of the third interspace displayed phasic inspiratory activity during resting breathing in each subject; this activity involved many motor units and was recorded in every breath. However, this activity declined rapidly in the caudal and the ventral direction, such that both the external intercostal in the dorsal portion of the seventh interspace and the external intercostal in the ventral portion of the third interspace were commonly silent.

CONCLUSIONS

The respiratory muscle pump in dogs and in humans appears to operate with remarkable precision, such that the muscles with the greatest respiratory effect also receive the greatest neural drive during breathing. In contrast, the muscles with a small respiratory effect receive little or no drive, and this raises several questions:

1. What is the mechanism of this matching? Although it is primarily related to central mechanisms in the case of the parasternal intercostals, proprioceptive mechanisms might play a significant role in the case of the external and internal intercostals. Indeed, in contrast to the parasternal intercostals, these are abundantly supplied with muscle spindles.
2. How does this matching evolve? Is it an evolutionary process? For example, if the inspiratory effect of the external intercostals in the dorsal portion of the rostral segments of the rib cage were abolished by locking the costovertebral articulations of the upper ribs, would these muscle areas continue indefinitely to receive a large inspiratory drive?
3. What are the potential benefits of this matching? Even though the muscle areas with a small respiratory effect receive only a small neural drive, activating these areas should be less effective with respect to lung volume than activating exclusively the muscle areas with the greatest effect. Is the existing distribution of neural drive the solution to expand the chest wall with minimum distortion and minimum energy expenditure (De Troyer and Wilson, 2000) ?
4. Do non-respiratory muscles obey the same principle of recruitment?

REFERENCES

Close, R.I., 1972, Dynamic properties of mammalian skeletal muscles, *Physiological Reviews*, 52, 129-197.

De Troyer, A., and Kelly, S., 1982, Chest wall mechanics in dogs with acute diaphragm paralysis, *Journal of Applied Physiology*, 53, 373-379.

De Troyer, A., and Legrand, A., 1995, Inhomogeneous activation of the parasternal intercostals during breathing, *Journal of Applied Physiology*, 79, 55-62.

De Troyer, A., and Legrand, A., 1998, Mechanical advantage of the canine triangularis sterni, *Journal of Applied Physiology*, 84, 562-568.

De Troyer, A., Legrand, A., Gevenois, P. A., and Wilson, T. A., 1998, Mechanical advantage of the human parasternal intercostal and triangularis sterni muscles, *Journal of Physiology* 513, 915-925.

De Troyer, A., Legrand, A., and Wilson, T. A., 1996, Rostrocaudal gradient of mechanical advantage in the parasternal intercostal muscles of the dog, *Journal of Physiology*, 495, 239-246.

De Troyer, A., Legrand, A., and Wilson, T. A., 1999, Respiratory mechanical advantage of the canine external and internal intercostal muscles, *Journal of Physiology*, 518, 283-289.

De Troyer, A., and Wilson, T. A., 2000, The canine parasternal and external intercostal muscles drive the ribs differently, *Journal of Physiology*, 523, 799-806.

Hamberger, E., 1749, *De Respiriationis Mechanismo et usu Geneino, Iena.*

Kirkwood, P. A., Sears, T.A., Stagg, D., and Westgaard, R. H., 1982, The spatial distribution of synchronization of intercostal motoneurones in the cat, *Journal of Physiology*, 327, 137-155.

Le Bars, P., and Duron, B., 1984, Are the external and internal intercostal muscles synergist or antagonist in the cat? *Neuroscience Letters*, 51, 383-386.

Legrand, A., Brancatissano, A., Decramer, M., and De Troyer, A., 1996, Rostrocaudal gradient of electrical activation in the parasternal intercostal muscles of the dog, *Journal of Physiology*, 495, 247-254.

Legrand, A., and De Troyer, A., 1999, Spatial distribution of external and internal intercostal activity in dogs, *Journal of Physiology*, 518, 251-300.

Sears, T. A., 1964, Efferent discharges in alpha and fusimotor fibres of intercostal nerves of the cat, *Journal of Physiology,* 174, 295-315.

Taylor, A., 1960, The contribution of the intercostal muscles to the effort of respiration in man, *Journal of Physiology,* 151, 390-402.

Wilson, T. A., and De Troyer, A., 1992, Effect of respiratory muscle tension on lung volume, *Journal of Applied Physiology,* 73, 2283-2288.

Wilson, T. A., Legrand, A., Gevenois, P.-A., and De Troyer, A., 2001, Respiratory effects of the external and internal intercostal muscles in humans, *Journal of Physiology,* 530, 319-330.

INDEX